Annals of the International Society of Dynamic Games
Volume 3

Series Editor
Tamer Başar

Associate Editors

M. Bardi, Padova
T. Başar, Urbana
R. P. Hämäläinen, Helsinki
A. Haurie, Geneva
A. A. Melikyan, Moscow
G. J. Olsder, Delft
T.E.S. Raghavan, Chicago
J. Shinar, Haifa
T.L. Vincent, Tucson
A. J. deZeeuw, Tilburg

Editorial Board

Leonard D. Berkovitz, Purdue University
P. Bernhard, INRIA, Sophia-Antipolis
R. P. Hämäläinen, Helsinki University of Technology
Alain Haurie, University of Geneva
N. N. Krasovskii, Academy of Sciences, Ekaterinburg
George Leitmann, University of California, Berkeley
G. J. Olsder, Delft University of Technology
T. E. S. Raghavan, University of Illinois, Chicago
Josef Shinar, Technion–Israel Institute of Technology
B. Tolwinski, Operations Research Experts, Black Hawk, Colorado
Klaus H. Well, Stuttgart University

Annals of the International Society of Dynamic Games

New Trends in Dynamic Games and Applications

Geert Jan Olsder

Editor

Birkhäuser
Boston • Basel • Berlin

Geert Jan Olsder
Delft University of Technology
Section of Applied Analysis
Mekelweg 4
Delft, The Netherlands

Library of Congress Cataloging-in-Publication Data

New trends in dynamic games and applications / Geert Jan Olsder,
    editor.
          p.   cm. -- (Annals of the International Society of Dynamic
    Games ; v. 3)
        Includes bibliographical references.
        ISBN 0-8176-3812-1
        1. Game theory.   I. Olsder, Geert Jan. II. Series.
    QA269.N48   1995   95-33319
    519.3-dc20   CIP

ISBN 0-8176-3812-1
ISBN 3-7643-3812-1
Reformatted from authors' diskettes by Texniques, Inc., Boston, MA
Printed and bound by Quinn-Woodbine, Woodbine, NJ
Printed in the United States of America

9 8 7 6 5 4 3 2 1

Table of Contents

Part III. Solution methods

Part IV. Nonzero-sum games, theory

Preface

The theory of dynamic games is very rich in nature and very much alive! If the reader does not already agree with this statement, I hope he/she will surely do so after having consulted the contents of the current volume. The activities which fall under the heading of 'dynamic games' cannot easily be put into one scientific discipline. On the theoretical side one deals with differential games, difference games (the underlying models are described by differential, respectively difference equations) and games based on Markov chains, with deterministic and stochastic games, zero-sum and nonzero-sum games, two-player and many-player games — all under various forms of equilibria. On the practical side, one sees applications to economics (stimulated by the recent Nobel prize for economics which went to three prominent scientists in game theory), biology, management science, and engineering.

The contents of this volume are primarily based on selected presentations made at the Sixth International Symposium on Dynamic Games and Applications, held in St Jovite, Quebec, Canada, 13-15 July 1994. Every paper that appears in this volume has passed through a stringent reviewing process, as is the case with publications for archival technical journals. This conference, as well as its predecessor which was held in Grimentz, 1992, took place under the auspices of the International Society of Dynamic Games (ISDG), established in 1990. One of the activities of the ISDG is the publication of these Annals.

The contributions in this volume have been grouped around five themes. The first three deal with zero-sum games and the specific themes are: minimax control (5 papers), pursuit evasion (8 papers) and solution methods (3 papers). The last two areas deal with nonzero-sum games and have been split up according to theory (5 papers) and applications (3 papers). The contents of these themes are as follows.

- *Minimax control*   The first paper by Bernhard sheds new light on minimax control by using a morphism between the conventional algebra and the so-called max-plus algebra. In this way, a remarkable parallel between stochastic control and minimax control becomes visible. In the next paper Fridman studies the H^∞ problem for nonlinear singularly perturbed systems and obtains solvability conditions in terms of invariant manifolds. The paper by Altman and Gaitsgory also studies a game with a fast and a slow part, but the controls of the players enter through a Markov chain which influences the continuous time dynamics of the game. Pan and Başar, consider the H^∞ control of a linear system in which the system and cost parameters also change according to a Markov chain process, but in this case the jumps in this Markov chain are determined stochastically and are not influenced by the players. The last paper on this theme, by Fristedt, Lapic and Sudderth, analyses a two-person zero-sum game analogous to the 'big match'. It is shown that this unusual game has a value and 'good' strategies are given.

- *Pursuit evasion*   The first paper, by Pesch, Gabler, Miesbach and Breitner, solves the well-known cornered rat game by means of neural network techniques. The authors make it plausible that this technique is more generally applicable. The paper by Lipman and Shinar studies ship defence scenarios against highly manoeuvrable attacking missiles. The existence of a state constraint leads to new singular phenomena. Lachner, Breitner and Pesch also consider an air combat, which they solve by means of the Isaacs' equation which then leads to multipoint boundary problems. Le Menec and Bernhard, in the following paper, explore AI techniques for the solution of aerial combats. Kumkov and Patsko study a pursuit game with incomplete information, defined by geometrical constraints, and where the pursuer has impulse control, the total 'amount of impulses' being a constraint. The game considered by Olsder and Pourtallier is also one in which the final miss distance should be minimized (as in the previous paper), but here the pursuer (a helicopter — the evader is a submarine) can choose the moments of observation. The paper by Neveu, Pignon, Raimondo, Nicolas and Pourtallier compares the results of an analytical treatment versus a realistic simulation, with expert rules of the same game treated in the previous paper. Surprisingly, the analytical results with the simpler model do not lead to worse results. In the last paper on this theme, Chikrii and Prokopovich study the interaction of n pursuers and m evaders. Various results on evasion are given.

- *Zero-sum games, solution methods*   The first contribution on this theme by Bardi, Bottacin and Falconi discusses a novel approximation scheme for the solution of general pursuit evasion games, based on recently obtained results for viscosity solutions. Tidball, in the next paper, considers discretization schemes for zero-sum differential games with stopping times. Chikrii and Rappoport suggest a solution method based on the resolving function method.

- *Nonzero-sum games, theory*   Li and Gajic derive an iterative algorithm leading to the nonnegative (positive) definite stabilizing solution of the coupled algebraic Riccati equations, resulting from a nonzero-sum differential game. As a byproduct, an algorithm is presented that solves the Riccati equation of H^∞ zero-sum differential games. Carlson and Haurie study a class of open-loop infinite horizon differential games; the so-called 'overtaking optimality' is used to deal with unbounded payoffs. Xu and Mizukami show that closed-loop no-memory information on the descriptor variables is sufficient for the leader to enforce his team solution in a Stackelberg game with linear dynamics (descriptor system) and quadratic criteria. Ehtamo and Ruusunen study bargaining games with the feasible set consisting of utility gains evaluated over multiple time periods. They provide various refinements of the optimality principle. As the last paper of this theme, Petrosjan extends the notion of Shapley value for cooperative games to dynamic games.

- *Nonzero-sum games, applications* Jørgenson's paper deals with a number of issues in dynamic game modelling, management strategy, the interface between the two and in particular pertaining to a firm's market strategy. Clemhout and Wan study endogenous growth as a dynamic game. One of their conclusions is that the underlying strategical complementarity equals the externality of the 'fish war'. Hines, in the last paper, applies game theory to biology. Recent results concerning the effects of sexual reproduction are discussed.

This volume would not have been made possible if

- the authors had not responded as positively and promptly as they did, when asked for changes, postcript files of the figures, ...;

- the associate editors, see for the list elsewhere, had not been very conscientious about their tasks regarding the reviewing process;

- the anonymous reviewers had not reviewed the papers so carefully and constructively;

- the co-chairs of the symposium in St Jovite, Michèle Breton and Georges Zaccour, had not arranged the excellent setting for the presentation of these papers;

- I had not gotten a tremendous hand from my secretary, Tatiana Tijanova, with the preparation of this volume, when she handled so diligently postscipt files, turned nonlatex files into latex files, etc.

I thank them all sincerely.

Geert Jan Olsder
Delft, April 1995

PART I
Minimax control

Expected Values, Feared Values,
and Partial Information Optimal Control

Pierre Bernhard

INRIA - BP 93

06902 Sophia-Antipolis Cedex

France

Abstract

We show how a morphism between the ordinary algebra $(+, \times)$ and the $(\max, +)$ algebra offer a completely parallel treatment of stochastic and minimax control of disturbed nonlinear systems with partial information.

1   Introduction

Minimax control, or worst case design, as a means of dealing with uncertainty is an old idea. It has gained a new popularity with the recognition, in 1988, of the fact that H_∞-optimal control could be cast into that concept. Although some work in that direction existed long before (see [8]), this viewpoint has vastly renewed the topic. See [3] and related work.

Many have tried to extend this work to a nonlinear setup. Most prominent among them perhaps is the work of Isidori, [15] [16] but many others have followed suit : [23] [4] [5] and more recently [18] [19]. This has contributed to a renewed interest in nonlinear minimax control.

We insist that the viewpoint taken here is squarely that of minimax control, *and not nonlinear H_∞-optimal control*. There are several reasons for this claim. For one thing, we only consider finite time problems, and therefore do not consider stability issues which are usually central in H_∞-optimal control. We do not stress quadratic loss functions. But more importantly, we claim that the minimax problem is only an intermediary step in H_∞ theory, used to insure existence of a fixed point to the feedback equations $z = P_K w$, $w = \Delta P z$ (P_K is the controlled plant, ΔP the model uncertainty). In that respect, the nonlinear equivalent is *not* the minimax problem usually considered, but rather the contraction problem independently tackled by [14].

If we decide that minimax is an alternative to stochastic treatment of disturbances (input uncertainties, rather than plant uncertainties), it makes sense to try to establish a parallel. In this direction, we have the striking morphism developed by Quadrat and coworkers, see [21] [2] [1]. We shall

review here recent work, mainly by ourselves, Baras, and James, in the light
of this parallel, or Quadrat's morphism. This paper is to a large extent based
on [7].

2   Quadrat's morphism

In a series of papers [21] [2] [1], giving credit to other authors for early
developments, Quadrat and coauthors have fully taken advantage of the morphism introduced between the ordinary algebra $(+, \times)$ and the $(\min, +)$, or
alternatively the $(\max +)$, algebra to develop a *decision calculus* parallel to
probability calculus. It has been pointed out by Quadrat and coauthors that
a possible way of understanding that morphism was through Cramer's transform. We shall not, however, develop that way of thinking here, but merely
rely on the algebraic similarity between the two calculi.

Let us briefly review some concepts, based on [1].

2.1   Cost measure

The parallel to a probability measure is a *cost measure*. Let Ω be a topological
space, $\mathcal{A}$ a σ-field of subsets, $K : \mathcal{A} \to \mathbb{R} \cup \{-\infty\}$ is called a cost measure if
it satisfies the following axioms :

- $K(\emptyset) = -\infty$

- $K(\Omega) = 0$

- for any family of (disjoint) elements A_n of $\mathcal{A}$,

$$K(\cup A_n) = \sup_n K(A_n) .$$

(It is straightforward to see that the word "disjoint" can be omitted
from this axiom).

One may notice the parallel with a probability measure. In the first two
axioms, the O of probability measures, the neutral element of the addition,
is replaced by the neutral element of the max operator : $-\infty$, and the 1,
the neutral element of the product, is replaced by the neutral element of the
sum, 0. In the third axiom, the sum of the measures of the disjoint sets is
replaced by the max.

The function $\bar{G} : \Omega \to \mathbb{R} \cup \{-\infty\}$ is called a *cost density* of K if we have

$$\forall A \in \mathcal{A}, \quad K(A) = \sup_{\omega \in \Omega} \bar{G}(\omega) .$$

One has the following theorem (Akian)

Theorem 2.1 *Every cost measure defined on the open sets of a Polish space Ω admits a unique maximal extension to 2^{Ω}; this extension has a density, which is a concave u.s.c. function.*

2.2 Feared values

The term *feared values* is introduced here to stress the parallel with *expected values*.

When a stochastic disturbance is introduced into a problem model, in order to derive a controller design for instance, it comes with a given probability distribution. We shall always assume here that these distributions have densities. Therefore let $w \in W$ be a stochastic variable. If its probability distribution is $\Pi(\cdot)$, let ψ be a function of w with values in $\mathbb{R}$. We define its expected value as

$$\mathbb{E}_w \psi := \int \psi(w) \Pi(w) \, dw$$

and we omit the subscript w to $\mathbb{E}$ when no ambiguity results. Similarly, let a disturbance w be given together with a cost distribution $\Gamma(\cdot)$. The *feared value* of a function ψ from W into $\mathbb{R}$ is defined as

$$\mathbb{F}_w \psi := \max_w [\psi(w) + \Gamma(w)]$$

which is the formula dual to that of the expected value in Quadrat's morphism.

The "fear" operator enjoys the linearity properties one would expect in the (max, +) algebra :

$$\mathbb{F}(\max\{\phi, \psi\}) = \max\{\mathbb{F}\phi, \mathbb{F}\psi\} \,,$$

and if λ is a constant,

$$\mathbb{F}(\lambda + \psi) = \lambda + \mathbb{F}\psi \,.$$

A sequence of stochastic variables $\{w_t\}$, $t = 0 \ldots T - 1$ also denoted $w_{[0,T]}$, are said to be *independent* if their joint probability density is the product of their individual probability densities Π_t:

$$\Pi(w_{[0,T]}) = \prod_{t=0}^{T-1} \Pi_t(w_t)$$

leading to the following formula, where J is a function of the whole sequence

$$\mathbb{E}J(w_{[0,T]}) = \int J(w_{[0,T]}) \prod_{t=0}^{T} \Pi_t(w_t) \, dw_{[0,T]} \,.$$

In a similar fashion, a sequence of independent decision variables $w_{[0,T]}$ with cost densities Γ_t will have a joint cost density Γ equal to the sum of their individual cost densities:

$$\Gamma(w_{[0,T]}) = \sum_{t=0}^{T-1} \Gamma_t(w_t)$$

leading to the dual formula

$$\mathbb{F} J(w_{[0,T]}) = \max_{w_{[0,T]}} \left[J(w_{[0,T]}) + \sum_{t=0}^{T} \Gamma_t(w_t) \right].$$

Conditioning Let a pair of decision variables (v, w) ranging over sets $\mathsf{V} \times \mathsf{W}$ have a joint cost density $r(v, w)$. We may define the marginal law for v as

$$p(v) = \max_{w \in W} r(v, w)$$

for which it is true that the feared value of the characteristic function $\mathbb{1}_A(v)$ of a set $A \subset \mathsf{V}$ is given by

$$\mathbb{F} \mathbb{1}_A = \max_{v \in V} p(v)$$

preserving the duality with the probabilistic formulas

$$p(v) = \int_W r(v, w)\, dw$$

and

$$\mathbb{E} \mathbb{1}_A = \int_V p(v)\, dv\,.$$

Similarily, we have the dual of Bayes formula, defining the *conditional cost measure* $q(w|v)$ as

$$q(w|v) = r(v, w) - p(v)$$

Let $\mathbb{F}_w^v$ denote the corresponding feared value, we have the "embedded algebra" formula:

$$\mathbb{F}_v \left[\mathbb{F}_w^v \psi(v, w) \right] = \mathbb{F} \psi(v, w)\,.$$

We shall often need a less simple form of conditioning such as (with transparent notations)

$$\mathbb{F}[\psi(w) \mid w \in A] = \max_{w \in A} [\psi(w) + \Gamma(w)]\,.$$

which should clearly be seen as the basic conditioning operation.

3 The discrete time control problem

3.1 The problem

We consider a partially observed two input control system

$$\begin{aligned}
x_{t+1} &= f_t(x_t, u_t, w_t)\,, & (1)\\
y_t &= h_t(x_t, w_t)\,, & (2)
\end{aligned}$$

where $x_t \in \mathrm{I\!R}^n$ is the state at time t, $u_t \in \mathsf{U}$ the (minimizer's) control, $w_t \in \mathsf{W}$ the disturbance input, and $y_t \in \mathsf{Y}$ the measured output. We shall call $\mathbf{U}$ the set of input sequences over the time horizon $[0, T]$: $\{u_t\}_{t\in[0,T]}$ usually written as $u_{[0,T]} \in \mathbf{U}$, and likewise for $w_{[0,T]} \in \mathbf{W}$. The initial state $x_0 \in \mathsf{X}_0$ is also considered part of the disturbance. We shall call $\omega = (x_0, w_{[0,T]})$ the combined disturbance, and $\Omega = \mathsf{X}_0 \times \mathbf{W}$ the set of disturbances.

The solution of (1) (2) above shall be written as

$$\begin{aligned}
x_t &= \phi_t(u_{[0,T]}, \omega)\,,\\
y_t &= \eta_t(u_{[0,T]}, \omega)\,.
\end{aligned}$$

Finally, we shall call u^t a partial sequence $(u_0, u_1, \ldots, u_t)$ and U^t the set of such sequences [1], likewise for $w^t \in \mathsf{W}^t$ and $y^t \in \mathsf{Y}^t$. Also, we write $\omega^t = (x_0, w^t) \in \Omega^t$.

The solution of (1) and (2) may alternatively be written as

$$\begin{aligned}
x_t &= \phi_t(u^{t-1}, \omega^{t-1})\,, & (3)\\
y_t &= \eta_t(u^{t-1}, \omega^t)\,. & (4)
\end{aligned}$$

We shall also write

$$\begin{aligned}
x^t &= \phi^t(u^{t-1}, \omega^{t-1})\,, & (5)\\
y^t &= \eta^t(u^{t-1}, \omega^t)\,, & (6)
\end{aligned}$$

to refer to the partial sequences solution of (1) and (2).

Admissible controllers will be *strictly causal output feedbacks* of the form $u_t = \mu_t(u^{t-1}, y^{t-1})$. We denote by $\mathcal{M}$ the class of such controllers.

A performance index is given. In general, it may be of the form

$$J(x_0, u_{[0,T]}, w_{[0,T]}) = M(x_T) + \sum_{t=0}^{T-1} L_t(x_t, u_t, w_t)\,.$$

---

[1] notice the slight inconsistency in notations, in that our U^t is the cartesian $(t+1)$ power of U. Other choices of notations have their drawbacks too.

However, we know that, to the expense of increasing the state dimension by one if necessary, we can always bring it back to a purely terminal payoff of the form

$$J(x_0, u_{[0,T]}, w_{[0,T]}) = M(x_T) = M \circ \phi_T(u_{[0,T]}, \omega). \tag{7}$$

The data of a strategy $\mu \in \mathcal{M}$ and of a disturbance $\omega \in \Omega$ generates through (1)(2) a unique pair of sequences $(u_{[0,T]}, w_{[0,T]}) \in \mathbf{U} \times \mathbf{W}$. Thus, with no ambiguity, we may also use the abusive notation $J(\mu, \omega)$. The aim of the control is to minimize J, in some sense, "in spite of the unpredictable disturbances".

We want to compare here two ways of turning this unprecise statement into a meaningful mathematical problem.

In the first approach, *stochastic control,* we model the unknown disturbance as a random variable, more specifically here a random variable x_0 with a probability density $N(x)$ and an independant white stochastic process $w_{[0,T]}$ of known instantaneous probability distribution Π_t. (We notice that nothing in the sequel prevents Π_t from depending on x_t and u_t.) The criterion to be minimized is then

$$H(\mu) := \mathbb{E}_\omega J(\mu, \omega). \tag{8}$$

This can be expanded into

$$H(\mu) = \int M(x_T) \left(\prod_{t=0}^{T-1} \Pi_t(w_t) \right) N(\xi) \, dw_{[0,T]} \, d\xi$$

In the second approach, we are given the cost density N of x_0, and the cost densities Γ_t of the w_t's. (Again, Γ_t might depend on x_t and u_t.) The criterion to be minimized is then

$$G(\mu) := \mathbb{F}_\omega J(\mu, \omega), \tag{9}$$

which can be expanded into

$$G(\mu) := \max_\omega [M(x_T) + \sum_{t=0}^{T-1} \Gamma_t(w_t) + N(x_0)]$$

Remark If all cost measures of the disturbances are taken as constant, (e.g. 0), then $G(\mu)$ is, if it exists, the *guaranteed value* given only over the sets on which the perturbations range. Therefore, minimizing it will insure the best possible guaranteed value.

3.2 Dynamic programming

3.2.1 Stochastic dynamic programming

We quickly recall here for reference purposes the classical solution of the stochastic problem via dynamic programming. One has to introduce the *conditional state probability measure*, and, assuming it is absolutely continuous with respect to the Lebesgue measure, its density W. Let, thus, $W_t(x)\,dx$ be the conditional probability measure of x_t given y^{t-1}, or *a priori* state probability distribution at time t, and $W_t^\eta(x)\,dx$ be the conditional state distribution given y^{t-1} and given that $y_t = \eta$, or *a posteriori* state probability distribution at time t.

Clearly, W_t is a function only of past measurements. As a matter of fact, we can give the filter that lets one compute it. Starting from

$$W_0(x) = N(x) \tag{10}$$

at each step, W_t^η can be obtained by Bayes' rule. A standard condition for this step to be well posed is that, for all (t, x, w), the map $w \mapsto h_t(x, w)$ be locally onto, and more specifically that the partial derivative $\partial h_t(x, w)/\partial w$ be invertible. It suffices here to notice that, because the information is increasing, (the information algebras are nested), we have, for any test function $\psi(\cdot) \in L^1(\mathbb{R}^n)$,

$$\mathbb{E}_y \int \psi(x) W_t^y(x)\,dx = \int \psi(x) W_t(x)\,dx\,. \tag{11}$$

Then W_{t+1} is obtained by propagating $W_t^{y_t}$ through the dynamics. It suffices for our purpose to define this propagation by the dual operator: for any test function ψ,

$$\int \psi(x) W_{t+1}(x)\,dx = \int \mathbb{E}_w \psi(f_t(x, u_t, w)) W_t^{y_t}(x)\,dx\,. \tag{12}$$

The above expression shows the dependence of the sequence $\{W_t\}$ on the control $u_{[0,T]}$ and the observation sequence $y_{[0,T]}$. Let this define the function F_t as

$$W_{t+1} = F_t(W_t, u_t, y_t)\,. \tag{13}$$

Let $\mathcal{W}$ be the set of all possible such functions W_t.

Via a standard dynamic programming argument, we can check that the Bellman return function U is obtained by the recurrence relation

$$\forall W \in \mathcal{W}, \quad U_T(W) \quad = \quad \int M(x) W(x)\,dx\,, \tag{14}$$

$$\forall W \in \mathcal{W}, \quad U_t(W) \quad = \quad \inf_u \mathbb{E}_y U_{t+1}\left(F_t(W, u, y)\right)\,. \tag{15}$$

Moreover, assume that the minimum in u is attained in (15) above at $u = \hat{\mu}_t(W)$. Then (13) and

$$u_t = \hat{\mu}_t(W_t) \tag{16}$$

define an optimal controller for the stochastic control problem. The optimal cost is $U_0(N)$.

3.2.2 Minimax dynamic programming

Let us consider now the problem of minimizing $G(\mu)$. We have to introduce the *conditional state cost measure* and its cost density W (according to the concepts introduced in section 2.1 following [1]). It is defined as the maximum possible past cost knowing the past information, as a function of current state. To be more precise, let us introduce the following subsets of Ω. Given a pair $(u^t, y^t) \in \mathsf{U}^t \times \mathsf{Y}^t$, and a subset A of $\mathrm{I\!R}^n$, let

$$\Omega_t(A \mid u^t, y^t) = \{\omega \in \Omega \mid y^t = \eta^t(u^{t-1}, \omega^t), \text{ and } \phi_{t+1}(u^t, \omega^t) \in A\}. \tag{17}$$

For any $x \in \mathrm{I\!R}^n$, we shall write $\Omega_t(x \mid u^t, y^t)$, or simply $\Omega_t(x)$ when no ambiguity results, for $\Omega_t(\{x\} \mid u^t, y^t)$. And likewise for $\Omega_{t-1}(x)$.

The conditional cost measure of A is $\sup_{\omega \in \Omega_{t-1}(A)}[N(x_0) + \Gamma(w_{[0,T]})]$, and hence the conditional cost density function is

$$W_t(x) = \sup_{\omega \in \Omega_{t-1}(x)} \left[\sum_{t=0}^{T-1} \Gamma_t(w_t) + N(x_0) \right].$$

Initialize this sequence with

$$W_0(x) = N(x).$$

It is a simple matter to write recursive equations of the form

$$W_{t+1} = F_t(W_t, u_t, y_t).$$

In fact, F_t is defined by the following. Let for ease of notations

$$Z_t(x \mid u, y) = \{(\xi, v) \in \mathrm{I\!R}^n \times \mathsf{W} \mid f_t(\xi, u, v) = x, \quad h_t(\xi, u, v) = y\},$$

then we have

$$W_{t+1}(x) = \sup_{(\xi,v) \in Z_t(x|u_t,y_t)} [W_t(\xi) + \Gamma_t(v)]. \tag{18}$$

It is worthwhile to notice that, for any function $\psi(x)$, (such that the max exists)

$$\max_x [W_{t+1}(x) + \psi(x)] = \mathrm{I\!F}_{w_t} \max_{x|h_t(x,w_t)=y} [W_t(x) + \psi(f_t(x, u_t, w_t))]$$

and that hence

$$\max_{y}\max_{x}[W_{t+1}(x) + \psi(x)] = \max_{x}\mathbb{F}_{w_t}[W_t(x) + \psi(f_t(x, u_t, w_t))],$$

the counterparts of (12) and (11) above.

As was probably first shown in [20], (also presented in a talk in Santa Barbara in July 1993), one can do simple dynamic programming in terms of this function W. The value function U will now be obtained through the following relation

$$\forall W \in \mathcal{W}, \quad U_T(W) \;=\; \sup_{x}(M(x) + W(x)), \tag{19}$$

$$\forall W \in \mathcal{W}, \quad U_t(W) \;=\; \inf_{u}\sup_{y} U_{t+1}(F_t(W, u, y)). \tag{20}$$

Moreover, assume that the minimum in u is attained in (20) above at $u = \hat{\mu}(W)$. Then it defines an optimal feedback (16), with W_t now defined by (18), for the minimax control problem. The optimal cost is $U_0(N)$.

Of course, all our setup has been arranged so as to stress the parallel between (14),(15) on the one hand, and (19),(20) on the other hand.

3.3 Separation theorem

3.3.1 Stochastic separation theorem

We are here in the stochastic setup. The performance criterion is H and W stands for the conditional state probability density.

We introduce the *full information* Bellman return function V_t defined by the classical dynamic programming recursion :

$$\forall x \in \mathbb{R}^n, \quad V_T(x) = M(x),$$

$$\forall x \in \mathbb{R}^n, \quad V_t(x) = \inf_{u}\mathbb{E}_{w_t} V_{t+1}(f_t(x, u, w_t)).$$

Then we can state the following result.

Proposition 3.1 *Let*

$$S_t(x, u) := \mathbb{E}_{w_t} V_{t+1}\left(f_t(x, u, w_t)\right) W_t(x).$$

If there exists a (decreasing) sequence of (positive) numbers R_t with $R_T = 0$ such that,

$\forall t \in [0, T-1], \forall u_{[0,T]} \in \mathbf{U}, \forall \omega \in \Omega,$

$$\int \min_{u} S_t(x, u)\,dx + R_t = \min_{u}\int S_t(x, u)\,dx + R_{t+1},$$

then the optimal control is obtained by minimizing the conditional expectation of the full information Bellman return function, i.e. choosing a minimizing u in the right hand side above.

Proof The proof relies on the following fact :

Lemma 3.1 *Under the hypothesis of the proposition, the function*

$$U_t(W) = \int V_t(x) W(x)\, dx + R_t \tag{21}$$

satisfies the dynamic programming equations (14)(15).

Let us check the lemma. Assume that

$$\forall W_{t+1} \in \mathcal{W}, \quad U_{t+1}(W_{t+1}) = \int V_{t+1}(x) W_{t+1}(x)\, dx + R_{t+1}$$

and apply (15), using (12)

$$U_t(W_t) = \min_u \mathbb{E}_y \int \mathbb{E}_{w_t} V_{t+1}(f_t(x, u, w_t)) W_t^y(x)\, dx + R_{t+1}$$

and, according to (11) this yields

$$U_t(W_t) = \min_u \int \mathbb{E}_{w_t} V_{t+1}(f_t(x, u, w_t)) W_t(x)\, dx + R_{t+1}\,.$$

Using the hypothesis of the proposition and Bellman's equation for V_t, we obtain

$$U_t(W_t) = \int V_t(x) W_t(x)\, dx + R_t\,,$$

and the recursion relation holds.

The hypothesis of the theorem sounds in a large extent like wishful thinking. It holds, as is easily checked, in the linear quadratic case. (In that case, symmetry properties result in the certainty equivalence theorem.) There is little hope of finding other instances. We state it here to stress the parallel with the minimax case.

3.3.2   Minimax separation theorem

This section is based upon [6] [7]. The same result is to appear independently in [17].

We are now in the minimax setup. The performance criterion is G, and W stands for the conditional state cost density.

We introduce the *full information* Isaacs Value function $V_t(x)$ which satisfies the classical Isaacs equation:

$$\forall x \in \mathbb{R}^n, \quad V_T(x) = M(x)\,,$$

$$\forall x \in \mathbb{R}^n, \quad V_t(x) = \inf_u \mathbb{F}_{w_t} V_{t+1}(f_t(x, u, w_t))\,.$$

Notice we do not need that the Isaacs condition holds, i.e., the existence of a saddle point in the right hand side above. If it does not, V is an upper value, which is what is needed in the context of minimax control.

It is convenient here to introduce a binary operation denoted $\oplus$ which can be either the ordinary addition or its dual in our morphism: the max operation.

Proposition 3.2 *Let*

$$S_t(x, u) = I\!\!F_{w_t}\left[V_{t+1}\left(f_t(x, u, w)\right) + W_t(x)\right].$$

If there exists a (decreasing) sequence of numbers R_t, such that,

$$\forall t \in [0, T-1], \forall u_{[0,T]} \in \mathbf{U}, \forall \omega \in \Omega,$$
$$\max_x \min_u S_t(x, u) \oplus R_t = \min_u \max_x S_t(x, u) \oplus R_{t+1},$$

then the optimal control is obtained by minimizing the conditional worst cost, future cost being measured according to the full information Isaacs Value function, i.e. taking a minimizing u in the right hand side above.

Proof The proof relies on the following fact :

Lemma 3.2 *Under the hypothesis of the proposition, the function*

$$U_t(W) = \max_x[V_t(x) + W(x)] \oplus R_t$$

satisfies the dynamic programming equations (19)(20).

Let us check the lemma. Assume that

$$\forall W_{t+1} \in \mathcal{W}, \quad U_{t+1}(W_{t+1}) = \max_x[V_{t+1}(x) + W_{t+1}(x)] \oplus R_{t+1}$$

and apply (20), using (18)

$$U_t(W) = \min_u \max_y \left(\max_x[V_{t+1}(x) + \max_{(\xi,v) \in Z_t(x|u,y)} (W_t(\xi) + \Gamma_t(v))] \oplus R_{t+1} \right).$$

The max operations may be merged into

$$U_t(W) = \min_u \left(\max_{\xi,v}[V_{t+1}(f_t(\xi, u, v)) + \Gamma_t(v) + W_t(\xi)] \oplus R_{t+1} \right).$$

Then, using the hypothesis of the proposition and Isaacs equation for V, we obtain

$$U_t(W) = \max_x[V_t(x) + W_t(x)] \oplus R_t,$$

thus establishing the recursion relation.

The hypothesis of the proposition is not as unrealistic as in the stochastic case. It is satisfied in the linear quadratic case, but more generally, it can be satisfied if S is convex-concave, for instance, with $\oplus$ the ordinary addition and $R_t = 0$ (or $\oplus$ the max operation and $R_t = -\infty$). Moreover, in that case, the same u provides the minimum in both sides, yielding a certainty equivalence theorem.

3.4   An abstract formulation

It is known that in the stochastic control problem, some results, including derivation of the separation theorem, are more easily obtained using a more abstract formulation of the observation process, in terms of a family of σ-fields $\mathcal{Y}_t$ generated in the disturbance space. The axioms are that

- the brownian motion w_t is *adapted* to the family $\mathcal{Y}_t$,

- the family $\mathcal{Y}_t$ is *increasing*.

The same approach can be pursued in the minimax case. Instead of an explicit observation through an output (2), one may define the observation process in the following way. To each pair $(u_{[0,T]}, \omega)$ the observation process associates a sequence $\{\Omega_t\}_{t \in [0,T]}$ of subsets of Ω. The axioms are that, for any $(u_{[0,T]}, \omega)$, the corresponding family Ω_t satisfies the following properties.

- The process is *consistent*, i.e. $\forall t, \quad \omega \in \Omega_t$.

- The process is *strictly non anticipative*, i.e. $\omega \in \Omega_t \Leftrightarrow \omega^{t-1} \in \Omega_t^{t-1}$ where Ω_t^{t-1} stands for the set of restrictions to $[0, t-1]$ of the elements of Ω_t.

- The process is *with complete recall*: $\forall (u_{[0,T]}, \omega), \quad t < t' \Rightarrow \Omega_t \supset \Omega_{t'}$.

In the case considered above, we have

$$\Omega_t = \Omega(\mathbb{R}^n \mid u^t, y^t)$$

but the abstract formulation suffices, and allows one, for instance, to extend the minimax certainty equivalence principle to a variable end time problem. See [7] for a detailed derivation.

One may think of the subsets Ω_t as playing the role of the measurable sets of the σ-field $\mathcal{Y}_t$.

4 The continuous time control problem

4.1 The problem

We now have a continuous time system, of the form

$$\dot{x} \;=\; f_t(x,u,w)\,, \tag{22}$$
$$y \;=\; h(x,w)\,. \tag{23}$$

The notations will be the counterpart of the discrete ones. In particular, u^t, will stand for the restriction to $[0,t]$ of the continuous time function $u_{[0,T]} : t \mapsto u_t$. We shall again let U^t designate the set of such segments of function. Likewise for $w^t \in \mathsf{W}^t$, $\omega^t \in \Omega^t$, and $y^t \in Y^t$. Notice however that (3) and (4) must be replaced by

$$x_t \;=\; \phi_t(u^t,\omega^t)\,, \tag{24}$$
$$y_t \;=\; \eta_t(u^t,\omega^t)\,, \tag{25}$$

and similarily for (5) and (6).

Admissible controllers will be of the form $u_t = \mu_t(u^t,y^t)$. This seems to be an implicit definition, since u_t is contained in u^t. In fact, it is hardly more so than any feedback control. In any event, we let $\mathcal{M}$ be the class of controllers of that form, such that they generate a unique trajectory for any $\omega \in \Omega$.

As in the discrete case, we may always bring a classical integral plus terminal cost to the form (7). The two problems we want to investigate are again the minimization of $H(\mu)$ given by (8) with a stochastic model for ω and that of $G(\mu)$ given by (9) with a cost density for ω (or its sole set membership description if we take this cost density constant).

4.2 Hamilton Jacobi theory

4.2.1 Stochastic Hamilton Jacobi theory

In the continuous time case, the technicalities of diffusion processes and Ito calculus make the stochastic problem much more complex than its discrete counterpart, or, for that matter, than its continuous minimax counterpart. As far as we know, the classical literature concentrates on simpler, technically tractable, particular cases of the system (22),(23). Typically, classical nonlinear stochastic control deals with the system

$$dx_t \;=\; b_t(x,u)\,dt + \sigma_t(x,u)\,dw_t\,, \tag{26}$$
$$dy_t \;=\; c_t(x)\,dt + dv_t\,, \tag{27}$$

where v_t and w_t are standard independent vector brownian motions, and the above equations are to be taken in the sense of stochastic integrals. We shall need the notation $\sigma\sigma' = a$ where the prime stands for transposed, i.e.

$$a_{ij} = \sum_k \sigma_{ik}\sigma_{jk}\,.$$

Under suitable regularity and growth assumptions, one may compute a conditional state probability distribution W_t through the stochastic PDE (which can be derived, for instance, from Zakai's equation, see [12]), the dual form of which may be written $W_0 = N$ and, for any function $\psi(\cdot) \in C^2(\mathbb{R}^n)$,

$$d \int \psi(\xi)W_t(\xi)\,d\xi =$$
$$\left(\int (L_t(u)\psi)(\xi)W_t(\xi)\,d\xi\right) dt + \left(\int \psi(\xi)W_t(\xi)[c_t'(\xi) - \bar{c}_t']\,d\xi\right)(dy_t - \bar{c}_t\,dt),$$

where

$$(L_t(u)\psi)(\xi) = \frac{\partial\psi}{\partial x}(\xi)b_t(\xi,u) + \frac{1}{2}\sum_{ij}\frac{\partial^2\psi}{\partial x_i \partial x_j}(\xi)a_{ij}(\xi,u)\,, \qquad (28)$$

and $\bar{c}_t$ stands for the conditional expectation of $c_t(x_t)$:

$$\bar{c}_t = \int c_t(z)W_t(z)\,dz\,.$$

A full information control problem can be written in terms of that probability density as a state. We refer to [12] for a complete treatment. The formal development is too intimately intermingled with the technical aspects to lend itself to a simple exposition of the kind given here. In particular, a nonlinear Hamilton Jacobi theory would imply the Ito calculus with an infinite dimensional state, which we have rather avoided writing.

4.2.2   Minimax Hamilton Jacobi theory

The minimax problem is not as complex as the stochastic one, at least to state formally, and as long as one only seeks sufficient conditions. It was independently developed in [6], and in [19] in a slightly less general context, but with a much more complete development in that it includes a first mathematical analysis of the resulting Isaacs equation.

We introduce the counterpart of (17) : for a given pair $(u^t, y^t) \in \mathsf{U}^t \times \mathsf{Y}^t$ and a subset A of $\mathbb{R}^n$, let

$$\Omega_t(A \mid u^t, y^t) = \{\omega \in \Omega \mid y^t = \eta^t(u^t, \omega^t), \text{ and } \phi_t(u^t, \omega^t) \in A\}$$

be the conditional disturbance subset of A, and again write $\Omega_t(\xi)$ instead of $\Omega_t(\{\xi\} \mid u^t, y^t)$. The conditional cost density function is now

$$W_t(x) = \sup_{\omega \in \Omega_t(x)} \left(N(x_0) + \int_0^T \Gamma_t(w_t)\,dt\right)\,.$$

If it is C^1, W_t satisfies a forward Hamilton Jacobi equation. Let

$$\mathsf{W}_t(x \mid y) = \{w \in \mathsf{W} \mid h_t(x, w) = y\}\,,$$

then this forward equation is, for u^t and y^t fixed:

$$\frac{\partial W_t(x)}{\partial t} = \sup_{w \in \mathsf{W}_t(x \mid y_t)} \left[-\frac{\partial W(x)}{\partial x} f_t(x, u_t, w) + \Gamma_t(w) \right] \qquad (29)$$

which we write as

$$\frac{\partial}{\partial t} W_t = \mathbb{F}_w^y \left[-\frac{\partial W(x)}{\partial x} f_t(x, u_t, w) \right] =: F_t(W_t, u_t, y_t)$$

and, together with the initial condition $W_0 = N$, it may define W_t along any trajectory.

Assume $\mathcal{W}$ is endowed with a topology for which U is absolutely continuous in W, and admits a Gâteaux derivative $D_W U$. Then, the value function $U_t(W)$ is obtained through the following Isaacs equation. U_T is again given by (19), and

$$\forall W \in \mathcal{W}, \quad \frac{\partial U_t(W)}{\partial t} + \inf_{u \in U} \sup_{y \in \mathsf{Y}} D_W U_t(W) F_t(W, u, y) = 0\,. \qquad (30)$$

Moreover, assume that the minimum in u is attained in (30) above at $u = \hat{\mu}_t(W)$, then (16) defines an optimal feedback for the minimax control problem. The optimal cost is $U_0(N)$.

Notice again that the easy task is to show a *sufficient* condition: if there exist C^1 functions W and U satisfying these equations, and if the feedback (16) is admissible, then we have a solution to the problem. It is worth noticing that the only existence result we are aware of is in [18], and is in a particular case somewhat similar to the set up we have outlined for the stochastic case.

A further remark is that in a case, say where $N = 0$ and $\Gamma = 0$, only the function W precisely characterizes the reachable set given the past information. Let $\mathsf{X}_t(u^t, y^t)$ be that set, then we have

$$W_t(x) = \left\{ \begin{array}{ll} 0 & \text{if } x \in \mathsf{X}_t\,, \\ -\infty & \text{if } x \notin \mathsf{X}_t\,. \end{array} \right.$$

This is of course highly nondifferentiable, an apparently serious drawback for this theory, since this is an important case.

There are two ways that may help resolve this problem. The first one is developed in [6]. It consists in using the Fenchel transform W^* of W, defined as

$$W^*(p) = \min_x [(p, x) - W(x)]\,. \qquad (31)$$

We show that, under some additional assumptions, W^* satisfies a dual forward Hamilton Jacobi equation:

$$\frac{\partial W_t^*(p)}{\partial t} + \sup_{w \in W_t(\xi_t|y_t)} [-pf_t(\xi_t, u_t, w) + \Gamma_t(w)] = 0 \,.$$

where

$$\xi_t = \frac{\partial W_t^*(p)}{\partial p} \,.$$

Now, U can be taken as a function of W^*. If W is a concave function, i.e. if X_t is convex in the case (31) above, the dual approach yields the exact minimax control. If W is not concave, the strategy thus computed yields a guaranteed cost $U_0(N^*)$.

Another possible way around the nondifferentiability of W is given by the following remark. One can replace W in the theory by a *parametrization* of W. Let $\mathcal{P}$ be a topological space, called the parameter space, and let $\pi : \mathcal{P} \to \mathcal{W}$ be a one-to-one map. Assume that to any pair (u^t, y^t) we can associate a time function p_t satisfying a differential equation

$$\dot{p}_t = \mathcal{F}_t(p_t, u_t, y_t)$$

such that $\pi(p_t)$ is the conditional cost density W_t of the process. Then it is clear that the Value function can be expressed in terms of p instead of U, and we recover the necessary differentiability to write the equivalent of (30), which becomes

$$\forall p \in \mathcal{P}, \quad \frac{\partial U_t(p)}{\partial t} + \inf_{u \in U} \sup_{y \in Y} D_p U_t(p) \mathcal{F}_t(p, u, y) = 0 \,.$$

In the case (31), p parametrizes as well $\mathsf{X}_t(u^t, y^t)$ as its characteristic function W. This is what we do in [22], where it is clear that W (or X_t) lies on a three dimensional manifold of $\mathcal{W}$ (or 2^{X}), so that we may take $\mathbb{R}^3$ for $\mathcal{P}$.

The $L_t(u)$ operator. We notice here a strange parallel. Assume that the dynamics are as in (26), or more precisely, since w is not a white noise anymore but a (deterministic) decision variable

$$f_t(x, u, w) = b_t(x, u) + \sigma_t(x, u)w \,.$$

Notice that the natural dual to the Gaussian law is the cost measure $\Gamma(w) = -1/2\|w\|^2$. Then, a counterpart of (28) is, for a function $\psi(x(t))$

$$\mathbb{F}\left[\frac{d\psi}{dt}(\xi, u)\right] = \frac{\partial \psi}{\partial x}(\xi)b_t(\xi, u) + \sum_{i,j} \frac{\partial \psi}{\partial x_i}(\xi)\frac{\partial \psi}{\partial x_j}(\xi)a_{ij}(\xi, u) \,.$$

At this stage, we do not know whether the similarity with (28) is anything more than a curiosity. Notice that the above operator is not linear. (Here we refer to (max, +) linearity.)

4.3   Separation theorem

4.3.1   Stochastic separation theorem

We may take advantage of the linear character of the equation (21) to write, at least formally, the continuous time counterpart to the stochastic separation principle of section 3.3.1. We need first to introduce the *full information* (state feedback) Bellman function $V_t(x)$ which satisfies the *stochastic Bellman equation* (see [13])

$$V_T = M \,,$$

$$\forall (t, x) \in [0, T] \times \mathbb{R}^n, \quad \frac{\partial V_t}{\partial t}(x) + \inf_u (L_t(u) V_t)(x) = 0 \,.$$

We can then state the following result.

Proposition 4.1 *Let*

$$S_t(x, u) = (L_t(u) V_t)(x) W_t(x) \,.$$

If there exists a real (positive) $L^1([0, T])$ function r_t such that,

$\forall t \in [0, T], \forall u_{[0,T]} \in \mathbf{U}$, almost surely

$$\int \min_u S_t(x, u) \, dx + r_t = \min_u \int S_t(x, u) \, dx \,,$$

then an optimal control is obtained by choosing the minimizing u, that we shall call $\hat{\mu}_t(W_t)$, on the above right hand side.

Proof. The proof relies on the following fact.

Lemma 4.1 *Under the hypothesis of the proposition, the stochastic process $\alpha_t = U_t(W_t)$ is a submartingale for any admissible control, and a martingale if $u_t = \hat{\mu}_t(W_t)$, where the function U_t is defined over the set $L^1(\mathbb{R}^n)$ by*

$$U_t(W) = \int_{\mathbb{R}^n} V_t(\xi) W(\xi) \, d\xi + R_t \,,$$

and

$$R_t = \int_t^T r_s \, ds \,. \tag{32}$$

Let us check the lemma. Consider the diffusion process $\alpha_t = U_t(W_t)$ where the system, and thus the filter, is driven by a control process u_t. It satisfies the stochastic differential equation

$$d\alpha_t = \int [L_t(u_t) V_t - \inf_u L_t(u) V_t](\xi) W_t(\xi) \, d\xi \, dt - r_t \, dt$$

$$+ \int V_t(\xi) W_t(\xi) [c'_t(\xi) - \bar{c}'_t] \, d\xi [dy_t - \bar{c}_t \, dt] \,.$$

Assume that u_t is *admissible*, i.e. measurable over the σ-field $\mathcal{Y}_t$ generated by the observation process y_t, and take the conditional expectation. One obtains, at least formally

$$\mathbb{E}^{\mathcal{Y}_t} d\alpha_t = \int [L_t(u_t)V_t - \inf_u L_t(u)V_t](\xi)W_t(\xi)\, d\xi\, dt - r_t\, dt\,.$$

It follows that if $u_t = \hat{\mu}_t(W_t)$, which is indeed admissible, the hypothesis of the proposition yields $\mathbb{E}d\alpha_t = 0$, and for any other admissible control $\mathbb{E}d\alpha_t \geq 0$. Thus under the feedback control $\hat{\mu}_t(W_t)$, $\mathbb{E}U_T(W_T) = U_0(W_0)$, hence, recalling that $V_T = M$ and $W_0 = N$, $\mathbb{E}M(x_T) = \mathbb{E}V_0(x_0) + R_0$. And for any other admissible control, $\mathbb{E}U_T(W_T) \geq U_0(W_0)$, hence $\mathbb{E}M(x_T) \geq \mathbb{E}V_0(x_0) + R_0$.

The above proof is formal in that we have not detailed the regularity and growth hypotheses under which these calculations are valid. But it can be made rigorous, and we provide a proof of the separation theorem for the linear quadratic case as an example.

4.3.2   Minimax separation theorem

Introduce as in the discrete time case the *full information* Isaacs' Value function V. This satisfies the Isaacs equation

$$\forall x \in \mathbb{R}^n, \quad V_T(x) = M(x),$$
$$\forall t, \forall x \in \mathbb{R}^n, \quad \frac{\partial V_t(x)}{\partial t} = \min_{u \in U} \mathbb{F}_{w_t}\left(\frac{\partial V_t(x)}{\partial x} f_t(x, u, w_t)\right).$$

The use of weak solutions, the viscosity solution, is now well understood. However, for our purpose here, which is to stress the formal duality according to Quadrat's morphism, we shall assume that V and W are C^1. We shall also assume that the full information game admits a unique state feedback solution $u_t = \phi_t^*(x_t)$, argument of the min above.

As in [3], introduce also the *auxiliary problem*

$$\max_{x \in \mathbb{R}^n} [V_t(x) + W_t(x)],$$

and assuming it has a (nonunique) solution, let $\hat{X}_t$ be the set of maximizing x's, or *conditional worst states*.

We have the following fact:

Proposition 4.2 *Let*

$$S_t(x, u) = \mathbb{F}_{w_t}\left[\frac{\partial V_t}{\partial x}(x) f_t(x, u, w_t)\right].$$

If there exists a real (positive) $L^1([0,T])$ function r_t such that

$$\forall t \in [0,T], \forall u_{[0,T]} \in \mathbf{U}, \forall \omega \in \Omega, \quad \min_{x \in \hat{X}_t} \min_u S_t(x,u) + r_t = \min_u \max_{x \in \hat{X}_t} S_t(x,u),$$

then an optimal control is obtained by minimizing the conditional worst rate of increase of the full information value function among the conditional worst states, i.e., taking the minimizing u on the right hand side above.

Proof The proof relies on the following fact :

Lemma 4.2 *Under the hypothesis of the proposition, the function*

$$U_t(W) = \max_x [V_t(x) + W_t(x)] + R_t$$

with R_t defined as in (32) satisfies the dynamic programming equations (19), and (30) replacing derivatives with right derivatives in time.

The lemma hinges on Danskin's theorem [11] to obtain for the right time derivative

$$\left(\frac{\partial U_t}{\partial t} \right)^+ (W) = \max_{x \in \hat{X}_t} \frac{\partial V_t}{\partial t}(x) - r_t$$

and for the directional derivative in a direction $dW \in \mathcal{W}$:

$$D_W U_t(W) \cdot dW = \max_{x \in \hat{X}_t} dW(x).$$

Then it is a simple matter to place this in (30); notice that because all $x \in \hat{X}_t$ maximize the auxiliary problem, then at these points $-\partial W_t / \partial x = \partial V_t / \partial x$, and that, as in the case of mathematical expectations, the cascade of the two max operators $\max_y \max_{w \in W(x|y)}$ collapses in $\max_w$, to get the result.

Remarks The condition of the proposition looks a bit odd. A first remark is that W_t, hence ω, *seems* not to enter it. This is of course *not* the case, because $\hat{X}_t$ depends on W_t.

The second remark is that we have quoted the proposition this way to stress a parallel with the other cases. (The parallel would have been better if we had not converted $- \max(-\bullet)$ in $\min(\bullet)$.) But its only reasonable use seems to be the following corollary, the now well known minimax certainty equivalence principle of [3],[9] :

Corollary 4.1 *If $\forall t \in [0,T]$, $\forall u_{[0,T]} \in \mathbf{U}$, $\forall \omega \in \Omega$, $\hat{X}_t$ is a singleton $\{\hat{x}_t\}$, then an optimal control is obtained by replacing x_t by $\hat{x}_t$ in the optimal state feedback of the full information problem, i.e. taking $u_t = \phi_t^*(\hat{x}_t)$.*

4.4　An abstract formulation

The abstract formulations of the observation process have indeed been originally introduced for the continuous time problems. The parallel here is exactly the same as in the discrete time case, the only difference for the minimax problem being that nonanticipativeness of the process is now written as

$$\omega \in \Omega_t \Leftrightarrow \omega^t \in \Omega_t^t.$$

This approach to proving the certainty equivalence theorem was first proposed in [9]. It allows one to extend the theorem to variable end time problems.

5　Conclusion

The parallel between stochastic and minimax control appears striking, even if some technicalities make it less clear in the continuous time case than in the discrete time case. Some more work probably remains to be done to fully explain and exploit it. But it is clear that "Quadrat's morphism" is at the root of the problem.

REFERENCES

[1] M. Akian, J-P. Quadrat and M. Viot: "Bellman Processes", 11th International Conference on Systems Analysis and Optimization, Sophia Antipolis, 1994.

[2] F. Baccelli, G. Cohen, J-P. Quadrat, G-J. Olsder: *Synchronization and Linearity*, Wiley, Chichester, 1992.

[3] T. Başar and P. Bernhard: H_∞-*Optimal Control and Related Minimax Design Problems, a Game Theory Approach*, Birkhäuser, Boston, 1991.

[4] J.A. Ball and J.W. Helton : "$\mathcal{H}_\infty$-Control for Nonlinear Plants: Connections with Differential Games", 28th CDC, Tampa, 1989

[5] J.A. Ball, J.W. Helton and M.L. Walker: "H_∞-Control For Nonlinear Systems with Output Feedback", IEEE Trans. Automatic Control, **AC-38**, 1993, 546–559.

[6] P. Bernhard "Sketch of a Theory of Nonlinear Partial Information Min-Max Control", Research Report 2020, INRIA, France, 1993, to appear in a revised form, IEEE Trans. A.C., 1994.

[7] P. Bernhard: "A Discrete Time Min-max Certainty Equivalence Principle", to appear, Systems and Control Letters, 1994.

[8] P. Bernhard and G. Bellec : "On the Evaluation of Worst Case Design, with an Application to the Quadratic Synthesis Technique", 3rd IFAC Symposium on Sensitivity, Adaptivity and Optimality, Ischia, 1973.

[9] P. Bernhard and A. Rapaport: "Min-max Certainty Equivalence Principle and Differential Games", revised version of a paper presented at the Workshop on Robust Controller Design and Differential Games, UCSB, Santa Barbara, 1993. Submitted for publication.

[10] G. Didinsky, T. Başar, and P. Bernhard: "Structural Properties of Minimax Policies for a Class of Differential Games Arising in Nonlinear H_∞-control", Systems and Control Letters, 1993,

[11] F. Clarke: *Optimization and Nonsmooth Analysis*, John Wiley, 1983.

[12] W.H. Fleming and E. Pardoux: "Existence of Optimal Controls for Partially Observed Diffusions", SIAM Jal on Control and Optimisation **20**, 1982, 261–283.

[13] W.H. Fleming and R.W. Rishel: *Deterministic and Stochastic Optimal Control*, Springer Verlag, New York, 1975.

[14] V. Fromion: "A Necessary and Sufficient Condition for Incremental Stability of Nonlinear Systems", to appear in this symposium.

[15] A. Isidori : "Feedback Control of Nonlinear Systems", 1st ECC, Grenoble, 1991

[16] A. Isidori and A. Astolfi : Disturbance Attenuation and $\mathcal{H}_\infty$ Control via Measurement Feedback in Nonlinear Systems", IEEE Trans. Automatic Control, **AC 37**, 1992 1283-1293,

[17] M. James: "On the Certainty Equivalence Principle and the Optimal Control of Partially Observed Dynamic Games", to appear IEEE Trans. on Automatic Control, 1994.

[18] M.R. James and J.S. Barras: "Partially Observed Differential Games, Infinite Dimensional HJI Equations, and Nonlinear H_∞ Control". To appear.

[19] M.R. James, J.S. Barras and R.J. Elliott: "Output Feedback Risk Sensitive Control and Differential Games for Continuous Time Nonlinear Systems", 32nd CDC, San Antonio, 1993.

[20] M.R. James, J.S. Barras and R.J. Elliott: "Risk Sensitive Control and Dynamic Games for Partially Observed Discrete Time Nonlinear Systems", IEEE Trans. on Automatic Control **AC-39**, 1994, 780–792

[21] J-P. Quadrat: "Théorèmes asymptotiques en programmation dynamique", *Compte Rendus de l'Académie des Sciences*, 311, 1990 745–748.

[22] A. Rapaport and P. Bernhard: "Un jeu de poursuite-evasion avec connaissance imparfaite des coordonnées", International Symposium on Differential Games and Applications, St Jovite, Canada, 1994.

[23] A.J. van der Schaft : "Nonlinear State Space $\mathcal{H}_\infty$ Control Theory", ECC 1993, Groningen.

H^∞-Control of Nonlinear Singularly Perturbed Systems and Invariant Manifolds

Emilia Fridman *

Tel-Aviv University, Department of Electrical Engineering - Systems,
Ramat-Aviv 69978, Tel-Aviv, Israel.

Abstract

We study the state feedback H^∞-suboptimal control problem for an affine nonlinear singularly perturbed system in the infinite horizon case. Game-theoretic approach to this problem leads to a Hamilton-Jacobi partial differential equation, which is solvable iff there exists a special invariant manifold of the corresponding Hamiltonian system. We get sufficient conditions for the solvability of the H^∞-suboptimal control problem in terms of a reduced-order slow submanifold, or in the hyperbolic case, in terms of a reduced-order slow Riccati equation. We construct an asymptotic expansion of the optimal controller in the powers of a small parameter by solving the reduced-order slow partial differential equations and algebraic equations. Under some assumptions a high-order accuracy controller achieves a performance with a high order of accuracy.

1 Introduction

Game-theoretic approach to the state-feedback singularly perturbed H^∞- suboptimal control problem leads to a high dimensional Hamilton-Jacobi partial differential equation, or inequality, of two-time-scales for an optimal controller evaluation [11]. To alleviate the difficulties caused by the high dimensionality and the stiffness that result from the interaction of slow and fast dynamical modes, Pan and Basar [10,11] have designed a composite controller based on the reduced-order slow and fast subproblems. This independence of the singular perturbation parameter ε controller is an $O(\varepsilon)$ -approximation, to the optimal one, that assures a given performance level. Under some assumptions, the composite controller achieves the desired performance level for the full-order system for all sufficiently small ε.

---

*Supported in part by the Ministry of Science and the Arts and by the Ministry of Absorption of Israel

However, for values of ε that are not too small, higher-order approximations to the optimal controller are needed to guarantee the desired performance level. In the linear case such approximations can be constructed on the basis of the exact decomposition of the full-order Riccati equation into the reduced-order Riccati and linear algebraic equations [5]. Moreover, it has been proved that the full-order Riccati equation has a stabilizing solution iff the reduced-order slow Riccati equation has such a solution.

In the present paper we get the nonlinear counterpart of [5]. We apply the geometric approach of [7,13] which relates Hamilton-Jacobi equations with special invariant manifolds of Hamiltonian systems. We obtain the exact decomposition of the special slow-fast manifold into the reduced-order slow submanifold of the Hamiltonian system and the fast manifold of an auxiliary system. Unlike the linear case, the fast manifold depends also on the slow variables, and there is no immediate order reduction. Still the fast manifold can be found in the form of asymptotic expansions with terms evaluated by algebraic operations. The special manifold exists iff the Hamiltonian system possesses the slow submanifold. Thus, we get the reduced-order sufficient conditions for the solvability of the H^∞-suboptimal control problem in terms of the slow submanifold or, in the hyperbolic case, in terms of a slow Riccati equation. We construct a higher-order approximation to the optimal controller in the form of expansion in the powers of ε by solving slow partial differential equations and algebraic equations. Under some assumptions the high-order accuracy controller achieves the performance with a high order of accuracy. We consider a numerical example that shows that the higher-order controller improves performance.

The present paper is organized as follows. In the next section we formulate the nonlinear singularly perturbed H^∞-control problem and the known [7,13] sufficient conditions for its solvability in terms of the special invariant manifold of the Hamiltonian system. In section 3 we express this manifold via slow and fast ones. We formulate sufficient conditions for the H^∞-suboptimal control problem to be solvable. In section 4 we study the hyperbolic case. In section 5 we construct an asymptotic expansion for the optimal controller and consider an illustrative example. The paper ends with an Appendix containing proofs of the main theorems.

2   Problem formulation

Consider the system

$$
\begin{aligned}
\dot{x}_1 &= a_1(x_1) + A_1(x_1)x_2 + B_1(x_1)u + D_1(x_1)w, &\qquad (1a)\\
\varepsilon\dot{x}_2 &= a_2(x_1) + A_2(x_1)x_2 + B_2(x_1)u + D_2(x_1)w, &\qquad (1b)\\
z &= \mathrm{col}\{k_1(x_1) + k_2(x_1)x_2, u\}, &\qquad (1c)
\end{aligned}
$$

where $x_1(t) \in \mathbf{R}^{n_1}$ and $x_2(t) \in \mathbf{R}^{n_2}$ are the state vectors, $x = \text{col}\{x_1, x_2\}$, $u(t) \in \mathbf{R}^m$ is the control input, $w \in \mathbf{R}^q$ is the disturbance and $z \in \mathbf{R}^s$ is the output to be controlled. The functions a_i, A_i, B_i, D_i and k_i ($i = 1, 2$) are all smooth. We assume also that $a_i(0) = 0$ and $k_1(0) = 0$. The system (1a-1c) has a standard singularly perturbed form being nonlinear only on the slow variable x_1 (see e.g.[3,9,11]). In fact the results will be true also for more general systems containing nonlinear on x_2 terms of the order $O(\varepsilon)$ (see Remark 2 below).

Denote by $|\cdot|$ the Euclidean norm of a vector. Let γ be a fixed positive constant. Then, the nonlinear H^∞- suboptimal control problem (for performance level γ) is to find a nonlinear state-feedback

$$u = \beta(x), \quad \beta(0) = 0, \tag{2}$$

such that the closed-loop system of (1a-1c) and (2) has a L_2-gain less than or equal to γ [13]. It means that the following inequality holds:

$$\int_0^\tau |z(t)|^2 dt \le \gamma^2 \int_0^\tau |w(t)|^2 dt \tag{3}$$

for all $w \in L_2[0, \tau]$ and all $\tau \ge 0$, where z denotes the response of the closed-loop system of (1a-1c) and (2) for $w \in L_2[0, \tau]$ and the initial condition $x(0) = 0$ [7,13]. The H^∞-suboptimal control problem is solvable on $\Omega \subset \mathbf{R}^{n_1} \times \mathbf{R}^{n_2}$ containing 0 as an interior point if (3) holds for every $\tau \ge 0$ and for every $w \in L_2[0, \tau]$ for which the state trajectory of the closed-loop system (1a-1c) and (2) starting from 0 remains in Ω for all $t \in [0, \tau]$.

Consider the Hamiltonian function

$$\mathcal{H}_\gamma(x_1, x_2, p_1, p_2) = p_1'(a_1 + A_1 x_2) + p_2'(a_2 + A_2 x_2)$$
$$-\frac{1}{2}(p_1' p_2') \begin{pmatrix} S_{11} & S_{12} \\ S_{21} & S_{22} \end{pmatrix} \begin{pmatrix} p_1 \\ p_2 \end{pmatrix} + \frac{1}{2}(k_1' + x_2' k_2')(k_1 + k_2 x_2), \tag{4}$$

where prime denotes the transposition of a matrix, p_1 and εp_2 play the role of the costate variables and $S_{ij} = B_i B_j' - 1/\gamma^2 D_i D_j'$. The corresponding Hamiltonian system has the form:

$$\dot{x}_1 = f_1(x_1, p_1, x_2, p_2), \qquad\qquad \dot{p}_1 = f_2(x_1, p_1, x_2, p_2), \tag{5a}$$
$$\varepsilon \dot{x}_2 = A_2 x_2 - S_{22} p_2 + f_3(x_1, p_1), \qquad \varepsilon \dot{p}_2 = -k_2' k_2 x_2 - A_2' p_2 + f_4(x_1, p_1), \tag{5b}$$

where $f_1 = a_1 + A_1 x_2 - S_{11} p_1 - S_{12} p_2$, $f_2 = -\nabla_{x_1}\mathcal{H}$, $f_3 = a_2 - S_{21} p_1$, $f_4 = -A_1' p_1 - k_1' k_2$.

For each $\varepsilon > 0$ the problem is solvable on $\Omega \subset \mathbf{R}^{n_1} \times \mathbf{R}^{n_2}$ if there exists a C^2 nonnegative solution $V : \Omega \to \mathbf{R}$ to the Hamilton-Jacobi partial differential equation

$$\mathcal{H}_\gamma(x_1, x_2, V_{x_1}', \varepsilon^{-1} V_{x_2}') = 0, \quad V(0) = 0, \tag{6}$$

with the property that the system of the first equation of (5a) and the first equation of (5b) with $p_1 = V'_{x_1}, p_2 = \varepsilon^{-1}V'_{x_2}$ has an asymptotically stable equilibrium at $x = 0$ [7], [13], where (V_{x_1}, V_{x_2}) denotes the Jacobian matrix of V. The latter is equivalent to the existence of the invariant manifold of (5a-5b)

$$p_1 = Z_1(x_1, x_2), \quad p_2 = Z_2(x_1, x_2), \tag{7}$$

where

$$V_{x_1} = Z'_1, \quad V_{x_2} = \varepsilon Z'_2, \tag{8}$$

with asymptotically stable flow

$$\dot{x}_1 = a_1 + A_1 x_2 - S_{11} Z_1 - S_{12} Z_2 \,, \quad \varepsilon \dot{x}_2 = a_2 + A_2 x_2 - S_{21} Z_1 - S_{22} Z_2 \tag{9}$$

and such that $V \geq 0$, $V(0) = 0$ (that implies $V_x(0) = 0$). The controller

$$u = -[B'_1, \ \varepsilon^{-1} B'_2] \, V'_x = -B'_1 Z_1 - B'_2 Z_2 \tag{10}$$

solves the problem. Further, by an optimal controller we shall mean (10). Note that the manifold (7) is not necessarily the stable manifold of the Hamiltonian system (5a-5b) because (9) needs not to be exponentially stable [7].

We shall reduce the analysis of the $(2n_1 + 2n_2)$-dimensional Hamiltonian system (5a-5b) to the slow $2n_1$-dimensional subsystem, that corresponds to the restriction of (5a-5b) to its slow (center) manifold. Namely, we shall show that the existence of (7) is equivalent to the existence of the reduced order invariant manifold of the slow subsystem. Moreover, we shall find the functions Z_1 and Z_2 from algebraic equations by means of the latter manifold and a fast manifold of an auxiliary system.

3 Decomposition of the slow-fast manifold

For each $x_1 \in \mathbf{R}^{n_1}$ consider the fast linear subproblem

$$\dot{x}_2 = A_2 x_2 + B_2 u + D_2 w, \quad z = \mathrm{col}\{k_2 x_2, u\}, \tag{11}$$

and the corresponding algebraic Riccati equation

$$A'_2 M + M A_2 + k'_2 k_2 - M S_{22} M = 0 \,. \tag{12}$$

We assume further

A1. *For a given γ (12) has a positive definite symmetric solution $M(x_1)$, continuous on $x_1 \in \mathbf{R}^{n_1}$, such that for each $x_1 \in \mathbf{R}^{n_1}$ the matrices $A_2 - B_2 B'_2 M$ and $\Lambda = A_2 - S_{22} M$ are Hurwitz.*

Consider the matrix

$$R = \begin{pmatrix} A_2 & -S_{22} \\ -k'_2 k_2 & -A'_2 \end{pmatrix} = \begin{pmatrix} I & 0 \\ M & I \end{pmatrix} \begin{pmatrix} \Lambda & -S_{22} \\ 0 & -\Lambda' \end{pmatrix} \begin{pmatrix} I & 0 \\ -M & I \end{pmatrix} \tag{13}$$

Under A1 R possesses the following property: it has n_2 stable eigenvalues λ, $\mathrm{Re}\lambda < -\alpha < 0$, and n_2 unstable ones λ, $\mathrm{Re}\lambda > \alpha$ for all $|x_1| \le m$. Then for any $m > 0$ there exists $\varepsilon_m > 0$, such that for all $\varepsilon \in (0, \varepsilon_m]$ and $|x_1| + |p_1| < m$ the system (5a-5b) has the slow manifold [2,6,15]

$$\begin{pmatrix} x_2 \\ p_2 \end{pmatrix} = \begin{pmatrix} L_3^*(x_1, p_1, \varepsilon) \\ L_4^*(x_1, p_1, \varepsilon) \end{pmatrix} = L^*(x_1, p_1, \varepsilon). \tag{14}$$

The subscripts of L^* correspond to the third and the fourth variables in the system of (5a-5b). To avoid cumbersome notation we shall omit ε argument in the functions below.

Setting (14) into (5a) and substituting u_1 and v_1 for x_1 and p_1 respectively, we get the $2n_1$-dimensional system for the flow on the slow manifold:

$$\dot{u}_1 = f_1[u_1, v_1, L_3^*(u_1, v_1), L_4^*(u_1, v_1)], \tag{15a}$$
$$\dot{v}_1 = f_2[u_1, v_1, L_3^*(u_1, v_1), L_4^*(u_1, v_1)]. \tag{15b}$$

The function L^* can be found in the form of expansion

$$L^*(x_1, p_1, \varepsilon) = \sum_{j=0}^{q} \varepsilon^j l_j^*(x_1, p_1) + O(\varepsilon^{q+1}). \tag{16}$$

The terms of (16) can be determined from the equation

$$\varepsilon \frac{\partial L^*}{\partial x_1} f_1 + \varepsilon \frac{\partial L^*}{\partial p_1} f_2 = \begin{pmatrix} A_2 L_3^* - S_{22} L_4^* + f_3(x_1, p_1) \\ -k_2' k_2 L_3^* - A_2' L_4^* + f_4(x_1, p_1) \end{pmatrix}, \tag{17}$$

where $f_i = f_i(x_1, p_1, L_3^*, L_4^*)$, $i = 1, 2$, by algebraic operations. Thus, $l_0^* = -R^{-1} f_0$, where $f_0 = col\{f_3, f_4\}$. Note that (17) can be derived by differentiating on t of (14), where $x_1 = u_1(t)$, $p_1 = v_1(t), x_2 = x_2(t)$, $p_2 = p_2(t)$, and by substituting for $\dot{u}_1$ and $\dot{v}_1$ the right sides of (15a-15b).

Consider the slow system (15a-15b). Denote by $\Omega_{m_i} = \{x_i \in \mathbf{R}^{n_i} : |x_i| < m_i\}$, $i = 1, 2$. Our next assumption is

A2. There exist $m_1 > 0$ and $\varepsilon_1 > 0$ such that for all $\varepsilon \in (0, \varepsilon_1]$ and $u_1 \in \Omega_{2m_1}$ the system (15a-15b) possesses the invariant manifold

$$v_1 = N(u_1), \tag{18}$$

where the function $N = N(u_1, \varepsilon)$ is continuous on both arguments and uniformly bounded together with its first derivative on u_1, and $N(0) = 0$.

The restriction of (15a-15b) to (18) is governed by the n_1- dimensional system

$$\dot{u}_1 = F_1(u_1), \tag{19}$$

where

$$F_i(u_1) = f_i[u_1, N(u_1), L_3^*(u_1, N(u_1)), L_4^*(u_1, N(u_1))] \tag{20}$$

and $i = 1$. Additionally we assume

A3. *For all $\varepsilon \in (0, \varepsilon_1]$ equation (19) is asymptotically stable.*

The theorem below states that A2 and A3 are necessary conditions for the existence of the invariant manifold (7) with asymptotically stable flow (9).

Theorem 3.1 *Let A1 hold, and for all small enough ε there exist $\overline{m}_1$ and $\overline{m}_2$ such that the $(2n_1+2n_2)$-dimensional Hamiltonian system (5a-5b) has an invariant on $\Omega_{\overline{m}_1} \times \Omega_{\overline{m}_2}$ manifold (7) with (9) asymptotically stable, where V has continuous and uniformly bounded derivatives on $(x_1, x_2) \in \Omega_{\overline{m}_1} \times \Omega_{\overline{m}_2}$ up to the second order, then A2 and A3 are valid.*

For proof of theorem see Appendix. Note that the stable solutions of (5a-5b) are exponentially approaching the solutions on the slow manifold (14) [8,15]. Under A1-A3 we shall construct the invariant manifold (7) with the stable flow by means of the slow submanifold (18) and a fast manifold of an auxiliary system. To get the latter system let us introduce the following change of variables:

$$\begin{pmatrix} u_2 \\ \overline{p}_2 \end{pmatrix} = \begin{pmatrix} x_2 \\ p_2 \end{pmatrix} - L^*(x_1, p_1), \quad \begin{pmatrix} \overline{x}_1 \\ \overline{p}_1 \end{pmatrix} = \begin{pmatrix} x_1 \\ p_1 \end{pmatrix} - \begin{pmatrix} u_1 \\ v_1 \end{pmatrix}, \qquad (21)$$

where u_1 and v_1 satisfy (15a-15b). For the new variables we get the system

$$\begin{aligned}
\dot{\overline{x}}_1 &= g_1(u_1, v_1, \overline{x}_1, \overline{p}_1, u_2, \overline{p}_2), \quad \dot{\overline{p}}_1 = g_2(u_1, v_1, \overline{x}_1, \overline{p}_1, u_2, \overline{p}_2), \\
\varepsilon \dot{u}_2 &= A_2(\overline{x}_1 + u_1)u_2 - S_{22}(\overline{x}_1 + u_1)\overline{p}_2 + \varepsilon g_3(u_1, v_1, \overline{x}_1, \overline{p}_1, u_2, \overline{p}_2), (22) \\
\varepsilon \dot{\overline{p}}_2 &= -k_2'(\overline{x}_1 + u_1)k_2(\overline{x}_1 + u_1)u_2 - A_2'(\overline{x}_1 + u_1)\overline{p}_2 \\
&+ \varepsilon g_4(u_1, v_1, \overline{x}_1, \overline{p}_1, u_2, \overline{p}_2),
\end{aligned}$$

where for $i = 1, 2$

$$g_i = f_i[\overline{x}_1 + u_1, \overline{p}_1 + v_1, u_2 + L_3^*(\overline{x}_1 + u_1, \overline{p}_1 + v_1), \overline{p}_2 + L_4^*(\overline{x}_1 + u_1, \overline{p}_1 + v_1)]$$

$$-f_i[u_1, v_1, L_3^*(u_1, v_1), L_4^*(u_1, v_1)],$$

and for $i = 3, 4$

$$g_i = -\frac{\partial L_i^*}{\partial x_1}\Delta f_1 - \frac{\partial L_i^*}{\partial p_1}\Delta f_2,$$

$$\Delta f_j = f_j[\overline{x}_1 + u_1, \overline{p}_1 + v_1, u_2 + L_3^*(\overline{x}_1 + u_1, \overline{p}_1 + v_1), \overline{p}_2 + L_4^*(\overline{x}_1 + u_1, \overline{p}_1 + v_1)]$$

$$-f_j[\overline{x}_1 + u_1, \overline{p}_1 + v_1, L_3^*(u_1, v_1), L_4^*(u_1, v_1)], \quad j = 1, 2.$$

Let $m_2 > 0$ be any positive. We choose m' such that

$$|x_2 - L_3^*(x_1, N(x_1))| \leq m'/2, \quad (x_1, x_2) \in \Omega_{2m_1} \times \Omega_{m_2}.$$

Then under A1 there exists ε' such that for all $\varepsilon \in (0, \varepsilon']$ the system of (15a-15b) and (22) has the fast (stable) manifold for $|u_2| < m'$ [15,4]

$$\begin{pmatrix} \overline{x}_1 \\ \overline{p}_1 \end{pmatrix} = \begin{pmatrix} \varepsilon L_1^+(u_1, v_1, u_2) \\ \varepsilon L_2^+(u_1, v_1, u_2) \end{pmatrix}, \quad \overline{p}_2 = L_4^+(u_1, v_1, u_2), \tag{23}$$

where $L_4^+ = M(u_1)u_2 + O(\varepsilon)$. The functions $L_i^+(i = 1, 2, 4)$ satisfy the inequalities

$$|L_i^+(u_1, v_1, u_2)| \le c|u_2|, \quad |L_i^+(u_1, v_1, u_2) - L_i^+(u_1, v_1, \widetilde{u}_2)| \le c|u_2 - \widetilde{u}_2|, \tag{24}$$
$$|L_i^+(u_1, v_1, u_2) - L_i^+(\widetilde{u}_1, \widetilde{v}_1, u_2)| \le c|u_2|(|u_1 - \widetilde{u}_1| + |v_1 - \widetilde{v}_1|).$$

The flow on this manifold is governed by the decoupled system of the slow equations (5a-5b) and the fast equation

$$\varepsilon \dot{u}_2 = A_2 u_2 - S_{22} L_4^+ + \varepsilon g_3(u_1, v_1, \varepsilon L_1^+, \varepsilon L_2^+, u_2, L_4^+), \tag{25}$$

where $L_i^+ = L_i^+(u_1, v_1, u_2)(i = 1, 2, 4)$, $A_2 = A_2(u_1 + \varepsilon L_1^+)$ and $S_{22} = S_{22}(u_1 + \varepsilon L_1^+)$. The solution of (25) with the initial value $u_2(0) = u_2^0$ satisfies the inequality

$$|u_2(t)| \le K \exp\left(-\frac{\alpha}{\varepsilon} t\right) \cdot |u_2(0)|, \quad K > 0, \ t > 0. \tag{26}$$

Hence, due to the first inequality of (24), the solutions of (22) lying on the fast manifold of (23) are rapidly exponentially decaying as t increases. Substituting (18) and (23) into (21) we get the algebraic equations for Z_1 and Z_2 determination:

$$\begin{aligned} x_1 &= u_1 + \varepsilon L_1^+[u_1, N(u_1), u_2], &\tag{27a} \\ x_2 &= u_2 + L_3^*[x_1, N(u_1) + \varepsilon L_2^+(u_1, N(u_1), u_2)], &\tag{27b} \end{aligned}$$

and

$$\begin{aligned} p_1 &= N(u_1) + \varepsilon L_2^+[u_1, N(u_1), u_2], &\tag{28a} \\ p_2 &= L_4^*[x_1, N(u_1) + \varepsilon L_2^+(u_1, N(u_1), u_2)] + L_4^+(u_1, N(u_1), u_2). &\tag{28b} \end{aligned}$$

Consider (27a-27b) as the system with respect to u_1 and u_2. Using the contraction principle argument, one can prove that there exists ε_2 such that for $\varepsilon \in (0, \varepsilon_2]$, the system (27a-27b) has a unique solution on $\Omega_{m_1} \times \Omega_{m_2}$

$$\begin{aligned} u_1 &= U_1(x_1, x_2) = x_1 + \varepsilon \overline{U}_1(x_1, x_2), &\tag{29a} \\ u_2 &= U_2(x_1, x_2) = x_2 - L_3^*(x_1, N(x_1)) + \varepsilon \overline{U}_2(x_1, x_2), &\tag{29b} \end{aligned}$$

where the functions $\overline{U}_1$ and $\overline{U}_2$ are Lipschitzian on x_1 and x_2, they vanish at $(x_1, x_2) = 0$, and satisfy the inequalities $\varepsilon_2|\overline{U}_1| \le m_1$, $\varepsilon_2|\overline{U}_2| \le m'/2$.

Further, applying the implicit function theorem one can show that U_1 and U_2 are continuously differentiable on x_1 and x_2. Substituting (29a-29b) into (28a) and (28b) we get (7), where

$$
\begin{aligned}
Z_1 &= N(U_1) + \varepsilon L_2^+[U_1, N(U_1), U_2], &\text{(30a)}\\
Z_2 &= L_4^*[x_1, N(U_1) + \varepsilon L_2^+(U_1, N(U_1), U_2)] + L_4^+(U_1, N(U_1), U_2). &\text{(30b)}
\end{aligned}
$$

In the Appendix we prove

Theorem 3.2 *Under A1-A3 for any $m_2 > 0$ there exists $\varepsilon_2 > 0$ such that for all $\varepsilon \in (0, \varepsilon_2]$*

(i) the $(2n_1+2n_2)$-dimensional Hamiltonian system (5a-5b) has the invariant on $\Omega_{m_1} \times \Omega_{m_2}$ manifold (7) with (9) asymptotically stable, where continuously differentiable on x_1 and x_2 functions Z_1 and Z_2 are defined by formula (30a-30b) from the algebraic systems of (27a-27b) and (28a-28b);

(ii) there exists a C^2 function $V : \Omega_{m_1} \times \Omega_{m_2} \to \mathbf{R}$, satisfying the Hamilton-Jacobi equation (6);

(iii) if additionally $V \geq 0$, then the H^∞-suboptimal control problem is solvable on $\Omega_{m_1} \times \Omega_{m_2}$ by the controller (10).

Note that the asymptotic stability of the closed-loop system (1a-1c) and (10) with $w = 0$ given by

$$
\dot{x}_1 = a_1 + A_1 x_2 - B_1 B_1' Z_1 - B_1 B_2' Z_2, \quad \varepsilon \dot{x}_2 = a_2 + A_2 x_2 - B_2 B_1' Z_1 - B_2 B_2' Z_2
$$

$$\text{(31)}$$

implies $V \geq 0$, and thus the validity of (3) [7,13]. Applying the center manifold theory [2,14,15] to (31) one can reduce its stability to the stability of a slow subsystem. Under A1 there exists $\alpha_0 > 0$ such that for all $|x_1| \leq m_1$ the eigenvalues of the matrix $A_2 - B_2 B_2' M$ satisfy the inequality $Re\lambda < -\alpha_0$. This implies the existence of a center manifold $x_2 = P^*(x_1, \varepsilon)$ of the system (31) for all small ε and $|x_1| < m_1$[3,6,15]. The system (31) is globally asymptotically stable on $\Omega_{m_1} \times \Omega_{m_2}$ iff the flow on its center manifold

$$
\dot{x}_1 = a_1 + A_1 P^* - B_1 B_1' Z_1(x_1, P^*) - B_1 B_2' Z_2(x_1, P^*) , \quad \text{(32)}
$$

is globally asymptotically stable on Ω_{m_1} (see e.g. [15]). Note also that analogously to L^* the function P^* can be easily found in the form of expansion. These observations and Theorem 3.2 imply the following

Corollary 3.1 *Under A1-A3 let for all $\varepsilon \in (0, \varepsilon_2]$ the functions Z_1 and Z_2 be defined by (28a-28b) on $\Omega_{m_1} \times \Omega_{m_2}$. Suppose that the system (32) is globally asymptotically stable on Ω_{m_1}. Then for $\varepsilon \in (0, \varepsilon_2]$ the H^∞-suboptimal control problem is solvable on $\Omega_{m_1} \times \Omega_{m_2}$ by a controller (10).*

4   Hyperbolic case

Assumption A2 is not easily verifiable. In this section we will consider a particular case when A2 holds. Consider the linearization of (1a-1c) at 0:

$$
\begin{aligned}
\dot{x}_1 &= A_{11}x_1 + A_{12}x_2 + B_{10}u + D_{10}w, \\
\varepsilon\dot{x}_2 &= A_{21}x_1 + A_{22}x_2 + B_{20}u + D_{20}w, \quad\quad (33a) \\
z &= \mathrm{col}\{C_1x_1 + C_2x_2, u\}, \quad\quad (33b)
\end{aligned}
$$

where $A_{i1} = \frac{\partial a_i}{\partial x_1}(0)$, $A_{i2} = A_{i2}(0)$, $D_{i0} = D_i(0)$, $C_1 = \frac{\partial k_1}{\partial x_1}(0)$, and $C_2 = k_2(0)$. The Hamiltonian matrix corresponding to (33a-33b) is similar to the matrix Ham_γ that corresponds to the linearization at 0 of the Hamiltonian system (5a-5b). It is easy to see that

$$
Ham_\gamma = \begin{pmatrix} R_{11} & R_{12} \\ \varepsilon^{-1}R_{21} & \varepsilon^{-1}R_{22} \end{pmatrix}, \quad\quad (34)
$$

where for $i = 1,2$, $j = 1,2$

$$
R_{ij} = \begin{pmatrix} A_{ij} & -S_{ij} \\ -Q_{ij} & -A'_{ji} \end{pmatrix}, \quad S_{ij} = B_{i0}B'_{j0} - 1/\gamma^2 D_{i0}D'_{j0}, \quad Q_{ij} = C'_iC_j.
$$

Under A1 the matrix R_{22} has no purely imaginary eigenvalues since $R_{22} = R(0)$. The matrix (34) has one group of $2n_1$ small eigenvalues $O(\varepsilon)$ close to those of $R_0 = R_{11} - R_{12}R_{22}^{-1}R_{21}$ and another group of $2n_2$ large eigenvalues $O(1)$ close to those of $\varepsilon^{-1}R_{22}$ [2,9]. To guarantee that for all small enough ε the matrix (34) has no purely imaginary eigenvalues, i.e. the vectorfield defined by (5a-5b) is hyperbolic we suppose

A4. *The matrix* $R_0 = R_{11} - R_{12}R_{22}^{-1}R_{21}$ *has no purely imaginary eigenvalues.*

It is known [13] that in the hyperbolic case for each ε the H^∞ -suboptimal control problem is solvable on a small enough neighborhood of $\mathbf{R}^{n_1} \times \mathbf{R}^{n_2}$ containing 0 if the linearized problem is solvable. The latter is equivalent to the existence of a nonnegative definite stabilizing solution to a corresponding $(n_1 + n_2) \times (n_1 + n_2)$-algebraic Riccati equation (ARE). We will get the reduced-order (in terms of $n_1 \times n_1$-ARE) sufficient conditions for the solvability of H^∞-suboptimal control problem on the domains containing large values of x_2 for all sufficiently small ε.

Under A4 the matrix R_0 has n_1 eigenvalues with negative real parts and n_1 with positive ones. This fact follows from the symmetry of the eigenvalues of Ham_γ and of R_{22}. Note that R_0 coincides with the linearization of the slow subsystem (15a-15b) on u_1, v_1 at $(u_1, v_1, \varepsilon) = 0$. Hence under A4 the invariant manifold (18) of stable solutions of (15a-15b)(if it exists) is a stable manifold of (15a-15b). To guarantee its existence we consider the linearized on u_1 and v_1 at $(u_1, v_1, \varepsilon) = 0$ system (15a-15b):

$$\dot{u}_1 = T_1 u_1 + T_2 v_1, \quad \dot{v}_1 = T_3 u_1 + T_4 v_1, \quad \begin{pmatrix} T_1 & T_2 \\ T_3 & T_4 \end{pmatrix} = R_0. \qquad (35)$$

Suppose that the stable manifold (i.e. the stable eigenspace) of (35) can be parameterized by u_1-coordinates in the form $v_1 = N^{(0)} u_1$. Then $N^{(0)}$ satisfies the following $n_1 \times n_1$-ARE:

$$N^{(0)}(T_1 + T_2 N^{(0)}) = T_3 + T_4 N^{(0)}, \qquad (36)$$

and the matrix $T_1 + T_2 N^{(0)}$ is Hurwitz. We suppose further

A5. *ARE (36) has a solution $N^{(0)}$ such that the matrix $T_1 + T_2 N^{(0)}$ is Hurwitz.*

From the theory of nonlinear differential equations it is known that under A4 and A5 the hypotheses A2 and A3 are valid. Then from Theorem 3.2 (i) follows that the Hamiltonian system (5a-5b) has the stable manifold (7) on $\Omega_{m_1} \times \Omega_{m_2}$, where Z_1 and Z_2 are continuously differentiable.

To guarantee the solvability of the H^∞-suboptimal control problem we have to assure the asymptotic stability of (32) (see Corollary 3.1). Let Ξ be the matrix that corresponds to the linearization of (32) on x_1 and x_2 at $(x_1, \varepsilon) = 0$. Direct computations show that Ξ is the matrix of the reduced system (33a), where $w = 0$, $u = u_l^{(0)}$ and

$$u_l^{(0)} = -B_{10}'\left\{ N^{(0)} - \begin{pmatrix} M(0) & I \end{pmatrix} R_{22}^{-1} R_{21} \begin{pmatrix} I \\ N^{(0)} \end{pmatrix} \right\} x_1 - B_{20}' M(0) x_2.$$

Assume that

A6. *The matrix Ξ is Hurwitz.*

Under A6 the system (32) is locally asymptotically stable for small enough ε . Thus we obtain

Theorem 4.1 *Under A1 and A4-A6 for any $m_2 > 0$ there exist $m_1 > 0$ and $\varepsilon_3 > 0$ such that for all $\varepsilon \in (0, \varepsilon_3]$ the H^∞-suboptimal control problem is solvable on $\Omega_{m_1} \times \Omega_{m_2}$ by the controller (10), where Z_1 and Z_2 are defined by (30a-30b).*

5   Asymptotic expansion of the optimal controller

We shall find an asymptotic approximation to the controller (10) by expanding Z_1 and Z_2, defined by (30a-30b), into the powers of ε. We assume further

A7. *For small enough ε and $|u_1| \le 2m_1$ the function N can be represented in the form:*

$$N(u_1, \varepsilon) = \sum_{j=0}^{q} \varepsilon^j N_j(u_1) + O(\varepsilon^{q+1}).$$ (37)

Note that the hypotheses A4 and A5 imply A7. The terms of (37) can be found from the partial differential equation:

$$\frac{\partial N}{\partial u_1} F_1(u_1) = F_2(u_1),$$ (38)

where F_i are defined by (20).

Analogously to (16) we construct the expansion of the function $L^+ = col\{L_1^+, L_2^+, L_4^+\}$

$$L^+ = \sum_{j=0}^{q} \varepsilon^j l_j^+ + O(\varepsilon^{q+1})$$ (39)

from the equation [15]

$$\varepsilon \frac{\partial L^+}{\partial u_1} f_1 + \varepsilon \frac{\partial L^+}{\partial v_1} f_2 + \frac{\partial L^+}{\partial u_2}[A_2 u_2 - S_{22} L_4^+ + \varepsilon g_3]$$
$$= col\{g_1, g_2, -k_2' k_2 u_2 - A_2' L_4^+ + \varepsilon g_4\},$$ (40)

where A_2, S_{22} and k_2 depend on $u_1 + \varepsilon L_1^+$, and $f_i = f_i(u_1, v_1, L_3^*, L_4^*)$, $i = 1, 2$, $g_k = g_k(u_1, v_1, \varepsilon L_1^+, \varepsilon L_2^+, u_2, \varepsilon L_4^+)$, $k = 1, ...4$, $L^+ = L^+(u_1, v_1, u_2)$. For the terms of (37) we get successively the equations of the form:

$$\frac{\partial l_j^+}{\partial u_2}[A_2(u_1) - S_{22}(u_1) M(u_1)] u_2 = G_j(u_1, v_1, u_2),$$ (41)

where $G_j(u_1, v_1, 0) = 0$. The equation (41) depends on u_1 and v_1 as on the parameters, and its solution is given by

$$l_j^+ = \int_0^{\infty} G_j \left[u_1, v_1, e^{[A_2(u_1) - S_{22}(u_1) M(u_1)]t} u_2 \right] dt.$$ (42)

In the case of the system of (1a-1c) G_j and, hence, l_j^+ is a (j+1)-order polynomial with respect to u_2. The coefficients of this polynomial can be found from (41) by algebraic operations. In the case of the system of Remark 2 l_j^+ has a more complicated structure.

Next we obtain from (27a-27b)

$$U_i = \sum_{j=0}^{q} \varepsilon^q U_{ij}(x_1, x_2) + O(\varepsilon^{q+1}), \quad i = 1, 2.$$ (43)

Thus, $U_{10} = x_1$, $U_{20} = x_2 - l_{30}^*[u_1, N(u_1)]$. Substituting the expansions (43), (37), (39) and (16) into (28a) and (28b) we get the asymptotic approximations to Z_1 and Z_2:

$$Z_i = \sum_{j=0}^{q} \varepsilon^j Z_{ij}(x_1, x_2, \varepsilon) + O(\varepsilon^{q+1}), \quad i = 1, 2, \tag{44}$$

where

$$Z_{1j} = N_j a + l_{2,j-1}^+(a, b, c), \quad Z_{2j} = l_{4j}^*\left(x_1, \sum_{k=0}^{q-j} \varepsilon^k Z_{1k}\right) + l_{4j}^+(a, b, c), \tag{45}$$

$$a = \sum_{k=0}^{q-j} \varepsilon^k U_{1k}, \quad b = \sum_{k=1}^{q-j} N_k\left(\sum_{p=0}^{q-j-k} \varepsilon^p U_{1p}\right), \quad c = \sum_{k=0}^{q-j} \varepsilon^k U_{2k}.$$

Note that,

$$Z_{10} = N_0(x_1), \quad Z_{20} = l_{40}^*[x_1, N_0(x_1)] + M(x_1)[x_2 - l_{30}^*(x_1, N_0(x_1))].$$

Substituting (44) into (10) we get the following $O(\varepsilon^{q+1})$-approximation to the optimal controller (if the latter exists):

$$u = u^{(q)} + O(\varepsilon^{q+1}), \quad u^{(q)} = -\sum_{k=1}^{2}\sum_{j=0}^{q} \varepsilon^j B_k' Z_{kj}(x_1, x_2, \varepsilon). \tag{46}$$

The higher-order terms in the approximation (46) lead to improved performance (see Theorem 5.1 and example below).

Theorem 5.1 *(i) Under A1-A3 and A7 (or A1, A4 and A5) for small enough ε and $|x_1| \le m_1$, $|x_2| \le m_2$ the invariant manifold (7) can be represented in the form (44);*

(ii) If additionally the optimal controller (10) exists (see Theorems 3.2, 4.1 and Corollary 3.1) it can be approximated by (46) for small enough ε;

(iii) Under A1, A4-A7 for any $m_2 > 0$ there exist m_1', ε_1' and ρ such that for all $|x_1| < m_1'$, $|x_2| < m_2$, $\varepsilon \in (0, \varepsilon_1']$ and $|w| \le \rho$ (i.e. $|w(t)| \le \rho$ for all $t \ge 0$) the controller $u^{(q)}$ achieves the performance level $\gamma + O(\varepsilon^{q+1})$.

For proof of theorem see Appendix.

Remark 1. The generating function V can be found from Hamilton-Jacobi equation (6) in the form of nested expansion introduced in [3] for the optimal control problem. It was shown in [3] that the truncated series of this expansion satisfies the Hamilton-Jacobi equation with a high order of accuracy. From Theorem 5.1 above the stronger result follows: under assumptions of (ii) the truncated series (46) leads to near-optimal controller and, in the hyperbolic case, to near-optimal performance.

Remark 2. All the results of the present paper are also valid for the systems containing nonlinear on x_2 terms of the order of $O(\varepsilon)$. In this case V cannot be found in the form of the expansion of [3] having a more complicated structure.

Remark 3. Let u_l be an optimal controller of (10) corresponding to the linear problem (33a-33b). It is well-known (see e.g.[1]) that for each ε the controller u_l can be found by solving $(n_1 + n_2) \times (n_1 + n_2)$-dimensional ARE. Theorem 3.2 reduces the full-order ARE to the lower order Riccati and linear algebraic equations. Clearly, the functions L^*, L^+ and N are linear operators, and L^+ does not depend on x_1 and p_1. Then the equations (17), (40) and (38) lead to the reduced-order algebraic equations for corresponding matrices (see [5] for details). Analogously to (46) the controller u_l can be found from the latter equations in the form of the expansion

$$u_l = u_l^{(q)} + O(\varepsilon^{q+1}), \quad u_l^{(q)} = \sum_{j=0}^{q} \varepsilon^j u_{lj}(x_1, x_2) \qquad (47)$$

with the linear operators u_{lj}.

In the hyperbolic case for each ε the nonlinear H^∞-suboptimal control problem can be locally solved by the optimal controller u_l of the linearized problem (33a-33b) since $u = u_l + O(|x_1|^2 + |x_2|^2)$ [7,13]. Similarly to (iii) of Theorem 5.1 it can be proved that u_l leads to the performance level $\gamma + O(|x_1|^2 + |x_2|^2)$ for small $|w|$. It is easy to check that for all small enough ε and $|x_1|$ the controller $u_l^{(q)}$ is $O(|x_1|^2 + \varepsilon|x_2|^2 + \varepsilon^{q+1})$ close to u and for small enough $|w|$ achieves the performance level $\gamma + O(|x_1|^2 + \varepsilon|x_2|^2 + \varepsilon^{q+1})$.

Example. Consider the system

$$\dot{x}_1 = u + f(x_1)w, \quad \varepsilon\dot{x}_2 = x_2 - u, \quad z = (x_1 \ x_2 \ u)'. \qquad (48)$$

The Hamilton-Jacobi equation (6) for this system can be written as

$$Z_2 x_2 + \frac{1}{2}Z_1^2\left(\frac{f^2(x_1)}{\gamma^2} - 1\right) + Z_1 Z_2 - \frac{1}{2}Z_2^2 + \frac{x_1^2}{2} + \frac{x_2^2}{2} = 0,$$

where $Z_1 = V_{x_1}$ and $\varepsilon Z_2 = V_{x_2}$. Here A4 holds and we have a hyperbolic case.

We suppose, that $0.5 - f^2\gamma^{-2} > 0$ for all $x_1 \in [-m_1, m_1]$. Neglecting terms of the order $O(\varepsilon^2)$, we get finally

$$Z_1 = \frac{x_1 + \varepsilon(\frac{\sqrt{2}}{2} + 1)[x_2 + 0.5x_1(0.5 - \gamma^{-2}f^2(x_1))^{-\frac{1}{2}}]}{\{0.5 - \gamma^{-2}f^2[x_1 + \varepsilon(\frac{\sqrt{2}}{2} + 1)(x_2 + 0.5x_1(0.5 - \gamma^{-2}f^2(x_1))^{-\frac{1}{2}})]\}^{\frac{1}{2}}},$$

$$Z_2 = (1 + \sqrt{2})x_2 + (\frac{\sqrt{2}}{2} + 1)Z_1 - \varepsilon\frac{1 + \sqrt{2}}{2}\left[x_1 + \frac{f(x_1)\dot{f}(x_1)x_1^2}{\gamma^2(0.5 - \gamma^{-2}f^2(x_1))}\right].$$

Thus, the controller

$$u^{(1)} = (1 + \sqrt{2})x_2 + \frac{\sqrt{2}}{2}Z_1 - \varepsilon\frac{1 + \sqrt{2}}{2}\left[x_1 + \frac{f(x_1)\dot{f}(x_1)x_1^2}{\gamma^2(0.5 - \gamma^{-2}f^2(x_1))}\right] \qquad (49)$$

is an $O(\varepsilon^2)$-approximation to an optimal one. Finally, A6 holds. Hence, for any $\gamma > 0$ and $m_2 > 0$ there exist $m_1 > 0$ and $\varepsilon_3 > 0$ such that for all $\varepsilon \in (0, \varepsilon_3]$ the H^∞-suboptimal control problem (48) is solvable on $\Omega_{m_1} \times \Omega_{m_2}$ and $u^{(1)}$ is an $O(\varepsilon^2)$-approximation to the controller of (10).

Consider $f = \cos x_1$ and constant and sinusoidal disturbance functions. We show here some simulations of the behavior of (48) under the nonlinear controllers: $u^{(1)}$ and $u^{(0)}$ - the $O(\varepsilon)$-approximation to (10) that can be obtained from (49) by substituting $\varepsilon = 0$, and under the linear controllers:

$$u_l^{(0)} = (1 + \sqrt{2})x_2 + \frac{2^{-\frac{1}{2}}x_1}{(0.5 - \gamma^{-2})^{\frac{1}{2}}}, \quad u_l^{(1)} = (1 + \sqrt{2})x_2 + \frac{\sqrt{2}Z_{1l}}{2} - \varepsilon\frac{1 + \sqrt{2}}{2}x_1,$$

where $Z_{1l} = \{x_1 + \varepsilon(\sqrt{2}/2 + 1)[x_2 + 0.5x_1(0.5 - \gamma^{-2})^{-\frac{1}{2}}]\}(0.5 - \gamma^{-2})^{-\frac{1}{2}}]$, and the optimal controller of the linearized problem u_l (if it exists).

Choose $\gamma = 1.5$ and consider two values of ε: $\varepsilon = 0.2$ and $\varepsilon = 0.4$. For $\varepsilon = 0.2$ the optimal controller of the linearized problem is given by: $u_l = 8.9326x_1 + 5.43882x_2$, while for $\varepsilon = 0.4$ it does not exist. We include plots of the functional cost

$$J = \int_0^T [x_1^2 + x_2^2 + u^2 - \gamma^2 w^2]ds$$

in Figure 1, where (1) is the plot under $u^{(1)}$, (2) - under $u^{(0)}$, (3) - under $u_l^{(1)}$, (4) - under $u_l^{(0)}$ and (5) - under u_l. For $\varepsilon = 0.2$ and $w = 2$ all the controllers lead to the negative functional cost (see Fig.1 (a)). Hence, under this particular disturbance input all the controllers achieve the performance bound $\gamma = 1.5$. Notice that the game cost incurred by the nonlinear controllers goes to negative infinity at a higher rate. Also, the cost incurred by $u^{(1)}$ and $u_l^{(1)}$ — the higher order approximations to the optimal nonlinear and linear controllers — goes to negative infinity at a higher rate than those incurred by $u^{(0)}$ and $u_l^{(0)}$.

For $\varepsilon = 0.4$ and $w = 2$ similar remarks can be made (see (c) of Fig.1). For $\varepsilon = 0.2$ and $w = \cos 2t$ the controller $u_l^{(0)}$ does not achieve the performance bound $\gamma = 1.5$ since it leads to a growing positive game cost, whereas all the other controllers under consideration achieve this performance bound (see (b) of Fig.1). For $\varepsilon = 0.4$ and $w = \cos 2t$ only the controller $u^{(1)}$ achieves the performance bound $\gamma = 1.5$ (see (d) of Fig.1). Thus the $O(\varepsilon^2)$-approximation to the optimal controller (10) achieves better performance bound than its $O(\varepsilon)$-approximation. The same conclusion can be reached about approximations to u_l - to the optimal controller of the linearized problem. The improvement is more significant for the values of ε that are not too small.

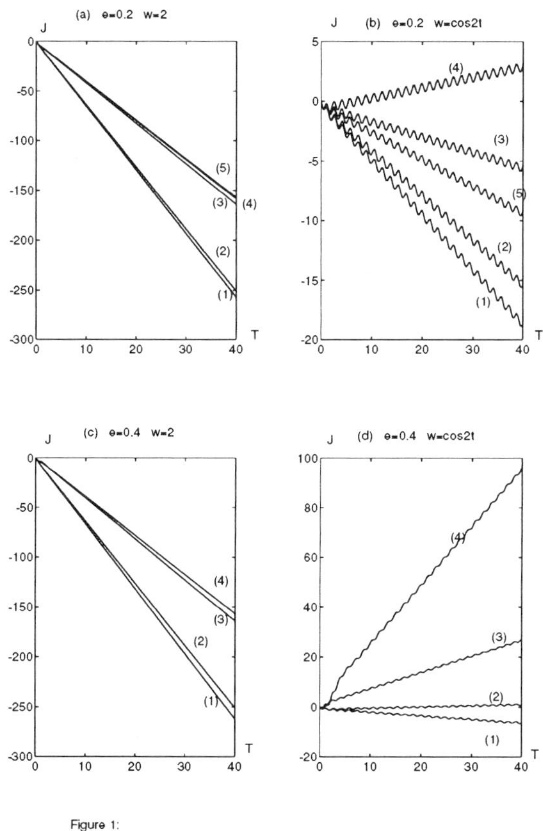

Figure 1:

6   Conclusions

We have developed a geometric approach of [7, 13] in the case of singularly perturbed H^∞-suboptimal control problem, nonlinear on the slow state variables. We have got the exact decomposition of the slow-fast invariant manifold of the Hamiltonian system into the reduced-order slow manifold and a fast manifold. As a result, sufficient conditions for the solvability of the H^∞-suboptimal control problem in terms of the slow manifold have been obtained. Also, an asymptotic expansion of the optimal controller have been constructed by solving partial differential equations, depending only on the slow variables. We have shown that a higher-order accuracy controller improves a performance. The results are valid for the domains containing large values of the fast variables.

A   Appendix

Proof of Theorem 3.1. Under A1 the system (5a-5b) has a center-stable manifold (for analogous derivations see [8,4,14])

$$p_2 = L^{*+}(x_1, p_1, x_2), \tag{A.1}$$

such that all the stable solutions of (5a-5b) belong to it. Let Z_1 and Z_2 be defined by (8). Then (7) determines an invariant on $\Omega_{\overline{m}_1} \times \Omega_{\overline{m}_2}$ manifold of stable solutions to (5a-5b), i.e. (7) is the submanifold of (A.1). Therefore p_1 and p_2, defined by (7), satisfy also (A.1), which implies the following relation:

$$Z_2(x_1, x_2) = L^{*+}[x_1, Z_1(x_1, x_2), x_2], \quad (x_1, x_2) \in \Omega_{\overline{m}_1} \times \Omega_{\overline{m}_2}. \tag{A.2}$$

Let $|Z_1(x_1, x_2)| \leq m_3$ for $(x_1, x_2) \in \Omega_{\overline{m}_1} \times \Omega_{\overline{m}_2}$, and $m = \overline{m}_1 + m_3$. Let further (14) determine a center manifold of (5a-5b) for $|x_1| + |p_1| \leq m$. We shall prove that for any $u_1^0 \in \Omega_{2m_1}$, where m_1 will be chosen below (from the solvability of (A.4a) for v_1), there exists $v_1^0 \in \mathbf{R}^{n_1}$ such that the solution of (5a-5b), lying on its center manifold,

$$x_1 = u_1, \; p_1 = v_1, \; x_2 = L_3^*(u_1, v_1), \; p_2 = L_4^*(u_1, v_1), \quad t \in \mathbf{R}, \tag{A.3a}$$
$$x_1(0) = u_1^0, \quad p_1(0) = v_1^0, \tag{A.3b}$$

lies also on the invariant manifold (7), i.e. satisfies for some $t_1 < 0 < t_2$ the equations:

$$v_1 = Z_1[u_1, L_3^*(u_1, v_1)], \tag{A.4a}$$
$$p_2 = Z_2[u_1, L_3^*(u_1, v_1)]. \tag{A.4b}$$

Note that (A.4b) follows from (A.2)-(A.4a). Clearly, substituting the first and the third of the relations (A.3a) into (A. 2) and applying further (A.4a) we have

$$\begin{aligned} Z_2(u_1, L_3^*(u_1, v_1)) &= L^{*+}[u_1, Z_1(u_1, L_3^*(u_1, v_1)), L_3^*(u_1, v_1)] \\ &= L^{*+}[u_1, v_1, L_3^*(u_1, v_1)]. \end{aligned} \tag{A.5}$$

The expression in the right side of (A.5) coincides with $L_4^*(u_1, v_1)$ since the center manifold is an invariant submanifold of the center-stable manifold. This, together with the last of (A.3a), implies (A.4b).

Consider (A.4a) as a system for v_1 evaluation. First, we shall show that Z_1 can be represented as follows:

$$Z_1(x_1, x_2) = C_1(x_1) + \varepsilon C_2(x_1, x_2, \varepsilon), \tag{A.6}$$

where C_1 and C_2 are Lipschitzian on $(x_1, x_2) \in \Omega_{\overline{m}_1} \times \Omega_{\overline{m}_2}$. Clearly, differentiating the first of the relations (8) on x_2, and the second on x_1, we get $Z_{1x_2} = \varepsilon Z_{2x_1} = V'_{x_1 x_2}$, which yields the representation (A.6). Substituting (A.6) into (A.4a) and applying to the latter equation the contraction principle, one can show that there exists $m_1 > 0$ such that (A.4a) has a solution (33a-33b) for $|u_1| \leq 2m_1$, where N is Lipschitzian and $N(0) = 0$. Further, applying the implicit function theorem one can prove that N is continuously differentiable for $|u_1| < 2m_1$.

Let $|u_1| < 2m_1$ for $t \in (t_1, t_2)$. Then from (A.4a) it follows that (18) is valid for (t_1, t_2), i.e. (18) defines an invariant manifold of (15a-15b). The solutions of the latter invariant manifold are asymptotically stable being at the same time the solutions of (7) with asymptotically stable (9). $\qquad\square$

Proof of Theorem 3.2. (i) The relations (7) define the invariant on $\Omega_{m_1} \times \Omega_{m_2}$ manifold of (5a-5b) if for any $(x_1^0, x_2^0) \in \Omega_{m_1} \times \Omega_{m_2}$ there exists $t_1 < 0 < t_2$ such that a solution of (5a-5b) with the initial values

$$x_1(0) = x_1^0, \;\; x_2(0) = x_2^0, \;\; p_1(0) = Z_1(x_1^0, x_2^0), \;\; p_2(0) = Z_2(x_1^0, x_2^0) \tag{A.7}$$

satisfies (7) for $t \in (t_1, t_2)$.

Let $(x_1^0, x_2^0) \in \Omega_{m_1} \times \Omega_{m_2}$ be any prechosen. Let u_1 and u_2 be solutions of (19) and (25), where v_1 is defined by (18), and with the following initial conditions

$$u_1(0) = U_1(x_1^0, x_2^0), \quad u_2(0) = U_2(x_1^0, x_2^0). \tag{A.8}$$

Denote by x_1, x_2, p_1, p_2 a solution of (5a-5b), (A.7). Note that the relations (A.7) and (A.8) imply (27a-27b) and (28a-28b) at $t = 0$. Let $t_1 < 0 < t_2$ be such an interval that for $t \in (t_1, t_2)$ we have $(x_1, x_2) \in \Omega_{m_1} \times \Omega_{m_2}$ and $|u_1| \leq 2m_1$. Then due to the uniqueness of the solution of (5a-5b), (A.7) the relations (27a-27b) and (28a-28b) are satisfied for all $t \in (t_1, t_2)$. This yields (7) for all $t \in (t_1, t_2)$. Hence, the relations (7) define an invariant on $\Omega_{m_1} \times \Omega_{m_2}$ manifold of (5a-5b). The asymptotic stability of (7) follows from the same property of u_1, u_2 and from the relations (27a-27b), which completes the proof of (i).

The invariant manifold (7) with asymptotically stable (9) is Lagrangian (it can be proved as Lemma 1 of [12]) and is projectable on the simply connected manifold $\Omega_{m_1} \times \Omega_{m_2}$, which implies the existence of the generating function V, satisfying (8) and (6) [13]. Finally, (ii) implies (iii) [7,13]. $\qquad\square$

Proof of Theorem 5.1. We have to prove only (iii). The controller (10) can be represented as follows (cf.(29a-29b), (30a-30b)):

$$u = \beta(x) = \beta_1(x_1) + M(x_1)x_2 + \varepsilon\beta_2(x,\varepsilon),$$

where β_1 and β_2 are smooth functions vanishing at 0. We use u to (1a-1c):

$$\begin{aligned}
\dot{x} &= f_1(x_1) + f_2(x_1)x_2 + \varepsilon g(x,\varepsilon) + D(x_1)w, & \text{(A.9a)} \\
z &= col\{k_1(x_1) + k_2(x_1)x_2; \beta(x)\}, & \text{(A.9b)}
\end{aligned}$$

where for $i = 1, 2$ $f_i = col\{f_{1i}, f_{2i}/\varepsilon\}$, $f_{1i} = a_i + B_i\beta_1$, $f_{2i} = A_i + B_iM$, $g = col\{B_1\beta_2, B_2\beta_2/\varepsilon\}$, $D = col\{D_1, D_2/\varepsilon\}$. We represent (A.9a) in the form:

$$\dot{x} = Ax + G(x,\varepsilon) + Dw, \tag{A.10}$$

where

$$A = \begin{pmatrix} A_{11} & A_{12} \\ \frac{A_{21}}{\varepsilon} & \frac{A_{22}}{\varepsilon} \end{pmatrix}, \quad A_{i1} = \frac{\partial f_i}{\partial x_1}\Big|_{x_1=0}, \quad A_{i2} = f_{i2}(0), \quad i = 1, 2.$$

The components of $G = col\{G_1, G_2/\varepsilon\}$ satisfy inequalities:

$$|G_i| \le c(|x_1| + \varepsilon)|x|, \quad |G_i(x,\varepsilon) - G_i(\tilde{x},\varepsilon)| \le c(|x| + |\tilde{x}| + \varepsilon)|x - \tilde{x}|. \tag{A.11}$$

Since $A_{22} = A_2 - B_2B_2'M\big|_{x=0}$ is Hurwitz there is a nonsingular transformation $y = T^{-1}x$ that block diagonalizes A [9,p.210]: $T^{-1}AT = diag\{A_s, A_f/\varepsilon\}$, $A_s = A_{11} - A_{12}L$, $A_f = A_{22} + \varepsilon LA_{12}$, where

$$T^{-1} = \begin{pmatrix} I - \varepsilon HL & -\varepsilon H \\ L & I \end{pmatrix}, \quad T = \begin{pmatrix} I & \varepsilon H \\ -L & I - \varepsilon LH \end{pmatrix},$$

L and H are defined by (12) and (13) from [9,p.210]. For y we get

$$\begin{aligned}
\dot{y}_1 &= A_sy_1 + G_s(y,\varepsilon) + D_s(y)w, & y_1(0) = 0, & \quad \text{(A.12a)} \\
\varepsilon\dot{y}_2 &= A_fy_2 + G_f(y,\varepsilon) + D_f(y)w, & y_2(0) = 0, & \quad \text{(A.12b)} \\
z &= col\{k(y), \xi(y), \} & & \quad \text{(A.12c)}
\end{aligned}$$

where $col\{G_s, G_f/\varepsilon\} = T^{-1}G(Ty)$, $col\{D_s, D_f/\varepsilon\} = T^{-1}D(y_1 + \varepsilon Hy_2)$, $k = k_1(y_1 + \varepsilon Hy_2) + k_2(y_1 + \varepsilon Hy_2)[-Ly_1 + (I - \varepsilon LH)y_2]$, $\xi = \beta(Ty)$. Obviously, G_s and G_f satisfy (A.11). Note that under A6 A_s is Hurwitz. Similarly substituting $u^{(q)}$ for u in (1a-1c) we get (A.10) with x_q, G_q and u_q substituted for x, G and β. It can be shown that $|u^{(q)} - u| \le c\varepsilon^{q+1}|x|$ since $u^{(q)} - u$ vanishes at $x = 0$. Therefore,

$$|G_q(x,\varepsilon) - G(x,\varepsilon)| \le c\varepsilon^{q+1}|x|. \tag{A.13}$$

Applying to the differential equation for x_q the block diagonalizing transformation $v = T^{-1}x_q$ we obtain

$$\dot{v}_1 = A_s v_1 + G_{sq}(v,\varepsilon) + D_s(v)w, \quad v_1(0) = 0, \qquad \text{(A.14a)}$$

$$\varepsilon\dot{v}_2 = A_f v_2 + G_{fq}(v,\varepsilon) + D_f(v)w, \quad v_2(0) = 0, \qquad \text{(A.14b)}$$

$$z = col\{k(v), \xi_q(v),\} \qquad \text{(A.14c)}$$

where $col\{G_{sq}, G_{fq}/\varepsilon\} = T^{-1}G_q(Tv)$, $\xi_q = u_q(Tv)$.

Denote by $||\cdot||$ the norm in $L_2[0,\tau]$. From (A.12b) and (A.14b) we get

$$\left|\,||z||^2 - ||z_q||^2\,\right| \le C\int_0^\tau (|\Delta y(t)| + \varepsilon^{q+1}|v|)(|y(t)| + |v|)dt, \qquad \text{(A.15)}$$

where $\Delta y = y - v$.

Applying to (A.12b) the variation of constants formula, using an exponential bound on $|exp\{A_f t/\varepsilon\}|$ and (A.11) we get

$$||y_2||^2 \le \int_0^\tau \int_0^t \int_0^t \frac{K}{\varepsilon^2} e^{-\frac{\alpha}{\varepsilon}(2t-s-p)}[\nu|y(p)| + |w(p)|][\nu|y(s)| + |w(s)|]dsdpdt,$$

where $\nu \to 0$ when $m_1 + \varepsilon \to 0$. Estimating from above the product of the square brackets by $\nu^2(|y(p)|^2 + |y(s)|^2) + |w(p)|^2 + |w(s)|^2$ and reversing the order of integration we deduce

$$||y_2||^2 \;\le\; \frac{2K}{\varepsilon^2}\int_0^\tau \int_p^\tau \int_0^t e^{-\frac{\alpha}{\varepsilon}(2t-s-p)}dsdt[\nu^2|y(p)|^2 + |w(p)|^2]dp \;\le$$

$$\le\; \frac{2K}{\alpha^2}[\nu^2||y||^2 + ||w||^2]. \qquad \text{(A.16)}$$

Analogously we get $|y_1||^2 \le 2K/\alpha^2[\nu^2||y||^2 + ||w||^2]$. Then for small m_1 and ε we have $||y||^2 \le c||w||^2$, where c does not depend on τ. Similarly one can derive

$$||y||^2 + ||v||^2 \le c_1||w||^2, \quad \int_0^\tau |y||v|dt \le c_1||w||^2. \qquad \text{(A.17)}$$

Applying to (A.12a), (A.12b) the variation of constants formula and using (A.11) we establish for small m_1 and ε the inequality: $\sup_{t\ge 0}|y| \le c_2\rho$. Then G with subindices satisfies the Lipschitz's condition with a small constant (cf.(A.11)). From (A.12a), (A.12b) and (A.14a), (A.14b) we get

$$|\Delta y(t)| \le K\int_0^t \left[\varepsilon^{-1}e^{-\frac{\alpha}{\varepsilon}(t-s)} + e^{-\alpha(t-s)}\right][\nu_1|\Delta y| + \varepsilon^{q+1}(|y|+|v|)]ds, \quad \text{(A.18)}$$

where $\nu_1 \to 0$ when $m_1 + \varepsilon + \rho \to 0$. From (A.18) analogously to (A.17) we obtain

$$\int_0^\tau |\Delta y|(|v| + |y|)dt \le c\varepsilon^{q+1}(||y||^2 + ||v||^2) \le c'\varepsilon^{q+1}||w||^2. \qquad \text{(A.19)}$$

Thus, from (A.15)-(A.17) and (A.20) we establish for small m_1, ρ, ε,

$$||z_q||^2 = ||z||^2 + O(\varepsilon^{q+1})||w||^2 .$$

By the condition $||z|| \leq \gamma^2||w||^2$. Hence $||z_q||^2 \leq [\gamma^2 + O(\varepsilon^{q+1})]||w||^2 = [\gamma + O(\varepsilon^{q+1})]^2||w||^2.$ $\qquad\square$

Acknowledgement

I would like to thank U. Shaked for very helpful discussions.

REFERENCES

[1] Basar T. and Bernhard P., H^∞-*Optimal Control and Related Minimax Design Problems: a Dynamic Game Approach.* Birkhäuser, Boston, 1991.

[2] Carr J., *Applications of centre manifold theory.* Springer-Verlag, New York, 1981.

[3] Chow J.H. and Kokotovic P.V., Near-optimal feedback stabilization of a class of nonlinear singularly perturbed systems. *SIAM J. Control Optim.* **16**, 756-770, 1978.

[4] Fridman E.M., Decomposition of boundary problems for singularly perturbed systems of neutral type in conditionally stable case. *Differential equations* (Moscow) **28** , no.6, 800-810, 1992.

[5] Fridman E.M., Exact slow-fast decomposition of linear singularly perturbed H^∞-optimal control problem. In: Proc. of 2 IEEE Mediterranean Symposium on New directions in Control Theory and Applications, Chania, Crete, 504-511, June 1994.

[6] Henry D., *Geometric theory of parabolic equations.* Springer- Verlag, New York, 1982.

[7] Isidori A. and Astolfi A., Disturbance attenuation and H^∞-control via measurement feedback in nonlinear systems. *IEEE Trans. Automat. Contr.* **37**, No 9, 1283-1293, 1992.

[8] Kelley A., The stable, center-stable, center, center-unstable, and unstable manifolds. *J. Diff. Eqns.*, **3**, 546-570, 1967.

[9] Kokotovic P., Khalil H. and O'Reilly J., *Singular Perturbation Methods in Control: Analysis and Design*. New York, Academic Press, 1986.

[10] Pan Z. and Basar T., H^∞-optimal control for singularly perturbed systems. Part I: Perfect State Measurements. *Automatica* **2**, 401-424, 1993.

[11] Pan Z. and Basar T., H^∞-optimal control for nonlinear singularly perturbed systems under perfect state measurements. CSL Report, University of Illinois, Urbana, May, 1993.

[12] Van der Schaft A., On a state space approach to nonlinear H_∞ control. *Systems and Control Letters*, No 16, 1-8, 1991.

[13] Van der Schaft A., L^2-gain analysis of nonlinear systems and nonlinear state feedback H^∞ control. *IEEE Trans. Automat. Contr.* **37** , no. 6, 770-784, 1992.

[14] Pliss M., *Integral sets of periodic systems of differential equations*. Moscow, Nauka, 1977(in Russian).

[15] Sobolev V. Integral manifolds and decomposition of singularly perturbed systems. *Systems and Control Letters* **4** , 169-179, 1984.

A Hybrid (Differential-Stochastic) Zero-Sum Game with a Fast Stochastic Part

Eitan Altman
Projet MISTRAL, INRIA, BP93
2004 Route des Lucioles
06902 Sophia Antipolis Cedex
France

Vladimir Gaitsgory
School of Mathematics, Univ. of South Australia
the Levels, Pooraka, South Australia 5095
Australia

Abstract

We consider in this paper a continuous time stochastic hybrid system with a finite time horizon, controlled by two players with opposite objectives (zero-sum game). Player one wishes to maximize some linear function of the expected state trajectory, and player two wishes to minimize it. The state evolves according to a linear dynamic. The parameters of the state evolution equation may change at discrete times according to a MDP, i.e., a Markov chain that is directly controlled by both players, and has a countable state space. Each player has a finite action space. We use a procedure similar in form to the maximum principle; this determines a pair of stationary strategies for the players, which is asymptotically a saddle point, as the number of transitions during the finite time horizon grows to infinity.

Keywords: Hybrid stochastic systems, stochastic games, asymptotic optimality, linear dynamics, Markov decision processes, finite horizon.

1   Introduction and statement of the problem

Consider the following hybrid stochastic controlled system. The state $Z_t \in \mathbb{R}^n$ evolves according to the following linear dynamics:

$$\frac{d}{dt}Z_t = AZ_t + BY_t, \qquad t \in [0,1], \qquad Z_0 = z \qquad (1)$$

where $Y_t \in \mathbb{R}^k$ is the "control" and $A(n \times n)$ and $B(n \times k)$ are matrices of real numbers. Y_t is not chosen directly by the controllers, but is obtained as a result of controlling the following underlying stochastic discrete event system. Let ϵ be the basic time unit. Time is discretized, i.e. transitions occur at times $t = n\epsilon$, $n = 0, 1, 2, ..., \lfloor \epsilon^{-1} \rfloor$, where $\lfloor x \rfloor$ stands for the greatest integer which is smaller or equal to x. There is a countable state space $\mathbf{X} = \mathbb{N}$ and two players having finite action spaces $\mathbf{A}_1$ and $\mathbf{A}_2$ respectively. Let $\mathbf{A} = \mathbf{A}_1 \times \mathbf{A}_2$. If the state is v and actions $a = (a_1, a_2)$ are chosen by the players, then the next state is w with probability P_{vaw}. Denote $\mathcal{P} = \{P_{vaw}\}$. A policy $u^i = \{u_0^i, u_1^i, ...\}$ in the set of policies U^i for player i, $i = 1, 2$ is a sequence of probability measures on $\mathbf{A}_i$ conditioned on the history of all previous states and actions of both players, as well as the current state. More precisely, define the set of histories: $\mathbf{H} := \cup_l \mathbf{H}_l$, where

$$\mathbf{H}_l := \{(x_0, a_0^1, a_0^2, x_1, a_1^1, a_1^2, \ldots, x_l)\}$$

are the sets of all sequences of $3l + 1$ elements describing the possible samples of previous states and actions prior to l as well as the current state at stage l (i.e. at time $l\epsilon$). (The range of l will be either $l = 0, 1, \ldots, \lfloor \epsilon^{-1} \rfloor$, or, in other contexts, all nonnegative integers, depending on whether we consider finite or infinite horizon problems). The policy at stage l for player i, u_l^i, is a map from $\mathbf{H}_l$ to the set of probability measures over the action space $\mathbf{A}_i$. (Hence at each time $t = l\epsilon$, player i, observing the history h_l, chooses action a_i with probability $u_l^i(a_i|h_l)$). Let $\mathcal{F}_l$ be the discrete σ-Algebra of subsets of $\mathbf{H}_l$. Each initial distribution ξ and policy pair u for the players uniquely define a probability measure P_ξ^u over the space of samples $\mathbf{H}$ (equipped with the discrete σ-algebra), see e.g. [4]. Denote by E_ξ^u the corresponding expectation operator. On the above probability space are now defined the random processes X_l and $A_l = (A_l^1, A_l^2)$, denoting the state and actions processes. When the initial distribution is concentrated on a single state x, we shall denote the corresponding probability measure and expectation by P_x^u and E_x^u.

Let $y^j : \mathbf{X} \times \mathbf{A} \to \mathbb{R}$, $j = 1, ..., k$ be some given bounded functions. Then Y_t in (1) is given by

$$Y_t = y(X_{\lfloor t/\epsilon \rfloor}, A_{\lfloor t/\epsilon \rfloor}). \tag{2}$$

Y_t and thus Z_t are well defined stochastic processes, and are both $\mathcal{F}_{\lfloor \epsilon^{-1} \rfloor}$ measurable.

We shall be especially interested in the following classes of policies.

(i) The Markov policies $\mathcal{M}_1, \mathcal{M}_2$: these are policies where u_l^i depends only on the current state (at time $t = l\epsilon$) and on l, and does not depend on previous states and actions. If a Markov policy $u^i \in \mathcal{M}_i$ is used by player i, we shall denote

$$u_l^i(a|x) : \text{ the probability under } u^i \text{ of choosing } a \in \mathbf{A}_i \text{ in state } x \text{ at stage } l. \tag{3}$$

Denote $\mathcal{M} = \mathcal{M}_1 \times \mathcal{M}_2$.

(ii) The stationary policies, denoted by $\mathcal{S}_1$, for player 1, and $\mathcal{S}_2$, for player 2. A policy u is called stationary if u_l depends only on the current state, and does not depend on previous states and actions nor on the time. Let $\mathcal{S} := \mathcal{S}_1 \times \mathcal{S}_2$. If a stationary policy f is used, we shall denote by $f_x(a)$ the probability under f of choosing action a when in state x. When stationary policies $f = (f^1, f^2)$ are used by the players, we set

$$P_{vfw} = P_{vf^1f^2w} = \sum_{a^1,a^2} P_{va^1a^2w} f_v^1(a^1) f_v^2(a^2),$$

$$y(v, f) = y(v, f^1, f^2) = \sum_{a^1,a^2} y(v, a^1, a^2) f_v^1(a^1) f_v^2(a^2).$$

Let $P_f = \{P_{vfw}\}$ be the transition probabilities of the Markov chain induced by a stationary policy pair f, and let $P_f^l = \{[P_f^l]_{vw}\}$ be the l step transition probabilities under f.

We make throughout the following assumption, which is a strong version of the Simultaneous Doeblin Condition, introduced in [5] Section 11.1, with a communicating condition.

(A1): There exists a state $x^* \in \mathbf{X}$ and a positive real number q_0 such that

$$P_{xfx^*} \geq q_0, \qquad \forall x \in \mathbf{X} f \in \mathcal{S}.$$

Let c be an n-dimensional vector representing the (linear) operating cost related to the process Z_t. Define the cost:

$$J_x^z(u^1, u^2) = E_x^{(u^1,u^2)} c^T Z_1, \qquad Z_0 = z$$

when policies u^1, u^2 are used by the players, and the initial state of the linear system is z, and the initial state of the controlled Markov chain is x. In our dynamic game, player 1 wishes to maximize $J_x^z(u^1, u^2)$ and player 2 wants to minimize it. More precisely, define the following problems:
$\mathbf{Q1}_I^\epsilon$: find a policy $u^1 \in U^1$ that achieves

$$F_I^\epsilon(x) = \sup_{u^1 \in U^1} \inf_{u^2 \in U^2} J_x^z(u^1, u^2)$$

where Z_1 is obtained through (1). If such a policy exists, then it is called optimal for $\mathbf{Q1}_I^\epsilon$. If for some δ and $u^1 \in U^1$,

$$F_I^\epsilon(x) \leq \inf_{u^2 \in U^2} J_x^z(u^1, u^2) + \delta$$

then u^1 is called δ-optimal for $\mathbf{Q1}_I^\epsilon$. One may consider also:
$\mathbf{Q1}_{II}^\epsilon$: find a policy $u^2 \in U^2$ that achieves

$$F_{II}^\epsilon(x) = \inf_{u^2 \in U^2} \sup_{u^1 \in U^1} J_x^z(u^1, u^2).$$

Similarly define optimality and δ-optimality of policies for $\mathbf{Q1}^\epsilon_{II}$. We clearly have $F^\epsilon_{II}(x) \geq F^\epsilon_I(x)$. If there exist some $u = (u^1, u^2)$ and δ such that

$$F^\epsilon_I(x) + \delta \geq E_x^{(u^1, u^2)} c^T Z_1 \geq F^\epsilon_{II}(x) - \delta,$$

then u is called δ-saddle point, or δ-equilibrium strategy pair for $\mathbf{Q1}^\epsilon$ (we need not specify $\mathbf{Q1}^\epsilon_I$ or $\mathbf{Q1}^\epsilon_{II}$). If this holds for $\delta = 0$, then u is called the saddle point or equilibrium strategy for $\mathbf{Q}^\epsilon$.

Remarks:

(i) $\mathbf{Q1}^\epsilon_I$ is equivalent to the problem: find a policy $u^1 \in U^1$ that achieves $\sup_{u^1 \in U^1} \inf_{u^2 \in U^2} c^T \overline{Z}_1$, where $\overline{Z}_t \in \mathbb{R}^n$ is given by

$$\frac{d}{dt}\overline{Z}_t = A\overline{Z}_t + BE_x^{(u^1, u^2)} Y_t, \qquad t \in [0, 1], \qquad \overline{Z}_0 = z \tag{4}$$

The same holds for $\mathbf{Q1}^\epsilon_{II}$.

(ii) By solving the problem $\mathbf{Q1}^\epsilon_I$, one can also solve a problem with an integral cost function, i.e. to find a policy u that achieves

$$\sup_{u^1 \in U^1} \inf_{u^2 \in U^2} E_x^{(u^1, u^2)} \int_0^1 c^T Z_t dt.$$

This is obtained by using a new variable R_t defined by $dR_t/dt = c^T Z_t$.

Note that the controllers do not require knowledge of the initial value z of Z_0, which may be assumed to be zero. More precisely, due to the linearity of the system (1), if a control strategy is optimal (or δ-optimal) for a given Z_0, then it is optimal (or δ-optimal, respectively) for any other value of Z_0.

Our model is characterized by the fact that ϵ is supposed to be a small parameter. We construct a set of Markov policies $\overline{u}^\epsilon = (\overline{u}^{1,\epsilon}, \overline{u}^{2,\epsilon})$ such that $\overline{u}^\epsilon$ is $\gamma(\epsilon)$-equilibrium for $\mathbf{Q1}^\epsilon$ where $\lim_{\epsilon \to 0} \gamma(\epsilon) = 0$. This implies, in particular, that the game has the value in the limit as $\epsilon \to 0$ and we call the above-mentioned sequence of Markov policies asymptotically saddle-point.

This paper is a continuation and generalization of our previous work [1] which solves a hybrid problem restricted to a single controller and to a finite state space. As in [1], the fact that ϵ is small means that the variables Y_t can be considered to be fast with respect to Z_t, since, by (2), they may have a finite (not tending with ϵ to zero) change at each interval of the length ϵ. This along with the linearity of the system (1) allows us to decompose the game into stochastic subgames in a sequence of intervals which are short with respect to the variables Z_t (in the sense that Z_t remain almost unchanged in these intervals) and which are long enough with respect to Y_t (so that the corresponding stochastic subgames show in these intervals their limit properties).

The type of model which we introduce is natural in the control of inventories or of production, where we deal with material whose quantity may change in a continuous (linear) way. Breakdowns, repairs and other control decisions yield the underlying controlled Markov chain. In particular, repair or preventive maintenance decisions are typical actions of a player that minimizes costs. If there is some unknown parameter (disturbance) of the dynamics of the system (e.g. the probability of breakdowns) which may change in a way that depends on the current and past states in a way that is unknown and unpredictable by the minimizer, we may formulate this situation as a zero-sum game, where the minimizer wishes to guarantee the best performance (lowest expected cost) under the worst case behavior of nature. Nature may then be modeled as the maximizing player. (This yields $\mathbf{Q1}_{II}^{\epsilon}$.)

Our model may also be used in the control of highly loaded queueing networks for which the fluid approximation holds (see Kleinrock [6] p. 56). The quantities Z_t may then represent the number of customers in the different queues whereas the underlying controlled Markov chain may correspond to routing, or flow control of, say, some on-off traffic, with again, nature controlling some disturbances in quantities such as service rates.

The structure of the paper is as follows. In Section 2 we present the main result; we construct the sequence of non-stationary policy for the hybrid control problems $\mathbf{Q1}^{\epsilon}$. We prove in Section 3 that the sequence of policies introduced in Section 2 is indeed asymptotically saddle-point as ϵ tends to zero. Proofs of some technical lemmas are left to the Appendix.

Below, B^T will denote the transpose of a matrix (or of a column vector) B, and $\|B\|$ will denote the sum of absolute values of the components of B.

2   Construction of ϵ-equilibrium Markov strategies

Consider a family of infinite horizon stochastic games, all with the same state and action spaces $\mathbf{X}$ and $\mathbf{A}$ as above, and the same transition probabilities $\mathcal{P}$, parametrized by a vector $\lambda \in \mathbb{R}^n$. Let $r : \mathbb{R}^n \times \mathbf{X} \times \mathbf{A} \to \mathbb{R}$ be the immediate cost, i.e. $r(\lambda, x, a)$ is the cost in the MDP λ, when at state x and the actions chosen are a. r is given by

$$r(\lambda, x, a) = \lambda^T B y(x, a).$$

The definition of policies $U = (U^1, U^2)$ is as in Section 1. Define the following cost functions. The finite horizon total expected cost:

$$\sigma^m(\lambda, \xi, u) := E_\xi^u \sum_{i=0}^{m-1} r(\lambda, X_i, A_i); \tag{5}$$

The infinite horizon expected average cost:

$$\overline{\sigma}(\lambda, \xi, u) := \varliminf_{m \to \infty} \frac{\sigma^m(\lambda, \xi, u)}{m}$$

Remark: The results of the paper are unchanged if the liminf is replaced by a limsup in the definition of the infinite horizon average cost.

A policy pair $u^\lambda = (u^{1,\lambda}, u^{2,\lambda}) \in U$ is said to be a saddle point or an equilibrium policy pair for problem λ with infinite horizon expected average cost criterion, if for all $u^1 \in U^1, u^2 \in U^2$,

$$\overline{\sigma}(\lambda, \xi, u^1, u^{2,\lambda}) \leq \overline{\sigma}(\lambda, \xi, u^{1,\lambda}, u^{2,\lambda}) \leq \overline{\sigma}(\lambda, \xi, u^{1,\lambda}, u^2). \tag{6}$$

Let $f^\lambda = (f^{1,\lambda}, f^{2,\lambda})$, where $f^{1,\lambda} \in \mathcal{S}_1, f^{2,\lambda} \in \mathcal{S}_2$, be some stationary equilibrium policy pair for the expected average problem. The existence of such a stationary equilibrium policy pair (under assumption (A1)) is well known, see e.g. [3].

$$\overline{\sigma}(\lambda) := \overline{\sigma}(\lambda, \xi, f^{1,\lambda}, f^{2,\lambda}) \tag{7}$$

is then defined to be the value of the λ stochastic game, and is known to be independent of ξ (which we shall thus omit from the notation). It can be computed using value iteration, see e.g. [11].

Let $\lambda(t) \in \mathbb{R}^n$, $t \in [0, 1]$ be the solution of

$$\frac{d}{dt}\lambda_t = -A^T\lambda_t, \quad \lambda_1 = -c \tag{8}$$

i.e.,

$$\lambda(t) = e^{A^T(1-t)}c.$$

Define the following:

- $\Delta(\epsilon):=$ a function of ϵ such that

$$\lim_{\epsilon \to 0} \Delta(\epsilon) = 0, \qquad \lim_{\epsilon \to 0} \frac{\Delta(\epsilon)}{\epsilon} = \infty$$

  $\Delta(\epsilon)$ will be the length of sub-intervals of $[0,1]$ during which we shall use fixed stationary policies. (In each new sub-interval, a new stationary policy has to be computed).

- $\tau_l := l\Delta(\epsilon)$, $l = 0, 1, 2, \ldots, \lfloor \Delta(\epsilon)^{-1} \rfloor$, is the instant at which the lth sub-interval begins.

- $M_\epsilon := \lfloor \Delta(\epsilon)^{-1} \rfloor$ is the number of sub-intervals.

- $\tau_{M_\epsilon+1} := 1$.

- $m_l := \lfloor (l+1)\Delta(\epsilon)\epsilon^{-1} \rfloor - \lfloor (l)\Delta(\epsilon)\epsilon^{-1} \rfloor$.

- $\overline{u}^\epsilon := (\overline{u}^{1,\epsilon}, \overline{u}^{2,\epsilon})$ is a pair of Markov policies defined by the players as follows: each player $i = 1,2$ defines $\overline{u}^{i,\epsilon}$ by applying $f^{i,\lambda(\tau_l)}$, $i = 1,2$ during $n = \lfloor \tau_l/\epsilon \rfloor, \lfloor \tau_l/\epsilon \rfloor + 1, \ldots, \lfloor \tau_{l+1}/\epsilon \rfloor - 1; l = 0, 1, \ldots, M_\epsilon$, where $f^{i,\lambda(\tau_l)}$ is defined in the paragraph above (7), and by choosing an arbitrary action at $\lfloor \epsilon^{-1} \rfloor$.

Theorem 2.1 $\overline{u}^\epsilon$ *is an asymptotically saddle point, i.e. for every ϵ there exists some $\gamma(\epsilon)$ with $\lim_{\epsilon \to 0} \gamma(\epsilon) = 0$, such that $\overline{u}^\epsilon$ is $\gamma(\epsilon)$-equilibrium for problem* **Q1**$^\epsilon$*:*

$$J_x^z(u^1, \overline{u}^{2,\epsilon}) - \gamma(\epsilon) \leq J_x^z(\overline{u}^{1,\epsilon}, \overline{u}^{2,\epsilon}) \leq J_x^z(\overline{u}^{1,\epsilon}, u^2) + \gamma(\epsilon), \qquad \forall u^1 \in U^1, u^2 \in U^2. \tag{9}$$

Moreover,

$$J_x^z(\overline{u}^{1,\epsilon}, \overline{u}^{2,\epsilon}) = \lambda^T(0)z + \int_0^1 \overline{\sigma}(\lambda(t))dt + O(\gamma(\epsilon)), \tag{10}$$

where $\overline{\sigma}(\lambda)$ was defined in (7).

Remark: As follows from the proof below, one can choose

$$\gamma(\epsilon) = O\left(\max\left\{\Delta(\epsilon), \frac{\epsilon}{\Delta(\epsilon)}\right\}\right),$$

so taking $\Delta(\epsilon) = \epsilon^{1/2}$, one obtains $\gamma(\epsilon) = O(\epsilon^{1/2})$.

3 Proof of main result

The proof is based on the following Lemmas, whose proof is provided in the appendix.

Lemma 3.1 *There exists some constant L such that for any initial distributions ξ, ζ and η on the initial state X_0, and any m,*

$$\sigma^m(\lambda, \xi, u^1, f^{2,\lambda}) - L \;\; \leq \;\; \sigma^m(\lambda, \zeta, f^{1,\lambda}, f^{2,\lambda}) \tag{11}$$
$$\leq \;\; \sigma^m(\lambda, \eta, f^{1,\lambda}, u^2) + L, \quad \forall u^1 \in U^1, u^2 \in U^2. \tag{12}$$

and

$$|\sigma^m(\lambda, \xi, f^\lambda) - m\overline{\sigma}(\lambda)| \leq L, \tag{13}$$

where f^λ are defined below (6), and λ belongs to a bounded set containing $\lambda(t)$, $t \in [0,1]$.

Lemma 3.2 *The value functions $\overline{\sigma}$, defined in (7), are continuous functions of λ.*

Proof of Theorem 2.1: We first note that for each fixed ϵ, the hybrid dynamic game problem can be formulated as a finite-horizon non-stationary zero-sum stochastic game (see e.g. Nowak [7, 8]), with bounded immediate cost, a countable state space and a finite number of actions. Although we do not pursue this direction, we conclude that both players may restrict to

Markov policies, so that it suffices in (9) to restrict to Markov policies u^1 and u^2 (this follows e.g. from Remark 2.1 in [7] or Lemma 3.5 in [2]).

Due to the linearity of the system, for any $u \in \mathcal{M}$, one can write the value of the hybrid game

$$J_x^z(u) = \lambda^T(0)z + \int_0^1 \lambda^T(t) B E_x^u Y(t) dt$$

which implies the inequality

$$\left| J_x^z(u) - \lambda^T(0)z - \sum_{l=0}^{M_\epsilon - 1} E_x^u \left\{ \lambda^T(\tau_l) B \int_{\tau_l}^{\tau_{l+1}} Y(t) dt \right\} \right| \leq L_1 \Delta(\epsilon), \qquad (14)$$

where L_1 is some constant (that does not depend on u, x and z). By (2) we have,

$$E_x^u \left| \lambda^T(\tau_l) B \int_{\tau_l}^{\tau_{l+1}} Y(t) dt - \epsilon \sum_{i=\lfloor \tau_l \epsilon^{-1} \rfloor}^{\lfloor \tau_{l+1} \epsilon^{-1} \rfloor - 1} \lambda^T(\tau_l) B y(X_i, A_i) \right| \leq L_2 \epsilon \qquad (15)$$

where L_2 is some constant (that does not depend on u, x and z).

We define for any Markov policy u^i for player i the s-step *shifted strategy* $\theta^j u^i$ by

$$\left(\theta^j u^i \right)_l (a|x) = u_{j+l}(a|x), \qquad \forall l, x, a \in \mathbf{A}_i$$

(we used (3) for the notation of a Markov policy). When both players use Markov policies $u = (u^1, u^2)$, we shall use the notation $\theta^j u = (\theta^j u^1, \theta^j u^2)$. For any Markov policy pair u,

$$E_x^u \left\{ \sum_{i=\lfloor \tau_l \epsilon^{-1} \rfloor}^{\lfloor \tau_{l+1} \epsilon^{-1} \rfloor - 1} \lambda^T(\tau_l) B y(X_i, A_i) \right\} = E_x^u \left\{ \sigma^{m_l} \left(\lambda(\tau_l), X(\lfloor \tau_l \epsilon^{-1} \rfloor), \theta^{\lfloor \tau_l \epsilon^{-1} \rfloor} u \right) \right\}$$

$$(16)$$

(where σ^m is defined in (5). Notice that by definition of the policies $\overline{u}^\epsilon$,

$$E_x^{\overline{u}^\epsilon} \left\{ \sum_{i=\lfloor \tau_l \epsilon^{-1} \rfloor}^{\lfloor \tau_{l+1} \epsilon^{-1} \rfloor - 1} \lambda^T(\tau_l) B y(X_i, A_i) \right\} = E_x^u \left\{ \sigma^{m_l} \left(\lambda(\tau_l), X(\lfloor \tau_l \epsilon^{-1} \rfloor), f^{\lambda(\tau_l)} \right) \right\}$$

$$(17)$$

By (11), for any distributions ξ, ζ and η on the state space, and $\forall u^1 \in U^1, u^2 \in U^2$,

$$\sigma^{m_l} \left(\lambda(\tau_l), \xi, \theta^{\lfloor \tau_l \epsilon^{-1} \rfloor} u^1, f^{2,\lambda} \right) - L$$

$$\leq \quad \sigma^{m_l} \left(\lambda(\tau_l), \zeta, f^{1,\lambda}, f^{2,\lambda} \right)$$

$$\leq \quad \sigma^{m_l} \left(\lambda(\tau_l), \eta, f^{1,\lambda}, \theta^{\lfloor \tau_l \epsilon^{-1} \rfloor} u^2 \right) + L, \quad \forall u^1 \in \mathcal{M}_1, u^2 \in \mathcal{M}_2.$$

which, along with (16)-(17) implies that

$$E_x^{(u^1,\overline{u}^{2,\epsilon})} \left\{ \sum_{i=\lfloor \tau_l \epsilon^{-1} \rfloor}^{\lfloor \tau_{l+1}\epsilon^{-1} \rfloor - 1} \lambda^T(\tau_l) By(X_i, A_i) \right\} - L$$

$$\leq E_x^{(\overline{u}^{1,\epsilon},\overline{u}^{2,\epsilon})} \left\{ \sum_{i=\lfloor \tau_l \epsilon^{-1} \rfloor}^{\lfloor \tau_{l+1}\epsilon^{-1} \rfloor - 1} \lambda^T(\tau_l) By(X_i, A_i) \right\}$$

$$\leq E_x^{(\overline{u}^{1,\epsilon},u^2)} \left\{ \sum_{i=\lfloor \tau_l \epsilon^{-1} \rfloor}^{\lfloor \tau_{l+1}\epsilon^{-1} \rfloor - 1} \lambda^T(\tau_l) By(X_i, A_i) \right\} + L$$

This, in turn, leads via (15) to

$$E_x^{(u^1,\overline{u}^{2,\epsilon})} \lambda^T(\tau_l) B \int_{\tau_l}^{\tau_{l+1}} Y(t)\,dt - (L + L_2)\epsilon$$

$$\leq E_x^{(\overline{u}^{1,\epsilon},\overline{u}^{2,\epsilon})} \lambda^T(\tau_l) B \int_{\tau_l}^{\tau_{l+1}} Y(t)\,dt$$

$$\leq E_x^{(\overline{u}^{1,\epsilon},u^2)} \lambda^T(\tau_l) B \int_{\tau_l}^{\tau_{l+1}} Y(t)\,dt + (L + L_2)\epsilon$$

and this, via (14), to

$$J_x^z(u^1,\overline{u}^{2,\epsilon}) - L_1 \Delta(\epsilon) - (L + L_2)\epsilon M_\epsilon$$
$$\leq J_x^z(\overline{u}^{1,\epsilon},\overline{u}^{2,\epsilon})$$
$$\leq J_x^z(\overline{u}^{1,\epsilon},u^2) + L_1 \Delta(\epsilon) + (L + L_2)\epsilon M_\epsilon$$

This proves (9) with

$$\gamma(\epsilon) = L_1 \Delta(\epsilon) + (L + L_2)\epsilon M_\epsilon = L_1 \Delta(\epsilon) + (L + L_2)[\Delta(\epsilon)^{-1}].$$

Now, from (17) and (13) it follows that

$$\left| E_x^{\overline{u}^\epsilon} \left\{ \sum_{i=\lfloor \tau_l \epsilon^{-1} \rfloor}^{\lfloor \tau_{l+1}\epsilon^{-1} \rfloor - 1} \lambda^T(\tau_l) By(X_i, A_i) \right\} - m_l \overline{\sigma}(\lambda(\tau_l)) \right| \leq L.$$

This, with (14) and (16), implies that

$$\left| J_x^z(\overline{u}^{1,\epsilon},\overline{u}^{2,\epsilon}) - \lambda^T(0)z - \sum_{l=0}^{M_\epsilon - 1} \overline{\sigma}(\lambda(\tau_l))\epsilon m_l \right| \leq L_1 \Delta(\epsilon) + (L + L_2)\epsilon M_\epsilon. \quad (18)$$

By definition of m_l, we have

$$|\epsilon m_l - \Delta(\epsilon)| \leq 2\epsilon. \quad (19)$$

On the other hand, since by Lemma 3.2, the function $\overline{\sigma}(\lambda)$ is continuous, it follows that it is uniformly continuous (since we need only consider a compact set of λ), so that

$$\left| \sum_{l=0}^{M_\epsilon - 1} \overline{\sigma}(\lambda(\tau_l))\Delta(\epsilon) - \int_0^1 \overline{\sigma}(\lambda(t))dt \right| = O(\Delta(\epsilon)).$$

This, along with (18) and (19) establishes (10).   ■

4   Appendix

Before proving Lemma 3.1, we introduce some definitions and quote some results from dynamic programming. Define the matrices $\Pi, D : \mathbf{X} \times \mathbf{X} \to \mathbb{R}$ parametrized by the stationary policies $f = (f^1, f^2)$:

$$\Pi^f(vw) := \lim_{l \to \infty} \frac{1}{l+1} \sum_{i=0}^{l} [P_f^i]_{vw}, \qquad D^f(v, w) = \sum_{l=0}^{\infty} \left([P_f^l]_{vw} - \Pi_{vw} \right).$$

Define the vector $h_\lambda^f = D^f r(\lambda, f)$. Consider a bounded vector of "terminating cost" $\alpha : \mathbf{X} \to \mathbb{R}$, and define the finite horizon expected average cost corresponding to α by

$$\sigma_\alpha^m(\lambda, \xi, u) := E_\xi^u \left[\left(\sum_{i=0}^{m-1} r(\lambda, X_i, A_i) \right) + \alpha(X_m) \right].$$

Define the optimal cost against policy $f^{2,\lambda}$:

$$\sigma_\alpha^m(\lambda, x, f^{2,\lambda}) = \sup_{u^1 \in U^1} \sigma_\alpha^m(\lambda, x, u^1, f^{2,\lambda}).$$

Lemma 4.1 *(i) Under any stationary policy pair f, Π^f is well defined and has identical rows equal to the unique steady state probability under f. Moreover,*

$$\overline{\sigma}(\lambda, f) := \overline{\sigma}(\lambda, v, f) = \sum_{w \in \mathbf{X}} \Pi^f(vw) r(\lambda, w, f),$$

and is independent of $v \in \mathbf{X}$.
(ii) D is well defined and $\sum_{w \in \mathbf{X}} |D^f(v, w)|$ are bounded by some constant $\mathcal{D}$, uniformly over all states v and all stationary policies of both players. Hence $|h_\lambda^f(v)|$ are bounded by some constant $\hat{h}$, uniformly in all stationary policies f, all states v and all λ in some compact set that contains $\lambda(t)$, $t \in [0, 1]$.

(iii) The pair $(\overline{\sigma}(\lambda, f^{\lambda}), h_{\lambda}^{f})$ is the unique bounded solution (the uniqueness of h_{λ}^{f} is up to an additive constant) of the dynamic programming equation

$$\overline{h}(v) + g = \max_{a^{1} \in \mathbf{A}_{1}} \left\{ r(\lambda, a^{1}, f^{2,\lambda}) + P_{va^{1}f^{2}w}\overline{h}(w) \right\}.$$

(iv) $\sigma_{\alpha}^{m}(\lambda, v, f^{2,\lambda})$ satisfies the following dynamic programming equation:

$$\sigma_{\alpha}^{0}(\lambda, v, f^{2,\lambda}) \quad := \quad \alpha(v)$$

$$\sigma_{\alpha}^{m}(\lambda, v, f^{2,\lambda}) \quad = \quad \max_{a_{1} \in \mathbf{A}_{1}} \left\{ r(\lambda, v, a_{1}, f^{2,\lambda}) + \sum_{w \in \mathbf{X}} P_{va^{1}f^{2,\lambda}w}\sigma_{\alpha}^{m-1}(\lambda, w, f^{2,\lambda}) \right\}$$

for all $v \in \mathbf{X}$.

Proof: The proof of (i), (ii) and (iii) are given in Proposition 5.1 in [10] (by choosing $\mu = 1$ there). (iv) are well known, see e.g. [9] (Note that when player two restricts to a stationary policy, i.e. to $f^{2,\lambda}$, then player 1 is faced with a standard Markov decision process (MDP)). ∎

Proof of Lemma 3.1: We prove the inequality

$$\sigma^{m}(\lambda, \xi, u^{1}, f^{2,\lambda}) - L \leq m\overline{\sigma}(\lambda, \xi, f^{1,\lambda}, f^{2,\lambda}).$$

The proof of the other one is the same. Consider the following terminating costs:

$$\alpha(v) = h_{\lambda}^{f}(v) + \hat{h}, \qquad v \in \mathbf{X},$$

where $\hat{h}$ is defined in Lemma 4.1 (ii). It follows from Lemma 4.1 (ii) that $\alpha \geq 0$. This implies that for any m,

$$\sigma_{\alpha}^{m}(\lambda, x, f^{2,\lambda}) \geq \sigma^{m}(\lambda, x, f^{2,\lambda}). \tag{20}$$

We now compute $\sigma_{\alpha}^{m}(\lambda, x, f^{2,\lambda})$ by Lemma 4.1 (iv):

$$\sigma_{\alpha}^{0}(\lambda, x, f^{2,\lambda}) \quad = \quad \alpha(x)$$

$$\sigma_{\alpha}^{1}(\lambda, x, f^{2,\lambda}) \quad = \quad \max_{a_{1} \in \mathbf{A}_{1}} \left\{ r(\lambda, x, a_{1}, f^{2,\lambda}) + \sum_{w \in \mathbf{X}} P_{xa^{1}f^{2,\lambda}w}\sigma_{\alpha}^{0}(\lambda, w, f^{2,\lambda}) \right\}$$

$$= \quad \max_{a_{1} \in \mathbf{A}_{1}} \left\{ r(\lambda, x, a_{1}, f^{2,\lambda}) + \sum_{w \in \mathbf{X}} P_{xa^{1}f^{2,\lambda}w}h_{\lambda}^{f}(w) \right\} + \hat{h}$$

$$= \quad h_{\lambda}^{f}(x) + \overline{\sigma}(\lambda, f^{\lambda}) + \hat{h},$$

where the last equality follows from Lemma 4.1 (iii). We can now establish by recursion that

$$\sigma_\alpha^m(\lambda, x, f^{2,\lambda}) = h_\lambda^f(x) + m\overline{\sigma}(\lambda, f^\lambda) + \hat{h} \leq m\overline{\sigma}(\lambda, f^\lambda) + 2\hat{h}. \qquad (21)$$

Combining (20) with (21), we obtain

$$\sigma^m(\lambda, \xi, u^1, f^{2,\lambda}) - 2\hat{h} \leq m\overline{\sigma}(\lambda)$$

for any ξ. The reverse inequality

$$m\overline{\sigma}(\lambda) \leq \sigma^m(\lambda, \xi, u^1, f^{2,\lambda}) + 2\hat{h}$$

is obtained similarly. This implies both (11) and (13).   ■

Proof of Lemma 3.2: From Lemma 4.1 (i), we have for any $f \in \mathcal{S}_1 \times \mathcal{S}_2$,

$$\overline{\sigma}(\lambda, f) = \sum_w \Pi^f(vw) r(\lambda, w, f)$$

(which in fact does not depend on v), so for any λ_1, λ_2 and any initial distribution ξ,

$$
\begin{aligned}
|\overline{\sigma}(\lambda_1, \xi, f) - \overline{\sigma}(\lambda_2, \xi, f)| \ &\leq\ \sum_v \sum_w \xi(v) \Pi^f(vw) |r(\lambda_1, w, f) - r(\lambda_2, w, f)| \\
&\leq\ \sup_{w,a} |r(\lambda_1, w, a) - r(\lambda_2, w, a)| \\
&\leq\ \|\lambda_1 - \lambda_2\| \sup_{w,a} \|By(w, a)\|
\end{aligned}
$$

Hence, for any initial distribution ξ,

$$
\begin{aligned}
\overline{\sigma}(\lambda_1) - \overline{\sigma}(\lambda_2) \ &=\ \overline{\sigma}(\lambda_1, \xi, f^{1,\lambda_1}, f^{2,\lambda_1}) - \overline{\sigma}(\lambda_2, \xi, f^{1,\lambda_2}, f^{2,\lambda_2}) \\
&\leq\ \overline{\sigma}(\lambda_1, \xi, f^{1,\lambda_1}, f^{2,\lambda_2}) - \overline{\sigma}(\lambda_2, \xi, f^{1,\lambda_1}, f^{2,\lambda_2}) \\
&\leq\ \|\lambda_1 - \lambda_2\| \sup_{w,a} \|By(w, a)\|
\end{aligned}
$$

and, in the same way we obtain

$$\overline{\sigma}(\lambda_2) - \overline{\sigma}(\lambda_1) \leq \|\lambda_1 - \lambda_2\| \sup_{w,a} \|By(w, a)\|.$$

Since y is bounded, we conclude that $\overline{\sigma}(\lambda)$ is continuous in λ.   ■

REFERENCES

[1] E. Altman and V. A. Gaitsgory, "Control of a hybrid Stochastic System", *Systems and Control Letters* **20**, 307-314, 1993.

[2] H.A.M. Couwenbergh, "Stochastic Games with metric state space", *J. of Game Theory* **9**, issue 1, 25-36, 1980.

[3] A. Federgruen, "On N-person stochastic Games with denumerable state space", *Adv. Appl. Prob.* **10**, 452-471, 1978.

[4] K. Hinderer, Foundations of Non-stationary Dynamic Programming with Discrete Time parameter, *Lecture Notes in Operations Research and Mathematical Systems* **33**, Springer-Verlag, 1970.

[5] A. Hordijk, *Dynamic Programming and Markov Potential Theory*, Second Edition, Mathematical Centre Tracts 51, Mathematisch Centrum, Amsterdam, 1977.

[6] L. Kleinrock, *Queuing Systems, Volume II: Computer Applications*, John Wiley, New York, 1976.

[7] A.S. Nowak, "Approximation Theorems for zero-sum nonstationary stochastic games", *Proc. of the American Math. Soc.*, **92**, No. 3, 418-424, 1984.

[8] U. Rieder, "Non-Cooperative Dynamic Games with General Utility Functions", *Stochastic Games and related topics*, T.E.S. Raghavan et al (eds), pp. 161-174, Kluwer Academic Publishers, 1991.

[9] S. Ross, *Applied Probability Models with Optimization Applications*, Holden-Day, 1970.

[10] F. M. Spieksma, *Geometrically Ergodic Markov Chains and the Optimal Control of Queues*, Ph.D. thesis, University of Leiden, 1990.

[11] J. Van der Wal, "Successive Approximations for Average Reward Markov Games", *Int. J. of Game Theory* **9**, issue 1, 13-24.

H^∞-Control of Markovian Jump Systems and Solutions to Associated Piecewise-Deterministic Differential Games *

Zigang Pan and Tamer Başar
Decision and Control Laboratory
Coordinated Science Laboratory and the
Department of Electrical and Computer Engineering
University of Illinois
1308 West Main Street
Urbana, IL 61801/USA

Abstract

A class of linear-quadratic piecewise deterministic soft-constrained zero-sum differential games is formulated and solved, where the minimizing player has access to perfect or imperfect (continuous) state measurements. Such systems are also known as jump linear-quadratic systems, and the underlying game problem can also be viewed as an H^∞ optimal control problem, where the system and cost matrices depend on the outcome of a Markov chain. Both finite- and infinite-horizon cases are considered, and a set of sufficient, as well as a set of necessary, conditions are obtained for the upper value of the game to be bounded. Policies for the minimizing player that achieve this upper value (which is zero) are piecewise linear on each sample path of the stochastic process, and are obtained from solutions of linearly coupled generalized Riccati equations. For the associated H^∞-optimal control problem, these policies guarantee an $\mathcal{L}_2$ gain type inequality on the closed-loop system.

1 Introduction

Systems prone to failure or abrupt structural changes can be modeled by the introduction of an auxiliary stochastic process, generally described as a Markov chain, the outcome of which determines the structure according to which (out of a finite number of alternatives) the system behaves at each

---

*Research supported in part by the U.S. Department of Energy under Grant DE-FG-02-88-ER-13939, and in part by the National Science Foundation under Grant NSF ECS 93-12807 and the Joint Services Electronics Program through the University of Illinois.

point in time. In the case of linear systems, this manifests itself as the system matrices depending on the current state of a Markov chain, leading to what are called *jump linear systems*. This class of systems has attracted much attention for a long time, with some of the representative books and papers being [1] [2] [3] [4] [5] [6] [7] [8] [9] [10] [11]. The optimal control problem for jump linear systems (with or without additive white noise inputs) has been solved completely when the state of the Markov chain is available ([12] [13] [14]). There has also been a major effort in studying the solution to the adaptive jump linear quadratic Gaussian optimal control, where the state of the Markov chain is not measured perfectly ([15] [16] [17] [18] [19]).

The worst-case design problem for jump linear systems has been studied earlier in [20], where it was assumed that the controller has access to both the system state x and the form process θ and the disturbance is restricted to time functions only. These results actually follow from a game theoretic approach, and using a result on Nash equilibria of differential games with jump linear dynamics [21]. The results obtained in [21] also cover nonlinear systems and different modes of equilibria for the case when both players have complete information about x and θ. More recently, these results have been extended to sampled state information, and relationship with "H^∞-optimal" control of piecewise-deterministic nonlinear systems has been established [22]. Additional results on piecewise-deterministic nonzero-sum differential games, especially from the viewpoint of overtaking feedback equilibria, can be found in [23].

In this paper, we revisit the worst-case controller design problem for jump linear systems, and obtain a complete set of solutions not only for the state feedback case, but also for the case when imperfect state measurements are available to the controller. This can be viewed as an H^∞-optimal control problem [24], where now the system and cost matrices depend on the outcome of a Markov chain. Solving this problem is equivalent to solving (for upper value of) a class of parametrized soft-constrained zero-sum differential games with piecewise deterministic linear dynamics, under a quadratic cost function. In this game, Player 1 (P1), which is the controller, strives to minimize the (expected) cost function based on an exact knowledge of the history of the state of the Markov process and a partial (noise corrupted) measurement on the system state. On the other hand, Player 2 (P2), the disturbance, tries to maximize the same (expected) cost function using progressively measurable strategies. As indicated earlier, we are primarily interested in the upper value of this zero-sum game, and especially in the question of whether the upper value is less than or equal to zero. A set of sufficient conditions, as well as a set of necessary conditions, are given to ensure satisfaction of this bound.

The balance of the paper is as follows. The next section, Section 2, provides a precise formulation of the problem under study, in both finite and infinite horizons. Section 3 presents results for the perfect state measurements case, which are subsequently extended to imperfect state measurements in Section 4. Section 5 contains three illustrative numerical examples, and the paper ends with the concluding remarks of Section 6 and an Appendix.

2   Problem Formulation

The jump linear system under consideration is described by:

$$
\begin{aligned}
\dot{x} &= A(t,\theta(t))x + B(t,\theta(t))u + D(t,\theta(t))w; \quad x(t_0) = x_0 & (1)\\
y &= C(t,\theta(t))x + E(t,\theta(t))w & (2)
\end{aligned}
$$

where x is the n-dimensional system state vector; u is the p-dimensional input, determined by the strategy of P1 (the controller); w is the q-dimensional input, determined by the strategy of P2 (the disturbance); y is the m-dimensional output vector; θ is a finite state Markov chain defined on the state space $\mathcal{S} = \{1,\ldots,s\}$ with the infinitesimal generator matrix

$$
\Lambda = (\lambda_{ij}(t))_{s \times s},
$$

and a positive initial distribution $\pi_0 := [\pi_{01},\ldots,\pi_{0s}]$ (i.e., $\pi_{0i} > 0 \ \forall i \in \mathcal{S}$). The underlying probability space is the triple $(\Omega, \mathcal{F}, \boldsymbol{P})$. The initial condition x_0 is unknown and is determined by the strategy of P2. Here $\theta(t)$ is the so-called *form* or *indicant process*, which determines the current form of the system at time t. The system state x, inputs u, y, and w each belong to appropriate $(\mathcal{L}^2)$ Hilbert spaces $\mathcal{H}_x$, $\mathcal{H}_u$, $\mathcal{H}_y$ and $\mathcal{H}_w$ respectively, defined on the time interval $[t_0, t_f]$. Let $\boldsymbol{E}$ denote the expectation with respect to the underlying probability space.

The input u is generated by a strategy μ_I of P1, according to

$$
u(t) = \mu_I(t, y_{[t_0,t]}, \theta_{[t_0,t]}) \tag{3}
$$

where $\mu_I : [t_0, t_f] \times \mathcal{H}_y \times \Omega \to \mathcal{H}_u$ is piecewise continuous in t and Lipschitz continuous in y and measurable in θ, further satisfying the given causality condition. Let us denote the class of all admissible strategies for P1 by $\mathcal{M}_I$.

The initial condition x_0 and the input w are generated by a strategy $\delta := (\delta_0, \nu)$ of P2, according to:

$$
\begin{aligned}
x_0 &= \delta_0(\theta(t_0)) & (4)\\
w(t) &= \nu(t, x_{[t_0,t]}, \theta_{[t_0,t]}) & (5)
\end{aligned}
$$

where $\delta_0 : \mathcal{S} \to \mathcal{R}^n$, and $\nu : [t_0, t_f] \times \mathcal{H}_x \times \Omega \to \mathcal{H}_w$ is piecewise continuous in t and Lipschitz continuous in x and measurable in θ, further satisfying the given causality condition. Let us denote the class of all admissible δ by $\mathcal{D}$.

Associated with this system is the finite-horizon quadratic performance index, parametrized by $\gamma > 0$:

$$
\begin{aligned}
J_\gamma(\mu_I, \delta) &= \boldsymbol{E}\{|x(t_f)|^2_{Q_f(\theta(t_f))} + \int_{t_0}^{t_f} (|x(t)|^2_{Q(t,\theta(t))} + |u(t)|^2_{R(t,\theta(t))} \\
&\quad -\gamma^2|w(t)|^2)\,dt - \gamma^2|x_0|^2_{Q_0(\theta(t_0))}\} \\
&\quad Q_f(.) \geq 0, \quad Q(.,.) \geq 0, \quad R(.,.) > 0, \quad Q_0(.) > 0
\end{aligned}
\tag{6}
$$

where P1 chooses his strategy μ_I to minimize the expected cost function (6), and P2 chooses the strategy δ to maximize the same expected cost function. The upper value of the game is determined by

$$
\bar{V} = \inf_{\mu_I \in \mathcal{M}_I} \sup_{\delta \in \mathcal{D}} J_\gamma(\mu_I, \delta)
\tag{7}
$$

As in standard H^∞-control [24], we will study a parametrization of the solution to this problem in terms of γ.

Both players are assumed to have access to the past state of the form process θ. The case when past values of θ are not completely available to the controller leads to a dual control problem, which is much more difficult (if not impossible) to solve. In practice, the form process is usually available to the controller with a small delay. Hence, the formulation here should provide a good approximation to what arises in practice.

It should be noted that the upper value of the game, $\bar{V}$, is bounded below by 0, which can be ensured for P2 by choosing x_0 and $w(t)$ to be zero. Because of the linear-quadratic nature of the problem, the upper value of the game will be infinite if $\bar{V} > 0$. Hence, the case of particular interest to us is $\bar{V} = 0$. In view of this, let us introduce a set Γ_I:

$$
\Gamma_I := \{\gamma > 0 : \quad \forall \gamma' > \gamma, \bar{V} = 0\}
\tag{8}
$$

and a threshold γ_I^*:

$$
\gamma_I^* := \inf \Gamma_I.
\tag{9}
$$

Note that γ_I^* is then the "smallest" value of γ such that the zero-sum differential game admits a bounded upper value.

This threshold is of particular relevance to H^∞-optimal control due to the following $\mathcal{L}_2$ type gain:

$$
\inf_{\mu_I \in \mathcal{M}_I} \sup_{\delta \in \mathcal{D}} \frac{J(\mu_I, \delta)^{1/2}}{(\|w\|^2 + \boldsymbol{E}\{|x_0|^2_{Q_0(\theta(t_0))}\})^{1/2}} = \gamma_I^*
$$

where

$$
J(\mu_I, \delta) = \boldsymbol{E}\{|x(t_f)|^2_{Q_f(\theta(t_f))} + \int_{t_0}^{t_f} (|x(t)|^2_{Q(t,\theta(t))} + |u(t)|^2_{R(t,\theta(t))})\,dt\}
$$

and $\|.\|$ denotes the norm:

$$\left[\boldsymbol{E}\{ \int_{t_0}^{t_f} |.|^2 \, dt \} \right]^{1/2}$$

The methodology developed in this paper can actually be applied to a more general class of problems where the system dynamics further admit a known bias input term $c(t, \theta(t))$:

$$\dot{x} \;=\; A(t, \theta(t))x + B(t, \theta(t))u + D(t, \theta(t))w + c(t, \theta(t))$$

and the general cost structure:

$$J_\gamma(\mu_I, \nu) = \boldsymbol{E}\{ |x(t_f)|^2_{Q_f(\theta(t_f))} + \int_{t_0}^{t_f} (|x(t)|^2_{Q(t,\theta(t))} + 2x(t)'P(t, \theta(t))u(t)$$

$$+ |u(t)|^2_{R(t,\theta(t))} + 2x(t)'p(t, \theta(t)) - \gamma^2|w(t)|^2) \, dt - \gamma^2|x_0|^2_{Q_0(\theta(t_0))} \}$$

with additional a cross term and a linear term. This general class of problems addresses wider practical issues, such as target tracking problem. But, we will only present the solution to the specific class of problems formulated by (1)–(6) to illustrate the basic idea in solving these problems without extra notational burdens.

We now make the following four basic assumptions for the problem formulated above:

Assumption 1 $A(t, i)$, $Q(t, i)$, $B(t, i)$, $D(t, i)$, $C(t, i)$, $E(t, i)$, $R(t, i)$ are piecewise continuous in t for each $i \in \mathcal{S}$.

Assumption 2 Matrix functions $R(t, i)$ and $Q_0(i)$ are positive definite for all $t \in [t_0, t_f]$ and $i \in \mathcal{S}$.

Assumption 3 The Markov chain θ is irreducible.

Assumption 4 The matrices $D(t, i)$ and $E(t, i)$ satisfy the conditions $D(t, i)E'(t, i) = 0$ and $N(t, i) := E(t, i)E'(t, i) > 0$ for all $t \in [t_0, t_f]$ and $i \in \mathcal{S}$.

To study the infinite horizon case (i.e. as $t_f \to \infty$ and $t_0 \to -\infty$, as well as when $t_f = \infty$ and $t_0 = -\infty$), we take A, B, D, C, E, Q, R, Λ to be time-invariant, $Q_f(.) = 0$, π_0 be the stationary distribution of the Markov chain, and further require $x(t) \to 0$ as $t \to -\infty$. Obviously, the weighting matrix $Q_0(.)$ does not play any role here since $x(-\infty) = 0$. In this case, P2's strategy only dictates the choice of the input $w(t)$.

As the counterparts of (8) and (9), let us introduce, respectively, the set

$$\Gamma_{I\infty} := \{ \gamma > 0 : \quad \forall \gamma' > \gamma, \bar{V} = 0 \}, \tag{10}$$

and the threshold

$$\gamma_{I\infty}^* := \inf \Gamma_{I\infty}. \tag{11}$$

Our objective, for both the finite and infinite horizon cases, is to obtain upper and lower bounds on γ_I^* and $\gamma_{I\infty}^*$, and to derive strategies for P1 that will lead to finite (i. e. zero) upper values for the game. We first study in the next section the perfect state measurements case, that is when $y(t) \equiv x(t)$. Subsequently, in Section 4, we study the original differential game, as formulated above, with imperfect state measurements, which is directly related to the H^∞-optimal control problem for jump linear systems.

3   Perfect State Measurements

The finite-horizon case

In this subsection, we study the problem for finite horizon and under perfect state measurements, i. e., $C(t, \theta(t)) \equiv I$ and $E(t, \theta(t)) \equiv 0$, under Assumptions 1–2. The strategy space for P1 is then denoted by $\mathcal{M}$, which consists of strategies $\mu(t, x(t), \theta(t))$ that are piecewise continuous in t and Lipschitz continuous in x, and measurable in θ. Furthermore, we denote the counterparts of (8) and (9) in this case by Γ and γ^*, respectively.

Let $\gamma > 0$ be fixed. Then it follows from the results of Section 5 of [21] that if the following set of coupled backward generalized Riccati differential equations (GRDE's):

$$\dot{Z}_i + A'(t,i)Z_i + Z_i A(t,i) - Z_i(B(t,i)R^{-1}(t,i)B'(t,i) - \frac{1}{\gamma^2}D(t,i)D'(t,i))$$

$$\cdot Z_i + Q(t,i) + \sum_{j=1}^{s} \lambda_{ij}(t)Z_j = 0; \quad Z_i(t_f) = Q_f(i) \quad i \in \mathcal{S} \tag{12}$$

admits nonnegative definite solutions $Z_i(t)$, $i = 1, \ldots, s$, on $[t_0, t_f]$, and furthermore

$$\gamma^2 Q_0(i) - Z_i(t_0) \geq 0, \qquad i = 1, \ldots, s, \tag{13}$$

then the game has a finite value, which is zero. This implies that $\gamma \geq \gamma^*$.

We will now show that the existence of a set of nonnegative definite solutions to (12) along with (13) is also necessary for $\gamma > \gamma^*$ (i. e., for the upper value to be bounded). Toward this end, we first introduce the matrix-valued function $Z(t, \theta(t))$ on $[t_0, t_f] \times \mathcal{S}$, defined as:

$$Z(t, \theta(t)) \quad := \quad Z_i(t) \qquad\qquad \text{if} \quad \theta(t) = i$$

when $Z_i(t)$, $i = 1, \ldots, s$, exist on $[t_0, t_f]$. Next, we introduce the notion of a conjugate point for the set (12).

Definition 3.1 *The set of coupled GRDE's (12) is said to have a conjugate point at T, where $T \in [t_0, t_f)$, if $Z_i(t)$, $i = 1, \ldots, s$, exists on $(T, t_f]$ and, for some $i \in \mathcal{S}$, one of the eigenvalues of $Z_i(t)$ goes to $+\infty$ as $t \downarrow T$.*

Now, we can state (and prove) the following theorem:

Theorem 3.1 *Consider the piecewise-deterministic soft-constrained zero-sum differential game of this subsection. Let Assumptions 1–2 hold. Then, $\gamma^* < +\infty$, and for $\gamma > \gamma^*$, the GRDE's (12) admit nonnegative definite solutions $Z_i(t)$, $i = 1, \ldots, s$, on $[t_0, t_f]$, satisfying (13), and hence the upper value $\bar{V}$ of the zero-sum game is 0. For $\gamma > \gamma^*$, a minimax strategy for P1 that attains this upper value is:*

$$u_\gamma^*(t) = \mu_\gamma^*(t, x(t), \theta(t)) = -R^{-1}(t, \theta(t))B'(t, \theta(t))Z(t, \theta(t))x(t). \qquad (14)$$

Furthermore, for $\gamma \leq \gamma^*$, condition (13) is violated for at least one $i \in \mathcal{S}$ and the upper value of the game is $+\infty$.

Proof In view of the discussions preceding the theorem, we need only show that if condition (13) is violated, then the upper value of the game is $+\infty$.

Choose a $\gamma > 0$ such that condition (13) is violated. Then, either the set of GRDE's (12) admits solutions on $[t_0, t_f]$, but the matrix $\gamma^2 Q_0(i) - Z_i(t_0)$ has a negative eigenvalue for at least one i, or the set (12) has a conjugate point at T for some $T \in [t_0, t_f)$. If we have the former case, then by conditioning (6) first on $\theta(t_0) = i$, and realizing that μ has x_0 as its argument,

$$\inf_{\mu \in \mathcal{M}} \sup_{(\delta_0, \nu) \in \mathcal{D}} J_\gamma(\mu, \delta_0, \nu) = \sup_{x_0 \in \mathcal{R}^n} E\{|x_0|^2_{Z_i(t_0)} - \gamma^2 |x_0|^2_{Q_0(i)}\}$$

which directly leads to the necessity of condition (13) for the boundedness (by zero) of the upper value.

The proof for the latter case is more involved. Again, choose a $\gamma > 0$ such that the set of GRDE's (12) does not admit a solution on $[t_0, t_f]$. Since the solution of (12) is always bounded below by 0 for each $i \in \mathcal{S}$ whenever it exists, the set of GRDE's (12) must have a conjugate point at T for some $T \in [t_0, t_f)$. Let $i_0 \in \mathcal{S}$ be the particular index associated with this conjugate point, as in Definition 3.1. Then, there exists a vector $x^* \in \mathcal{R}^n$ and a sequence $\{T_k\}_{k=0}^\infty$ such that $t_f = T_0 > T_1 > \cdots$, $\lim_{k \to \infty} T_k = T$ and

$$|x^*|^2_{Z_{i_0}(T_k)} \to +\infty \qquad \text{as} \quad k \to \infty.$$

Suppose that there is a strategy $\mu \in \mathcal{M}$ for P1 that guarantees a finite upper value $\bar{V}$. Since μ is locally Lipschitz continuous in x, there exists an integer

p_0 such that, for each $k = p_0, p_0 + 1, \ldots$, there is a corresponding vector $x_0^k \in \mathcal{R}^n$ such that

$$\dot{x} = A(t, i_0)x + B(t, i_0)\mu; \qquad x(t_0) = x_0^k$$

leads to $x(T_k) = \left(|x^*|_{Z_{i_0}(T_k)}^2\right)^{-1/4} x^*$.

For any integer $k \geq p_0$, the maximum value of the game with P1 using strategy μ and system starting at x_0^k is bounded below by $J_\gamma(\mu, x_0^k, \nu_k)$, where $\nu_k \in \mathcal{D}$ is defined by

$$\begin{aligned}
w(t) &= 0; \qquad t \in [t_0, T_k) \\
w(t) &= \frac{1}{\gamma^2} D'(t, \theta(t)) Z(t, \theta(t)) x(t); \qquad t \in [T_k, t_f]
\end{aligned}$$

Then,

$$\sup_{\delta \in \mathcal{D}} J_\gamma(\mu, \delta) \geq \left(|x^*|_{Z_{i_0}(T_k)}^2\right)^{1/2} \boldsymbol{P}(r_{[t_0, T_k]} \equiv i_0) - \gamma^2 \boldsymbol{E}\{|x_0^k|_{Q_0(\theta(t_0))}^2\}$$

As $k \to \infty$ the left-hand-side (LHS) should be bounded by $\bar{V}$, while the right-hand-side (RHS) converges to $+\infty$. This leads to a contradiction. Therefore, there is no strategy $\mu \in \mathcal{M}$ that can guarantee a finite cost for P1.

This completes the proof of the Theorem. $\qquad\qquad\square$

Remark 3.1 Let $\bar{\gamma}_i^* > 0$ be the H^∞-optimal performance level of the deterministic linear system with coefficient matrices $\{A_i + (1/2)\lambda_{ii}I, B_i, D_i, Q_i, R_i\}$, $i = 1, \ldots, s$. Then, we have $\gamma^* \geq \bar{\gamma}_i^*$, $i = 1, \ldots, s$. $\qquad\square$

The infinite-horizon case

We now turn to the infinite-horizon case, where we take $t_0 = 0$, $t_f \to \infty$ (as well as the case when $t_f = \infty$), the system matrices A, B, D, Q, R, Λ to be time-invariant, $Q_f(.) = 0$, and π_0 be the stationary distribution of the Markov chain.

Let $\bar{V}$ denote the upper value of the game, and introduce the set (as the counterpart of (10))

$$\Gamma_\infty := \{\gamma > 0 : \quad \forall \gamma' > \gamma, \bar{V} = 0\} \tag{15}$$

Further introduce the counterpart of (11) for the perfect state measurements case:

$$\gamma_\infty^* := \inf \Gamma_\infty. \tag{16}$$

Introduce the following set of coupled generalized algebraic Riccati equations (GARE's):

$$A'(i)Z_i + Z_iA(i) - Z_i(B(i)R^{-1}(i)B'(i) - \frac{1}{\gamma^2}D(i)D'(i))Z_i$$

$$+Q(i) + \sum_{j=1}^{s}\lambda_{ij}Z_j = 0; \quad i = 1,\ldots,s \qquad (17)$$

and introduce the condition:

$$\gamma^2Q_0(i) - Z_i \geq 0; \qquad i = 1,\ldots,s. \qquad (18)$$

Let $Z(\theta(t))$ be defined as:

$$Z(\theta(t)) \quad := \quad Z_i \qquad\qquad \text{if} \quad \theta(t) = i$$

when Z_i, $i = 1,\ldots,s$, exist.

The following two assumptions will be needed for the ensuing analysis.

Assumption 5 The pair $(A(\theta(t)), B(\theta(t)))$ is stochastically stabilizable ([9]).

Assumption 6 The pair $(A(i), Q(i))$ is observable for each $i \in \mathcal{S}$.

Then, we have the following counterpart of Theorem 3.1.

Theorem 3.2 *Consider the piecewise-deterministic soft-constrained zero-sum differential game with perfect state measurements in the infinite-horizon case, as defined in this subsection. Let Assumptions 2, 3, 5 and 6 hold. Then, $\gamma^*_\infty < +\infty$, and for any $\gamma > \gamma^*_\infty$, there exist a set of minimal positive definite solutions Z_i, $i = 1,\ldots,s$, to GARE's (17), which further satisfy the condition (18), and a strategy $\mu^*_{\gamma\infty}$ for P1 that guarantees the zero upper value:*

$$u^*_{\gamma\infty}(t) = \mu^*_{\gamma\infty}(t, x(t), \theta(t)) = -R^{-1}(\theta(t))B'(\theta(t))Z(\theta(t))\,x(t), \qquad (19)$$

The jump linear system driven by this control, and without any disturbance,

$$\dot{x} \quad = \quad (A(\theta(t)) - B(\theta(t))R^{-1}(\theta(t))B'(\theta(t))Z(\theta(t)))\,x(t), \qquad (20)$$

is mean-square stable, i. e., $\lim_{t\to\infty} \mathbf{E}\{|x(t)|^2\} = 0$.

Furthermore, for almost all $\gamma > \gamma^*_\infty$, the jump linear system driven by both the optimal control and the optimal disturbance,

$$\dot{x} \quad = \quad (A(\theta(t)) - (B(\theta(t))R^{-1}(\theta(t))B'(\theta(t))$$

$$-\frac{1}{\gamma^2}D(\theta(t))D'(\theta(t)))Z(\theta(t)))\,x(t) \qquad (21)$$

is also mean-square stable.

For $\gamma < \gamma_\infty^*$, on the other hand, either condition (18) is not satisfied, or the set of GARE's (17) does not admit nonnegative definite solutions, and in both cases the upper value of the game is $+\infty$.

Proof   First, we will show that for any $\gamma > 0$ such that the set of GARE's (17) does not admit any nonnegative definite solutions, the upper value of the game is $+\infty$.

Fix any $\gamma > 0$ such that the set of GARE's (17) does not admit any nonnegative definite solutions. Consider the following set of coupled GRDE's indexed by $T > 0$:

$$\dot{Z}_i^T + A'(i)Z_i^T + Z_i A(i) - Z_i^T (B(i)R^{-1}(i)B'(i) - \frac{1}{\gamma^2}D(i)D'(i))Z_i^T$$

$$+ Q(i) + \sum_{j=1}^{s} \lambda_{ij} Z_j^T = 0 \quad Z_i^T(T) = 0 \quad i = 1, \ldots, s. \quad (22)$$

Denote the solutions to the above GRDE's by $Z_i^T(t)$, $i = 1, \ldots, s$. The following relationship holds for all $i \in \mathcal{S}$:

$$Z_i^{T_1}(t) \geq Z_i^{T_2}(t) \qquad \forall\ 0 \leq t \leq T_2 \leq T_1.$$

Since the set of GARE's (17) does not admit any nonnegative definite solutions, there exists a sequence of strictly increasing positive scalars $\{T_k\}_{k=0}^{\infty}$ and an index $i_0 \in \mathcal{S}$ such that the maximum eigenvalue of $Z_{i_0}^{T_k}(0)$ goes to ∞ as $k \to \infty$. Without loss of generality, we can assume there exists an $x^* \in \mathcal{R}^n$ such that

$$\lim_{k \to \infty} |x^*|^2_{Z_{i_0}^{T_k}(0)} = +\infty.$$

(Such an x^* must exist for some subsequence of $\{T_k\}_{k=0}^{\infty}$. In the proof below we work directly with that subsequence.)

For any integer k, the maximum value of the game with P1 using strategy μ and system starting at x^* is bounded below by $J_\gamma(\mu, x^*, \nu_k)$, where $\nu_k \in \mathcal{D}$ is defined by

$$\begin{aligned} w(t) &= 0; & (T_k, \infty) \\ w(t) &= \frac{1}{\gamma^2}D'(\theta(t))Z^{T_k}(t, \theta(t))x(t); & t \in [0, T_k] \end{aligned}$$

where $Z^{T_k}(t, \theta(t))$ is defined on $[0, T_k]$ by

$$Z^{T_k}(t, \theta(t)) = Z_i^{T_k}(t); \qquad \text{if} \qquad \theta(t) = i$$

Then,

$$\sup_{\delta \in \mathcal{D}} J_\gamma(\mu_\infty, \delta) \geq (|x^*|^2_{Z_{i_0}^{T_k}(0)})\pi_{0i_0} - \gamma^2 \boldsymbol{E}\{|x^*|^2_{Q_0(\theta(0))}\}$$

As $k \to \infty$ RHS converges to $+\infty$. Therefore, there is no strategy $\mu \in \mathcal{M}$ that can guarantee a finite cost for P1. This implies that $\gamma \le \gamma_\infty^*$.

For the case where (17) admits a nonnegative definite solution but the condition (18) is violated, again first condition (6) on $\theta(t_0) = i_0$, where $i_0 \in \mathcal{S}$ is such that $\gamma^2 Q_0(i_0) - Z_{i_0}$ has at least one negative eigenvalue, and note that

$$\inf_{\mu \in \mathcal{M}} \sup_{(\delta_0,\nu) \in \mathcal{D}} J_\gamma = \sup_{x_0 \in \mathcal{R}^n} \boldsymbol{E}\{|x_0|^2_{Z_{i_0}} - \gamma^2 |x_0|^2_{Q_0(i_0)}\}$$

which directly leads to the conclusion that the upper value is $+\infty$.

Next, we fix a $\gamma > 0$ such that the set of GARE's (17) admits nonnegative definite solutions $\bar{Z}_i$, $i = 1, \ldots, s$. Then the set of GRDE's (22) must admit nonnegative definite solutions $Z_i^T(t)$, $i \in \mathcal{S}$, on $[0,T]$ for any $T > 0$, which further satisfy the following bound:

$$Z_i^T(t) \le \bar{Z}_i; \qquad \forall i \in \mathcal{S}, \ \forall T > 0, \ \forall t \in [0,T].$$

By the monotonicity of the $Z_i^T(t)$'s, we can define

$$Z_i \ := \ \lim_{T \to \infty} Z_i^T(0); \qquad \forall i \in \mathcal{S}.$$

Hence, Z_i, $i \in \mathcal{S}$, satisfy the set of GARE's (17). Obviously, $Z_i \le \bar{Z}_i$, $i = 1 \ldots, s$. Thus, Z_i, $i = 1, \ldots, s$ is the set of minimal nonnegative definite solutions to the GARE's (17).

For each $i \in \mathcal{S}$, (17) can be rewritten in the following form:

$$(A'(i) + \frac{1}{2}\lambda_{ii}I)Z_i + Z_i(A(i) + \frac{1}{2}\lambda_{ii}I) - Z_i(B(i)R^{-1}(i)B'(i)$$

$$-\frac{1}{\gamma^2}D(i)D'(i))Z_i + Q(i) + \sum_{\substack{j=1 \\ j \ne i}}^{s} \lambda_{ij}Z_j = 0$$

By Assumption 6, we have the observability of the pair $(A(i) + \frac{1}{2}\lambda_{ii}I, Q(i)$ $+ \sum_{\substack{j=1 \\ j \ne i}}^{s} \lambda_{ij}Z_j)$, $i = 1, \ldots, s$. This implies, by the existing theory on the solution to GARE's [24], that $Z_i > 0$, $i = 1, \ldots, s$. Hence, $Z_i > 0$, $i = 1, \ldots, s$, is the set of minimal positive definite solutions to the GARE's (17).

For this value of γ, a "completion of squares" yields the following equivalent expression for J_γ:

$$J_\gamma(\mu,\delta) \ = \ \boldsymbol{E}\{\int_0^T (|u(t) + R^{-1}(\theta(t))B'(\theta(t))Z(\theta(t))x(t)|^2_{R(\theta(t))}$$

$$-\gamma^2 |w(t) - \frac{1}{\gamma^2}D'(\theta(t))Z(\theta(t))x(t)|^2)\,dt$$

$$+|x_0|^2_{Z(\theta(0))-\gamma^2 Q_0(\theta(0))} - |x(T)|^2_{Z(\theta(t))}\},$$

for all $T > 0$. Hence, if further the condition (18) is satisfied, the strategy μ_∞^* is minimax, and it guarantees a cost 0 for P1. This further implies that $\gamma \geq \gamma_\infty^*$.

Obviously, the set of GARE's (17) admits nonnegative definite solutions also at $\gamma = \infty$ under Assumption 5. Hence, we have the threshold $\gamma_\infty^* < +\infty$.

Thus, we have shown that γ_∞^* is the infimum of γ's such that the set of GARE's (17) admits positive definite solutions, which further satisfy condition (18). Rewrite the set of GARE's (17) in the following form:

$$(A(i) - B(i)R^{-1}(i)B'(i)Z_i)'Z_i + Z_i(A(i) - B(i)R^{-1}(i)B'(i)Z_i) + Z_i(B(i)$$
$$\cdot R^{-1}(i)B'(i) + \frac{1}{\gamma^2}D(i)D'(i))Z_i + Q(i) + \sum_{j=1}^{s} \lambda_{ij}Z_j = 0; \quad i = 1,\ldots,s.$$

Since the pairs $(A(i) - B(i)R^{-1}(i)B'(i)Z_i, Z_i(B(i)R^{-1}(i)B'(i) + (1/\gamma^2)D(i)$ $D'(i))Z_i + Q(i))$, $i = 1,\ldots,s$ are observable, the mean-square stability of the jump linear system (20) readily follows.

To complete the proof of the Theorem, we only need to show that, for almost all $\gamma > \gamma_\infty^*$, the set of GARE's (17) admits nonnegative definite solutions Z_i, $i = 1,\ldots,s$, such that the jump linear system (21) is mean-square stable.

Denote Z_i by $Z_{\gamma i}$ to emphasize its dependence on the parameter γ. Since $Z_{\gamma i}$ is nonincreasing in γ for each $i \in \mathcal{S}$, the derivative $\dfrac{d}{d\gamma}Z_{\gamma i}$ exists for all $i \in \mathcal{S}$ almost everywhere in the interval $(\gamma_\infty^*, \infty)$.

Fix any $\gamma > \gamma_\infty^*$ such that $\dfrac{d}{d\gamma}Z_{\gamma i}$ exists for all $i \in \mathcal{S}$. Let

$$-\Delta_{\gamma i} := \frac{d}{d\gamma}Z_{\gamma i}; \qquad \forall i \in \mathcal{S}.$$

Then, $\Delta_{\gamma i} \geq 0$ for each $i \in \mathcal{S}$, and they further satisfy the following coupled Lyapunov equations:

$$A'_{F\gamma}(i)\Delta_{\gamma i} + \Delta_{\gamma i}A_{F\gamma}(i) + \frac{2}{\gamma^3}Z_{\gamma i}D(i)D'(i)Z_{\gamma i} + \sum_{j=1}^{s}\lambda_{ij}\Delta_{\gamma j} = 0; \quad i \in \mathcal{S},$$

where

$$A_{F\gamma}(i) := A(i) - (B(i)R^{-1}(i)B'(i) - \frac{1}{\gamma^2}D(i)D'(i))Z_{\gamma i}; \quad i = 1,\ldots,s.$$

This, coupled with the mean-square stability of the jump linear system (20), implies the mean-square stability of the jump linear system (21).

This completes the proof of the theorem.   □

Remark 3.2 In the proof of the above theorem, Assumption 3 may be relaxed to the condition that the initial distribution π_0 has only positive elements. □

The above theorem further implies (with some additional work) the following stronger conclusion on the mean-square stability of the closed-loop system:

Corollary 3.1 *Consider the piecewise-deterministic soft-constrained zero-sum differential game covered by Theorem 3.2. For any $\gamma > \gamma_\infty^*$, there exists a set of positive definite solutions $\bar{Z}_i$, $i = 1, \ldots, s$, to the following GARI's:*

$$A'(i)\bar{Z}_i + \bar{Z}_i A(i) - \bar{Z}_i(B(i)R^{-1}(i)B'(i) - \frac{1}{\gamma^2}D(i)D'(i))\bar{Z}_i$$

$$+Q(i) + \sum_{j=1}^{s} \lambda_{ij}\bar{Z}_j < 0 \quad i = 1, \ldots, s \qquad (23)$$

such that the jump linear systems

$$\dot{x} = (A(\theta(t)) - B(\theta(t))R^{-1}(\theta(t))B'(\theta(t))\bar{Z}(\theta(t)))x(t) \qquad (24)$$

and

$$\dot{x} = (A(\theta(t)) - (B(\theta(t))R^{-1}(\theta(t))B'(\theta(t))$$

$$-\frac{1}{\gamma^2}D(\theta(t))D'(\theta(t)))\bar{Z}(\theta(t)))x(t) \qquad (25)$$

are mean-square stable. The strategy $\bar{\mu}_{\gamma\infty}^*$ defined by:

$$\bar{\mu}_{\gamma\infty}^*(t, x(t), \theta(t)) = -R^{-1}(\theta(t))B'(\theta(t))\bar{Z}(\theta(t))x(t). \qquad (26)$$

guarantees the zero upper value for P1.

Proof   Fix a $\gamma > \gamma_\infty^*$. Then, by Theorem 3.2, there exists a $\gamma_1 \in (\gamma_\infty^*, \gamma)$ such that a set of minimal positive definite solutions $Z_{\gamma_1 i}$ to GARE's (17) exists and the jump linear system:

$$\dot{x} = (A(\theta(t)) - (B(\theta(t))R^{-1}(\theta(t))B'(\theta(t)) - \frac{1}{\gamma_1^2}D(\theta(t))D'(\theta(t)))Z_{\gamma_1}(\theta(t)))x(t)$$

is mean-square stable. Then, by an application of the Implicit Function Theorem, there exists a $\rho_0 > 0$ such that $\forall \rho \in [0, \rho_0]$, the following set of coupled GARE's:

$$(A'(i) + 1/2\lambda_{ii}I)\bar{Z}_i + \bar{Z}_i(A(i) + 1/2\lambda_{ii}I) - \bar{Z}_i(B(i)R^{-1}(i)B'(i)$$

$$-\frac{1}{\gamma_1^2}D(i)D'(i))\bar{Z}_i + Q(i) + \sum_{\substack{j=1 \\ j \neq i}}^{s}(1 + \rho)\lambda_{ij}\bar{Z}_j = 0 \quad i = 1, \ldots, s \qquad (27)$$

admits minimal positive definite solutions $\bar{Z}_{\gamma_1\rho i}$, $i = 1, \ldots, s$ (see the Appendix for a proof).

This further implies that $\forall \rho \in [0, \rho_0]$, the following set of coupled GARE's:

$$(A'(i) + 1/2\lambda_{ii}I)\bar{Z}_i + \bar{Z}_i(A(i) + 1/2\lambda_{ii}I) - \bar{Z}_i(B(i)R^{-1}(i)B'(i)$$

$$-\frac{1}{\gamma^2}D(i)D'(i))\bar{Z}_i + Q(i) + \sum_{\substack{j=1 \\ j \neq i}}^{s}(1 + \rho)\lambda_{ij}\bar{Z}_j = 0 \quad i = 1, \ldots, s$$

admits minimal positive definite solutions $\bar{Z}_{\gamma\rho i}$, $i = 1, \ldots, s$. Furthermore, $\bar{Z}_{\gamma\rho i}$ is nondecreasing in ρ for each $i \in \mathcal{S}$. Hence, the derivative $\frac{d}{d\rho}\bar{Z}_{\gamma\rho i}$ exists for all $i \in \mathcal{S}$ almost everywhere on $[0, \rho_0]$.

Fix a $\rho \in (0, \rho_0]$ such that $\frac{d}{d\rho}\bar{Z}_{\gamma\rho i}$ exists for all $i \in \mathcal{S}$. Obviously, $\bar{Z}_{\gamma\rho i}$, $i = 1, \ldots, s$, satisfy the GARI's (23). Let

$$\bar{\Delta}_{\gamma\rho i} := \frac{d}{d\rho}\bar{Z}_{\gamma\rho i}; \qquad \forall i \in \mathcal{S}.$$

Then, $\bar{\Delta}_{\gamma\rho i} \geq 0$ for each $i \in \mathcal{S}$, and they further satisfy the following coupled Lyapunov equations, for $i = 1, \ldots, s$:

$$\bar{A}'_{F\gamma}(i)\bar{\Delta}_{\gamma\rho i} + \bar{\Delta}_{\gamma\rho i}\bar{A}_{F\gamma}(i) + \sum_{\substack{j=1 \\ j \neq i}}^{s}(\lambda_{ij}\bar{Z}_j + \rho\lambda_{ij}\bar{\Delta}_{\gamma\rho j}) + \sum_{j=1}^{s}\lambda_{ij}\bar{\Delta}_{\gamma\rho j} = 0,$$

where

$$\bar{A}_{F\gamma}(i) := A(i) - (B(i)R^{-1}(i)B'(i) - \frac{1}{\gamma^2}D(i)D'(i))\bar{Z}_{\gamma\rho i}; \quad i = 1, \ldots, s.$$

This implies the mean-square stability of the jump linear system (25).

Note that $\gamma_1^2 Q_0(i) \geq Z_{\gamma_1 i}$ since $\gamma_1 > \gamma_\infty^*$. Then, by choosing ρ sufficiently small, the solutions $\bar{Z}_{\gamma\rho i}$, $i = 1, \ldots, s$, can be made to satisfy condition (18). For this value of ρ, by an argument similar to that used in the proof of Theorem (3.2), we have that the jump linear system (24) is mean-square stable. The optimality of the strategy $\bar{\mu}_{\gamma\infty}^*$ follows easily from a "completion of squares" argument as in the case of the preceding theorem, and this completes the proof of the corollary. $\qquad\qquad\square$

Remark 3.3 Let $\bar{\gamma}_{i\infty}^* > 0$ be the H^∞-optimal performance level of the deterministic linear system with coefficient matrices $\{A_i + (1/2)\lambda_{ii}I, B_i, D_i, Q_i, R_i\}$, $i = 1, \ldots, s$. Then, we have $\gamma_\infty^* \geq \bar{\gamma}_{i\infty}^*$, $i = 1, \ldots, s$. $\qquad\square$

4 Imperfect State Measurements

The finite-horizon case

Before presenting the solution to the imperfect state measurements case, as formulated in Section 2, we introduce a set of coupled generalized Riccati differential inequalities (GRDI's) defined in forward time:

$$\dot{\Theta}_i + A'(t,i)\Theta_i + \Theta_i A(t,i) + \Theta_i D(t,i)D'(t,i)\Theta_i - C'(t,i)N^{-1}(t,i)C(t,i)$$

$$+\frac{1}{\gamma^2}Q(t,i) + \sum_{j=1}^{s}\lambda_{ij}(t)\Theta_j \le 0; \quad \Theta_i(t_0) = Q_0(i) \quad i = 1,\ldots,s \quad (28)$$

Furthermore, we introduce a set $\hat{\Gamma}_I \subset \mathcal{R}$ by:

$$\hat{\Gamma}_I := \{\gamma > 0 : \quad \text{the set of coupled GRDE's (12) admits a set of}$$

nonnegative definite solutions $Z_i(t)$, $i = 1,\ldots,s$, on $[t_0, t_f]$,

and the set of coupled GRDI's (28) admits a set of nonnegative

definite solutions $\Theta_i(t)$, $i = 1,\ldots,s$, on $[t_0, t_f]$, such that

$$\gamma^2\Theta_i(t) - Z_i(t) > 0 \text{ for all } t \in [t_0, t_f] \text{ and } i \in \mathcal{S}.\} \quad (29)$$

Let $\hat{\gamma}_I$ be the infimum of the set Γ_I:

$$\hat{\gamma}_I := \inf \hat{\Gamma}_I \quad (30)$$

Let $\Theta(t, \theta(t))$ be defined on $[t_0, t_f]$ as:

$$\Theta(t, \theta(t)) \quad := \quad \Theta_i(t) \qquad \qquad \text{if } \theta(t) = i$$

when $\Theta_i(t)$, $i = 1,\ldots,s$, exist on $[t_0, t_f]$.

Then, we have the following sufficiency result:

Theorem 4.1 *Consider the piecewise-deterministic soft-constrained zero-sum differential game formulated by (1)–(7), and let Assumptions 1–4 hold. If $\hat{\gamma}_I < +\infty$, then for $\gamma > \hat{\gamma}_I$, the upper value $\bar{V}$ of the zero-sum game is 0, and a minimax strategy for P1 that attains this upper value is $\mu^*_{I\gamma}$:*

$$u^*_{I\gamma}(t) = \mu^*_{I\gamma}(t, y_{[t_0,t]}, \theta_{[t_0,t]}) = -R^{-1}(t,\theta(t))B'(t,\theta(t))Z(t,\theta(t))\,\hat{x}(t) \quad (31)$$

where $\hat{x}$ is generated by the filter (observer):

$$\dot{\hat{x}} = (A(t,\theta(t)) - (B(t,\theta(t))R^{-1}(t,\theta(t))B'(t,\theta(t)) - \frac{1}{\gamma^2}D(t,\theta(t))$$

$$\cdot D'(t,\theta(t)))Z(t,\theta(t)))\hat{x} + \gamma^2(\gamma^2\Theta(t,\theta(t)) - Z(t,\theta(t)))^{-1}C'(t,\theta(t))$$

$$\cdot N^{-1}(t,\theta(t))(y - C(t,\theta(t))\,\hat{x}); \qquad \hat{x}(t_0) = 0 \quad (32)$$

Hence, we have the relationship $\gamma^*_I \le \hat{\gamma}_I$.

Proof   Fix a $\gamma > \hat{\gamma}_I$. Substitution of the minimax strategy $\mu^*_{I\gamma}$ into the original system (1) and the cost function (6), leads to the following system and cost function, in terms of $\xi := [x', \hat{x}']'$:

$$\dot{\xi} = \begin{bmatrix} A(t,\theta(t)) & -B(t,\theta(t))K(t,\theta(t)) \\ G(t,\theta(t))C(t,\theta(t)) & F(t,\theta(t)) \end{bmatrix} \xi$$

$$+ \begin{bmatrix} D(t,\theta(t)) \\ \gamma^2 \Xi(t,\theta(t))C'(t,\theta(t))N^{-1}(t,\theta(t))E(t,\theta(t)) \end{bmatrix} w$$

$$=: \quad \bar{A}(t,\theta(t))\xi + \bar{D}(t,\theta(t))w \tag{33}$$

$$J_\gamma(\mu^*_{I\gamma},\delta) = \boldsymbol{E}\{\int_{t_0}^{t_f} (\xi(t)' \begin{bmatrix} Q(t,\theta(t)) & 0 \\ 0 & K'(t,\theta(t))R(t,\theta(t))K(t,\theta(t)) \end{bmatrix} \xi(t)$$

$$-\gamma^2 w'(t)w(t))\, dt + |\xi(t_f)|^2 \begin{bmatrix} Q_f(\theta(t_f)) & 0 \\ 0 & 0 \end{bmatrix}$$

$$-\gamma^2 |\xi(t_0)|^2 \begin{bmatrix} Q_0(\theta(t_0)) & 0 \\ 0 & 0 \end{bmatrix} \tag{34}$$

where

$$\begin{aligned}
\Xi(t,\theta(t)) &:= \gamma^2\Theta(t,\theta(t)) - Z(t,\theta(t)) \\
K(t,\theta(t)) &:= R^{-1}(t,\theta(t))B'(t,\theta(t))Z(t,\theta(t)) \\
G(t,\theta(t)) &:= \gamma^2\Xi^{-1}(t,\theta(t))C'(t,\theta(t))N^{-1}(t,\theta(t)) \\
F(t,\theta(t)) &:= A(t,\theta(t)) - (B(t,\theta(t))R^{-1}(t,\theta(t))B'(t,\theta(t)) \\
&\qquad -\frac{1}{\gamma^2}D(t,\theta(t))D'(t,\theta(t)))Z(t,\theta(t)) - G(t,\theta(t))C(t,\theta(t))
\end{aligned}$$

To complete the proof of the theorem, we only need to show that

$$\sup_{\delta \in \mathcal{D}} J_\gamma(\mu_{I\gamma},\delta) \leq 0.$$

Introduce the matrix-valued function

$$\Sigma(t,i) := \begin{bmatrix} \gamma^2\Theta(t,i) & -\Xi(t,i) \\ -\Xi(t,i) & \Xi(t,i) \end{bmatrix}; \qquad t \in [t_0, t_f],\ i \in \mathcal{S}, \tag{35}$$

and denote $\Sigma(t,i)$ by $\Sigma_i(t)$, for $i \in \mathcal{S}$. It can be shown, by extensive but straightforward algebraic manipulations, that the matrices Σ_i, $i = 1,\ldots,s$, satisfy the following differential inequalities:

$$\dot{\Sigma}_i + \Sigma_i\bar{A}(t,i) + \bar{A}'(t,i)\Sigma_i + \frac{1}{\gamma^2}\Sigma_i\bar{D}(t,i)\bar{D}'(t,i)\Sigma_i + \sum_{j=1}^{s}\lambda_{ij}\Sigma_j$$

$$+ \begin{bmatrix} Q(t,i) & 0 \\ 0 & Z(t,i)B(t,i)R^{-1}(t,i)B'(t,i)Z(t,i) \end{bmatrix} \leq 0;\quad i \in \mathcal{S}. \tag{36}$$

Let $\mathcal{A}$ be the extended generator [25] associated with the jump linear system (33). Then, we have

$$\mathcal{A}(\xi'(t)\Sigma(t,\theta(t)=i)\xi(t)) = \xi'(t)\dot{\Sigma}(t,i)\xi(t) + \sum_{j=1}^{s}\lambda_{ij}\xi'(t)\Sigma(t,j)\xi(t)$$
$$+2\xi'(t)\Sigma(t,i)(\bar{A}(t,i)\xi + \bar{D}(t,i)w).$$

Thus, by a "completion of squares" argument, the cost function $J_\gamma(\mu^*_{I\gamma},\delta)$ can be bounded above by zero, for any strategy δ of P2:

$$J_\gamma(\mu^*_{I\gamma},\delta) = J_\gamma(\mu^*_{I\gamma},\delta) + \boldsymbol{E}\{\xi'(t_0)\Sigma(t_0,\theta(t_0))\xi(t_0) - \xi'(t_f)\Sigma(t_f,\theta(t_f))\xi(t_f)$$
$$+ \int_{t_0}^{t_f}\mathcal{A}(\xi'(t)\Sigma(t,\theta(t))\xi(t))\,dt\} = \boldsymbol{E}\{\int_{t_0}^{t_f} -\gamma^2|w(t) - \frac{1}{\gamma^2}\bar{D}'\Sigma(t,\theta(t))$$
$$\cdot\xi(t)|^2\,dt + |\xi(t_f)|^2\begin{bmatrix} Q_f(\theta(t_f)) & 0 \\ 0 & 0 \end{bmatrix} - \gamma^2|\xi(t_0)|^2\begin{bmatrix} Q_0(\theta(t_0)) & 0 \\ 0 & 0 \end{bmatrix}$$
$$+\xi'(t_0)\Sigma(t_0,\theta(t_0))\xi(t_0) - \xi'(t_f)\Sigma(t_f,\theta(t_f))\xi(t_f)\}$$
$$\leq \boldsymbol{E}\{-|\xi(t_f)|^2\begin{bmatrix} \Xi(t_f,\theta(t_f)) & -\Xi(t_f,\theta(t_f)) \\ -\Xi(t_f,\theta(t_f)) & \Xi(t_f,\theta(t_f)) \end{bmatrix}$$
$$+\gamma^2|\xi(t_0)|^2\begin{bmatrix} 0 & -\Xi(t_0,\theta(t_0)) \\ -\Xi(t_0,\theta(t_0)) & \Xi(t_0,\theta(t_0)) \end{bmatrix}\} \leq 0$$

The last inequality follows from the fact that $\hat{x}(t_0) = 0$. It, in turn, implies that the proposed strategy $\mu^*_{I\gamma}$ guarantees the upper value 0 for P1, and hence it is minimax. This completes the proof of the theorem. $\square$

Remark 4.1 Let $\bar{\gamma}^*_{Ii} > 0$ be the finite-horizon (imperfect-state) H^∞-optimal performance level of the deterministic linear system with coefficient matrices $\{A_i + (1/2)\lambda_{ii}I, B_i, D_i, C_i, E_i, Q_i, R_i\}$, $i = 1,\ldots,s$. Then, we have $\gamma^*_I \geq \bar{\gamma}^*_{Ii}$, $i = 1,\ldots,s$. $\square$

Theorem 4.1 provides only a sufficient condition for a minimax strategy for P1 when $\gamma > \gamma^*_I$. In other words, it leaves open the possibility that $\hat{\gamma}_I$ might be strictly larger than γ^*_I. In the next theorem, we present a necessary condition for the upper value of the game to be 0.

Theorem 4.2 *Consider the piecewise-deterministic soft-constrained zero-sum differential game formulated by (1)–(7), and let Assumptions 1–4 hold. For any $\gamma > \gamma^*_I$, if there exists a minimax strategy μ_I that guarantees a cost of 0 for P1, then the set of coupled GRDE's (12) admits nonnegative definite* *solutions $Z_i(t)$, $i = 1, \ldots, s$, on $[t_0, t_f]$. Furthermore, for each $T \in [t_0, t_f]$, the following set of coupled backward GRDE's:*

$$\dot{\Theta}_i^T + A'(t,i)\Theta_i^T + \Theta_i^T A(t,i) + \Theta_i^T D(t,i)D'(t,i)\Theta_i^T - C'(t,i)N^{-1}(t,i)$$

$$\cdot C(t,i) + \frac{1}{\gamma^2}Q(t,i) + \sum_{j=1}^{s}\lambda_{ij}(t)\Theta_j^T = 0; \quad \Theta_i^T(T) = \frac{1}{\gamma^2}Z_i(T) \; i \in \mathcal{S} \quad (37)$$

admits symmetric solutions $\Theta_i^T(t)$, $i = 1, \ldots, s$, on $[t_0, T]$ such that

$$\Theta_i^T(t_0) < Q_0(i); \qquad i = 1, \ldots, s$$

Proof   Fix a $\gamma > \gamma_I^*$; then the set of coupled GRDE's (12) must admit nonnegative definite solutions on $[t_0, t_f]$, since this is a necessary condition for the game to have upper value 0 when P1 has access to full state information.

Note that the minimax strategy μ_I must satisfy the following condition:

$$\mu_I(t, y_{[t_0,t]}, \theta_{[t_0,t]}) = 0, \qquad \text{if} \quad y_{[t_0,t]} \equiv 0.$$

Otherwise, P2 can simply choose $x_0 = 0$ and $w(t) \equiv 0$ to achieve a positive cost.

By the assumption of the theorem, we must have

$$\sup_{\delta \in \mathcal{D}} J_\gamma(\mu_I, \delta) \leq 0.$$

Fix any $T \in [t_0, t_f]$, and introduce a subclass, $\mathcal{D}_1$, of strategies for P2 that satisfy the following condition:

$$w(t) \;\; = \;\; \frac{1}{\gamma^2}D'(t, \theta(t))Z(t, \theta(t))x(t); \qquad t \in [T, t_f]$$

$$w(t) \;\; = \;\; -E'(t, \theta(t))N^{-1}(t, \theta(t))C(t, \theta(t))x(t)$$
$$+D'(t, \theta(t))v(t, x_{[t_0,t]}, \theta_{[t_0,t]}); \quad t \in [t_0, T]$$

where x_0 and v are to be determined later. Obviously, any strategy in $\mathcal{D}_1$ results in $y_{[t_0,T]} \equiv 0$, and thus $u_{[t_0,T]} \equiv 0$ when $u(t)$ is generated by the strategy μ_I.

Fix any $\delta_1 \in \mathcal{D}_1$. Then, we have

$$J_\gamma(\mu_I, \delta_1) = \boldsymbol{E}\Big\{ \int_{t_0}^{T} (|x(t)|^2_{Q(t,\theta(t))-\gamma^2 C'(t,\theta(t))N^{-1}(t,\theta(t))C(t,\theta(t))}$$

$$-\gamma^2|v(t)|^2_{D'(t,\theta(t))D(t,\theta(t))}) \, dt - \gamma^2|x_0|^2_{Q_0(\theta(t_0))} + |x(t_f)|^2_{Q_f(\theta(t_f))}$$

$$+ \int_{T}^{t_f} (|x(t)|^2_{Q(t,\theta(t))-(1/\gamma^2)Z(t,\theta(t))D'(t,\theta(t))D(t,\theta(t))Z(t,\theta(t))} \; +$$

$$|u(t)|^2_{R(t,\theta(t))}) \, dt\} = E\{\int_{t_0}^T (|x(t)|^2_{Q(t,\theta(t))-\gamma^2 C'(t,\theta(t))N^{-1}(t,\theta(t))C(t,\theta(t))}$$

$$-\gamma^2 |v(t)|^2_{D'(t,\theta(t))D(t,\theta(t))}) \, dt - \gamma^2 |x_0|^2_{Q_0(\theta(t_0))} + |x(T)|^2_{Z(T,\theta(T))}$$

$$+ \int_T^{t_f} |u(t) + B'(t,\theta(t))Z(t,\theta(t))x(t)|^2_{R(t,\theta(t))} \, dt\}$$

$$\geq E\{\int_{t_0}^T (|x(t)|^2_{Q(t,\theta(t))-\gamma^2 C'(t,\theta(t))N^{-1}(t,\theta(t))C(t,\theta(t))}$$

$$-\gamma^2 |v(t)|^2_{D'(t,\theta(t))D(t,\theta(t))}) \, dt - \gamma^2 |x_0|^2_{Q_0(\theta(t_0))} + |x(T)|^2_{Z(T,\theta(T))}\}$$

where the second equality follows from a "completion of squares" argument.

The following inequalities then follow:

$$0 \geq \sup_{\delta \in \mathcal{D}} J_\gamma(\mu_I, \delta) \geq \sup_{\delta_1 \in \mathcal{D}_1} J_\gamma(\mu_I, \delta_1) \geq \sup_{\delta_1 \in \mathcal{D}_1} E\{|x(T)|^2_{Z(T,\theta(T))}$$

$$-\gamma^2 |x_0|^2_{Q_0(\theta(t_0))} + \int_{t_0}^T (|x(t)|^2_{Q(t,\theta(t))-\gamma^2 C'(t,\theta(t))N^{-1}(t,\theta(t))C(t,\theta(t))}$$

$$-\gamma^2 |v(t)|^2_{D'(t,\theta(t))D(t,\theta(t))}) \, dt\}$$

Note that the solution $\Theta_i^T(t)$, $i = 1, \ldots, s$, to GRDE's (37) cannot explode to $-\infty$ on $[t_0, T]$, because if they do then the supremum in the last inequality would be $+\infty$, which leads to a clear contradiction.

When the GRDE's (37) admit symmetric solutions on $[t_0, t_f]$, the following inequality holds (again by completion of squares):

$$0 \geq \sup_{\delta_1 \in \mathcal{D}_1} E\{\int_{t_0}^T -\gamma^2 |D(t,\theta(t))(v(t) + \Theta^T(t,\theta(t))x(t))|^2) \, dt$$

$$-\gamma^2 |x_0|^2_{Q_0(\theta(t_0))-\Theta^T(t_0,\theta(t_0))}\}.$$

Then, necessarily, $Q_0(i) \geq \Theta_i^T(t_0)$ for each $i \in \mathcal{S}$. The strict inequality follows because the above holds for all $\gamma > \gamma_I^*$.

This completes the proof of the theorem. $\qquad\square$

The infinite-horizon case

Now, we turn to study the infinite-horizon case as formulated in Section 2.

The following results are the counterparts of those in the finite-horizon case. Introduce a set of coupled generalized algebraic Riccati inequalities (GARI's):

$$A'(i)\Theta_i + \Theta_i A(i) + \Theta_i D(i) D'(i) \Theta_i - C'(i) N^{-1}(i) C(i) + \frac{1}{\gamma^2} Q(i)$$

$$+ \sum_{j=1}^s \lambda_{ij} \Theta_j \leq 0 \quad i = 1, \ldots, s \qquad (38)$$

Define a set $\hat{\Gamma}_{I\infty} \subset \mathcal{R}$ by:

$$\hat{\Gamma}_{I\infty} := \{\gamma > 0 : \quad \text{the set of coupled GARE's (17) admits a set of}$$
$$\text{nonnegative definite solutions } Z_i, \; i = 1, \ldots, s, \text{ and the set of}$$
$$\text{coupled GARI's (38) admits a set of nonnegative definite solutions}$$
$$\Theta_i, \; i = 1, \ldots, s, \text{ such that } \gamma^2 \Theta_i - Z_i > 0 \text{ for all } i \in \mathcal{S}.\} \tag{39}$$

Furthermore, let $\hat{\gamma}_{I\infty}$ be defined as:

$$\hat{\gamma}_{I\infty} := \inf \hat{\Gamma}_{I\infty}, \tag{40}$$

and $\Theta(\theta(t))$ be defined as:

$$\Theta(\theta(t)) \quad := \quad \Theta_i \qquad\qquad \text{if} \quad \theta(t) = i$$

whenever $\Theta_i \; i = 1, \ldots, s$, exist.

Then, we have the following sufficiency result as the counterpart of Theorem 4.1:

Theorem 4.3 *Consider the piecewise-deterministic soft-constrained zero-sum differential game formulated by (1)–(7) in the infinite-horizon case, with $t_f \to \infty$ and $t_0 \to -\infty$, as well as when $t_f = \infty$ and $t_0 = -\infty$, and with system matrices A, B, D, C, E, Q, R, Λ being time-invariant, $Q_f(.) = 0$, and $x(t) \to 0$ as $t \to -\infty$. Let Assumptions 2–6 hold. If $\hat{\gamma}_{I\infty} < +\infty$, then for $\gamma > \hat{\gamma}_{I\infty}$ the upper value, $\bar{V}$, of the zero-sum game is 0, and a minimax strategy for P1 that attains this upper value is $\mu^*_{I\gamma\infty}$:*

$$u^*_{I\gamma\infty}(t) = \mu^*_{I\gamma\infty}(t, y_{(-\infty,t]}, \theta_{(-\infty,t]}) = -R^{-1}(\theta(t))B(\theta(t))Z(\theta(t))\,\hat{x}(t) \tag{41}$$

where $\hat{x}$ is generated by the filter (observer):

$$\dot{\hat{x}} = (A(\theta(t)) - (B(\theta(t))R^{-1}(\theta(t))B'(\theta(t)) - \frac{1}{\gamma^2}D(\theta(t))D'(\theta(t)))Z(\theta(t)))\hat{x}$$
$$+\gamma^2(\gamma^2\Theta(\theta(t)) - Z(\theta(t)))^{-1}C'(\theta(t))N^{-1}(\theta(t))(y - C(\theta(t))\hat{x}) \tag{42}$$

with initial condition $\hat{x}(-\infty) = 0$. Hence, we have the relationship $\gamma^*_{I\infty} \leq \hat{\gamma}_{I\infty}$.

Proof Fix a $\gamma > \hat{\gamma}_{I\infty}$. Substitution of the minimax strategy $\mu^*_{I\gamma\infty}$ into the original system (1) and the cost function (6), leads to system dynamics and cost function that are the time-invariant versions of (33) and (34), expressed in terms of $\xi := [x', \hat{x}']'$.

Introduce a function $\Sigma(i)$, $i \in \mathcal{S}$, as the time-invariant counterpart of (35). Then, the matrices Σ_i, $i = 1, \ldots, s$, satisfy the following GARI's:

$$\Sigma_i \bar{A}(i) + \bar{A}'(i)\Sigma_i + \frac{1}{\gamma^2}\Sigma_i \bar{D}(i)\bar{D}'(i)\Sigma_i + \sum_{j=1}^{s} \lambda_{ij}\Sigma_j$$

$$+\begin{bmatrix} Q(i) & 0 \\ 0 & Z(i)B(i)R^{-1}(i)B'(i)Z(i) \end{bmatrix} \leq 0; \qquad i = 1, \ldots, s. \qquad (43)$$

An application of the "completion of squares" argument similar to that in the finite-horizon case completes the proof of this theorem. $\qquad\square$

Remark 4.2 Let $\bar{\gamma}^*_{Ii\infty} > 0$ be the infinite-horizon (imperfect-state) H^∞-optimal performance level of the deterministic linear system with coefficient matrices $\{A_i + (1/2)\lambda_{ii}I, B_i, D_i, C_i, E_i, Q_i, R_i\}$, $i = 1, \ldots, s$. Then, we have $\gamma^*_{I\infty} \geq \bar{\gamma}^*_{Ii\infty}$, $i = 1, \ldots, s$. $\qquad\square$

A set of necessary conditions for $\gamma > \gamma^*_{I\infty}$ is stated below as the counterpart of Theorem 4.2.

Theorem 4.4 *Consider the piecewise-deterministic soft-constrained zero-sum differential game covered by Theorem 4.3. Let Assumptions 2–6 hold, and assume that $\gamma^*_I < +\infty$. For any $\gamma > \gamma^*_I$, if there exists a minimax strategy $\mu_{I\infty}$ that guarantees a cost of 0 for P1, then the set of coupled GARE's (17) admits nonnegative definite solutions Z_i, $i = 1, \ldots, s$. Furthermore, the following set of coupled backward GRDE's:*

$$\dot{\Theta}^\infty_i + A'(i)\Theta^\infty_i + \Theta^\infty_i A(i) + \Theta^\infty_i D(i)D'(i)\Theta^\infty_i - C'(i)N^{-1}(i)C(i)$$

$$+\frac{1}{\gamma^2}Q(i) + \sum_{j=1}^{s}\lambda_{ij}\Theta^\infty_j = 0; \quad \Theta^\infty_i(0) = \frac{1}{\gamma^2}Z_i \quad i = 1, \ldots, s \quad (44)$$

admits symmetric solutions $\Theta^\infty_i(t)$, $i = 1, \ldots, s$, on $(-\infty, 0]$.

Proof Fix a $\gamma > \gamma^*_{I\infty}$; then the set of coupled GARE's (17) must admit nonnegative definite solutions, since this is a necessary condition for the game to have upper value 0 when P1 has access to full state information.

Note that the minimax strategy $\mu_{I\infty}$ must satisfy the following condition:

$$\mu_{I\infty}(t, y_{(-\infty,t]}, \theta_{(-\infty,t]}) = 0; \quad \text{if} \quad y_{(-\infty,t]} \equiv 0.$$

By the hypothesis of the theorem, we must have

$$\sup_{\delta \in \mathcal{D}} J_\gamma(\mu_I, \delta) \leq 0.$$

Take a subclass, $\mathcal{D}_1$, of strategies for P2 that satisfy the following condition on $(-\infty, 0]$:

$$w(t) = -E'(\theta(t))N^{-1}(\theta(t))C(\theta(t))x(t) + D(\theta(t))v(t, x_{(-\infty,t]}, \theta_{(-\infty,t]}).$$

Obviously, any strategy in $\mathcal{D}_1$ results in $y_{(-\infty,0]} \equiv 0$, and thus $u_{(-\infty,0]} \equiv 0$ when $u(t)$ is generated by the strategy μ_I.

Fix any $\delta_1 \in \mathcal{D}_1$; then we have

$$
\begin{aligned}
J_\gamma(\mu_I, \delta_1) = E\Big\{ &\int_{-\infty}^0 \big(|x(t)|^2_{Q(\theta(t))-\gamma^2 C'(\theta(t))N^{-1}(\theta(t))C(\theta(t))} \\
&-\gamma^2|v(t)|^2_{D'(\theta(t))D(\theta(t))}\big)\, dt + \int_0^\infty \big(|x(t)|^2_{Q(\theta(t))} + |u(t)|^2_{R(\theta(t))} \\
&-\gamma^2|w(t)|^2\big)\, dt\Big\} \geq E\Big\{ \int_{-\infty}^0 \big(|x(t)|^2_{Q(\theta(t))-\gamma^2 C'(\theta(t))N^{-1}(\theta(t))C(\theta(t))} \\
&-\gamma^2|v(t)|^2_{D'(\theta(t))D(\theta(t))}\big)\, dt + |x(0)|^2_{Z(\theta(0))}\Big\}
\end{aligned}
$$

where the inequality follows from a "completion of squares" argument. Hence,

$$
\begin{aligned}
0 \;\geq\; \sup_{\delta_1 \in \mathcal{D}_1} E\Big\{ &\int_{-\infty}^0 \big(|x(t)|^2_{Q(\theta(t))-\gamma^2 C'(\theta(t))N^{-1}(\theta(t))C(\theta(t))} \\
&-\gamma^2|v(t)|^2_{D'(\theta(t))D(\theta(t))}\big)\, dt + |x(0)|^2_{Z(\theta(0))}\Big\}
\end{aligned}
$$

Note that the solution $\Theta_i^\infty(t)$, $i = 1, \ldots, s$, to GRDE's (44) cannot explode to $-\infty$ on $[t, 0]$ for any $t < 0$, because if they do, then the supremum in the last inequality above must be $+\infty$, which is a contradiction. Hence, GRDE's (44) must admit solutions on $(-\infty, 0]$.

This completes the proof of the theorem. $\qquad\qquad\qquad\qquad\square$

Remark 4.3 The well-known results for the deterministic H^∞-optimal control problem [24] can be obtained as a special case of the results in this paper by choosing $\mathcal{S}$ to be a singleton. Note that when $\mathcal{S}$ is a singleton there is no gap between the sufficiency and necessity conditions of Theorems 4.1 and 4.2, respectively (Theorems 4.3 and 4.4, in the infinite-horizon case). $\qquad\square$

Remark 4.4 *Computational issues.* As shown in Theorem 4.3, the solution to the H^∞-control problem for jump linear systems involves the solution of a set of coupled GARE's and a set of coupled GARI's. The former can be solved using iteratively any standard Riccati solver, in view of the observation that the solution to (17) can be obtained as

$$
Z_i = \lim_{k \to \infty} Z_i^{(k)}, \qquad i \in \mathcal{S},
$$

where $Z_i^{(k)}$, $k = 0, 1, \ldots$, are recursively generated from:

$$
\begin{aligned}
(A'(i) + \tfrac{1}{2}\lambda_{ii}I)Z_i^{(k+1)} &+ Z_i^{(k+1)}(A(i) + \tfrac{1}{2}\lambda_{ii}I) - Z_i^{(k+1)}(B(i)R^{-1}(i)B'(i) \\
&- \tfrac{1}{\gamma^2}D(i)D'(i))Z_i^{(k+1)} + Q(i) = -\sum_{\substack{j=1 \\ j \neq i}}^s \lambda_{ij}Z_j^{(k)}; \quad Z_i^{(0)} = 0 \quad i \in \mathcal{S}.
\end{aligned}
$$

Note that the sequence thus generated is monotonically nondecreasing. For the latter, on the other hand, i.e. (38), along with the spectral radius condition $\gamma^2 \Theta_i - Z_i > 0$, $i \in \mathcal{S}$, it is possible to obtain an equivalent linear matrix inequality (LMI), and thereby use an existing software package for LMI's [26]. □

We note that, as $\gamma \to \infty$, the filters (32) and (42) do not yield the conditional mean estimators for the case when the disturbances are white Gaussian noises. [1] This observation is in contrast with the fact that the deterministic H^∞-optimal controller yields the LQG controller as $\gamma \to +\infty$.

5 Examples

In this section, we consider three numerical examples in the infinite-horizon case to illustrate the theoretical findings of the previous two sections.

Example 1

Take a scalar piecewise deterministic system with a two-state Markov chain, where the system matrices are specified as follows:

$$A(1) = 1; \quad B(1) = 1; \quad D(1) = \begin{bmatrix} 1 & 0 \end{bmatrix}; \quad C(1) = 1;$$
$$E(1) = \begin{bmatrix} 0 & 1 \end{bmatrix}; \quad Q(1) = 1; \quad R(1) = 1; \quad Q_0(1) = 100;$$
$$A(2) = -2; \quad B(2) = -1; \quad D(2) = \begin{bmatrix} 1/2 & 0 \end{bmatrix}; \quad C(2) = -1;$$
$$E(2) = \begin{bmatrix} 0 & 1 \end{bmatrix}; \quad Q(2) = 1; \quad R(2) = 1; \quad Q_0(2) = 100.$$

The infinitesimal generator of the Markov chain is given as

$$\Lambda = \begin{bmatrix} -6 & 6 \\ 1 & -1 \end{bmatrix}.$$

Consider first the perfect state measurements case. (In this case, the matrices $C(1)$, $C(2)$, $E(1)$ and $E(2)$ are irrelevant.) Using a particular search algorithm, we find that $\gamma_\infty^* = 0.7154$.

The quantities $\bar{\gamma}_{1\infty}^*$ and $\bar{\gamma}_{2\infty}^*$, on the other hand, are found to be

$$\bar{\gamma}_{1\infty}^* = 0.4472 \qquad \bar{\gamma}_{2\infty}^* = 0.1836.$$

Clearly, γ_∞^* is larger than both $\bar{\gamma}_{1\infty}^*$ and $\bar{\gamma}_{2\infty}^*$, which corroborates the statement of Remark 3.3.

---

[1] When the disturbances are white Gaussian noises, the conditional mean estimator is given by a standard Kalman filter whose gain is determined from the solution of a piecewise deterministic forward Riccati differential equation.

Let $\gamma^*_{1\infty}$ denote the optimal performance level of system 1 under perfect state measurements without any switching, and $\gamma^*_{2\infty}$ denote the same for system 2. Then, these quantities can be computed to be

$$\gamma^*_{1\infty} = 1 \qquad \gamma^*_{2\infty} = 0.2236.$$

Note that here γ^*_{∞} lies between $\gamma^*_{1\infty}$ and $\gamma^*_{2\infty}$, but this is not a general rule (as we shall observe a different pattern in the next example).

Choose the desired performance level as $\gamma = 0.75$; then,

$$Z_1 = 1.061; \quad Z_2 = 0.3949.$$

A minimax strategy for P1 (or equivalently a controller that guarantees the performance level γ) is therefore

$$\begin{aligned}
\mu^*_{\infty}(x(t), \theta(t)) &= -1.061x & \theta(t) = 1 \\
&= 0.3949x & \theta(t) = 2
\end{aligned}$$

Next, we consider the same system under imperfect state measurements. Using a particular search algorithm (that involves iterative Riccati and LMI solvers), we compute $\hat{\gamma}_{I\infty} = 0.8004$. Obviously, we have the relationship $0.8004 \geq \gamma^*_{I\infty} \geq 0.7154$, where the latter value is the threshold level above which the set of GARE's (17) admits positive solutions.

The quantities $\bar{\gamma}^*_{I1\infty}$ and $\bar{\gamma}^*_{I2\infty}$ are found to be

$$\bar{\gamma}^*_{I1\infty} = 0.4495 \qquad \bar{\gamma}^*_{I2\infty} = 0.1944,$$

which shows that $\gamma^*_{I\infty}$ is larger than both $\bar{\gamma}^*_{I1\infty}$ and $\bar{\gamma}^*_{I2\infty}$ — an observation that corroborates the statement of Remark 4.2.

Let $\gamma^*_{I1\infty}$ denote the optimal performance level of system 1 under imperfect state measurements without any switching, and $\gamma^*_{I2\infty}$ denote the same for system 2. Then, these quantities can be computed to be

$$\gamma^*_{I1\infty} = 2.732 \qquad \gamma^*_{I2\infty} = 0.22425,$$

where again $\hat{\gamma}_{I\infty}$ lies in between these two values.

Choose a desired performance level $\gamma = 0.85$; then,

$$Z_1 = 0.8495; \quad Z_2 = 0.3535; \quad \Theta_1 = 2.309; \quad \Theta_2 = 0.5723.$$

A minimax strategy for P1 (or equivalently a controller that guarantees the performance level γ) is

$$\begin{aligned}
\mu^*_{I\infty}(y_{(-\infty,t]}, \theta_{(-\infty,t]}) &= -0.8495\hat{x} & \theta(t) = 1 \\
&= 0.3535\hat{x} & \theta(t) = 2
\end{aligned}$$

where $\hat{x}$ is generated by:

$$\begin{aligned}
\dot{\hat{x}} &= 0.4438\hat{x} + 0.8825y; & \theta(t) = 1 \\
&= -14.28\hat{x} - 12.05y; & \theta(t) = 2
\end{aligned}$$

Example 2

We now have a two-dimensional piecewise deterministic system with a two-state Markov chain. The system matrices are chosen as

$$A(1) = \begin{bmatrix} 0 & 1 \\ 0 & 0 \end{bmatrix}; \quad B(1) = \begin{bmatrix} 0 \\ 1 \end{bmatrix}; \quad D(1) = \begin{bmatrix} 1 & 0 & 0 \\ 0 & 1 & 0 \end{bmatrix};$$

$$C(1) = \begin{bmatrix} 2 & 1 \end{bmatrix}; \quad E(1) = \begin{bmatrix} 0 & 0 & 1 \end{bmatrix};$$

$$Q(1) = \begin{bmatrix} 1 & 0 \\ 0 & 1 \end{bmatrix}; \quad R(1) = 1; \quad Q_0(1) = \begin{bmatrix} 100 & 0 \\ 0 & 100 \end{bmatrix};$$

$$A(2) = \begin{bmatrix} 1 & -1 \\ 2 & 0 \end{bmatrix}; \quad B(2) = \begin{bmatrix} 1 \\ -1 \end{bmatrix}; \quad D(2) = \begin{bmatrix} 1 & 1 & 0 \\ -1 & 2 & 0 \end{bmatrix};$$

$$C(2) = \begin{bmatrix} -1 & 2 \end{bmatrix}; \quad E(2) = \begin{bmatrix} 0 & 0 & 1 \end{bmatrix};$$

$$Q(2) = \begin{bmatrix} 2 & 1 \\ 1 & 2 \end{bmatrix}; \quad R(2) = 1; \quad Q_0(2) = \begin{bmatrix} 100 & 0 \\ 0 & 100 \end{bmatrix},$$

and infinitesimal generator of the Markov chain is taken to be

$$\Lambda = \begin{bmatrix} -3 & 3 \\ 1 & -1 \end{bmatrix}.$$

In the perfect state measurements case, using a particular search algorithm, we find that $\gamma_\infty^* = 2.736$. It is again strictly larger than the quantities $\bar\gamma_{1\infty}^* = 0.7559$ and $\bar\gamma_{2\infty}^* = 2.052$.

The quantities $\gamma_{1\infty}^*$ and $\gamma_{2\infty}^*$ are computed to be

$$\gamma_{1\infty}^* = 1.414 \qquad \gamma_{2\infty}^* = 2.384.$$

Note that here γ_∞^* is larger than both $\gamma_{1\infty}^*$ and $\gamma_{2\infty}^*$, which is different from the observation made in the previous example.

Pick the desired performance level as $\gamma = 2.8$; then,

$$Z_1 = \begin{bmatrix} 26.98 & 11.38 \\ 11.38 & 6.289 \end{bmatrix}; \quad Z_2 = \begin{bmatrix} 33.35 & 10.14 \\ 10.14 & 4.368 \end{bmatrix}.$$

A minimax strategy for P1 (guaranteeing the chosen performance level γ) is

$$\begin{aligned}
\mu_\infty^*(x(t), \theta(t)) &= \begin{bmatrix} -11.38 & -6.289 \end{bmatrix} x \qquad \theta(t) = 1 \\
&= \begin{bmatrix} -23.20 & -5.774 \end{bmatrix} x \qquad \theta(t) = 2
\end{aligned}$$

In the imperfect state measurements case, using a particular search algorithm (that involves iterative Riccati and LMI solvers), the quantity $\hat\gamma_{I\infty}$ is found to be $\hat\gamma_{I\infty} = 6.094$. Obviously, $6.094 \geq \gamma_{I\infty}^* \geq 2.736$, where again the latter value is the threshold level above which the set of GARE's (17) admits positive solutions.

The quantities $\bar{\gamma}^*_{I1\infty}$ and $\bar{\gamma}^*_{I2\infty}$ are

$$\bar{\gamma}^*_{I1\infty} = 0.7559 \qquad \bar{\gamma}^*_{I2\infty} = 4.072,$$

and by Remark 4.2, we should have $\gamma^*_{I\infty} \geq 4.072$.

Furthermore, the quantities $\gamma^*_{I1\infty}$ and $\gamma^*_{I2\infty}$ are:

$$\gamma^*_{I1\infty} = 2.0242 \qquad \gamma^*_{I2\infty} = 6.110.$$

Choose the desired performance level as $\gamma = 6.5$; then,

$$Z_1 = \begin{bmatrix} 5.331 & 2.060 \\ 2.060 & 1.912 \end{bmatrix}; \quad Z_2 = \begin{bmatrix} 6.155 & 1.477 \\ 1.477 & 1.361 \end{bmatrix};$$

$$\Theta_1 = \begin{bmatrix} 0.8288 & -1.104 \\ -1.104 & 1.979 \end{bmatrix}; \quad \Theta_2 = \begin{bmatrix} 0.6572 & -0.4838 \\ -0.4838 & 0.5781 \end{bmatrix}.$$

A minimax strategy for P1 (guaranteeing the performance level γ) is

$$\begin{aligned} \mu^*_{I\infty}(y_{(-\infty,t]}, \theta_{(-\infty,t]}) &= \begin{bmatrix} -2.060 & -1.912 \end{bmatrix} \hat{x} \qquad \theta(t) = 1 \\ &= \begin{bmatrix} -4.677 & -0.1165 \end{bmatrix} \hat{x} \qquad \theta(t) = 2 \end{aligned}$$

where $\hat{x}$ is generated by:

$$\begin{aligned} \dot{\hat{x}} &= \begin{bmatrix} -336.7 & -167.4 \\ -203.8 & -102.8 \end{bmatrix} \hat{x} + \begin{bmatrix} 168.4 \\ 100.9 \end{bmatrix} y; \qquad \theta(t) = 1 \\ &= \begin{bmatrix} 45.14 & -97.99 \\ 56.74 & -99.17 \end{bmatrix} \hat{x} + \begin{bmatrix} 48.49 \\ 49.74 \end{bmatrix} y; \qquad \theta(t) = 2 \end{aligned}$$

Example 3

We now have a three-dimensional piecewise deterministic system with a three-state Markov chain. The system matrices are in this case

$$A(1) = \begin{bmatrix} 1 & 2 & 0 \\ -1 & -0.5 & 1 \\ 0 & 0 & 0 \end{bmatrix}; \quad D(1) = \begin{bmatrix} 1 & 0 & 0 & 0 \\ 0 & 1 & 0 & 0 \\ 0 & 0 & 1 & 0 \end{bmatrix};$$

$$C(1) = \begin{bmatrix} 3 \\ 1 \\ 0 \end{bmatrix}'; \quad E(1) = \begin{bmatrix} 0 \\ 0 \\ 0 \\ 1 \end{bmatrix}'; \quad Q(1) = \begin{bmatrix} 1 & 0 & 0 \\ 0 & 1 & 0 \\ 0 & 0 & 1 \end{bmatrix};$$

$$B(1) = \begin{bmatrix} 0 \\ 0 \\ 1 \end{bmatrix}; \quad R(1) = 1; \quad Q_0(1) = \begin{bmatrix} 100 & 0 & 0 \\ 0 & 100 & 0 \\ 0 & 0 & 100 \end{bmatrix};$$

$$A(2) = \begin{bmatrix} 3 & 1 & -2 \\ 2 & 0 & 0 \\ 1 & 2 & 1 \end{bmatrix}; \quad D(2) = \begin{bmatrix} 0 & 0 & 0 & 0 \\ 2 & 1 & 2 & 0 \\ -1 & 2 & -3 & 0 \end{bmatrix};$$

$$C(2) = \begin{bmatrix} 4 \\ -1 \\ 0 \end{bmatrix}'; \quad E(2) = \begin{bmatrix} 0 \\ 0 \\ 0 \\ 1 \end{bmatrix}'; \quad Q(2) = \begin{bmatrix} 2 & 0 & 0 \\ 0 & 2 & 0 \\ 0 & 0 & 2 \end{bmatrix};$$

$$B(2) = \begin{bmatrix} 1 \\ 2 \\ 0 \end{bmatrix}; \quad R(2) = 1; \quad Q_0(2) = \begin{bmatrix} 100 & 0 & 0 \\ 0 & 100 & 0 \\ 0 & 0 & 100 \end{bmatrix};$$

$$A(3) = \begin{bmatrix} 0 & -1 & -2 \\ 3 & 1 & 1 \\ 2 & -2 & 0 \end{bmatrix}; \quad D(3) = \begin{bmatrix} 2 & 1 & 0 & 0 \\ 1 & -1 & 0 & 0 \\ -2 & -2 & 0 & 0 \end{bmatrix};$$

$$C(3) = \begin{bmatrix} 2 \\ 0 \\ 4 \end{bmatrix}'; \quad E(3) = \begin{bmatrix} 0 \\ 0 \\ 0 \\ 1 \end{bmatrix}'; \quad Q(3) = \begin{bmatrix} 1 & 0 & 0 \\ 0 & 1 & 0 \\ 0 & 0 & 1 \end{bmatrix};$$

$$B(3) = \begin{bmatrix} 2 \\ 0 \\ -1 \end{bmatrix}; \quad R(3) = 1; \quad Q_0(3) = \begin{bmatrix} 100 & 0 & 0 \\ 0 & 100 & 0 \\ 0 & 0 & 100 \end{bmatrix},$$

and the infinitesimal generator of the Markov chain is taken to be

$$\Lambda = \begin{bmatrix} -2 & 1 & 1 \\ 3 & -4 & 1 \\ 1 & 0 & -1 \end{bmatrix}.$$

In the perfect state measurements case, using a particular search algorithm, we find that $\gamma_\infty^* = 24.55$. It is again strictly larger than the quantities $\bar{\gamma}_{1\infty}^* = 1.464$, $\bar{\gamma}_{2\infty}^* = 4.754$ and $\bar{\gamma}_{3\infty}^* = 1.848$.

The quantities $\gamma_{1\infty}^*$, $\gamma_{2\infty}^*$ and $\gamma_{3\infty}^*$ are found to be

$$\gamma_{1\infty}^* = 3.113 \qquad \gamma_{2\infty}^* = 10.26 \qquad \gamma_{3\infty}^* = 1.959.$$

Note that here γ_∞^* is larger than all the three threshold levels, which is different from the observation made in Example 1, but in line with the observation made in Example 2.

Pick a desired performance level $\gamma = 25$; then, a minimax strategy for P1 (guaranteeing the performance level γ) is

$$\begin{aligned} \mu_\infty^*(x(t), \theta(t)) &= \begin{bmatrix} 2.448 & -2.774 & -7.456 \end{bmatrix} x & \theta(t) = 1 \\ &= \begin{bmatrix} -2.726 & -9.017 & -9.080 \end{bmatrix} x & \theta(t) = 2 \\ &= \begin{bmatrix} -5.302 & -4.286 & 0.4695 \end{bmatrix} x & \theta(t) = 3 \end{aligned}$$

In the imperfect state measurements case, using a particular search algorithm (that involves iterative Riccati and LMI solvers), the quantity $\hat{\gamma}_{I\infty}$ is found to be $\hat{\gamma}_{I\infty} = 28.87$. Obviously, $28.87 \geq \gamma_{I\infty}^* > \gamma_\infty^* = 24.49$, where again the latter value is the threshold above which the set of GARE's (17) admits positive solutions.

The quantities $\bar{\gamma}_{I1\infty}^*$, $\bar{\gamma}_{I2\infty}^*$ and $\bar{\gamma}_{I3\infty}^*$ are

$$\bar{\gamma}_{I1\infty}^* = 1.522 \qquad \bar{\gamma}_{I2\infty}^* = 6.409 \qquad \bar{\gamma}_{I3\infty}^* = 5.580\,.$$

Again, $\gamma_{I\infty}^*$ is strictly larger than these three quantities, which corroborates the statement of Remark 4.2.

Furthermore, the quantities $\gamma_{I1\infty}^*$, $\gamma_{I2\infty}^*$ and $\gamma_{I3\infty}^*$ are:

$$\gamma_{I1\infty}^* = 5.445 \qquad \gamma_{I2\infty}^* = 57.13 \qquad \gamma_{I3\infty}^* = 8.722.$$

Note that in this case the optimal performance level $\gamma_{I\infty}^*$ is larger than the performance levels of first and third form systems, but less than that of the second form system.

Choose a desired performance level $\gamma = 29$; then, a minimax strategy for P1 (guaranteeing the performance level γ) is

$$
\begin{aligned}
\mu_{I\infty}^*(y_{(-\infty,t]}, \theta_{(-\infty,t]}) &= \begin{bmatrix} 1.719 & -2.450 & -5.922 \end{bmatrix} \hat{x} & \theta(t) = 1 \\
&= \begin{bmatrix} -3.694 & -7.680 & -6.010 \end{bmatrix} \hat{x} & \theta(t) = 2 \\
&= \begin{bmatrix} -4.965 & -3.945 & 0.4184 \end{bmatrix} \hat{x} & \theta(t) = 3
\end{aligned}
$$

where $\hat{x}$ is generated by:

$$
\begin{aligned}
\dot{\hat{x}} &= \begin{bmatrix} -25980 & -8658 & 0 \\ -58490 & -19500 & 1.003 \\ -10850 & -3619 & -5.916 \end{bmatrix} \hat{x} + \begin{bmatrix} 8660 \\ 19500 \\ 3616 \end{bmatrix} y; & \theta(t) = 1 \\
&= \begin{bmatrix} -1934 & 476.7 & 8.010 \\ 4344 & 1103 & 12.17 \\ 12760 & -3187 & 1.556 \end{bmatrix} \hat{x} + \begin{bmatrix} 483.4 \\ -1087 \\ -3190 \end{bmatrix} y; & \theta(t) = 2 \\
&= \begin{bmatrix} 11780 & 8.871 & 23580 \\ 75330 & 1.016 & 150700 \\ 32950 & 1.930 & -65910 \end{bmatrix} \hat{x} + \begin{bmatrix} -5895 \\ -37660 \\ 16480 \end{bmatrix} y; & \theta(t) = 3
\end{aligned}
$$

Note that in this example, even though the measurement is one-dimensional, there is not much difference between the achievable performance levels of the perfect and imperfect state cases, and hence the sufficiency result of Theorem 4.3 is almost necessary.

6 Conclusion

In this paper, we have studied a class of soft-constrained zero-sum differential game problems for jump linear systems with perfect as well as imperfect state measurements, in both finite and infinite horizons. Sufficient, as well as some necessary, conditions are presented to ensure the existence of a minimax strategy for P1 in all cases. Under these conditions, closed-form expressions for minimax strategies for P1 are obtained and are shown to attain the upper value 0 for the game. The results obtained have immediate applications to the H^∞-optimal control problem for jump linear systems, where an $\mathcal{L}_2$ gain type inequality is guaranteed on the closed-loop system.

Some challenging extensions of these results still remain. One of these would be to remove the existing gap between necessary and sufficient conditions presented here. Another one would be the problem where the form process $\theta(t)$ is not completely available for measurement. Yet another problem would be to study model simplification in this context, when some of the transitions are rare. All these are topics currently under study.

A Appendix

Here, we provide some details that were omitted in the proof of Corollary 3.1. Let us define $\mathrm{vec}(M)$, for any symmetric matrix M, to be a vector whose elements are the lower triangular elements of M, or in simple mathematical terms:

$$\mathrm{vec}(M) = \begin{bmatrix} m_{11} \\ m_{21} \\ m_{22} \\ m_{31} \\ \vdots \\ m_{nn} \end{bmatrix} ; \qquad M = \begin{bmatrix} m_{11} & \cdots & m_{1n} \\ \vdots & \ddots & \vdots \\ m_{n1} & \cdots & m_{nn} \end{bmatrix}.$$

Consider a jump linear system with state dynamics

$$\dot{x} = A(\theta(t))x. \tag{A.1}$$

This system is mean-square stable if the following set of coupled Lyapunov equations admits a unique positive definite solution

$$A(i)'M_i + M_i A(i) + I + \sum_{j=1}^{s} \lambda_{ij} M_j = 0; \qquad i \in \mathcal{S}. \tag{A.2}$$

The above set of coupled Lyapunov equation can equivalently be viewed as a set of algebraic equations:

$$F(A(1), \ldots, A(s), \Lambda)\vec{m} + \vec{b} = 0,$$

where $F(A(1), \ldots, A(s), \Lambda)$ denotes the Jacobian of the LHS of (A.2), and

$$
\vec{m} := \begin{bmatrix} \mathrm{vec}(M_1) \\ \vdots \\ \mathrm{vec}(M_s) \end{bmatrix}; \quad \vec{b} = \begin{bmatrix} \mathrm{vec}(I) \\ \vdots \\ \mathrm{vec}(I) \end{bmatrix}.
$$

Then, the following lemma holds:

Lemma A.1 *The jump linear system is mean-square stable if and only if the matrix $F(A(1), \ldots, A(s), \Lambda)$ is Hurwitz.*

Proof   First, assume that the matrix $F(A(1), \ldots, A(s), \Lambda)$ is Hurwitz. Consider the following set of coupled Lyapunov differential equations:

$$
\dot{M}_i^T + A(i)' M_i^T + M_i^T A(i) + I + \sum_{j=1}^{s} \lambda_{ij} M_j^T = 0; \quad M_i^T(T) = 0, \ i \in \mathcal{S} \tag{A.3}
$$

for any $T > 0$. Then, clearly the solutions $M_i^T(t)$, $i = 1, \ldots, s$, are monotonically nondecreasing in T. Let the vector $\vec{m}^T(t)$ be defined as

$$
\vec{m}^T(t) := \begin{bmatrix} \mathrm{vec}(M_1^T(t)) \\ \vdots \\ \mathrm{vec}(M_s^T(t)) \end{bmatrix}.
$$

It satisfies the following differential equation:

$$
\dot{\vec{m}}^T + F(A(1), \ldots, A(s), \Lambda)\vec{m}^T + \vec{b} = 0; \qquad \vec{m}^T(T) = 0.
$$

By the stability of the matrix $F(A(1), \ldots, A(s), \Lambda)$, we have

$$
\lim_{T \to \infty} \vec{m}^T(0) = -F(A(1), \ldots, A(s), \Lambda)^{-1} \vec{b}.
$$

Then, clearly the matrices M_i, $i = 1, \ldots, s$, such that $\vec{m} = -F(A(1), \ldots, A(s), \Lambda)^{-1} \vec{b}$ provide the unique solutions to (A.2) and are nonnegative definite. This implies that the jump linear system (A.1) is mean-square stable.

On the other hand, if the jump linear system (A.1) is mean-square stable, then the set of coupled Lyapunov equations (A.2) admits unique positive definite solutions M_i, $i = 1, \ldots, s$. In turn, the set of coupled Lyapunov differential equations (A.3) must admit solutions $M_i^T(t)$, $i = 1, \ldots, s$, for each $T > 0$, which further satisfy the limits:

$$
\lim_{T \to \infty} M_i^T(0) = M_i.
$$

The above limit must also hold for any solution to the set of differential equations (A.3) with a terminal condition that is positive definite and upper bounded by M_i. Now introduce the matrices $\tilde{M}_i^T$, $i = 1, \ldots, s$, which satisfy the set of differential equations (A.3) but with a different terminal condition:

$$\tilde{M}_i^T(T) = (1/2)M_i; \qquad i \in \mathcal{S}.$$

Suppose that the matrix $F(A(1), \ldots, A(s), \Lambda)$ is not Hurwitz. Let $\vec{p}_0$ be a nonzero vector such that the following differential equation

$$\dot{\vec{p}}^T + F(A(1), \ldots, A(s), \Lambda)\vec{p}^T = 0; \quad \vec{p}^T(T) = \vec{p}_0$$

admits a solution $\vec{p}^T(t)$ which does not converge to 0 as $T \to \infty$. Let $\vec{p}_0$ and $\vec{p}^T$ correspond (through the vec operation) to symmetric matrices P_{i0}, $i \in \mathcal{S}$, and P_i^T, $i \in \mathcal{S}$, respectively. Clearly, the matrices P_i^T, $i \in \mathcal{S}$, satisfy the following set of coupled Lyapunov differential equations:

$$\dot{P}_i^T + A(i)'P_i^T + P_i^T A(i) + \sum_{j=1}^{s} \lambda_{ij} P_j^T = 0; \qquad P_i^T(T) = P_{i0}, \quad i \in \mathcal{S}.$$

For any scalar α, the matrices $\alpha P_i^T + \tilde{M}_i^T$, $i \in \mathcal{S}$, satisfy the set of coupled Lyapunov differential equations (A.3) with the terminal condition:

$$\alpha P_i^T(T) + \tilde{M}_i^T(T) = \alpha P_{i0} + (1/2)M_i, \qquad i = 1, \ldots, s.$$

For sufficiently small α, we have $0 \leq \alpha P_{i0} + (1/2)M_i \leq M_i$ for each $i \in \mathcal{S}$. This then implies that $\alpha P_i^T(t) + \tilde{M}_i^T(t)$ converges to M_i as $T \to \infty$, and hence, $P_i^T(t) \to 0$ for each $i \in \mathcal{S}$. This contradicts the choice of the vector $\vec{p}_0$. Hence, the matrix $F(A(1), \ldots, A(s), \Lambda)$ is Hurwitz.

This completes the proof of the lemma. $\qquad\square$

Given the above result, we introduce the following notation:

Definition A.1 The matrix $F(A(1), \ldots, A(s), \Lambda)$ is the *stability matrix* for the jump linear system

$$\dot{x} = A(\theta(t))x$$

where $\theta(t)$ is a continuous time finite state Markov chain process with generator Λ and state space $\{1, \ldots, s\}$. $\qquad\diamond$

Now, given the mean-square stability of the jump linear system

$$\dot{x} = A_{F\gamma_1}(\theta(t))x,$$

where

$$A_{F\gamma_1}(i) := A(i) - (B(i)R^{-1}(i)B'(i) - \frac{1}{\gamma_1^2}D(i)D'(i))Z_{\gamma_1}(i),$$

its stability matrix $F(A_{F\gamma_1}(1), \ldots, A_{F\gamma_1}(s), \Lambda)$ is Hurwitz. A straightforward evaluation of the Jacobian of the RHS of (27) with respect to $(\text{vec}(\bar{Z}_1)', \ldots, \text{vec}(\bar{Z}_s)')$ at the point $(\rho, \bar{Z}_1, \ldots, \bar{Z}_s) = (0, Z_{\gamma_1}(1), \ldots, Z_{\gamma_1}(s))$ gives exactly the matrix $F(A_{F\gamma_1}(1), \ldots, A_{F\gamma_1}(s), \Lambda)$. Thus the Implicit Function Theorem can be applied to establish the existence of a $\rho_0 > 0$, such that the set of coupled GARE's (27) admits a minimal positive definite solution.

REFERENCES

[1] N. N. Krassovskii and E. A. Lidskii, "Analytical design of controllers in systems with random attributes I, II, III," *Automation and Remote Control*, vol. 22, 1961, 1021–1025, 1141–1146, 1289–1294.

[2] J. J. Florentin, "Optimal control of continuous-time Markov, stochastic systems," *J. Electron. Control*, vol. 10, 1961.

[3] H. J. Kushner, "On the stochastic maximum principle: Fixed time of control," *J. Math. Anal. Appl.*, vol. 11, 1965, 78–92.

[4] H. J. Kushner, *Stochastic stability and control*. Academic Press, New York, 1967.

[5] F. Kozin, "A survey of stability of stochastic systems," *Automatica*, vol. 5, 1969, 95–112.

[6] R. Rishel, "Dynamic programming and minimum principles for systems with jump Markov disturbances," *SIAM Journal on Control and Optimization*, vol. 13, 338–371, February 1975.

[7] M. Mariton, "Almost sure and moment stability of jump linear systems," *Systems & Control Letters*, vol. 11,1988, 393–397.

[8] M. Mariton, *Jump Linear Systems in Automatic Control*. Marcel Dekker, New York, 1990.

[9] Y. Ji and H. J. Chizeck, "Controllability, stabilizability, and continuous-time Markovian jump linear quadratic control," *IEEE Transactions on Automatic Control*, vol. AC-35, July 1990, 777–788.

[10] X. Feng, K. A. Loparo, Y. Ji, and H. J. Chizeck, "Stochastic stability properties of jump linear systems," *IEEE Transactions on Automatic Control*, vol. AC-37, January 1992, 38–53.

[11] M. Mariton, "Control of non-linear systems with Markovian parameter," *IEEE Transactions on Automatic Control*, vol. AC-36, 1991, 233–238.

[12] W. M. Wonham, "Random differential equations in control theory," in *Probabilistic Methods in Applied Mathematics* (A. T. Bharucha-Reid, ed.), Academic Press, New York, 1969, 131–212.

[13] M. Mariton and P. Bertrand, "Output feedback for a class of linear systems with stochastic jump parameters," *IEEE Transactions on Automatic Control*, vol. AC-30, no. 9, 1985, 898–900.

[14] D. D. Sworder, "Feedback control of a class of linear systems with jump parameters," *IEEE Transactions on Automatic Control*, vol. AC-14, 1969, 9–14.

[15] B. Griffiths and K. A. Loparo, "Optimal control of jump linear Gaussian systems," *Int. J. Contr.*, vol. 43, no. 4, 1983, 792–819.

[16] P. E. Caines and H. F. Chen, "Optimal adaptive LQG control for systems with finite state process parameters," *IEEE Transactions on Automatic Control*, vol. AC-30, 1985, 185–189.

[17] K. A. Loparo, Z. Roth, and S. J. Eckert, "Non-linear filtering for systems with random structure," *IEEE Transactions on Automatic Control*, vol. AC-31, 1986, 1064–1068.

[18] R. J. Elliott and D. D. Sworder, "Control of a hybrid conditionally linear Gaussian processes," *Journal of Optimization Theory and Applications*, vol. 74, 1992, 75–85.

[19] J. Ezzine and A. H. Haddad, "Error bounds in the averaging of hybrid systems," *IEEE Transactions on Automatic Control*, vol. AC-34, 1990, 1188–1192.

[20] C. E. de Souza and M. Fragoso, "H^∞ control of linear systems with Markovian jumping parameters," *Control Theory and Technology*, vol. 9, no. 2, 1993, 457–466.

[21] T. Başar and A. Haurie, *Feedback equilibria in differential games with structural and modal uncertainties*, vol. 1 of *Advances in Large Scale Systems*, (J. B. Cruz, Jr., ed.), JAI Press Inc., Connecticut, May 1984, 163–201. .

[22] T. Başar, "Minimax control of switching systems under sampling," in *Proceedings of the 33rd IEEE Conference on Decision and Control*, Orlando, 1994, 716–721.

[23] D. Carlson, A. Haurie, and A. Leizarowitz, "Overtaking equilibria for switching regulator and tracking games," in *Advances in Dynamic Games and Applications* (T. Başar and A. Haurie, eds.), Birkhäuser, Boston, 1994, 247–268.

[24] T. Başar and P. Bernhard, *H^∞-Optimal Control and Related Minimax Design Problems: A Dynamic Game Approach.* Birkhäuser, Boston, 1991.

[25] M. Davis, *Markov Models and Optimization.* Monographs on Statistics and Applied Probability, Chapman and Hall, London, October 1992.

[26] S. Boyd, L. E. Ghaoui, E. Feron, and V. Balakrishnan, *Linear Matrix Inequalities in System and Control Theory.* Philadelphia, PA: SIAM, 1994.

The Big Match on the Integers

Bert Fristedt, Stephan Lapic and William D. Sudderth*

School of Mathematics, University of Minnesota
Minneapolis, MN 55455 USA
Department of Mathematics, Case Western Reserve University
Cleveland, OH 44106 USA
School of Statistics, University of Minnesota
Minneapolis, MN 55455 USA

Abstract

A two-person, zero-sum game analogous to the big match is introduced. The game does not fall within any of the classes of games known, by general theorems, to have values. We prove directly that our game has a value and find good strategies for the players.

AMS 1991 subject classification 90D15, 60G40.

Keywords: stochastic games, the big match, martingales.

1 Introduction

The big match of Blackwell and Ferguson (1968) is one of the best known and most intriguing stochastic games. Because the big match is an average reward stochastic game with finite state and action spaces, it now follows from a theorem of Mertens and Neyman (1981) that the game has a value which can be calculated from that of the associated discounted games. These facts were far from clear in 1968 when the big match appeared. So Blackwell and Ferguson found the value by a clever direct argument, in the process proving its existence.

Consider now another stochastic game which is a variation on the big match and seems to lie just beyond the reach of theorems currently available. The state space S consists of the set of integers Z together with two terminal states λ and ρ. Every day each player chooses a 0 or a 1. If the current state is an integer x and player I chooses a 0, then the next state is $x - 1$ if player II chooses 0 and is $x + 1$ if player II chooses 1. If the current state is an integer x and player I chooses 1, then the next state is λ if player II chooses 1

*Research supported by National Science Foundation Grants DMS-9123358 and DMS-9423009.

and is ρ if player II chooses 0. The states λ and ρ are absorbing. The payoff from player II to player I is 1 if the process of states either converges to $-\infty$ or ever reaches λ, and the payoff is 0 otherwise. We regard a choice of action 1 by player I at any integer x as resulting in a "big match" because player I wins the game immediately if player II also chooses 1 and loses immediately otherwise.

Call this game the *big match on the integers*. It is the stochastic game with state space $S = Z \cup \{\lambda, \rho\}$, action sets $A = B = \{0, 1\}$ for the two players and law of motion q given by

$$q(x + 1|x, 0, 1) = q(x - 1|x, 0, 0) = q(\lambda|x, 1, 1) = q(\rho|x, 1, 0) = 1$$

for each $x \in Z$ and

$$q(\lambda|\lambda, a, b) = q(\rho|\rho, a, b) = 1$$

for all $a \in A, b \in B$. The payoff function from II to I is the indicator function of the following set of possible histories for the process $X_0, X_1, \ldots$ of states

$$G = \{(x_0, x_1, \ldots) : x_n \to -\infty \text{ or, for some } n, x_n = \lambda\}. \tag{1}$$

At each stage of the game, a player chooses an action, possibly at random, and may do so based on the knowledge of the current state and all previous states. A *strategy* for either player specifies the entire sequence of that player's choices of actions. (See Raghavan (1991) or Maitra and Sudderth (1992) for an introduction to stochastic games, precise definitions, and further references.)

Here is our main theorem which mirrors Theorem 1 of Blackwell and Ferguson (1968).

Theorem 1.1 *If the initial state is an integer x, then the big match on the integers has value $1/2$. An optimal strategy for player II is to toss a fair coin every day. Player I has no optimal strategy but for any positive integer N can get an expected payoff of at least $N/(2N + 2)$ by choosing action 1 with probability $1/(n + 1)^2$ whenever the current state is $x + N - n$ for some $n \geq 0$.*

Our proof involves some auxiliary big match games with essentially the same dynamics but with Z replaced by Z^+, Z^-, or a finite interval of integers. Endpoints are absorbing with a payoff of 1 from II to I when a left hand endpoint is reached. The game on Z^+ was studied by Nowak and Raghavan (1991) whose paper along with "The Big Match" led to this one.

The next section treats a difference equation which arises in all the auxiliary games as well as the basic game. Section 3 has some general lemmas on optimal strategies. The solutions of the auxiliary games are in Sections 4, 5, and 6. The proof of Theorem 1.1 is in Section 7. Some remarks on more general games are in the final section.

2   A difference equation

Consider the big match on Z or any one of the three auxiliary games: *the finite game* with state space $\{0, 1, \ldots, N\} \cup \{\lambda, \rho\}$, *the game on Z^+* with state space $\{0, 1, 2, \ldots\} \cup \{\lambda, \rho\}$, or *the game on Z^-* with state space $\{0, -1, -2, \ldots\} \cup \{\lambda, \rho\}$. In this section, we *assume* that the value $v(x)$ is well-defined for every initial state x. If the state x is an integer and is not an endpoint (i.e. $x \neq 0$ and in the finite game $x \neq N$), then clearly

$$v(x) = \text{ value } \begin{pmatrix} v(x-1) & v(x+1) \\ 0 & 1 \end{pmatrix}. \tag{2}$$

In each of the games player I is no worse off at $x - 1$ than at x. So v is a (weakly) decreasing function.

Assume further that $0 < v(x) < 1$ at integers x which are not endpoints. Let $p(x)$ denote the probability that I plays 0 when using some optimal strategy in the *one-day matrix game* of (2). (Actually, this optimal strategy is unique.) Then

$$v(x) = p(x)v(x-1) = p(x)v(x+1) + (1 - p(x)). \tag{3}$$

Eliminating $p(x)$ we get

$$v(x)v(x-1) - v(x-1) = v(x+1)v(x) - v(x).$$

Hence there is a constant k such that

$$v(x-1)(1 - v(x)) = k$$

for all integers x and $x-1$ in the state space. Since $v(x) < 1$ and $v(x-1) > 0$ by assumption, the constant k must be strictly positive. Rewrite the previous equation as

$$v(x) = \frac{v(x-1) - k}{v(x-1)}$$

and set $v(x) = a(x)/b(x)$ to get a system of linear homogeneous difference equations

$$\begin{aligned} a(x) &= a(x-1) - k\,b(x-1) \\ b(x) &= a(x-1) \end{aligned}$$

equivalent to our equation for v. The difference equations can be solved by standard methods yielding the following result for v:

$$v(x) =$$

$$\begin{cases} \dfrac{2(1+x)v(0) - x}{4xv(0) - 2(x-1)} & \text{if } k = \tfrac{1}{4} \\[2ex] \dfrac{(2v(0)-1+\sqrt{1-4k})(1+\sqrt{1-4k})^{x+1} - (2v(0)-1-\sqrt{1-4k})(1-\sqrt{1-4k})^{x+1}}{2[(2v(0)-1+\sqrt{1-4k})(1+\sqrt{1-4k})^{x} - (2v(0)-1-\sqrt{1-4k})(1-\sqrt{1-4k})^{x}]} & \text{if } 0 < k \neq \tfrac{1}{4}. \end{cases} \tag{4}$$

Before using this formula for v, we must, of course, check our assumptions that v exists and $0 < v(x) < 1$ at integers x which are not endpoints.

If v is a strictly decreasing function so that $v(x-1) > v(x+1)$ in (2), then both players have unique optimal strategies in the one-day matrix game at x. Let $p(x)$ and $q(x)$ denote the probabilities that I and II, respectively, play 0 when using these optimal strategies. Since $p(x)$ satisfies (3), it is given by

$$p(x) = \frac{v(x)}{v(x-1)}. \tag{5}$$

It is equally easy to calculate

$$q(x) = 1 - v(x). \tag{6}$$

3   Lemmas on optimal strategies

There is a nice interplay between optimal strategies in a stochastic game and optimal actions in the corresponding one-day matrix games. A complete treatment would be out of place here, but we will present three lemmas and a sketch of their proofs. (The impatient reader can skip this section and refer back to it if necessary.)

Although the lemmas hold quite generally, we continue to assume that the stochastic game under consideration is the big match on the integers or one of the three auxiliary big match games. As in the previous section we also assume the stochastic game has a value $v(x)$ for each initial state x.

Lemma 3.1 *An optimal strategy for either player in the stochastic game with initial state x must begin with an optimal action for that player in the matrix game at x.*

Sketch of proof: If player I (say) begins in the stochastic game with an action which is less than optimal in the matrix game, then player II can begin with an action which makes the expected value of $v(X_1)$ strictly less than $v(x)$. By continuing with a nearly optimal strategy in the stochastic game starting from X_1, player II can hold player I to an expected payoff less than $v(x)$ in the original stochastic game. $\square$

Let α be a strategy for one of the players in the stochastic game starting from state x. Suppose that on the first play, player I chooses action a_1, player II chooses action b_1 and the next state is x_1. Write $\alpha[a_1, b_1, x_1]$ for the *conditional strategy* corresponding to the continuation of play governed by α.

Lemma 3.2 *Suppose α is an optimal strategy for player I in the stochastic game with initial state x and assume that player II has an optimal randomized action in the matrix game at x which assigns positive probability to both members of II's action set B. Then, if I plays α and II plays any strategy, the conditional strategy $\alpha[a_1, b_1, x_1]$ is almost surely optimal for I in the stochastic game starting from x_1. (The lemma remains true if I and II are interchanged.)*

Sketch of proof: It follows from an elementary fact about matrix games (Lemma 2.1.2 in Karlin (1959)) that when I plays α and II any strategy, the expectation of $v(X_1)$ must equal $v(x)$. So to obtain a payoff of $v(x)$, I must continue to play optimally starting from X_1. $\square$

We consider two strategies for a player to be the same if they choose the same actions almost surely regardless of how the opponent plays. Also we regard all actions to be the same at the absorbing states λ, ρ and any endpoints. With these conventions our final lemma gives sufficient conditions for an optimal strategy to be unique.

Lemma 3.3 *Assume the following:*

(i) Player I has an optimal strategy in the stochastic game with initial state x.

(ii) For any state y, player I has a unique optimal action $\mu(y)$ in the matrix game at y.

(iii) For every state y, player II has an optimal action in the matrix game at y which gives positive probability to every element of B.

Then player I has a unique optimal strategy in the stochastic game starting from x; namely, play action $\mu(y)$ whenever the current state is y. (The lemma remains true if I and II are interchanged.)

Sketch of proof: Let α be an optimal strategy for I in the stochastic game starting from x. By (i), (ii), and Lemma 3.1, α must begin with action $\mu(x)$. By (iii) and Lemma 3.2, the conditional strategy $\alpha[a_1, b_1, x_1]$ is almost surely optimal. Apply Lemma 3.1 again to see that $\mu(x_1)$ is almost surely the next action dictated by α. And so forth. $\square$

4   The finite game

For the finite game with state space $\{0, 1, \ldots, N\} \cup \{\lambda, \rho\}$, it follows from any one of several well-known results that the value exists. (See, for example, Orkin (1972).) Let $v_N(x)$ be the value when the initial state is

x. Fair coin-tossing strategies for each player show that $0 < v_N(x) < 1$ for $x = 1, 2, \ldots, N-1$. Thus the assumptions of Section 2 hold and putting $v_N(0) = 1$ in (4) gives

$$v_N(x) = \begin{cases} \dfrac{2+x}{2+2x} & \text{if } k = \frac{1}{4} \\[2ex] \dfrac{(1+\sqrt{1-4k})^{x+2} - (1-\sqrt{1-4k})^{x+2}}{2[(1+\sqrt{1-4k})^{x+1} - (1-\sqrt{1-4k})^{x+1}]} & \text{if } 0 < k \neq \frac{1}{4}. \end{cases}$$

The case $k = 1/4$ does not apply because it gives $v_N(N) > 0$ when in fact $v_N(N) = 0$. (Recall that state N is absorbing and there is no payoff there.) Likewise $k < 1/4$ does not apply for the same reason.

It remains to consider $k > 1/4$, in which case

$$1 \pm \sqrt{1-4k} = se^{\pm i\theta}$$

for some $s > 0$ and $\theta \in (0, \pi/2)$ for which $s = \sec\theta$. Then

$$\begin{aligned} v_N(x) &= \frac{\sec\theta \sin((x+2)\theta)}{2\sin((x+1)\theta)} \\[1ex] &= \tfrac{1}{2}(1 + \tan\theta \cot((x+1)\theta)) \end{aligned} \tag{7}$$

for $x = 0, 1, \ldots, N$. The condition $v_N(N) = 0$ implies $\theta = k\pi/(N+2)$ for some integer $k \in [1, (N+2)/(2\pi))$. However, if such a k is larger than 1, formula (7) will give negative values for some x. So it must be the case that

$$\theta = \frac{\pi}{N+2}. \tag{8}$$

Proposition 4.1 *The finite game on* $\{0, 1, \ldots, N\}$ *has the value function*

$$\begin{aligned} v_N(x) &= \frac{\sec\frac{\pi}{N+2} \sin\frac{(x+2)\pi}{N+2}}{2\sin\frac{(x+1)\pi}{N+2}} \\[1ex] &= \frac{1}{2}\left(1 + \tan\frac{\pi}{N+2} \cot\frac{(x+1)\pi}{N+2}\right) \end{aligned}$$

for $x = 0, 1, 2, \ldots, N$. *Furthermore, the players have unique optimal strategies which are stationary and in which player I plays 0 with probability*

$$p_N(x) = \frac{\sin\frac{(x+2)\pi}{N+2} \sin\frac{x\pi}{N+2}}{\sin^2\left(\frac{(x+1)\pi}{N+2}\right)}$$

and player II plays 0 with probability

$$q_N(x) = \tfrac{1}{2}\left(1 - \tan\frac{\pi}{N+2} \cot\frac{(x+1)\pi}{N+2}\right)$$

for $x = 1, 2, \ldots, N-1$.

Proof: Substitute the θ of (8) into (7) to get the formula for v_N. The formula for p_N comes from (5) which gives the unique optimal strategy for I in the matrix game

$$\begin{pmatrix} v_N(x-1) & v_N(x+1) \\ 0 & 1 \end{pmatrix}$$

which has value $v_N(x)$ for $x = 1, \ldots, N-1$. So if I plays p_N, then, against any strategy for II, the stochastic process $v_N(x), v_N(X_1), v_N(X_2), \ldots$ corresponding to the value function evaluated at the successive states $x, X_1, X_2, \ldots$ is a bounded submartingale. Thus $v_N(X_n)$ converges almost surely to a limit variable L as $n \to \infty$. Furthermore the process of states $\{X_n\}$ must reach one of the absorbing states $0, N, \lambda$, or ρ with probability one. Consequently L is a $\{0, 1\}$ - valued variable and the payoff from II to I is

$$\begin{aligned} P[X_n \text{ reaches } 0 \text{ or } \lambda] &= EL \\ &= E \lim v_N(X_n) \\ &\geq v_N(x). \end{aligned}$$

This shows p_N is optimal. That it is uniquely so follows from Lemma 3.3. The proof that q_N, given by (6), is uniquely optimal for player II is similar. $\square$

It is amusing to observe that the optimal strategies for players I and II remain the same if player I's goal is to reach N or ρ rather than 0 or λ. This is related to the fact that if player I (say) plays her optimal strategy, then the process $v_N(X_n)$ is a martingale regardless of player II's strategy. This surprising symmetry is reflected by the formula

$$v_N(x) = 1 - v_N(N-x), \quad x = 0, 1, \ldots, N,$$

which also holds for p_N and q_N.

5　The game on Z^+

The game with state space $Z^+ \cup \{\lambda, \rho\} = \{0, 1, 2, \ldots\} \cup \{\lambda, \rho\}$ can be viewed as a stochastic game with a positive daily reward function. Namely, II pays I one dollar whenever there is a transition to either 0 or λ and pays nothing otherwise. Indeed this is essentially the game studied by Nowak and Raghavan (1991) who showed that player I has no uniformly ϵ-optimal stationary plan for $0 < \epsilon < 1$. It has long been known that positive games have a value (Frid (1973), Kamerud (1975)). A recent reference is Nowak (1985) which gives a very general form of the result. However, we will give a simple direct proof for the big match on Z^+.

Proposition 5.1 *The game on Z^+ has the value function v^+ where $v^+(\lambda) = 1, v^+(\rho) = 0$, and*

$$v^+(x) = \frac{x+2}{2x+2}, \quad x \in Z^+.$$

Player I has no optimal strategy while player II's unique optimal strategy is to play 0 with probability

$$q^+(x) = \frac{x}{2x+2}$$

for $x = 1, 2, \ldots$.

Proof: We need to show that $\underline{v}(x) \geq v^+(x) \geq \overline{v}(x)$ for each x, where $\underline{v}$ and $\overline{v}$ are the lower and upper value functions for the game on Z^+.

To get the first inequality, observe that for $N > x$, the finite game on $\{0, 1, \ldots, N\}$ is clearly no better for I than the game on Z^+. Hence,

$$\underline{v}(x) \geq \lim_{N \to \infty} v_N(x) = \frac{x+2}{2x+2} = v^+(x)$$

for each $x \in Z^+$.

To prove the other inequality, suppose that II plays according to q^+ and I plays any strategy. Now $q^+(x)$ is the unique optimal strategy for II in the matrix game

$$\begin{pmatrix} v^+(x-1) & v^+(x+1) \\ 0 & 1 \end{pmatrix}$$

which has value $v^+(x)$. Thus the process $v^+(x), v^+(X_1), v^+(X_2), \ldots$ is a bounded supermartingale when $x, X_1, X_2, \ldots$ is the process of successive states. Since $v^+(0) = v^+(\lambda) = 1$, the payoff from II to I is no larger than the limit of $v^+(X_n)$. Thus the expected payoff is no larger than

$$E \lim_n v^+(X_n) \leq v^+(x).$$

This proves that $\overline{v} \leq v^+$ and also shows q^+ to be optimal for II. Uniqueness follows from Lemma 3.3.

The only candidate for an optimal strategy for I is to play 0 at each $x \in Z^+$ with probability

$$p^+(x) = \frac{x(x+2)}{(x+1)^2}$$

and to play 1 with probability

$$1 - p^+(x) = \frac{1}{(x+1)^2},$$

since these are uniquely optimal for the matrix game. Now if I plays according to p^+ and II always chooses action 1, then player I wins if and only if she ends the game by playing 1 at some stage. But the probability that I ever plays 1 starting from x is no larger than

$$\sum_{n=x}^{\infty} \frac{1}{(n+1)^2} \to 0 \text{ as } x \to \infty. \quad \square$$

Notice that the value function v^+ is just the solution of the difference equation of section 2 when $v^+(0) = 1$ and $k = 1/4$. So we could have derived the formula for v^+ from (4) if we had had an argument to show directly that $k = 1/4$. Such an argument is given by Nowak and Raghavan (1991).

6   The game on Z^-

Consider next the game with state space $Z^- \cup \{\lambda, \rho\} = \{\ldots, -2, -1, 0\} \cup \{\lambda, \rho\}$. Since there is no left-hand endpoint, the payoff from II to I is the indicator function of the set G defined by (1). This game does not fall into any of the general categories known to have a value. Nevertheless, it is natural to conjecture a value by first translating the formula for the finite game to the interval $\{-N, -N+1, \ldots, 0\}$, setting

$$w_N(x) = v_N(x + N) = \frac{\sec \frac{\pi}{N+2} \sin \frac{x\pi}{N+2}}{2 \sin \frac{(x-1)\pi}{N+2}},$$

and taking the limit to get

$$\lim_{N \to \infty} w_N(x) = \frac{x}{2x - 2}$$

for $x \in Z^-$.

Proposition 6.1 *The game on Z^- has the value function v^- where $v^-(\lambda) = 1, v^-(\rho) = 0$, and*

$$v^-(x) = \frac{x}{2x - 2}, \quad x \in Z^-.$$

Player II has no optimal strategy while player I's unique optimal strategy is to play 0 with probability

$$p^-(x) = 1 - (x - 1)^{-2}$$

for $x = -1, -2, \ldots$.

Proof: For $-N < x \le 0$, the finite game on $\{-N, -N+1, \ldots, 0\}$ is at least as good for I as is the game on Z^-. So, if $\overline{v}(x)$ is the upper value of the game on Z^-, then

$$\overline{v}(x) \le \lim_{N \to \infty} w_N(x) = v^-(x), \quad x \in Z^-.$$

To get the opposite inequality for the lower value $\underline{v}(x)$, fix $x \leq -1$, suppose I plays the strategy corresponding to p^- and II plays any strategy. Now $p^-(x)$ is the unique optimal strategy for I in the matrix game

$$\begin{pmatrix} v^-(x-1) & v^-(x+1) \\ 0 & 1 \end{pmatrix}$$

which has the value $v^-(x)$. Hence, the process $v^-(x), v^-(X_1), v^-(X_2), \ldots$ is a bounded submartingale and must converge almost surely. The function v^- is strictly decreasing on Z^-. So it follows that with probability one either $X_n \to -\infty$ or X_n reaches one of the absorbing states λ, ρ, or 0. Since $v^-(X_n) \to 1/2$ when $X_n \to -\infty$ in which case the payoff is 1, we see that the expected payoff is at least

$$E \lim_n v^-(X_n) \geq v^-(x).$$

Hence, $\underline{v}(x) \geq v^-(x)$ and $v^-(x)$ is the value of the game. This argument also shows the optimality of p^-. Its uniqueness follows from Lemma 3.3.

The only candidate for an optimal strategy for II is to play 0 with probability

$$q^-(x) = \frac{x-2}{2x-2} \quad \text{at } x = -1, -2, \ldots,$$

since these are the uniquely optimal plays in the matrix games. Suppose II plays according to q^- and I always plays 0 starting from an initial state $x \in \{-1, -2, \ldots\}$. The process $v^-(x), v^-(X_1), v^-(X_2), \ldots$ is then a bounded martingale and must converge almost surely. Since v^- is strictly increasing on Z^- and I never terminates the game by playing 1, the process X_n either converges to $-\infty$ or reaches 0 almost surely. Hence,

$$v^-(x) = E \lim_n v^-(X_n) = \tfrac{1}{2} P[X_n \to -\infty]$$

while the expected payoff to I is $P[X_n \to -\infty]$ and is strictly greater than $v^-(x)$. $\qquad\square$

The proof of Proposition 6.1 shows slightly more than is stated. Suppose we choose the payoff to be $1/2$ if $X_n \to -\infty$ and keep it the same otherwise. Obviously player I is no better off in this *modified game*. Nevertheless, the proof shows that, by playing p^-, player I still obtains an expected payoff of at least

$$E \lim_n v^-(X_n) \geq v^-(x)$$

starting from an $x \in \{-1, -2, \ldots\}$. Hence, the modified game also has value function v^-. We will use this fact to connect the game on Z with the original big match in the next section. (By the way, in the modified game it is optimal for II to play 0 with probability $q^-(x)$ at each negative x. Player I's optimal strategy is unchanged.)

7 The proof of Theorem 1.1

We are finally ready for the big match on the integers. If player II uses a fair coin tossing strategy, then, regardless of the strategy used by player I, the process $x, X_1, X_2, \ldots$ will not converge to $-\infty$ and the state λ will be reached with probability at most $1/2$. Hence, the upper value $\bar{v}(x)$ is at most $1/2$ for $x \in Z$. To see that the lower value $\underline{v}(x)$ is at least $1/2$, consider the game on Z^- translated to $\{\ldots, x + N - 2, x + N - 1, x + N\}$ for a positive integer N. That is, suppose $x + N$ is absorbing, yields payoff 0, and the game is otherwise the same. This new game is clearly no better for player I than the original game and it follows from Proposition 6.1 that I can obtain a payoff of at least $v^-(-N)$ by playing according to p^- translated by N. Hence

$$\underline{v}(x) \geq \lim_{N \to \infty} v^-(-N) = 1/2,$$

and, therefore, the game has value $v(x) = 1/2$.

Player I has no optimal strategy since, at any $x \in Z$, such a strategy would necessarily make an optimal play in the matrix game

$$\begin{pmatrix} v(x-1) & v(x+1) \\ 0 & 1 \end{pmatrix} = \begin{pmatrix} \frac{1}{2} & \frac{1}{2} \\ 0 & 1 \end{pmatrix}$$

at each stage. That is, I must play 0 with probability one at every x. But II can then play 1 at every x and the payoff to I is zero.

This completes the proof of Theorem 1.1.

Notice that the coin toss strategy is not the only optimal strategy for II because II has many optimal strategies in the matrix game

$$\begin{pmatrix} \frac{1}{2} & \frac{1}{2} \\ 0 & 1 \end{pmatrix}.$$

For example, player II can choose action 0 at states $x \geq 0$ and when $x < 0$ make the straightforward choice of tossing a fair coin or can even use coins biased in favor of 0 provided that the bias tends to zero sufficiently fast as $x \to -\infty$.

In a certain sense, Theorem 1.1 includes the corresponding result for the classical big match. To see this, consider a *modified game* in which the payoff is changed to be $1/2$ if $X_n \to -\infty$. The remark of the previous section on a modified game on Z^- together with the proof just completed shows that this modified game also has value $1/2$ at each $x \in Z$. However, the payoff to I in the modified game is no greater than in the classical big match which gives an alternative proof that the lower value for the big match is at least $1/2$. Coin tossing again shows the upper value to be no greater than $1/2$.

8 Stochastic games with Borel payoff functions

The big match on the integers like the classical big match is a two-person, zero-sum stochastic game with a countable state space, finite action sets, and a Borel measurable payoff. A challenging open question is whether every such game has a value. The question is open also in the special case when the payoff function is the indicator of some Borel set G. Blackwell (1969, 1989) proved the existence of a value for an interesting class of stochastic games in which the set G is a G_δ or, equivalently an F_σ. His result was extended by Maitra, Purves and Sudderth (1992) to stochastic games with a payoff function g such that $[g \geq c]$ is a G_δ for every c.

It is natural to consider next indicators of sets G which are at the next level of the Baire hierarchy, namely $G_{\delta\sigma}$'s or $F_{\sigma\delta}$'s. For example, the payoff set defined in (1) for the big match on the integers is an $F_{\sigma\delta}$.

REFERENCES

[1] Blackwell, D. (1969). Infinite G_δ games with imperfect information. *Zastosowania Matematyki Applicationes Matematicae* **10**, 99–101.

[2] Blackwell, D. (1989). Operator solution of infinite G_δ games of imperfect information. *Probability, Statistics, and Mathematics: Papers in Honor of Samuel Karlin.* ed. T.W. Anderson, K. Athreya, and D.L. Iglehart. Academic Press, New York.

[3] Blackwell, D. and Ferguson, T.S. (1968). The big match. *Annals of Mathematical Statistics* **39**, 159–163.

[4] Frid, E.B. (1973). On stochastic games. *Theory of Probability and its Applications* **18**, 389–393.

[5] Kamerud, Dana B. (1975). *Repeated Games and Positive Stochastic Games.* Ph.D. dissertation. University of Minnesota.

[6] Karlin, Samuel (1959). *Mathematical Methods and Theory in Games, Programming, and Economics,* volume I. Addison-Wesley, Reading.

[7] Maitra, A. and Sudderth, W. (1992). An operator solution of stochastic games. *Israel Journal of Mathematics* **78**, 33–49.

[8] Maitra, A., Purves, R. and Sudderth, W. (1992). Approximation theorems for gambling problems and stochastic games. *Game Theory and Economic Applications: Lecture Notes in Economics and Mathematical Systems* **389** Springer-Verlag, Berlin, 114–131.

[9] Mertens, J.-F. and Neyman, A. (1981). Stochastic games. *International Journal of Game Theory* **10**, 53–66.

[10] Nowak, Andrzej (1985). Universally measurable strategies in zero-sum stochastic games. *Annals of Probability* **13**, 269–287.

[11] Nowak, Andrzej and Raghavan, T.E.S. (1991). Positive stochastic games and a theorem of Ornstein. *Stochastic Games and Related Topics*, ed. T.E.S. Raghavan et al., Kluwer Academic Press, Dordrecht, 127–134.

[12] Orkin, Michael (1972). Recursive matrix games. *Journal of Applied Probability* **9**, 813–820.

[13] Raghavan, T.E.S. (1991). Stochastic games—an overview. *Stochastic Games and Related Topics*, ed. T.E.S. Raghavan et al., Kluwer Academic Press, Dordrecht, 1–9.

PART II
Pursuit evasion

Synthesis of Optimal Strategies for Differential Games by Neural Networks

H.J. Pesch[*]
Department of Mathematics, Munich University of Technology,
D–80290 Munich, Germany.

I. Gabler [†]
Bayerische Hypotheken- und Wechsel-Bank AG,
D–80278 Munich, Germany.

S. Miesbach [‡]
Corporate Research and Development,
ZFE ST SN 41, Siemens AG, Munich, D–81730 Munich, Germany.

M. H. Breitner [§]
Department of Mathematics, Munich University of Technology,
D–80290 Munich, Germany.

Abstract

The paper deals with the numerical approximation of optimal strategies for two-person zero-sum differential games of pursuit-evasion type by neural networks. Thereby, the feedback strategies can be computed in real-time after the training of appropriate neural networks. For this purpose, sufficiently many optimal trajectories and their associated open-loop representations of the optimal feedback strategies must be computed, to provide data for training and cross-validation of the neural networks. All the precomputations can be carried through in a highly parallel way. This approach turns out to be applicable for differential games of more general type.

The method is demonstrated for a modified cornered rat game where a pursuing cat and an evading rat, both moving in simple motion, are constrained to a rectangular arena. Two holes in the walls

[*]Professor of Mathematics, Presently, Institute of Mathematics, Clausthal University of Technology, Erzstr. 1, D–38678 Clausthal-Zellerfeld, Germany.
E-mail: pesch@math.tu-clausthal.de.

[†]Software Engineer.

[‡]Research Scientist.
E-mail: stefan.miesbach@zfe.siemens.de.

[§]Assistant Professor, Presently, Institute of Mathematics, Clausthal University of Technology, Erzstr. 1, D–38678 Clausthal-Zellerfeld, Germany.
E-mail: breitner@math.tu-clausthal.de.

surrounding the arena enable the rat to evade once and for all, if the rat is not too far from these holes. The optimal trajectories in the escape zone can be computed analytically. In the capture zone, a game of degree is employed with terminal time as payoff. To compute optimal trajectories for this secondary game, the time evolution of the survival region for the rat is determined via a sequence of discretized games.

The combination of these methods permits the computation of more than a thousand trajectories leading to some ten thousand sample patterns which relate the state variables to the values of the optimal strategies. These data exhibit characteristic properties of the optimal strategies. It is shown that these properties can be extracted from the data by use of neural networks. By means of the trained networks, about 200 trajectories are finally simulated. The pursuer as well as the evader acts according to the controls proposed by the neural networks. Despite the simple structure of the neural networks used in this study, the strategies based upon them show a reasonable, close to optimal performance in a large variety of simulations of the pursuit-evasion game under consideration.

Keywords: Differential games, neural networks, real-time computation, pursuit-evasion games, optimal strategies, numerical solution of differential games, cornered rat game.

1 Introduction

Considering the stage of development of numerical solution methods for both optimal control and differential game problems, one finds that even complicated optimal control problems can today be solved by means of sophisticated numerical methods. Fields of application are, for example, aeronautics, astronautics, robotics, economics, and chemical engineering. See for example [28] and [29] for a review of methods and applications and the references cited therein. In contrast, the differential game problems solved so far are mostly rather simple and therefore do not describe the underlying real problems as realistically as those optimal control problems mentioned above.

The main reason is that, for optimal control problems, one often needs to know only the optimal controls as time-dependent functions, i. e., as open-loop controls, in particular when only the optimal processes are to be investigated. However, if these optimal open-loop controls, which are valid only along their associated optimal trajectories, are to be applied to real processes, trajectories may occur, which deviate considerably from the precomputed optimal trajectories due to inaccuracies of the model or unpredictable disturbances. Consequently, optimal open-loop controls cannot be applied in real processes directly. This is even more important for differential game problems, where optimal closed-loop or feedback controls, which depend on the state variables, are vital for the nature of differential games. The actors or players of a differential game usually have full information about the equations of motion, the actual state, the constraints imposed on the problem, the information available for the other players, and about the other players' goals. However, no player usually has any information about the present or future controls of his opponents. Therefore, each player tries to reach his goal against the unknown controls of the other players. This confirms the necessity to compute closed-loop controls.

Since optimal closed-loop controls generally cannot be calculated analytically, one has to approximate the optimal closed-loop controls numerically. The methods developed so far for optimal control processes are different for slow and fast processes. For a slow process, e.g., for an optimal control problem in economics, significant deviations from a precomputed optimal trajectory may appear only after a period of time long enough to allow a re-optimization. The optimal open-loop controls due to the actual initial conditions can then be recomputed by means of direct methods (see, e.g., the direct collocation method described in von Stryk [36] and von Stryk and Bulirsch [37]) whenever it is necessary; see, e.g., [19]. Online and offline methods are the same here. For a fast process, e.g., the flight of a space shuttle through the Earth's atmosphere, or for a very fast process, e.g., the motion of a robot arm, special methods must be applied, in order to provide approximations of the optimal closed-loop controls in real-time. Different approaches for the computation of closed-loop controls, all capable of real-time applications, can be found, e.g., in Bryson and Ho [10] and [7], [20], and [27]. All these approaches can be applied to optimal control problems as well as differential game problems. However, only [7] deals with differential games.

Here, we are going to develop an alternative method which can also be applied both to optimal control and differential game problems. This method is based on the approximation property of neural networks for multivariate transformations. If it is possible to compute the optimal open-loop controls and their associated optimal trajectories for various initial conditions, the optimal controls associated with all points along these trajectories can be used for the training of neural networks. The trained neural networks then enable an approximation of the optimal closed-loop controls in the area of the state space filled with those sample trajectories. First applications of neural networks to optimal control problems can be found in Järmark and Bengtsson [18], Nguyen and Widrow [26], and recently, for an application in robotics, in Goh and Edwards [14].

Contrary to optimal control theory, different concepts of optimality compete with each other in differential game theory; see, e.g., Başar and Olsder [2]. Common to all those concepts is that the computation of the optimal closed-loop controls, the so-called optimal strategies of the players, is much more involved than for optimal control problems. This is due to the fact that, for complicated differential games, optimal trajectories can today be computed only by indirect methods (for example by the multiple shooting method, see Stoer and Bulirsch [35]); see, e.g., [6], [8], [9], [21], and [22]. Along these optimal trajectories, the optimal controls of each player are so-called open-loop representations of their optimal strategies; see Isaacs [17] and [2]. Since optimal behavior of all players is extremely unlikely for realistic problems, the optimal strategies of the players must be approximated globally, i.e., the feedback controls must be valid in the entire relevant part of the state space. Additionally, the approximation has to be applicable in real time, too. The employment of neural networks seems to be one of the most promising approaches for the near future. Data for training and cross-validation of the neural networks are taken from the open-loop representations of the optimal strategies along suitable optimal trajectories.

The well-known *Cornered Rat Game*, a modified version of which is presented in this paper, will serve as an illustrative example. The major difficulties for the global approximation of optimal strategies can already be seen from this rather simple problem: The optimal strategies are global functions of four state variables; the optimal strategies are nondifferentiable and even discontinuous at certain hypermanifolds, the so-called singular surfaces; training and cross-validation data are concentrated only on a few optimal trajectories. Using thoroughly the modified cornered rat game as a driving example, it is demonstrated how to construct the neural feedback controller and how to apply it to more general problems of optimal control or differential games.

The present paper is an abridged version of the thesis [12] and the report [13].

2 Illustrative Example: A Modified Cornered Rat Game

2.1 Original Cornered Rat Game

The cornered rat game, firstly formulated by Isaacs [17] and also investigated by Breakwell [5], is a classical pursuit-evasion game. The rat E, the evader in the game, can run faster than the cat P, the pursuer, but E is entrapped in a corner, see Fig. 1.

Both players P and E can change their directions of motion instantaneously. The range d_P of the cat determines the circular terminal set Λ_P, by which capture of E by P is defined, if E penetrates Λ_P for the first time. The hypermanifold of the state space separating capture zone from escape

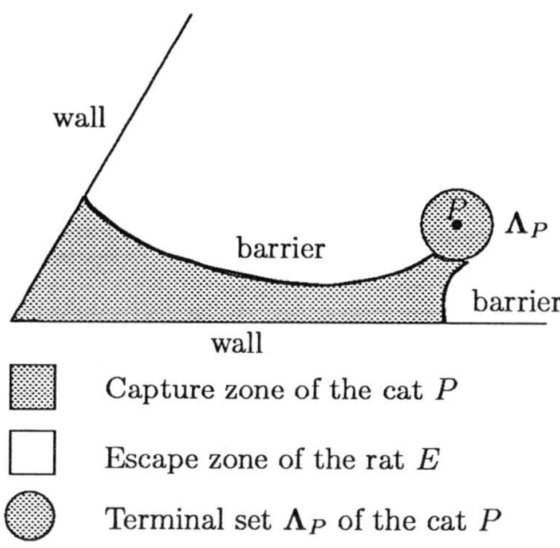

Figure 1: The original cornered rat game.

zone is the so-called barrier. If P uses an optimal pursuit strategy in the capture zone, P can force capture against all strategies of E. If E uses an optimal evasion strategy in the escape zone, P can in no way capture E.

2.2 Modified Cornered Rat Game

Inspired by the original cornered rat game, the following modified cornered rat game is introduced. This game includes additional phenomena typical for two-player zero-sum differential games of pursuit-evasion type. The state space S is four-dimensional, and a state $z \in S$, $z := (x_P, y_P, x_E, y_E)^\top$, represents permissible positions of the pursuing cat P and the evading rat E in a rectangular arena[1]; see Fig. 2. The players P and E are therefore responsible for obeying their respective constraints:

$$0\,[\mathrm{m}] \leq x_P \leq 10\,[\mathrm{m}], \quad 0\,[\mathrm{m}] \leq y_P \leq 6\,[\mathrm{m}], \tag{1}$$

$$0\,[\mathrm{m}] \leq x_E \leq 10\,[\mathrm{m}], \quad 0\,[\mathrm{m}] \leq y_E \leq 6\,[\mathrm{m}]. \tag{2}$$

Note that, because of these state constraints, the dimension of the problem cannot be reduced as in the unconstrained case.

---

[1]It is assumed that the cat enters the arena through the entrance, that she or he closes the door immediately because of supposing a rat to be in the arena, but she or he does not become aware of the rat directly.

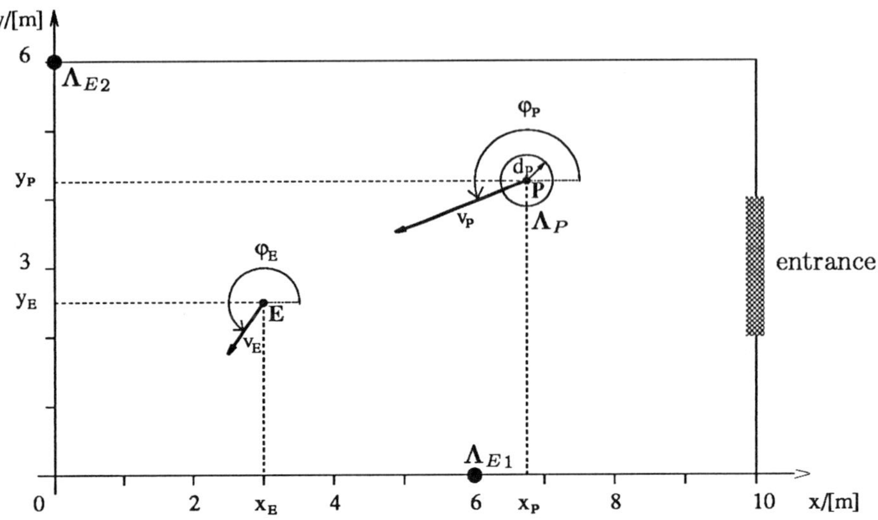

Figure 2: The Modified Cornered Rat Game.

The cat P runs with constant maximum velocity $v_P = 2 \left[\mathrm{m\,s^{-1}}\right]$ in the direction given by the angle φ_P. The angle φ_P is the control variable of P. The rat E runs either with constant maximum velocity $v_E = 1 \left[\mathrm{m\,s^{-1}}\right]$ or sits still in a corner of the capture zone and waits to be eaten well rested.[2] The player E uses the direction φ_E as his control variable, see Fig. 2. Both angles φ_P and φ_E can be changed instantaneously. Note that the cat is faster than the rat in the differential game investigated here. In order to compensate for this, we find ourself obliged to give the rat a chance to escape once and for all through one of two holes in the walls. The holes are located at $(x, y) = (6\,[\mathrm{m}], 0\,[\mathrm{m}])$ and at $(x, y) = (0\,[\mathrm{m}], 6\,[\mathrm{m}])$; see Fig. 2. Hence, capture of E by P takes place, if

$$d(z) := \sqrt{(x_P - x_E)^2 + (y_P - y_E)^2} < d_P \,, \qquad (3)$$

before E can reach one of the holes in the wall. Here, d_P denotes the range of P, also called the capture radius.

Thus, the target set Λ_P of P is defined by

$$\Lambda_P := \{z \in S : d(z) \le d_P\} \,, \qquad (4)$$

---

[2] Here, with the view of achieving a unique strategy whenever possible, we ignore the rat's natural desire to survive.

and the target sets $\boldsymbol{\Lambda}_{Ei}$, $i = 1, 2$, of E are defined by

$$\boldsymbol{\Lambda}_{E1} := \{\boldsymbol{z} \in \boldsymbol{S} : (x_E, y_E) = (6\,[\mathrm{m}], 0\,[\mathrm{m}]) \text{ and } d(\boldsymbol{z}) > d_P\}\,, \quad (5)$$

$$\boldsymbol{\Lambda}_{E2} := \{\boldsymbol{z} \in \boldsymbol{S} : (x_E, y_E) = (0\,[\mathrm{m}], 6\,[\mathrm{m}]) \text{ and } d(\boldsymbol{z}) > d_P\}\,. \quad (6)$$

The final time t_f is determined by

$$t_f := \min_{t \in [0, \infty[} \{t : \boldsymbol{z}(t) \in \boldsymbol{\Lambda}_P \cup \boldsymbol{\Lambda}_{E1} \cup \boldsymbol{\Lambda}_{E2}\}\,. \quad (7)$$

The game starts in an initial state $\boldsymbol{z}_0$ at time $t = 0$ and is governed by the following equations of motion, known as simple motion,

$$\frac{\mathrm{d}}{\mathrm{d}t} x_P = v_P \cos \varphi_P\,, \quad \frac{\mathrm{d}}{\mathrm{d}t} y_P = v_P \sin \varphi_P\,, \quad (8)$$

$$\frac{\mathrm{d}}{\mathrm{d}t} x_E = v_E \cos \varphi_E\,, \quad \frac{\mathrm{d}}{\mathrm{d}t} y_E = v_E \sin \varphi_E\,. \quad (9)$$

The players P and E have exact information about the actual state and the state history, but no information about the future evolution of the game. The modified cornered rat game is formulated now via two nested differential games.

Primary Game

In the primary game, both P and E try to find an optimal permissible strategy $\hat{\gamma}_P{}^*(\boldsymbol{z})$ and $\hat{\gamma}_E{}^*(\boldsymbol{z})$, respectively, in compliance with the above-defined information structure, to guarantee that their own target sets are attained against all permissible strategies of the antagonistic player. Here, we have a so-called game of kind; see, e. g., [2] and [17]. If E can enforce escape by choosing $\hat{\gamma}_E{}^*(\boldsymbol{z})$ for $\boldsymbol{z} \in \boldsymbol{S}$, then $\boldsymbol{z}$ is an element of the escape zone $\boldsymbol{S}_E$. The set $\boldsymbol{S}_P := \boldsymbol{S} \setminus \boldsymbol{S}_E$, where P can enforce capture using $\hat{\gamma}_P{}^*(\boldsymbol{z})$, is called the capture zone. The separating hypermanifold is the barrier $\boldsymbol{B}$, compare Fig. 3. Since the state space $\boldsymbol{S}$ is convex and P runs faster than E, both in simple motion, optimal strategies $\hat{\gamma}_P{}^*(\boldsymbol{z})$ and $\hat{\gamma}_E{}^*(\boldsymbol{z})$ can be easily obtained in the entire escape zone. The barrier can be calculated analytically as

$$\boldsymbol{B} := \left\{\boldsymbol{z} \in \boldsymbol{S} : \frac{\sqrt{(x_P - 6)^2 + y_P{}^2} - d_P}{v_P} = \frac{\sqrt{(x_E - 6)^2 + y_E{}^2}}{v_E}\right\} \bigcup$$

$$\left\{\boldsymbol{z} \in \boldsymbol{S} : \frac{\sqrt{x_P{}^2 + (y_P - 6)^2} - d_P}{v_P} = \frac{\sqrt{x_E{}^2 + (y_E - 6)^2}}{v_E}\right\}\,. \quad (10)$$

Secondary Game

If E cannot prevent capture for an initial state $z_0 \in S_P$, the problem arises of which performance index shall be used for the two players in the capture zone, to get a certain degree of uniqueness for the players' actions. Assuming that P tries to capture E efficiently, one can take the capture time t_f as the performance index of a secondary game leading to a game of degree; see [2] and [17].

Thus, both P and E try to find an optimal permissible strategy $\check{\gamma}_P{}^*(z)$ and $\check{\gamma}_E{}^*(z)$ which minimizes or maximizes, respectively, the final time t_f against the unknown present and future actions of the antagonistic player. Cumbersome to the general solution of the secondary game is a rich variety of different optimal scenarios depending on the starting point $z_0 \in S_P$. All these cases have to be investigated separately (see Fig. 3): **(i)** E runs straight away from P, and P runs directly behind E. This is known as tail chase. E is captured before any wall is reached. Both $\check{\gamma}_P{}^*(z^*(t))$ and $\check{\gamma}_E{}^*(z^*(t))$ are unique for all $t \in [0, t_f]$. **(ii)** E runs away from P on a straight line, but is constrained by one of the walls. The end position of E is on that wall, when capture occurs. Both $\check{\gamma}_P{}^*(z^*(t))$ and $\check{\gamma}_E{}^*(z^*(t))$ may not be unique for all $t \in [0, t_f[$, since dispersal surfaces may occur; see [17]. **(iii)** E runs into the most advantageous corner and eventually has to wait there, until P strikes. Generally, $\check{\gamma}_P{}^*(t)$ is unique for all $t \in [0, t_f]$, but $\check{\gamma}_E{}^*(z^*(t))$ and $z^*(t)$ are not unique for all $t \in [0, t_f]$.

Survival Region Algorithm

Although optimal strategies of both players can be computed analytically for all $z_0 \in S_P$ (see [12]), the distinction of the different cases is tedious and strongly dependent on the shape of the convex state space. Therefore, the following survival region method is introduced. This method is more generally applicable and can be used in the entire state space of the modified cornered rat game. The method approximates an optimal trajectory $z^*(t)$, $t \in [0, t_f]$, with $z^*(0) = z_0$, $z_0 \in S_P$, and thereby provides optimal strategies $\check{\gamma}_P{}^*(z^*(t))$ and $\check{\gamma}_E{}^*(z^*(t))$ along that trajectory.

The survival region $R(t, z_0)$ is defined as the subset of S, which can be reached only by E but not by P in time t, see Fig. 4. With $r_E := t\, v_E$ and $r_P := t\, v_P$ there holds

$$R(t, z_0) := \left\{ z(t) \in S : \sqrt{(x_P(t) - x_P(0))^2 + (y_P(t) - y_P(0))^2} > r_P + d_P \right.$$

$$\left. \text{and} \ \ \sqrt{(x_E(t) - x_E(0))^2 + (y_E(t) - y_E(0))^2} \le r_E \right\} . \qquad (11)$$

In the following, a numerical discretization scheme is presented by which the survival region can be easily computed. Defining a time discretization Δt

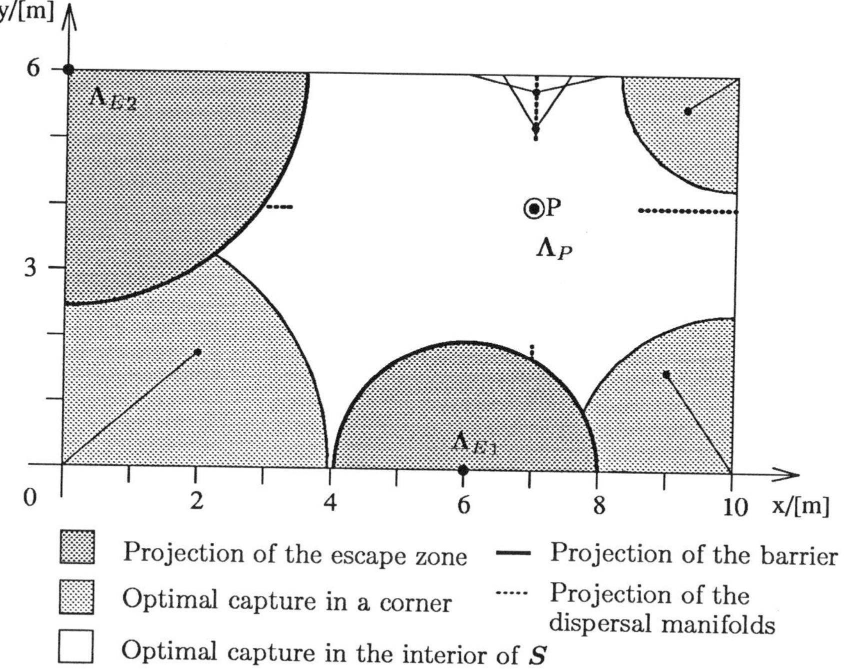

Figure 3: Partition of the state space into escape and capture zone and the dispersal manifolds for the position ($x_P(0) = 7\,[\text{m}]$, $y_P(0) = 4\,[\text{m}]$) of P. Thin lines describe different optimal trajectories of E.

and an associated circular spatial discretization according to Fig. 4, the discretized survival region $\tilde{R}(i\,\Delta t, z_0)$ is computed for $i = 1, 2, 3, \ldots$, until the first index i is found for which the set $\tilde{R}$ is empty. By bisection in time, the set $\tilde{R}$ is then computed which contains one and only one element. This element $\tilde{z}^*(\tilde{t_f})$ approximates the exact final point $z^*(t_f)$ of an optimal trajectory. The accuracy of the method can be controlled, e. g., by choosing $\alpha = \frac{\pi}{3}$ and $\Delta t = \frac{3\varepsilon}{\pi}$; see Fig. 4. This choice guarantees that $\|\tilde{z}^*(\tilde{t_f}) - z^*(t_f)\|_2 \le \varepsilon$. By this procedure, an approximation of optimal strategies $\check{\gamma}_P^*(z^*(t))$ and $\check{\gamma}_E^*(z^*(t))$ along the optimal trajectory is also obtained.

For the illustration of the survival region algorithm, we choose the initial state $z_0 = (10\text{ m}, 3\text{ m}, 4\text{ m}, 2\text{ m})$. The survival regions $R(t_i, z_0)$ are shown in Figs. 5 and 6 for $t_i = 0$ s, 0.19 s, 1.14 s, 2.67 s, 4.01 s, 4.77 s, and 5.01 s. After the increase of the survival region, its decrease and its splitting into two parts can be seen. An optimal trajectory $z^*(t)$, $t \in [0, t_f]$, with $z^*(0) = z_0$ and the projection of the barrier can be seen also in Fig. 6.

For the global computation of optimal strategies $\gamma_P^*(z)$ and $\gamma_E^*(z)$

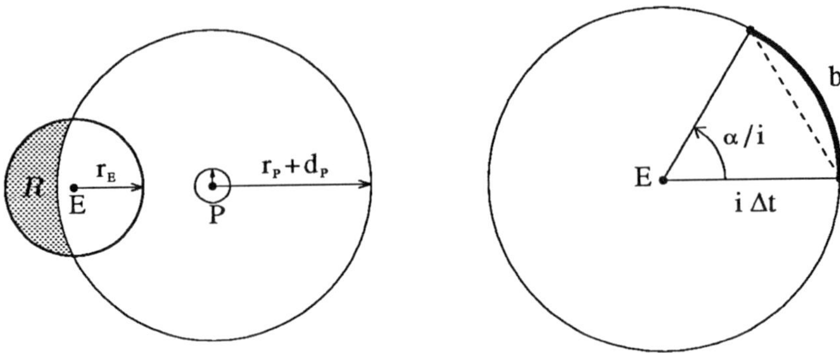

Figure 4: Construction (left) and discretization (right) of the survival region $\mathbf{R}(t, \mathbf{z}_0)$.

for all $\mathbf{z} \in \mathbf{S}$, it is necessary to combine the aforementioned methods. For $\mathbf{z} \in \mathbf{S}_E$, optimal strategies $\hat{\gamma}_P{}^*(\mathbf{z})$ and $\hat{\gamma}_E{}^*(\mathbf{z})$ are computed analytically, whereas optimal strategies $\check{\gamma}_P{}^*(\mathbf{z})$ and $\check{\gamma}_E{}^*(\mathbf{z})$ are computed by means of the survival region algorithm for $\mathbf{z} \in \mathbf{S}_P$.

First Step: Constructing the Neural Feedback Controller

Now, we are able to perform the first preparatory step for the construction of the neural feedback controller: By means of the above procedure, optimal trajectories and their associated optimal strategies can be computed efficiently, but, of course, not in real-time. These trajectories may then span the entire state space. In case of a more general differential game problem, an appropriate numerical method for the computation of the optimal trajectories must be employed. Thereby, it is sufficient to compute only the open-loop representations of the optimal feedback strategies. These computations can be done by means of an indirect method such as multiple shooting; see [6], [8], [9], [21], and [22]. Note that the modified cornered rat game as presented here as driving example is so simple that the optimal feedback strategies themselves can be computed in the entire state space even analytically.

We now leave the field of differential games and make a leap to neural networks.

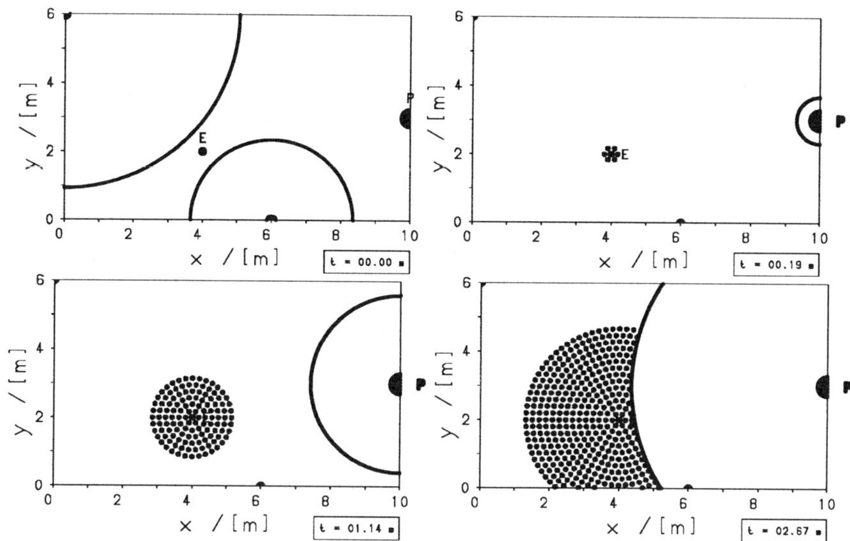

Figure 5: Computation of an optimal trajectory for the initial state $x_P(0) = 10\,[\mathrm{m}]$, $y_P(0) = 3\,[\mathrm{m}]$, $x_E(0) = 4\,[\mathrm{m}]$, and $y_E(0) = 2\,[\mathrm{m}]$ by means of the survival region algorithm. (Part 1)

3   Neural Networks for Multivariate Approximations

3.1   Introduction

Apart from their biological and connectional interpretations artificial neural networks, abbreviated NN, may be regarded as "black boxes" realizing nonlinear multivariate transformations $\mathrm{NN}\colon \mathbb{R}^n \to \mathbb{R}^m$, $y = \mathrm{NN}(x)$, which are parametrized by a set of n_w weight parameters $w \in \mathbb{R}^{n_w}$. These weights become specified by training procedures that are designed to extract regularities from measured or precomputed data sets.

Neural networks are built up by a large number of identical processing elements, which are called neurons due to analogies with biological nerve cells. The neurons are interconnected with each other and with input nodes. The interconnections conduct signals between the processing elements and are associated with weights w_{Nj} where the indices N and j refer to the signal-receiving and emitting neuron, respectively. The receiving neuron N collects signals a_j from several other neurons or input nodes and processes them into a scalar emission signal a_N. There is a vast body of literature

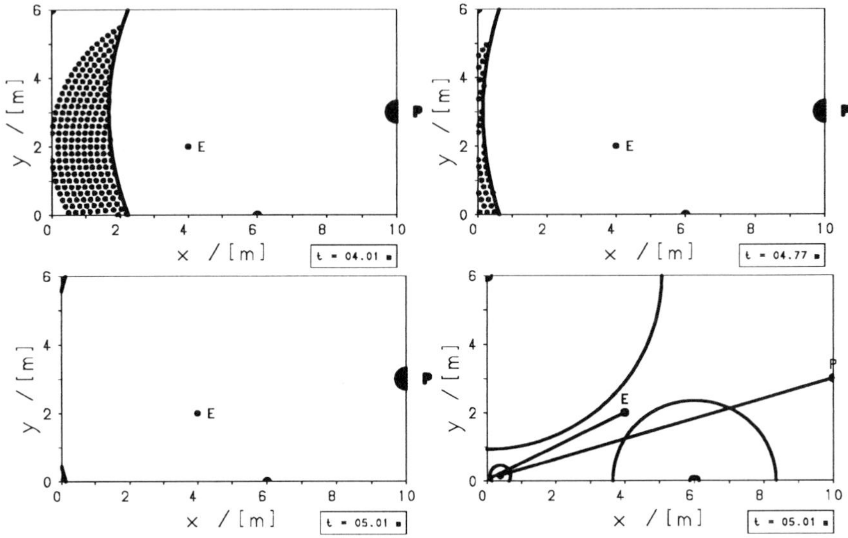

Figure 6: Computation of an optimal trajectory for the initial state $x_P(0) = 10\,[\text{m}]$, $y_P(0) = 3\,[\text{m}]$, $x_E(0) = 4\,[\text{m}]$, and $y_E(0) = 2\,[\text{m}]$ by means of the survival region algorithm. (Part 2)

concerning neural network architectures.[3]

For a first approach in applying neural networks in differential games, we concentrate in this paper on neural networks of the so-called multilayer-perceptron architecture.

3.2 Multilayer Perceptron

In a multilayer perceptron the neurons are arranged in several layers as shown in Fig. 6 for a 4-layer perceptron.[4] The layer 0 contains the input nodes which are supplied with the input variables x of the transformation to be extracted from a data set. Then, several so-called hidden layers follow, here layers 1 and 2, the neurons of which perform nonlinear operations on their

---

[3]The most important journals in the field of neural networks include *Neural Networks*, *Neural Computation*, and the *IEEE Transactions on Neural Networks*. In addition, the proceedings of the NIPS-Conference on *Neural Information Processing Systems* [30] provide an excellent source of information on recent developments. For readers especially interested in neural networks for control applications, the books of Ritter, Martinetz, and Schulten [31] and Miller, Sutton, and Werbos [25] are recommended.

[4]Note the enormous reduction in the number of neurons in a 4-layer perceptron compared to the brains of a cat and a rat.

received signals. Finally, the output signals of the network are computed in the output layer, here layer 3, by linear neurons which simply sum up the signals of the preceeding hidden layer. In more detail, the signal processing of the layers is as follows.

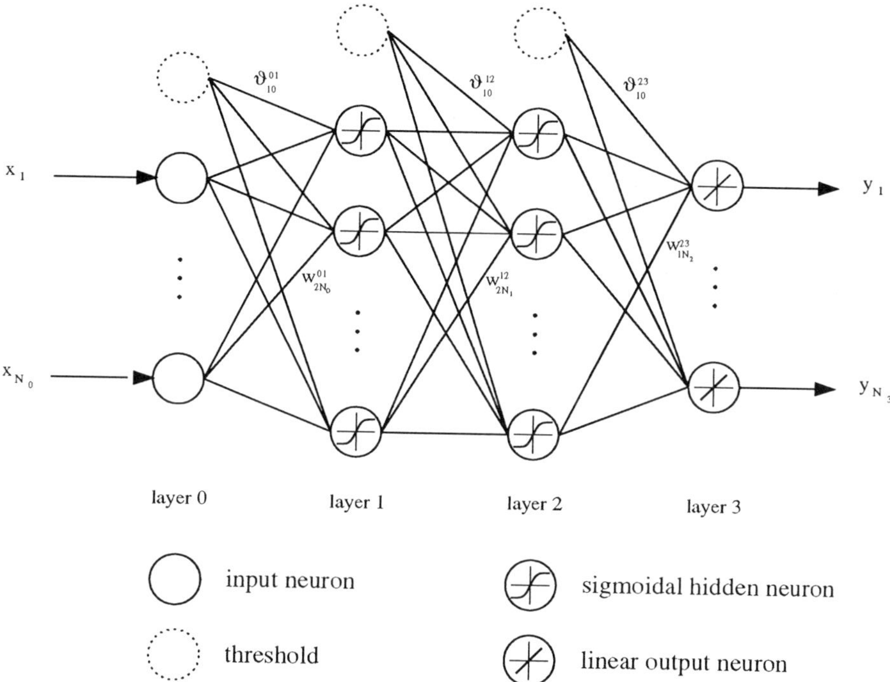

Figure 7: A four-layer neural network with N_0 neurons in the input layer 0, N_1 and N_2 neurons in the hidden layer 1 and 2, respectively, and N_3 neurons in the output layer.

A neuron N in a hidden layer computes its emission signal a_N by linearly combining the signals a_j it receives and by processing the result by a squashing function s,

$$a_N = s\left(\sum_j w_{Nj}\, a_j - \vartheta_N\right). \tag{12}$$

Here a_j denotes the input x_j or the emitted signal of neuron j in layer 1, if the neuron N lies in layer 1 or 2, respectively. The quantity ϑ_N represents a threshold signal which is visualized by the dashed neurons of Fig. 7. The squashing function s is usually realized by a scalar sigmoidal function, e. g., $s(u) = \tanh(u)$ or $s(u) = 1/(1 + e^{-u})$.

A neuron N in the output layer simply sums up the emitted signals of the preceeding hidden layer,

$$y_N = a_N = \sum_j w_{Nj}\, a_j - \vartheta_N \,. \tag{13}$$

Note that the weight parameters w_{Nj} and ϑ_N in (12) and (13) are different; the superscripts referring to the layer are suppressed to lighten the notation.

The construction of the multilayer perceptron originates in early works on the connectional approach to detecting some of the structure of human brains; see Rosenblatt [33]. From a mathematical point of view, there seems to be little motivation to use a multilayer perceptron to build up multivariate transformations. Recently a lot of effort, however, has been made to prove approximation properties. As an example, Hornik, Stinchcombe, and White [16] found the following theorem: *Let $S \subset \mathbb{R}^n$ be a Lebesgue-measurable subset in the input space and let $L_2[S]$ be the space of square-integrable functions defined on S. Then, any $f \in L_2[S]$ can be arbitrarily closely approximated by a three- or four-layer perceptron.* A more constructive theorem can be found in Blum and Li [4]. It should be mentioned that these theorems are mainly of theoretical importance here, as they give no answer to the question of how many neurons are to be used in the hidden layers or how to find an appropriate set of weights. For the latter, we take a glance at the so-called training algorithms.

3.3　Training by Incremental Gradient Descent

Most training algorithms for neural networks are variations of the incremental gradient descent, which may be regarded as the basic algorithm in neural computation. More details and more sophisticated versions of this method can be found in Hertz, Krogh, and Palmer [15].

The situation is as follows. Assume that we have to approximate, say, a piecewise continuous mapping $f\colon \mathbb{R}^n \to \mathbb{R}^m$, the analytical representation of which is not available, but we are provided with a set of measured or precomputed input and output vectors $(\boldsymbol{x}^j, \boldsymbol{f}^j)$, $j = 1, \ldots, n_p$, the so-called training patterns. Assume furthermore that, for this purpose, we choose a neural network $\mathrm{NN}(\cdot, \boldsymbol{w})$ with a fixed topology. Then, we have to seek an appropriate set of weights $\boldsymbol{w}$ which optimizes the network transformation with respect to the known data. The most straightforward approach is to define a least-squares functional

$$\mathcal{E}(\boldsymbol{w}) = \frac{1}{2} \sum_{j=1}^{n_p} \left\| \boldsymbol{e}^j(\boldsymbol{w}) \right\|^2 \tag{14}$$

with the error signal $e^j(w) := \mathrm{NN}(x^j, w) - f^j = y^j - f^j$ measuring the accuracy of the approximation by the network on the set of input patterns x^j. For applications in differential games, as we have in mind, we can proceed on the assumption that the data are free of noise. Then, the set of optimal weights is given by the global minimum of the error function[5] $\mathcal{E}(w)$. Any nonlinear least-squares method may now be applied to solve this minimization problem. Unfortunately, the special construction of the multilayer perceptron leads to extremely ill-conditioned optimization problems. Therefore, higher-order optimization methods, in general, do not exhibit their superior convergence properties. These problems are discussed in Saarinen, Bramley, and Cybenko [34]. A second difficulty one is often faced with arises from comparably large data sets. One has to solve large scale optimization problems with typically several hundreds of unknown parameters and several thousands of data. For neural training problems, methods based on the incremental gradient descent have turned out to produce reasonably good solutions.

If all n_p patterns have been drawn at least once, one training epoch of the incremental gradient descent is said to be completed. Concerning the convergence behaviour of the cumulative, i. e. the ordinary gradient descent, and the incremental gradient descent, one can show that the updates yielded by one cumulative gradient iteration and one epoch of incremental gradient descent will differ only by $O(\epsilon)$, where $\epsilon > 0$ denotes the learning rate, i. e. the step size of the gradient descent. This holds under the assumption that the iterations are started with the same initial weights and that each pattern (x^j, f^j) is drawn exactly once per training epoch. As the iteration proceeds, both methods may diverge exponentially. For more details, see [13].

For the convergence behaviour of the incremental gradient descent in case of a network having only linear weights or with the nonlinear weights fixed, respectively, see Anderson [1] and [23]. In general, the incremental gradient method does not precisely converge but it is captured by stable bassins around the optimal set of weights; see Benveniste, Metivier, and Priouret [3]. In case of a number of local minima, cumulative and incremental gradient descent may exhibit substantially different behavior. The cumulative gradient descent and any higher order technique as well may get stuck in a flat local minimum far from a reasonable network performance. Due to the above-mentioned non-convergence property, the incremental method usually cannot get captured by a flat local minimum.

The last topic which should be brought up in this subsection is the need to perform cross-validation tests between the training epochs. This is because the goal of the training is actually not to minimize the error

---

[5]Note that in most other applications of neural networks one has to deal with more or less noisy or inconsistent data. In those cases, the minimization of the error function (14) may not lead to meaningful results; see Finnoff [11].

function, i. e., to optimize the network on the training data, but to find an approximation of the unknown function f being valid over some compact set $X \subset \mathbb{R}^n$, e. g., the convex hull of the data. Therefore, the interpolation and extrapolation capabilities of the network have to be tested on a separate set of so-called cross-validation data which become excluded from the actual training. After each training epoch, the corresponding error function over these additional data is computed, too. Only if both error functions decrease during the learning phase, can a reasonable approximation of the function f be expected.

Finally, it should be mentioned that the incremental gradient descent yields an algorithm appropriate for parallel processing.

Second Step: Constructing the Optimal Feedback Controller

Now, we are able to perform the second step for the construction of the neural feedback controller: Identifying the input vectors x^j with the state variables $z \in S$ and the output vectors f^j with the optimal feedback strategies $\gamma_P{}^*(z)$ or $\gamma_E{}^*(z)$, respectively, of either player along the precomputed optimal trajectories of our driving example, a neural network for both players can be established and trained. Thereafter, one has completed the construction of the neural feedback controller which is assumed to approximate the optimal feedback strategies. Those two steps close the necessary off-line computations, and the neural feedback controller can be applied for on-line computations.

We are now going to apply this neural network technique in detail to synthesize optimal strategies for the modified cornered rat game.

4 Numerical Synthesis of Optimal Strategies

The computation of optimal strategies $\gamma_P{}^*(z)$ and $\gamma_E{}^*(z)$ for a $z \in S_P$ is performed by means of the survival region algorithm, which needs up to 10 minutes computing time on a SUN SPARC 10 work station. Nonoptimal behavior of the players now forces the recomputation of optimal strategies $\gamma_P{}^*(z)$ and $\gamma_E{}^*(z)$ at all times. However, the duration of the pursuit-evasion game usually takes only less than 10 seconds. This gap in time scales exhibits the main obstacle for real-time computations of optimal strategies. For complicated pursuit-evasion games, for example those investigated in [8], [9], and [21], the computation of optimal strategies may even need considerably more computing time. A well trained neural network can now close this gap. The evaluation of the multivariate transformation NN needs a few arithmetic operations to approximate the optimal strategies and is therefore real-time applicable.

4.1 Training Data and Cross-Validation Data

By means of the survival region method, a pool of 1365 optimal trajectories $z^*(t)$, $t \in [0, t_f]$, and their associated optimal strategies $\gamma_P{}^*(z)$ and $\gamma_E{}^*(z)$ have been computed based on initial conditions $z^*(0) = z_0$ given by any combination from

$$(x_P(0), y_P(0)) \in \{(10, 3), (9, 2), (7, 1), (9, 5), (7, 4)\} \tag{15}$$

and

$$(x_E(0), y_E(0)) := \left(\frac{i}{2}, \frac{j}{2}\right) \text{ with } i \in \{0, 1, \ldots, 20\}, \ j \in \{0, 1, \ldots, 12\} . \tag{16}$$

For all initial positions $(x_E(0), y_E(0))$ of E taken into consideration in (16), while holding the position $(x_P(0), y_P(0))$ of P fixed at $(7, 4)$ [see (15)], the values of the optimal strategies $\gamma_P{}^*(z_0)$ and $\gamma_E{}^*(z_0)$ and the optimal capture time $t_f{}^*(z_0)$ are depicted in the left parts of Figs. 8–10 as functions of those two variables $x_E(0)$ and $y_E(0)$. Note that by definition there holds $t_f{}^*(z_0) = \infty$ in the entire escape zone. These three (left) figures together with equivalent figures for all other positions of P given by (15) condense the entire information of the pool of trajectories at $t = 0$. Virtually all these trajectories can be attached at the seven-dimensional vector $(z_0, \gamma_P{}^*(z_0), \gamma_E{}^*(z_0), t_f{}^*(z_0))$. Finally, all vectors $(z, \gamma_P{}^*(z), \gamma_E{}^*(z), t_f{}^*(z))$ along these trajectories are potential input-output patterns for the training of the networks as described below.

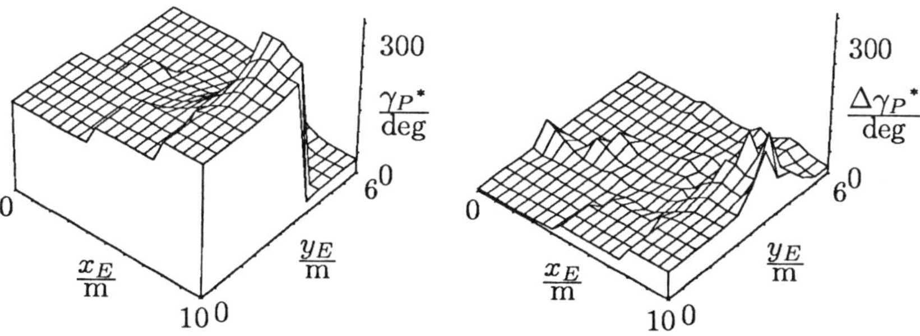

Figure 8: Optimal strategy $\gamma_P{}^*(x_E(0), x_P(0))$ (left) and absolute error $\Delta\gamma_P{}^*(x_E(0), x_P(0))$ of $\gamma_P{}^*$ (right) for $(x_P, y_P) = (7, 4)$.

Figures 11 and 12 illustrate a subset of those optimal trajectories for initial positions given by $(x_P(0), y_P(0)) = (7, 4)$, while $(x_E(0), y_E(0))$ is running through the data set (16). Simultaneously, the optimal feedback strategies can be seen from the slopes of the trajectories being constant along a specific trajectory. Note that the optimal strategies $\gamma_P^*(z_0)$ and $\gamma_E^*(z_0)$ have discontinuities at the barrier and for $(x_E(0), y_E(0)) \rightarrow (x_P(0), y_P(0))$; see again the left parts of Figs. 8, and 9. The value function $t_f^*(z_0)$ of the secondary game has also discontinuities at the barrier; see the left part of Fig. 10.

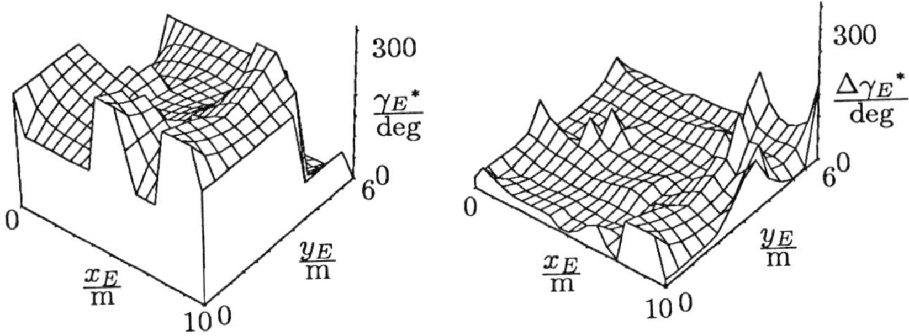

Figure 9: Optimal strategy $\gamma_E^*(x_E(0), x_P(0))$ (left) and absolute error $\Delta\gamma_E^*(x_E(0), x_P(0))$ of γ_E^* (right) for $(x_P, y_P) = (7, 4)$.

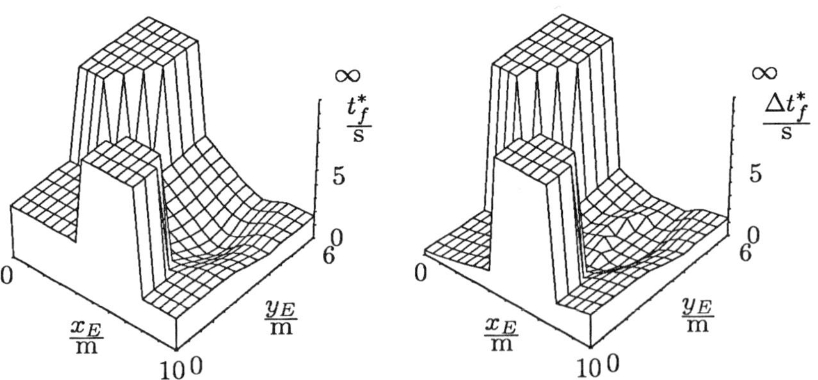

Figure 10: Optimal capture time $t_f^*(x_E(0), y_E(0))$ and absolute error $\Delta t_f^*(x_E(0), y_E(0))$ of t_f^* for $(x_P(0), y_P(0)) = (7, 4)$.

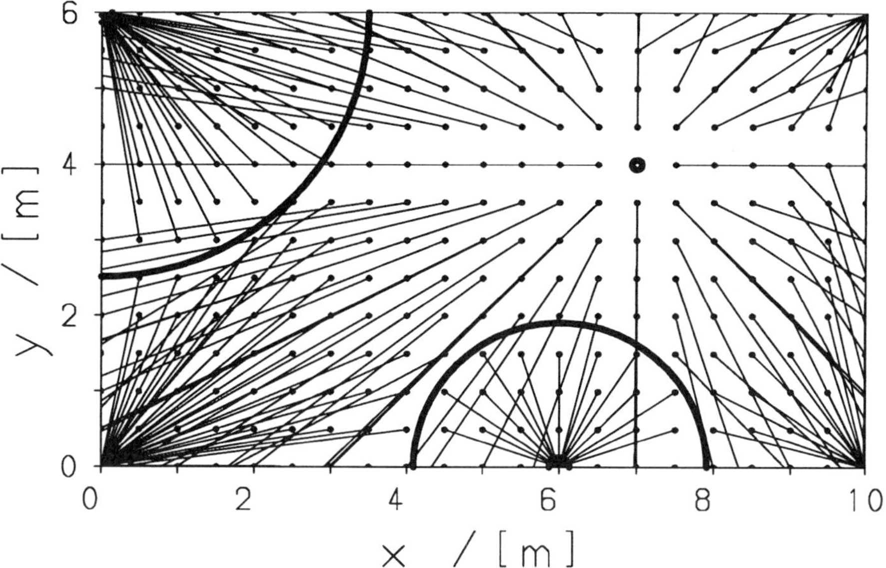

Figure 11: Optimal trajectories of the rat E for $(x_P(0), y_P(0)) = (7, 4)$.

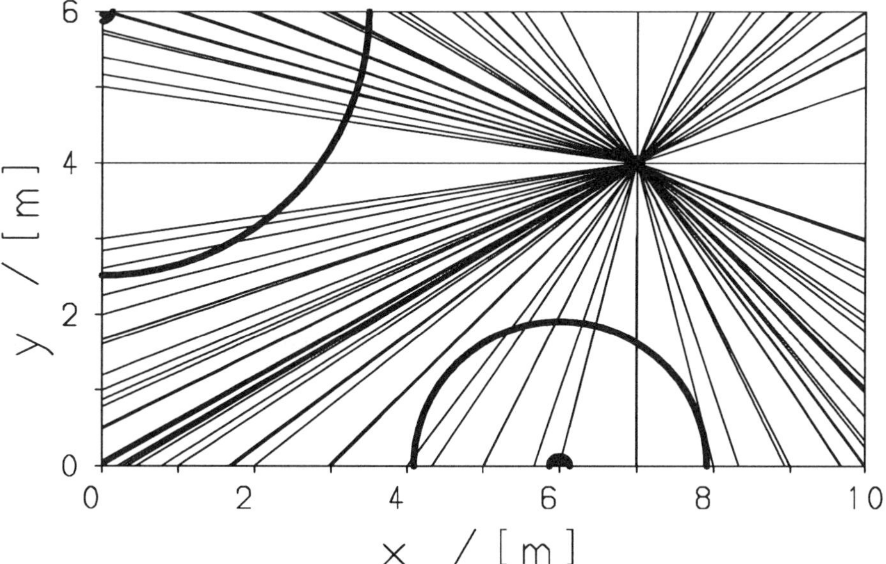

Figure 12: Optimal trajectories of the cat P for $(x_P(0), y_P(0)) = (7, 4)$.

4.2   Network Configuration

The subdivision of the state space S into three parts, the capture zone S_P and the two parts of the escape zone S_E associated with the holes in the wall, must be represented by the neural network configuration. A single neural network for the entire state space S is considered in [12], but cannot be recommended, since the topology of a single 4-layer perceptron seems to be unsuitable for modelling the discontinuities at the singular surfaces. For the modified cornered rat game, different neural networks are therefore combined for the capture and the escape zone, since the separating barrier B can be computed analytically via Eq. (10). Thus, six four-layer perceptrons with 20 sigmoidal neurons in each hidden layer have been used to approximate γ_E^* and γ_P^* in the capture zone as well as in the two parts of the escape zone. The seventh neural network of the same type is used to approximate the secondary value function t_f^* in the capture zone. Common to all networks are the input variables x_E, y_E, x_P, and y_P and a single output variable, either γ_E^*, γ_P^*, or t_f^*. For the sigmoidal neurons the squashing function $s(u) = \tanh(u)$ has been used.

4.3   Training Procedure and Approximation Error

A suitable subset of optimal trajectories according to the aforementioned initial positions of P and E [see Eqs. (15) and (16)] generates the training and cross-validation data for the seven neural networks. Choosing a grid of size $\Delta t = 0.1\,[\mathrm{s}]$ or $\Delta t = 0.01\,[\mathrm{s}]$, input data $z(i\,\Delta t) = (x_P(i\,\Delta t), y_P(i\,\Delta t), x_E(i\,\Delta t), y_E(i\,\Delta t))^{\mathsf{T}}$ are obtained for each trajectory by

$$x_P(i\,\Delta t) = x_P(0) + v_P\, i\,\Delta t\,\cos(\gamma_P^*)\,,$$

$$y_P(i\,\Delta t) = y_P(0) + v_P\, i\,\Delta t\,\sin(\gamma_P^*)\,, \tag{17}$$

$$x_E(i\,\Delta t) = x_E(0) + v_E\, i\,\Delta t\,\cos(\gamma_E^*)\,,$$

$$y_E(i\,\Delta t) = y_E(0) + v_E\, i\,\Delta t\,\sin(\gamma_E^*) \tag{18}$$

with $i = 0, 1, 2, \ldots$, as long as $d(z(i\,\Delta t)) > 1\,[\mathrm{m}]$. The related output data are given by $\gamma_E^*(z(i\,\Delta t)) \equiv \gamma_E^*(z(0))$, $\gamma_P^*(z(i\,\Delta t)) \equiv \gamma_P^*(z(0))$ and $t_f^*(z(i\,\Delta t)) = t_f^*(z(0)) - i\,\Delta t$. The total number of training and cross-validation data, the grid sizes, and the learning rates for the seven neural networks are shown in Table 1. To illustrate the convergence properties of the training algorithm, the error functions for the training data and for the cross-validation data are depicted in Fig. 13. The neural networks have been trained until the error functions versus the number of training epochs have been decreased sufficiently, here after about 300 training epochs.

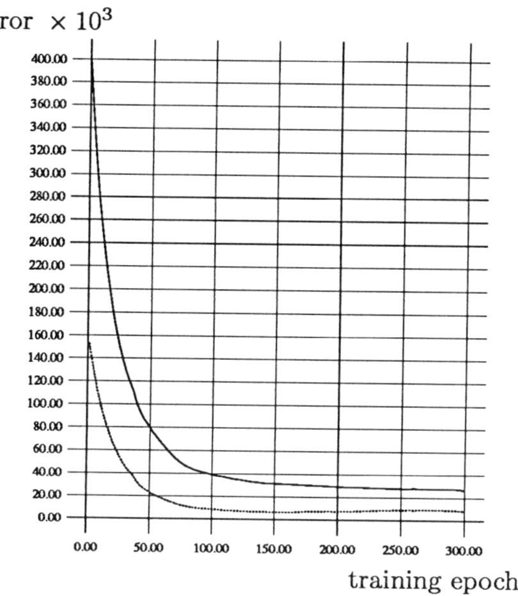

Figure 13: Error functions for training (upper curve) and cross-validation (lower curve) data versus training epochs for γ_P^* in the capture zone.

Table 1. Training data of the neural networks.

Approximation of t_f^*			
Network in the	Learning rate	Δt	Number of data
capture zone	10^{-5}	0.1	878

Approximation of γ_E^*			
Network in the	Learning rate	Δt	Number of data
capture zone	10^{-6}	0.1	2166
escape zone of Λ_{E1}	10^{-6}	0.01	6422
escape zone of Λ_{E2}	10^{-5}	0.1	1123

Approximation of γ_P^*			
Network in the	Learning rate	Δt	Number of data
capture zone	10^{-5}	0.1	2166
escape zone of Λ_{E1}	10^{-5}	0.01	6422
escape zone of Λ_{E2}	10^{-5}	0.1	1123

The right parts of Figs. 8–10 show the absolute errors $\Delta\gamma_P{}^*(z_0) := |\gamma_P{}^*(z_0) - \mathrm{NN}_{\gamma_P}(z_0; w)|$, $\Delta\gamma_E{}^*(z_0) := |\gamma_E{}^*(z_0) - \mathrm{NN}_{\gamma_E}(z_0; w)|$, and $\Delta t_f{}^*(z_0) := |t_f{}^*(z_0) - \mathrm{NN}_{t_f}(z_0; w)|$ of the global approximation for the networks NN_{γ_P}, NN_{γ_E}, and NN_{t_f} associated with $\gamma_P{}^*$, $\gamma_E{}^*$, and $t_f{}^*$, respectively. The networks NN_{γ_P} and NN_{γ_E} are split each into three nets according to the configuration described in Section 4.2 using the barrier condition (10) to decide which partial net is to be used at a certain $z \in S$. The approximation errors take their maximum values near the dispersal lines due to the initial condition $(x_P(0), y_P(0)) = (9, 5)$ for P; compare also Fig. 3.

Computer experiments have shown that a total number of less than 200 optimal trajectories suffice for a sufficiently accurate approximation of the optimal strategies for the problem under discussion. It should be mentioned that today the computation of some thousand optimal trajectories can be carried through even for very complicated differential games; see, e. g., [9].

4.4 Simulations

Finally, some simulations have been carried through, to demonstrate the performance of the trained neural networks for the approximation of the optimal strategies $\gamma_P{}^*$ and $\gamma_E{}^*$. Choosing $(x_P(0), y_P(0)) = (9, 5)$ and $(x_E(0), y_E(0))$ according to (16), 237 discretized pursuit-evasion games have been simulated. The grid size $\Delta t = 0.1\,[\mathrm{s}]$ has been chosen for the constant time steps. In each time interval, constant controls have been applied for both players. The discretized pursuit-evasion game terminates if one of the following events occurs: **(i)** $d(z(i\,\Delta t)) \leq 0.3\,[\mathrm{m}]$, i. e., if capture occures [cp. Eq. (3)]; **(ii)** $[x_E(i\,\Delta t) - 6]^2 + [y_E(i\,\Delta t)]^2 \leq 0.01\,[\mathrm{m}]$ and $d(z(i\,\Delta t)) > 0.3\,[\mathrm{m}]$, i. e., E is said to escape in the simulation [cp. Eq. (5)]; **(iii)** $[x_E(i\,\Delta t)]^2 + [y_E(i\,\Delta t) - 6]^2 \leq 0.01\,[\mathrm{m}]$ and $d(z(i\,\Delta t)) > 0.3\,[\mathrm{m}]$, i. e., E is said to escape in the simulation here, too [cp. Eq. (6)]; **(iv)** $t = i\,\Delta t > 10.0\,[\mathrm{s}]$, i. e., the simulation exhibits irregular behavior of both players. Additionally, controls violating the state constraints (1) and (2) have been suppressed and replaced by controls that are admissible with respect to the state constraints. In particular, if player E has reached one of the corners, E has been forced to stay there for the duration of the game.

Figures 14 and 15 present the results of the neural network controller for the 273 trajectories due to the initial condition $(x_P(0), y_P(0)) = (9, 5)$ for P and are fairly similar to those shown in Figs. 11 and 12. Moreover, various simulations in the escape zone S_E and in the capture zone S_P are shown in Figs. 16–21. Both P and E use the controls provided by the easy-to-evaluate multivariate transformation NN, except when $d(z(t)) \leq 1.0\,[\mathrm{m}]$. In this case, tail-chase pursuit has been employed for P. The final time t_f is indicated in the boxes. These simulations demonstrate the good performance of the controller. A typical example of a failure is given by Fig. 21. There, the optimal capture trajectory terminates at the wall $x = 10$. The vicinity of a dispersal surface causes the simulation to fail.

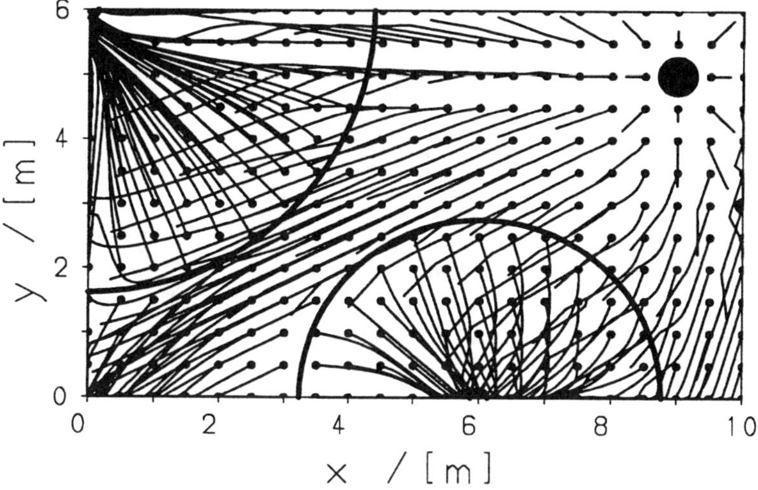

Figure 14: Trajectories of E with the controls of both players computed by the neural networks for $(x_P(0), y_P(0)) = (9, 5)$.

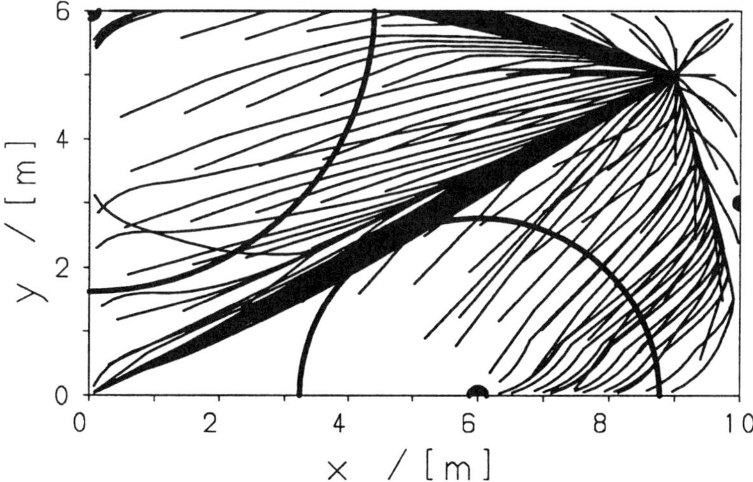

Figure 15: Trajectories of P with the controls of both players computed by the neural networks for $(x_P(0), y_P(0)) = (9, 5)$.

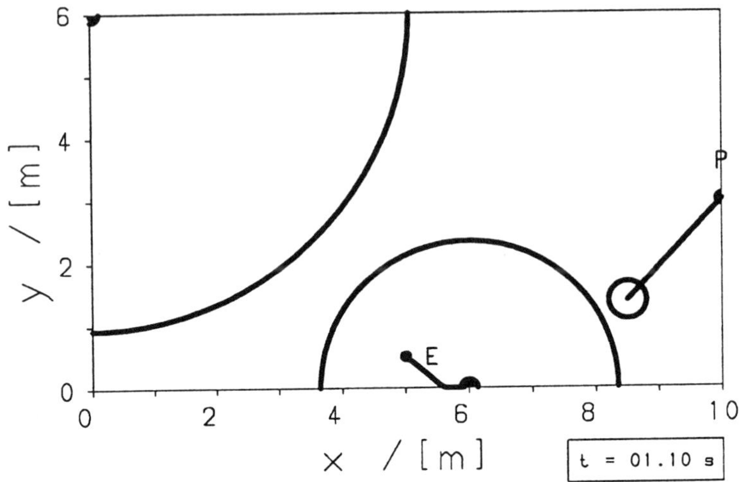

Figure 16: E can escape. Both players run to the nearest hole.

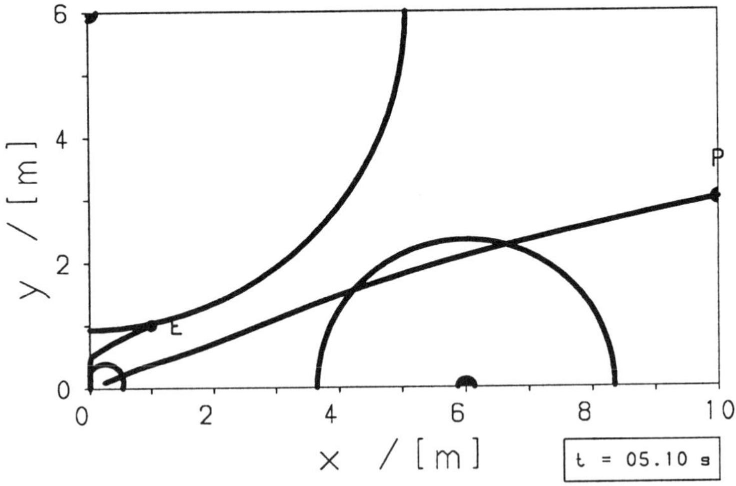

Figure 17: E cannot escape. Capture occurs in the corner next to E.

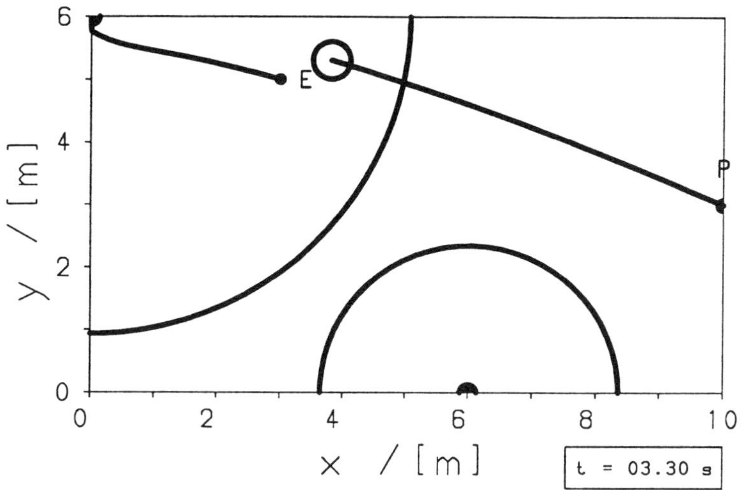

Figure 18: *E* can escape. Both players run to the hole next to *E*.

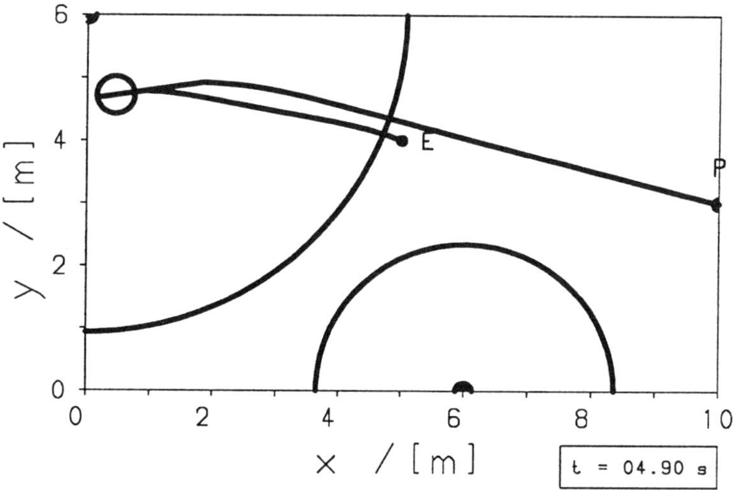

Figure 19: *E* cannot escape. Here, *P* uses final tail-chase pursuit.

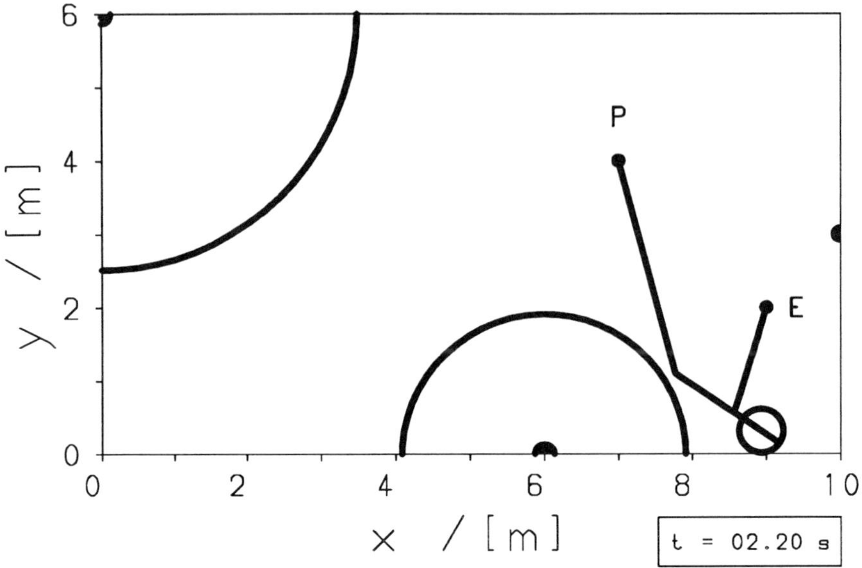

Figure 20: E cannot escape. Capture occurs in the interior of **S**.

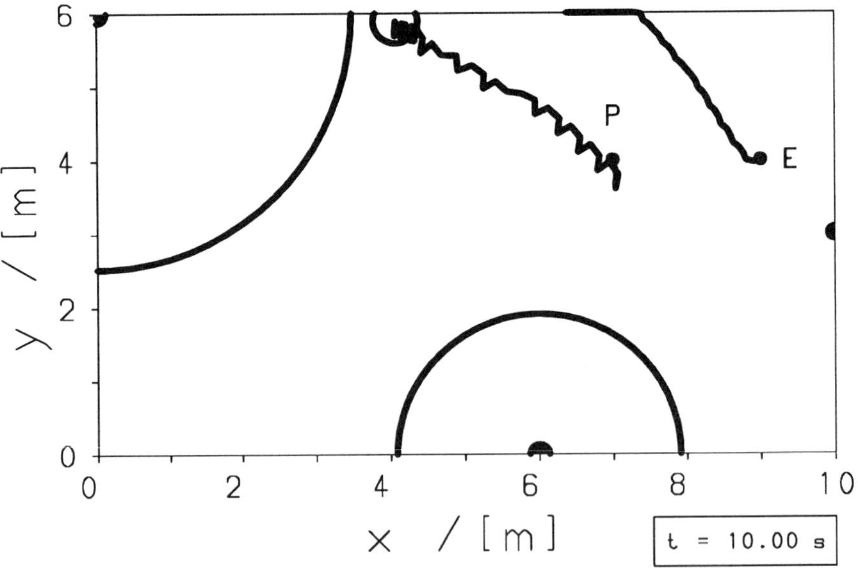

Figure 21: E would not be able to escape if P pursues optimally. Failure occurs because of the near dispersal surface.

5 Conclusions

The generalization of the neural-network-based approach presented in this paper, its real-time applicability, and its convenient easy-to-use black-box property make this approach very attractive for applications to more complicated differential games. Appropriate candidates are, for example, game-against-nature problems, such as the reentry of a space-shuttle orbiter under atmospheric uncertainty [6], [7], [8], or complex pursuit-evasion game problems, such as air combat problems [9], [21], [22]. If such problems of higher dimension are to be investigated, the following open problems have to be solved or have to be scrutinized: **(i)** how can a large number of optimal trajectories be efficiently computed, so that they span the relevant regions of the state space appropriately, e. g., without leaving holes; **(ii)** how can the topology and the size of the neural networks be designed appropriately, i. e., finding the right balance between a network too inadequate for the complexity of the problem and a network leading to overparametrization, which, in both cases, yields poor approximations; **(iii)** how can the training of the networks be performed efficiently on high performance computers; **(iv)** how can the global performance of the controller be evaluated efficiently, since neural network techniques are inherently experimental.

To summarize neural networks can only be the second choice in cases where methods more mathematically involved are not available or not applicable. This might be just the case in differential games.

Acknowledgements

The authors would like to thank Professor R. Bulirsch of Munich University of Technology, Department of Mathematics, and Professor B. Schürmann of Siemens AG, Corporate Research and Development, Munich, for their encouragement. This work has been partly granted by the German National Science Foundation within the Project "Applied Optimization and Control", and by FORTWIHR, the Bavarian Consortium on High Performance Scientific Computing.

REFERENCES

[1] ANDERSON, B. D. O., *Exponential Stability of Linear Equations Arising in Adaptive Identification*, IEEE Transactions on Automatic Control **22**, 1977, 83–88.

[2] BAŞAR, T. , and OLSDER, G. J., *Dynamic Noncooperative Game Theory*, Academic Press, London, 1982, 2nd ed., 1994.

[3] BENVENISTE, A., METIVIER, M., and PRIOURET, P., *Algorithmes adaptatifs et approximations stochastique*, Masson, Paris, 1987.

[4] BLUM, E. K., and LI, L. K., *Approximation Theory and Feedforward Networks*, Neural Networks **4**, 1990, 511–515.

[5] BREAKWELL, J. V., *Zero-Sum Differential Games with Terminal Pay-off*, in: P. Hagedorn, H. W. Knobloch, and G. J. Olsder (Eds.), Differential Games and Applications, Lecture Notes in Control and Information Sciences **3**, Springer, Berlin, 1977, 70–95.

[6] BREITNER, M. H., *Construction of the Optimal Feedback Controller for Constrained Optimal Control Problems with Unknown Disturbances*, in: R. Bulirsch, and D. Kraft (Eds.), Computational Optimal Control, International Series of Numerical Mathematics **115**, Birkhäuser, Basel, 1994, 147–162.

[7] BREITNER, M. H., *Real-Time Capable Approximation of Optimal Strategies in Complex Differential Games*, in: M. Breton, G. Zaccour (Eds.), Proceedings of the Sixth International Symposium on Dynamic Games and Applications, St-Jovite, Québec, GERAD, École des Hautes Études Commerciales, Montreal, 1994, 370–374.

[8] BREITNER, M. H., and PESCH H. J., *Reentry Trajectory Optimization under Atmospheric Uncertainty as a Differential Game*, in: T. Başar, and A. Haurie (Eds.), Advances in Dynamic Games and Applications, Annals of the International Society of Dynamic Games **1**, Birkhäuser, Boston, 1994, 70–88.

[9] BREITNER, M. H., PESCH, H. J., and GRIMM, W., *Complex Differential Games of Pursuit-Evasion Type with State Constraints, Part 1: Necessary Conditions for Optimal Open-Loop Strategies, Part 2: Numerical Computation of Optimal Open-Loop Strategies*, Journal of Optimization Theory and Applications **78**, 1993, 419–441, 443–463.

[10] BRYSON, A. E., and HO, Y.-C., *Applied Optimal Control*, Hemisphere, New York, 1975.

[11] FINNOFF, W., *Diffusion Approximations for the Constant Learning Rate Backpropagation Algorithm and Resistance to Local Minima*, Neural Computation **6**, 285–295, 1994.

[12] GABLER, I., *Numerische Berechnung optimaler Strategien eines Differentialspiels und Approximation durch Neuronale Netze*, Diploma thesis, Department of Mathematics, Munich University of Technology, Munich, 1993.

[13] GABLER, I., MIESBACH, S., BREITNER, M. H., and PESCH, H. J., *Synthesis of Optimal Strategies for Differential Games by Neural Networks*, Report No. 468, Deutsche Forschungsgemeinschaft, Schwerpunkt "Anwendungsbezogene Optimierung und Steuerung", Department of Mathematics, Munich University of Technology, Munich, 1993.

[14] GOH, C. J., and EDWARDS, N. J., *Feedback Control of Minimum-Time Optimal Control Problems Using Neural Networks*, Optimal Control Applications & Methods **14**, 1993, 1–16.

[15] HERTZ, J., KROGH, A., and PALMER, R. G., *Introduction to the Theory of Neural Computation*, Addison-Wesley, Redwood City, 1991.

[16] HORNIK, K., STINCHCOMBE, M., and WHITE, H., *Multi-Layer Feedforward Networks are Universal Approximators*, Neural Networks **2**, 1989, 359–366.

[17] ISAACS, R., *Differential Games*, John Wiley & Sons, New York, 1965, Krieger, New York, 1975.

[18] JÄRMARK, B., and BENGTSSON, H., *Near-Optimal Flight Trajectories Generated by Neural Networks*, Internal Report, Saab/Scania Aircraft Division, Linköping, 1989.

[19] KOSLIK, B., PESCH, H. J., BREITNER, M. H., and VON STRYK, O., *Optimal Control of a Complex Concern Model by Direct and Indirect Optimization Methods*, to appear in: H. Engl, and H. Neunzert (Eds.), Proceedings of the 8th Conference of the European Consortium for Mathematics in Industry, ECMI Series, Teubner, Stuttgart, 1995.

[20] KUGELMANN, B., and PESCH, H. J., *New General Guidance Method in Constrained Optimal Control, Part 1: Numerical Method, Part 2: Application to Space Shuttle Guidance*, Journal of Optimization Theory and Applications **67**, 1990, 421–435, 437–446.

[21] LACHNER, R., BREITNER, M. H., and PESCH H. J., *Three-Dimensional Air Combat: Numerical Solution of Complex Differential Games*, in: G. J. Olsder (Ed.), New Trends in Dynamic Games and Applications, this volume, Birkhäuser, Boston, 1995, 165–190.

[22] LACHNER, R., BREITNER, M.H., and PESCH H.J., *Optimal Strategies of a Complex Pursuit-Evasion Game*, to appear in Journal of Computing and Information, 1995.

[23] MIESBACH, S., *Bahnführung von Robotern mit Neuronalen Netzen*, PhD thesis, Department of Mathematics, Munich University of Technology, Munich, 1995.

[24] MIESBACH, S., and SCHÜRMANN, B., *Wenn Roboter arbeiten lernen: Ideen und Methoden der Neuroinformatik zur Regelung und Steuerung*, Informationstechnik **33**, 1991, 300–309.

[25] MILLER, W. T., SUTTON, R. S., and WERBOS, P. J. (Eds.), *Neural Networks for Control*, MIT Press, Cambridge, 1990.

[26] NGUYEN, D., and WIDROW, B., *The Truck Backer-Upper: An Example of Self-Learning in Neural Networks*, in: W. T. Miller, R. S. Sutton, and P. J. Werbos (Eds.), Neural Networks for Control, The MIT Press, Cambridge, 1990, 287–299.

[27] PESCH, H. J., *Real-Time Computation of Feedback Controls for Constrained Optimal Control Problems, Part 1: Neighboring Extremals, Part 2: A Correction Method Based on Multiple Shooting*, Optimal Control Applications & Methods **10**, 1989, 129–145, 147–171.

[28] PESCH, H. J., *Offline and Online Computation of Optimal Trajectories in the Aerospace Field*, in: A. Miele and A. Salvetti (Eds.), Applied Mathematics in Aerospace Science and Engineering, Plenum, New York, 1994, 165–219.

[29] PESCH, H. J., *Solving Optimal Control and Pursuit-Evasion Game Problems of High Complexity*, in: R. Bulirsch, and D. Kraft (Eds.), Computational Optimal Control, International Series of Numerical Mathematics **115**, Birkhäuser, Basel, 1994, 43–64.

[30] *Proceedings of the NIPS Neural Information Processing Systems*, Morgan Kaufmann, San Mateo, CA, annually since 1987.

[31] RITTER, H., MARTINETZ, T., and SCHULTEN, K., *Neuronale Netze: Eine Einführung in die Neuroinformatik selbstorganisierender Netzwerke*, Addison-Wesley, Bonn, 1991.

[32] ROJAS, R., *Theorie der neuronalen Netze*, Springer, Berlin, 1993.

[33] ROSENBLATT, F., *Principles of Neurodynamics*, Spartan, New York, 1962.

[34] SAARINEN, S., BRAMLEY, R., and CYBENKO, G., *Ill-Conditioning in Neural Network Training Problems*, SIAM Journal of Scientific and Computing **14**, 693–714, 1993.

[35] STOER, J., and BULIRSCH, R., *Introduction to Numerical Analysis*, Springer, New York, 2nd ed., 1993.

[36] VON STRYK, O., *Numerical Solution of Optimal Control Problems by Direct Collocation*, in: R. Bulirsch, A. Miele, J. Stoer, and K.-H. Well (Eds.), Optimal Control, International Series of Numerical Mathematics **111**, Birkhäuser, Basel, 1993, 129–143.

[37] VON STRYK, O., and BULIRSCH, R., *Direct and Indirect Methods for Trajectory Optimization*, Annals of Operations Research **37**, 1992, 357–373.

A Linear Pursuit-Evasion Game with a State Constraint for a Highly Maneuverable Evader [*]

Y. Lipman[†] and J. Shinar[‡]
Faculty of Aerospace Engineering
Technion, Israel Institute of Technology
Haifa, Israel

Abstract

Motivated to analyze anti-ballistic point defense - as well as ship defense scenarios - against highly maneuverable attacking missiles, a zero-sum pursuit-evasion game with a state constraint imposed on the evader is formulated. The performance index of the game is the miss distance. A perfect information structure, the worst case from the point of view of the defense, is assumed. The analysis is performed using a planar kinematic model linearized around a nominal *collision course*, leading to a fixed horizon, terminal pay-off linear differential game with bounded controls. Compared to the unconstrained game, the existence of the state constraint creates new singular phenomena which substantially modify the solution. The results of the analysis serve as the basis for the investigation of challenging realistic (imperfect information) future defense problems.

1   Introduction

This study is motivated by two future point defense scenarios: one is against a maneuvering tactical ballistic missile and the other is ship defense from anti-ship missiles. In both scenarios the objective of the defense system is to launch guided missiles to intercept a high speed and highly maneuverable attacking threat at a safe distance from the target. These scenarios can be formulated as zero-sum pursuit-evasion games with a state constraint imposed on the evader (the attacking missile). Respecting this constraint allows the attacking missile to hit the designated target if the interception fails. Though in a real scenario the attacking missile has no information on the interceptor, in this study a perfect information structure (the worst case from the point of view of the defense) is assumed. The investigation of the

---

[*] Research partly supported by AFOSR Grant No. F49620-92-J-0527.
[†] Doctoral Student
[‡] Professor, Max and Lottie Dresher Chair in Aerospace Performance and Propulsion

realistic imperfect information scenario requires the solution of the perfect information game as a basis.

The present analysis uses a planar constant speed kinematic model, which allows trajectory linearization around a nominal collision course. In this formulation the duration of the game is fixed. In the mathematical model of the resulting linear game it is assumed that both players have bounded controls, the pursuer (the interceptor missile) is represented by a first-order transfer function, while the evader is assumed to have instantaneous dynamics. The performance index of the game is the miss distance, to be minimized by the pursuer and maximized by the evader.

The unconstrained version of this linear game was formulated and solved in the past [1]. That solution, motivated to analyze anti-aircraft missile engagements (air-to-air and surface-to-air), served as the basis of several further studies [2]–[5], such as extensions to three dimensional and imperfect information analysis. The main parameter of the game solution is the pursuer/evader maneuver ratio (the ratio of the maximum lateral acceleration of the pursuer and the maximum lateral acceleration of the evader). If this ratio is sufficiently high the solution indicates that for most practical initial conditions a very small (but non zero) miss distance can be guaranteed. However, if the maneuverabilities of the evader and the pursuer are similar (moreover, for an evader with a maneuverability advantage) the game solution predicts very large miss distances. In anti-aircraft engagements a considerable maneuver advantage of the interceptor missile is easily achieved. Indeed, in most anti-aircraft missile designs the missile/target maneuver ratio has been kept at the order of 3 or higher. In future anti-ship defense scenarios, as well as in anti-ballistic missile defense, such a favorably high maneuver ratio cannot be guaranteed. In effect a maneuver ratio of the order of unity, or even smaller, is likely to be expected.

The objective of the present study is to derive the complete game solution of the perfect information linearized game with the state constraint, representing a point defense missile/anti-missile scenario. Such a solution consists of the decomposition of the game space to regions of different optimal strategies and outcomes. As a result, one can evaluate the effect of the state constraint, - imposed on the evader by the point defense scenario, on the game solution. The attention in this study is concentrated on the range of maneuver ratios near to unity, which may be expected in future anti-ballistic and ship defense scenarios. In order to obtain results of generalized validity, the analysis is carried out using non-dimensional variables.

2 Problem formulation

The analysis of the perfect information point defense scenario is based on the following set of assumptions:

(A-1). The engagement between the attacking missile "A"(evader) and the interceptor missile "D" takes place in a plane.
(A-2). The engagement starts ($t = 0$) when the point defense system launches "D" against "A".
(A-3). Both missiles have respective constant velocities V_j and bounded lateral acceleration capabilities $|a_j| \leq (a_j)_{\max}$, $(j = A, D)$.
(A-4). The velocity vectors of both missiles are pointed near to the line of sight, allowing trajectory linearization.
(A-5). The target "T", the object of the point defense, is collocated with the initial position of "D". Moreover, it is either stationary, or its velocity is negligible compared to each of the missile velocities.
(A-6). "A" has instantaneous dynamics, while the dynamics of "D" is represented by a first order transfer function with the time constant τ.
(A-7). Both missiles have perfect information on each other, as well as on the position of "T".
(A-8). The interception of "A" by "D" must take place between the minimum and maximum ranges of the point defense system.
(A-9). If the interception fails, "A" turns to hit and destroy "T".

The origin of the engagement's coordinate system is located at the target "T" and its x axis is aligned with the initial line of sight. The assumptions (A-3) and (A-4) imply that the velocity components parallel to the line of sight are approximately constant and as a consequence the final time of the engagement t_f and the interception range x_f are determined by initial range x_0. Therefore, the equations of motion in the x direction are trivially solved:

$$
\begin{aligned}
x_A(t) &= x_0 - V_A t & (1) \\
x_D(t) &= V_D t & (2) \\
\Delta x(t) &\triangleq x_A(t) - x_D(t) = x_0 - (V_A + V_D)t & (3) \\
t_f &= x_0/(V_A + V_D) & (4) \\
x_f &= x_0 - V_A t_f = V_D t_f & (5)
\end{aligned}
$$

Based on assumptions (A-4) and (A-6) the equations of motion normal to the initial line of sight and the corresponding initial conditions are

$$
\begin{aligned}
\dot{y}_A &= y_1 & y_A(0) &= 0 & (6) \\
\dot{y}_1 &= a_A^c & y_1(0) &= V_A \alpha_0 & (7) \\
\dot{y}_D &= y_2 & y_D(0) &= 0 & (8)
\end{aligned}
$$

$$\dot{y}_2 = y_3 \qquad y_2(0) = V_D\delta_0 \tag{9}$$

$$\dot{y}_3 = (a_D^c - y_3)/\tau \quad y_3(0) = 0 \tag{10}$$

where a_A^c and a_D^c are the commanded lateral accelerations of "A" and "D" respectively.

$$a_A^c = (a_A)_{\max}\, v \quad |v| \le 1 \tag{11}$$

$$a_D^c = (a_D)_{\max}\, u \quad |u| \le 1 \tag{12}$$

The non zero initial conditions $V_A\alpha_0$ and $V_D\delta_0$ represent the respective initial velocity components not aligned with the line of sight. By assumption (A-4) these components are small compared to the components along the line of sight.

The performance index of the engagement is the *miss distance*:

$$J = |y_A(t_f) - y_D(t_f)| \tag{13}$$

The objective of "D" is to minimize J, while "A" wants to maximize it, but still being able to hit "T". This constraint is expressed by

$$|y_A(t_f) + t_c\dot{y}_A(t_f)| \le (a_A)_{\max}\, t_c^2/2 \tag{14}$$

where t_c is the time needed for "A" to reach "T" after the interception:

$$t_c \triangleq x_f/V_A = t_f(V_D/V_A) \tag{15}$$

The linear differential equations (6)–(10) can be reduced to a set of two only, by using the following non dimensional variables:

$$\nu \triangleq V_D/V_A \tag{16}$$

$$\mu \triangleq (a_D)_{\max}/(a_A)_{\max} \tag{17}$$

$$\Theta \triangleq (t_f - t)/\tau\,, \qquad \Theta(0) = t_f/\tau \triangleq \Theta_0 \tag{18}$$

$$\Theta_c \triangleq t_c/\tau = \nu\Theta_0 \tag{19}$$

$$Z(\Theta) \triangleq \frac{(y_A - y_D) + \tau\Theta(y_1 - y_2) - \tau^2(e^{-\Theta} + \Theta - 1)y_3}{\tau^2(a_A)_{\max}} \tag{20}$$

$$W(\Theta) \triangleq \frac{y_A + \tau(\Theta + \Theta_c)y_1}{\tau^2(a_A)_{\max}} \tag{21}$$

$Z(\Theta)$ is the normalized *zero-effort miss distance* in the "D" against "A" engagement, while $W(\Theta)$ has a similar interpretation between "A" and a static "T". Substitution of the new variables (16)–(21) into (6)–(10) yields

$$\frac{dZ}{d\Theta} = \mu(e^{-\Theta} + \Theta - 1)u - \Theta v \quad Z(\Theta_0) = Z_0 \tag{22}$$

$$\frac{dW}{d\Theta} = -(\Theta_c + \Theta)v \qquad W(\Theta_0) = W_0 \tag{23}$$

with the following normalized initial conditions

$$Z_0 \;=\; (V_A\alpha_0 - V_D\delta_0)\Theta_0/\tau(a_A)_{\max} \tag{24}$$

$$W_0 \;=\; V_A\alpha_0(\Theta_c + \Theta_0)/\tau(a_A)_{\max} \tag{25}$$

The payoff function (13) and the constraint (14) can be written in the following non-dimensional form

$$J = |Z(\Theta = 0)| \overset{\Delta}{=} |Z_f| \tag{26}$$

$$|W(\Theta = 0)| \overset{\Delta}{=} |W_f| \leq \Theta_c^2/2 \tag{27}$$

Whether a trajectory reaches the constraint or not depends in general on the initial conditions. As long as a trajectory satisfies the constraint as a strict inequality, the solution is identical with the solution of the unconstrained game [1]. Thus, It seems useful to recall its solution prior to a detailed analysis of the game with state constraint.

3 Unconstrained Game Solution

In the non-dimensional formulation of the unconstrained game [1] there is only a single state variable Z and its dynamics are described by (22). The solution is governed by a single parameter μ, defined in (17). The Θ, Z game space is decomposed into two regions D_1 and its complement D_0 (see Fig. 1).

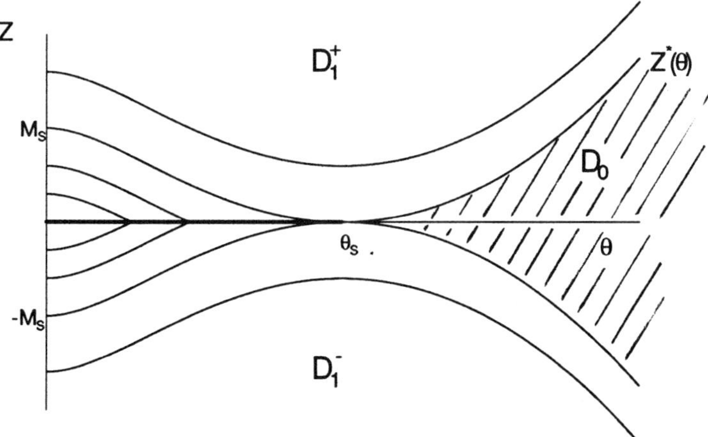

Figure 1: The unconstrained game solution.

The first region is defined by

$$D_1 = \{\Theta, Z \mid \ \Theta \le \Theta_s \ \cup \ |Z| \ge Z^*(\Theta)\} \tag{28}$$

where

$$Z^*(\Theta) = \Theta_s + (\mu - 1)(\Theta^2 - \theta_s^2)/2 - \mu(e^{-\Theta} + \Theta - 1) \tag{29}$$

and $\Theta_s(\mu)$ is the non vanishing solution of the equation

$$\Theta = \mu(e^{-\Theta} + \Theta - 1) \overset{\Delta}{=} \mu\Psi(\Theta) \tag{30}$$

The optimal strategies in D_1 are

$$v^*(\Theta, Z) = u^*(\Theta, Z) = sign\{Z\}, \quad Z \ne 0 \tag{31}$$

The segment $0 < \Theta \le \Theta_s$ of the Θ axis ($Z = 0$) is a *dispersal line* of "A" (the evader), which may select either a positive or negative maneuver. "D" (the pursuer) must maneuver to the same direction. The optimal outcome, the *value* of the game is a unique function of the initial condition (Θ_0, Z_0) in D_1.

In D_0 the optimal strategies are arbitrary. All the trajectories starting in this region must go through the same point, $Z = 0$ at $\Theta = \Theta_s$, which belongs to the *dispersal line*. As a consequence, the optimal outcome, called the *value* of the game, is constant (a function of μ only) for all these trajectories:

$$J_0^* = \Theta_s - (\mu - 1)\Theta_s^2/2 \overset{\Delta}{=} M_s(\mu) \tag{32}$$

Since in most anti-aircraft missile engagements the initial conditions are in D_0, the major conclusion drawn from this analysis has been that if μ is sufficiently large (at least $\mu > 2$), then the *guaranteed miss distance*, expressed in its non-dimensional form by (32), is negligibly small. This conclusion has implied a practical victory of guided anti-aircraft interceptor missiles in perfect information scenarios.

However, in future anti-ship defense scenarios, as well as in anti-ballistic missile defense, maneuver ratios of the order of unity are likely to be expected. Therefore the results of [1] have to be reexamined for smaller values of μ.

In Figs. 2 and 3 the values of Θ_s and M_s are shown for the range of $0.8 < \mu < 1.5$ and the following initial conditions $Z_0 = 0.0$ (*collision course*), $\Theta_0 = 50$.

Note that according to (30) for $\mu \le 1$ the value of Θ_s becomes infinite, which means that the entire game space is D_1. As a consequence, equation (32), predicting an infinite *miss distance*, is not valid. The finite value of the non-dimensional *miss distance*, displayed in Fig. 3, is obtained by integrating (22) with the appropriate initial conditions. These results show extremely large *miss distances*, clearly unacceptable for an effective point defense.

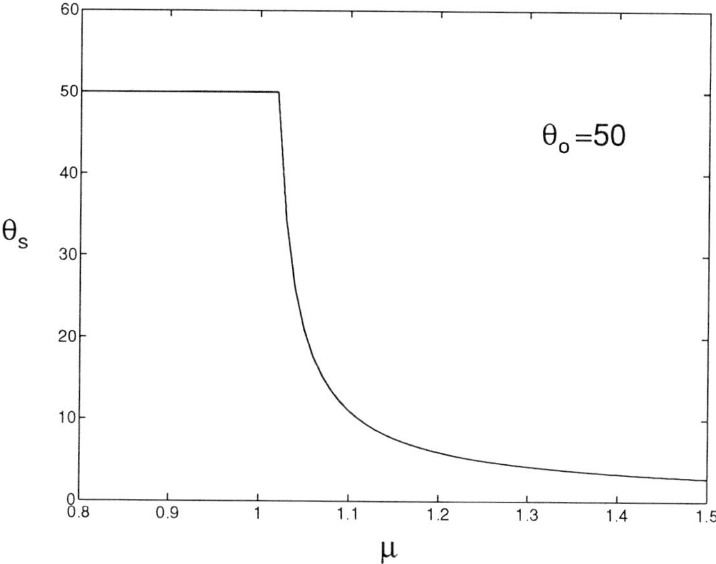

Figure 2: Normalized critical time to go (Θ_s) in the unconstrained game.

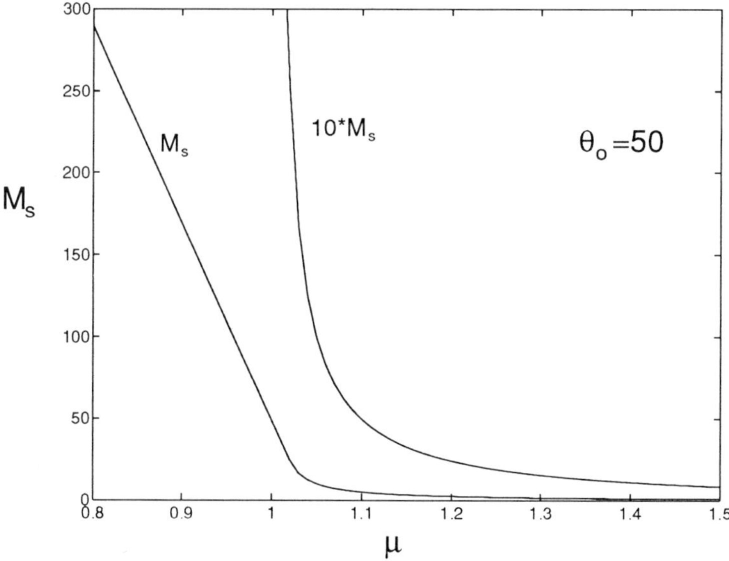

Figure 3: Normalized guaranteed miss distance (M_s) in the unconstrained game.

4   Constrained Game: Primary Solution

The game with constraint, the very subject of the present paper, has two state variables: Z and W (Θ being the independent variable). The non-dimensional formulation presented in (22)–(27) indicates dependence on two parameters, μ (the pursuer/evader maneuver ratio) and Θ_c (the normalized time of flight of the evader for reaching its target after an unsuccessful interception). Since Θ_c defined by (19) is the product of ν and Θ_0, for any given initial condition (Θ_0, Z_0, W_0) the game solution is determined by two independent parameters μ and ν, both having a physical significance.

For those initial conditions for which the constraint imposed by the point defense scenario does not become active, the solution of [1] described in the previous section remains valid. Such unconstrained optimal trajectories, though they are inherent parts of the solution, do not require additional analysis. Therefore, in the sequel it is assumed that along an optimal trajectory the constraint becomes active (at least at the end) and the solution of the corresponding pursuit-evasion game is derived. The game space, as well as the set of admissible initial conditions, is restricted by the state constraint

$$|W| \le W_c(\Theta), \quad \Theta \in [0, \Theta_0] \tag{33}$$

where

$$W_c(\Theta) \triangleq (\Theta_c + \Theta)^2/2 \tag{34}$$

is obtained by integrating (23) backwards from (27), using v_c the appropriate control on the constraint:

$$v_c = -sign\ W \tag{35}$$

This constraint is represented by two surfaces C^+ and C^- according to the sign of W. It is easy to see that if a trajectory reaches one of these surfaces it must stay on it until the game ends, satisfying the terminal constraint (27) as an equality. This observation is used in the derivation of the primary solution of the game.

The Hamiltonian of the game is

$$- H = p_\theta + p_z\{\mu\Psi(\Theta)u - \Theta v\} - p_w(\Theta + \Theta_c)v + \lambda\{W_c(\Theta) - |W|\} \tag{36}$$

where $\lambda \ne 0$ only when the trajectory is on the constraint. (The minus sign in the left hand side of (36) is the consequence of Θ being a normalized *inverse* time.)

The costate variables p_θ, p_z and p_w are the components of the vector **grad** J^* (whenever such gradient exists) and are obtained by solving the adjoint differential equations with the corresponding transversality conditions.

$$\frac{dp_\theta}{d\Theta} = \partial H/\partial\Theta\ , \qquad\qquad p_\theta(\Theta = 0) = p_{\theta f} \tag{37}$$

$$\frac{dp_z}{d\Theta} = \partial H/\partial Z = 0 \ , \qquad\qquad p_z(\Theta = 0) = \partial J/Z_f = sign\ Z_f \quad (38)$$

$$\frac{dp_w}{d\Theta} = \partial H/\partial W = -\lambda\ sign\ W \ , \qquad p_w(\Theta = 0) = p_{wf} \quad (39)$$

p_{wf} being an undetermined constant on each optimal trajectory reaching the constraint (because $|W_f| = \Theta_c^2/2$). The optimal strategies of the players are determined by

$$u^* \quad = \quad arg \min_u H = sign\ p_z \quad (40)$$

$$v^* \quad = \quad arg \max_v H = sign\ S_v \overset{\Delta}{=} sign\ [p_z\Theta + p_w(\Theta_c + \Theta)] \quad (41)$$

Substitution of (38) yields

$$u^* \quad = \quad sign\ Z_f \quad (42)$$

$$v^* \quad = \quad sign\ S_v = sign\ [\Theta(sign\ Z_f + p_w) + p_w\Theta_c] \quad (43)$$

Applying these strategies, optimal trajectories can be generated by retrograde integration from any endpoint (Z_f, W_f). In order for a trajectory to leave the constraint (in retro sense) one must have

$$v^*(\Theta = 0) = sign\ W_f \quad (44)$$

requiring

$$p_{wf} = 0 \quad (45)$$

Optimal trajectories may also remain (in retro sense) on one of the constraint surfaces (C^+ or C^-) creating in each surface two trajectory families, with $Z_f > 0$ ($u^* = 1$) and $Z_f < 0(u^* = -1)$, as shown for C^+ in Fig. 4. Note that in the zone between the two trajectory families with uniquely defined u^* the optimal strategy of the pursuer is arbitrary.

At any point (Θ^+, Z) along those "constraint trajectories" that are characterized by sign $Z_f = sign\ W = 1$, a yet unconstrained optimal trajectory reaches (in forward time) the surface C^+. At this point the switch function S_v changes sign, implying

$$p_w(\Theta^+) = -[\Theta^+/(\Theta_c + \Theta^+)] \quad (46)$$

which allows us to define the value of λ on the constraint by comparing the derivative of (46) with (39)

$$\lambda = \Theta_c/(\Theta_c + \Theta)^2 = \lambda(\Theta) \quad (47)$$

Since along the unconstrained part of each trajectory $\lambda = 0$, the value of p_w computed from (46) remains constant for $\Theta \in [\Theta^+, \Theta_0]$ and consequently

$$u^* = v^* = sign\ Z_f \quad (48)$$

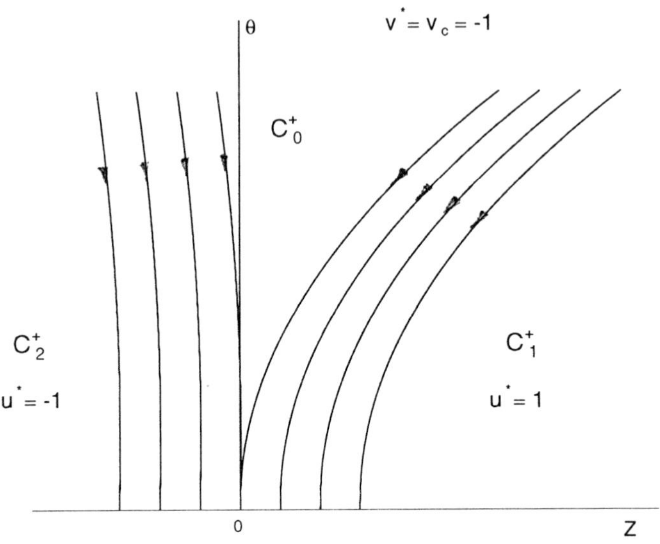

Figure 4: Optimal trajectories on the constraint C^+ .

Similar trajectory families can be generated by leaving (in retro sense) the other constraint surface C^- at Θ^-. A part of these trajectories can be integrated backwards without intersecting other optimal trajectories creating two fields of primary trajectories D^+ and D^-.

5   Dispersal Surface and Aggregation Line

However, other optimal trajectories generated by retrograde integration from the surface C^+, may intersect at some point (Θ_j, Z_j, W_j) another trajectory emanating (in retro sense) from C^-, leading to the same normalized miss distance $(|Z_f|)$ but generated by a different optimal strategy pair. These intersections generate a *dispersal surface* (DS) of the game. This DS is dominated by the evader, who decides at (Θ_j, Z_j, W_j) whether to go (in forward time) towards C^+ by using $v^* = 1$ or towards C^- by using $v^* = -1$. Each choice of the evader must be followed by a similar strategy of the pursuer $(u^* = v^*)$. Starting at (Θ_j, Z_j, W_j) the respective values of Θ^+ and Θ^- can be computed by integrating (23), which leads to

$$W(\Theta^+) \; \overset{\Delta}{=} \; W^+ = W_j + (\Theta_c + \Theta_j)^2/2 - (\Theta_c + \Theta^+)^2/2 > 0 \quad (49)$$

$$W(\Theta^-) \triangleq W^- = W_j - (\Theta_c + \Theta_j)^2/2 + (\Theta_c + \Theta^-)^2/2 < 0 \qquad (50)$$

Equating W^+ and W^- with W_c^+ and with W_c^- (the values of W on C^+ and C^- respectively)

$$W_c^+ \triangleq (\Theta_c + \Theta^+)^2/2 \qquad (51)$$

$$W_c^- \triangleq -(\Theta_c + \Theta^-)^2/2 \qquad (52)$$

yields (by using (34) and omitting the subscript "j")

$$\begin{aligned}
\Theta^+(\Theta, W) \triangleq \Theta^+ &= -\Theta_c + \sqrt{W + (\Theta_c + \Theta)^2/2} \\
&= -\Theta_c + \sqrt{W + W_c(\Theta)} \qquad (53) \\
\Theta^-(\Theta, W) \triangleq \Theta^- &= -\Theta_c + \sqrt{-W + (\Theta_c + \Theta)^2/2} \\
&= -\Theta_c + \sqrt{-W + W_c(\Theta)} \qquad (54)
\end{aligned}$$

If $W = 0$ both Θ^+ and Θ^- vanish for

$$\Theta = \Theta_N \triangleq (\sqrt{2} - 1)\Theta_c = 0.4142\Theta_c \qquad (55)$$

Depending on the value of W, either Θ^+ or Θ^- may become negative, which implies that the respective constraint is not reached before the end of the game ($\Theta = 0$). If both are negative then the optimal trajectories are unconstrained and the solution of [1] applies. Therefore in the sequel the following "extended" notations will be used.

$$\Theta^+ = \max\,[0, -\Theta_c + \sqrt{W + W_c(\Theta)}] \qquad (56)$$

$$\Theta^- = \max\,[0, -\Theta_c + \sqrt{-W + W_c(\Theta)}] \qquad (57)$$

The equation of the dispersal surface is obtained by integrating both trajectories that start at (Θ_j, Z_j, W_j) using the different optimal strategy pairs (one with $Z_f^+ \triangleq Z_f > 0$ and the other with $Z_f^- \triangleq Z_f < 0$) leading to

$$Z_f^+ = Z_j + \mu\Psi(\Theta) - (\mu - 1)\Theta^2/2 - (\Theta^+)^2 \qquad (58)$$

$$Z_f^- = Z_j - \mu\Psi(\Theta) + (\mu - 1)\Theta^2/2 + (\Theta^-)^2 \qquad (59)$$

Imposing the equality $Z_f^- = -Z_f^+$ and denoting Z_j as Z_{ds} yields

$$Z_{ds}(\Theta, W) \triangleq Z_{ds} = [(\Theta^+)^2 - (\Theta^-)^2]/2 \qquad (60)$$

The equation of the DS can be thus written as

$$DS(\Theta, Z, W) = Z - \{[\Theta^+(\Theta, W)]^2 - [\Theta^-(\Theta, W)]^2\}/2 = 0 \qquad (61)$$

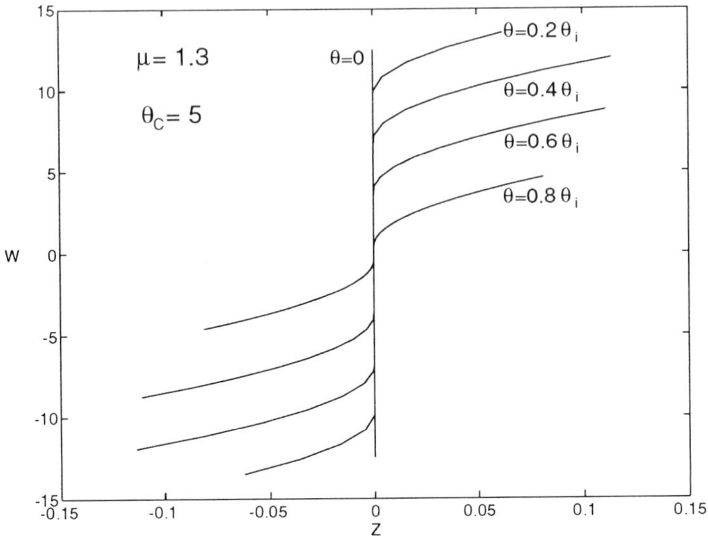

Figure 5: The dispersal surface.

It is easy to see that for $W = 0$ one has $\Theta^+ = \Theta^-$ and consequently $Z_{ds} = 0$.

In Fig. 5 several cuts of the DS are shown at different fixed values of Θ.

The definition of Z_{ds} allows us to conclude that in the primary domain of the game D (the union of D^+ and D^-)

$$sign\ Z_f = sign\ [Z - Z_{ds}] \tag{62}$$

leading to the expression of the optimal strategies in the following feedback form:

$$u^* = sign\ [Z - Z_{ds}] \tag{63}$$
$$v^* = sign\ [Z - Z_{ds}]\ , \quad |W| < W_c(\Theta) \tag{64}$$
$$v^* = -sign\ W\ , \quad |W| = W_c(\Theta) \tag{65}$$

The existence and optimality of the DS have to be validated by testing that the strategy corresponding to the field D^+ (or D^-) indeed leads the state of the game to the same field. This test requires us to satisfy the following inequality on the DS.

$$\max\ \{\Theta^+, \Theta^-\} < \Theta_c[\Theta - \mu\Psi(\Theta)]/[\Theta_c + \mu\Psi(\Theta)] \tag{66}$$

Since for $W > 0$ one has $\Theta^+ > \Theta^-$, on this side of the DS the condition with respect to Θ^+ will be violated first, indicating that optimal trajectories with $u^* = v^* = -1$ reach the surface (in retro sense) tangentially. Similarly, for $W < 0$ the inequality with respect to Θ^- will be first violated by "grazing" optimal trajectories with $u^* = v^* = 1$. All the "grazing" trajectories can be continued by retrograde integration beyond the *DS*.

The ensemble of points for which the inequality (66) becomes an equality defines a line which serves as the boundary of the *DS*. The coordinates of such points can be obtained by transforming (66) to the relevant equality

$$\max\{\Theta^+, \Theta^-\} = \Theta_c[\Theta - \mu\Psi(\Theta)]/[\Theta_c + \mu\Psi(\Theta)] \tag{67}$$

and using (53), (54) and (60). The projections of this *aggregation line (AL)* (so-called for reasons to be explained later) on the planes $(\Theta, W), (\Theta, Z)$ and (Z, W) are shown in Fig. 6 for $\Theta_N \leq \Theta_s$. The *AL* intersects the Θ axis at the point "*I*" $(\Theta_i, 0, 0)$. Note that Θ_i is the largest value of Θ for which the *DS* exists.

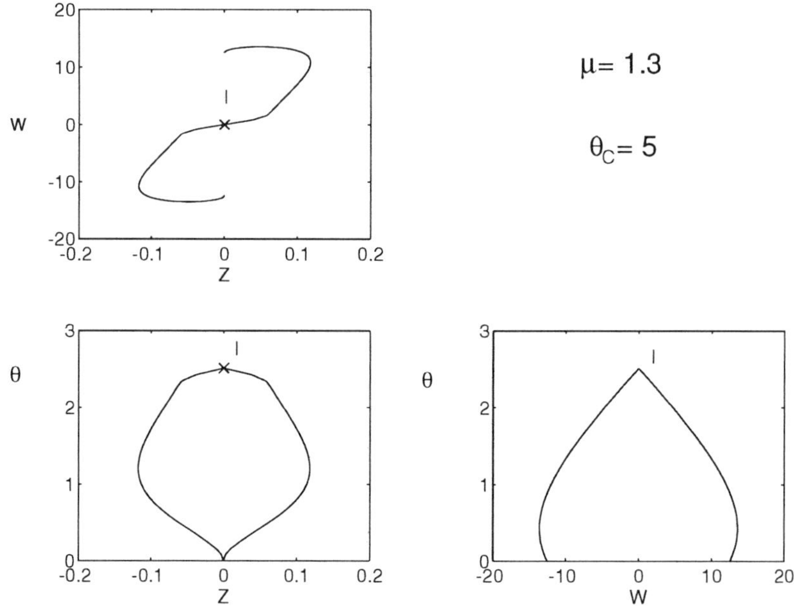

Figure 6: The aggregation line, $\Theta_N \leq \Theta_s$.

The value of Θ_i can be computed from (67) taking into account that $\Theta^+ = \Theta^-$. By equating the two different expressions for Θ^+ from (67) and (53) an expression for W on the *AL* is obtained. Evaluating this expression for $W = 0$ leads to

$$\mu\Psi(\Theta_i) = (\sqrt{2} - 1)\Theta_c = \Theta_N \tag{68}$$

This situation occurs only if $\Theta_N \leq \Theta_s$, $\Theta_s(\mu)$ being defined in (30). In this case it is easy to see that Θ_i satisfies $\Theta_N \leq \Theta_i \leq \Theta_s$. If, according to (53) and (54), $\Theta^+ = \Theta^- = 0$ then (67) is satisfied by $\Theta_i = \Theta_N = \Theta_s$. If $\Theta_N > \Theta_s$, there are points on the AL from which optimal trajectories do not reach the constraint and (67) yields trivially $\Theta_i = \Theta_s$. In such a case there is an entire segment of the AL with $|W(\Theta_s)| \leq W_B$, where W_B is given for $\Theta_N > \Theta_s$ by

$$W_B = [(\Theta_c^2/\Theta_N) + \Theta_s](\Theta_N - \Theta_s)/2 \tag{69}$$

that satisfies (67), as can be seen in Fig. 7

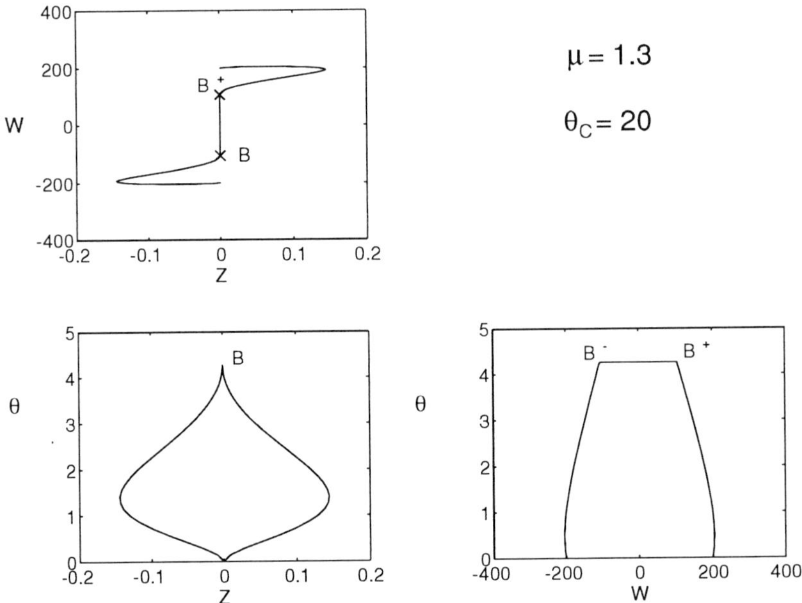

Figure 7: Aggregation line, $\Theta_N > \Theta_s$.

If this occurs, then two families of optimal trajectories from the fields D^+ and D^- respectively will be tangent to each other at $\Theta = \Theta_s$, similarly to the unconstrained game solution [1] shown in Fig. 1.

The ensemble of "grazing" trajectories on both sides of the AL creates two boundary surfaces for the primary domain of the game D, which can be filled by optimal trajectories generated by retrograde integration using the strategies given in (63)–(65). In order to complete the game solution the rest of the game space has also to be analyzed.

6 Equivocal Surface and Region E

The construction of optimal primary trajectories described in the previous sections leaves an open region of the game space, starting at the AL, yet to be filled with optimal trajectories. Since regular (primary) trajectories cannot be constructed in this region, the existence of some singular surface starting at the AL has to be assumed. Based on the fact that along the AL only one trajectory family arrives tangentially, it seems to be useful to assume and verify the existence of an *equivocal surface* (ES) starting there. This surface has to separate two regions of the game space with different optimal strategies, the primary domain D and the region to be called E. When an optimal trajectory reaches an ES, one of the players (the *dominant*) can switch strategy and in this case the trajectory traverses the ES and enters the other region. If the *dominant* player does not switch strategy, the trajectory continues along the ES. The *dominant* player can decide at any point on the ES to switch his strategy and enter the other region. The optimal strategy of the second (non *dominant*) player depends on the decision of the opponent.

In the present game, since the DS is dominated by the evader, it is reasonable to assume the same for the ES. If the assumptions on the existence of an optimal ES are valid, then the evader has the option to lead all trajectories originating in the respective open region E to the *aggregation line*, which explains the origin of this notion. Let us assume the existence of two *equivocal surfaces*, ES^+ and ES^-, leading to parts of the AL with negative or positive values of W respectively. The surface ES^+ separates the open region to be called E^+ and the primary domain D^-, while ES^- separates the open region E^- and the primary domain D^+. Optimal trajectories reach ES^+ with $v^* = 1$ and ES^- with $v^* = -1$. As can be expected, the costate variables p_θ, p_z and p_w become discontinuous along both parts of the ES.

If at any point on ES^- the evader switches strategy, the trajectory traverses the surface to D^+ and continues with $u^* = v^* = 1$ towards the constraint C^+. All such optimal trajectories reach the C^+ surface along the same "constrained trajectory" yielding the same *miss distance*. If the evader continues to play $v^* = -1$ the trajectory will remain on ES^- and move parallel to the surface C^+ until the AL is reached. The same is true on ES^+ with the signs $(+)$ and $(-)$ reversed.

The optimal strategy of the pursuer, keeping the trajectory on ES^-, varies as a function of Θ and W:

$$u^*_{ES^-} = 1 - 2\Theta_c(\Theta - \Theta^+)/[(\Theta_c + \Theta^+)\mu\Psi(\Theta)] \tag{70}$$

This last equation was derived using the Projection Lemma and the other necessary conditions that apply on an ES, as quoted by Bernhard [6]. For infinitely large values of Θ the second term in (70) becomes negligible and

158   *Y. Lipman and J. Shinar*

u^*_{ES-} tends to be $+1$, while on the AL its value (obtained by substituting from (67) the relevant expression for Θ^+) is -1. This indicates that on the AL the ES becomes tangent to the DS. From this point on the optimal trajectory may continue with $u^* = v^* = -1$ along one of the "grazing" primary trajectories towards C^-, as shown schematically in Fig. 8. Thus, the surface ES^- can be generated by backward integration from all points of the AL with $W > 0$, by using $v^* = -1$ and $u^* = u^*_{ES-}$.

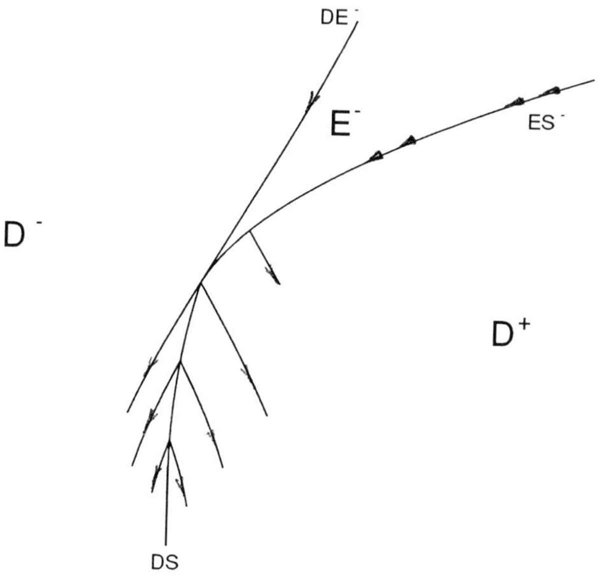

Figure 8: The equivocal surface.

If the pursuer does not follow the optimal (singular) strategy, the trajectory leaves the ES. If on ES^- the pursuer uses $u < u^*_{ES-}$, the trajectory enters the domain D^+ and the resulting *miss distance* will be larger than the value predicted for the ES. If the pursuer uses $u > u^*_{ES-}$, the trajectory moves back to the region E^-, but either returns later to the ES or reaches the AL. A similar description holds for ES^+ with the signs $(+)$ and $(-)$ reversed. Derivation of (70) and verification of equal *miss distances* along all alternative trajectories are detailed in [7].

In the regions E^+ and E^-, the optimal strategy of the evader ($v^* = 1$, or $v^* = -1$) is to reduce $|W|$ and the optimal trajectories follow a surface parallel to the nearby constraint surface. This is also true on the respective ES (serving as one of the boundaries of E) if the evader decides to keep the trajectory there. The very point on the AL (Θ_a, Z_a, W_a) where all these trajectories will finally *"aggregate"* (as well as the resulting value of Z_f) depends only on $W(\Theta)$ being controlled by the evader. The value of Θ_a can be computed by integrating (23) with the optimal strategy of (65) from any point (Θ_k, Z_k, W_k) in E and using (67) together with (53) or (54). This leads to the following implicit equation.

$$|W_k| - (\Theta_c + \Theta_k)^2/2 = (\Theta_c + \Theta_a)^2\{[\Theta_c/\{\Theta_c + \mu\Psi(\Theta_a)\}]^2 - 1\} = f(\Theta_a) \quad (71)$$

Once Θ_a is obtained the values of W_a and Z_a can be found, using (53) or (54) and (67), and (60) respectively, as functions of Θ_k and W_k. Knowing the coordinates (Θ_a, Z_a, W_a) the optimal cost in the region E can be derived, based on (26) and (58) or (59), as a function of Θ_k and W_k only. This leads us to determine that in the entire region $(p_z)_E = 0$ identically and as a consequence the optimal pursuer strategy is arbitrary. Note that even on the ES the pursuer can choose $u > u^*_{ES-}$ (or $u < u^*_{ES+}$) without affecting the outcome, a quite particular ES behavior not yet encountered (to the knowledge of the authors) in other examples.

Since away from the constraints $\lambda = 0$, along any optimal trajectory $(p_w)_E$ must be constant (a function of Θ_k and W_k) and equal to the value of p_w in the neighborhood of the AL. Note that on the AL itself, which is the boundary of the DS, p_w (as well as the other components of **grad** J^*) is discontinuous. The derivation (detailed in [7]) is based on (53), (54) and (67), yielding

$$(p_w)_E = -sign \, W\{[\Theta_a^+/\Theta_c + \Theta_a^+)] + [\Theta_a^-/(\Theta_c + \Theta_a^-)]\}/2 \quad (72)$$

confirming the assumption that the optimal strategy of the evader in E is the same as on the nearby constraint, i.e.,

$$(v^*)_E = -sign \, W \quad (73)$$

and compatible with (35).

One can summarize that each of the two disjoint subregions (E^+ or E^-) have three boundaries, namely an *equivocal surface* (ES^+ or ES^-), a surface generated by the family of "grazing" primary trajectories from the AL - (to be denoted as DE^+ or DE^-), and a constraint surface (C^- for E^+ and C^+ for E^-). Note that DE^+ and ES^+ (as well as DE^- and ES^-) are tangent to each other along the AL, as mentioned earlier.

7 "Neutral" Zone

The new subregions E^+ and E^- do not yet fill completely the entire game space. The trajectories with opposite optimal strategy pairs that reach the

AL, either at the point "I" $(\Theta_i, 0, 0)$ if $\Theta_N \leq \Theta_s$, or on the segment $B(\Theta = \Theta_s, Z = 0, |W| \leq W_B)$ if $\Theta_N > \Theta_s$, arrive there tangentially and create between them an empty region. This region, denoted here O, is separated from the others by the boundaries DO^+, DO^-, EO^+ and EO^- respectively. The boundary $DO^+(DO^-)$ is generated by backward integration from the trajectory on $ES^-(ES^+)$ that leads to "I" (or to the segment B), using the optimal strategy pair of $D^+(D^-)$, while the surfaces EO^+ and EO^- are defined by

$$|W_{EO}(\Theta)| = \{(\Theta + \Theta_c)^2 - (\Theta_i + \Theta_c)^2\}/2 \quad \Theta_N \leq \Theta_s \quad (74)$$
$$|W_{EO}(\Theta) - W_B| = \{(\Theta + \Theta_c)^2 - (\Theta_s + \Theta_c)^2\}/2 \quad \Theta_N > \Theta_s \quad (75)$$

and taking the sign of W. Region O is similar to the region D_0 of the unconstrained game [1]. If the initial conditions are in this region and $\Theta_N \leq \Theta_s$, the trajectory (if it did not reach the ES) must go through "I" leading to the same terminal cost M_i. If $\Theta_N > \Theta_s$, all trajectories (if they did not reach the ES) must go through the segment B and the terminal cost will be M_s. Therefore inside this region, which may be called the *neutral zone* of the game, the *value* is constant and the optimal strategies of both players are arbitrary. Unique optimal strategies have to be selected only when a trajectory reaches one of the boundary surfaces.

Nevertheless, in spite the similarity there is a substantial difference between the region D_0 of the unconstrained game and the region O. If $\mu \leq 1$, Θ_s becomes infinite. Therefore, in the unconstrained game [1] D_0 disappears and the entire game space is D_1. In the present game Θ_i, the solution of (68) is always finite. Thus, the region O exists for all values of μ if $\Theta_0 > \Theta_i$.

The importance of the *neutral zone* lies in the fact that in most problems of practical interest (such as in missile defense scenarios) the initial conditions are selected by the players. The "attacking" missile (the evader) can be easily aimed initially towards its target (i.e., with a rather small, or even zero, value of W_0). Similarly the interceptor missile can be launched with a small deviation from *collision course* (i.e. with a small value of Z_0). Therefore, most practically important engagements will start in the region O, where the *guaranteed miss distance* is constant.

For the case where $\Theta_N \leq \Theta_s$ and $\Theta_i \leq \Theta_0$, the values of Θ_i and M_i are the functions of Θ_c and the maneuver ratio μ, as shown in Figs. 9 and 10. It can be seen in Fig. 10 that for small values of the maneuver ratio the normalized *guaranteed optimal miss distance* M_i, is monotonically increasing with Θ_c. Therefore, the interest of the defense system is to keep Θ_c (which is obtained, as (19) indicates, by the selection of Θ_0) as small as possible, as long as it is compatible with the defense system's *minimum range*, taking into account the "safe distance from the target" requirement (see A-8). Therefore,

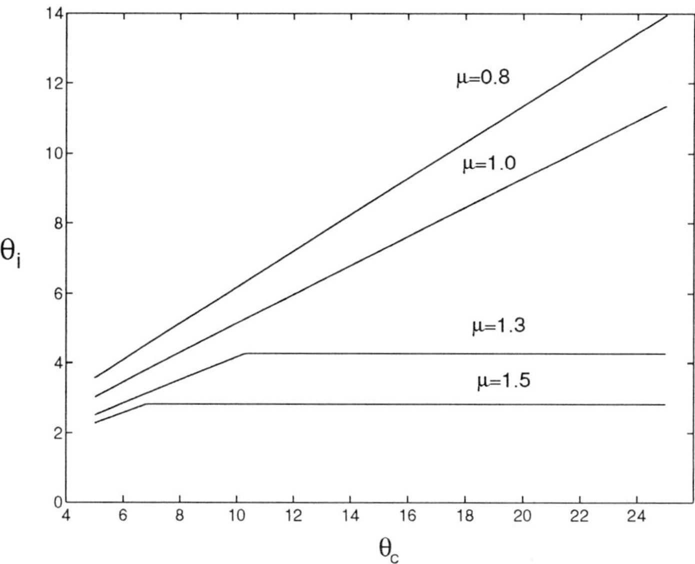

Figure 9: Normalized critical time to go (Θ_i) in the game with constraint.

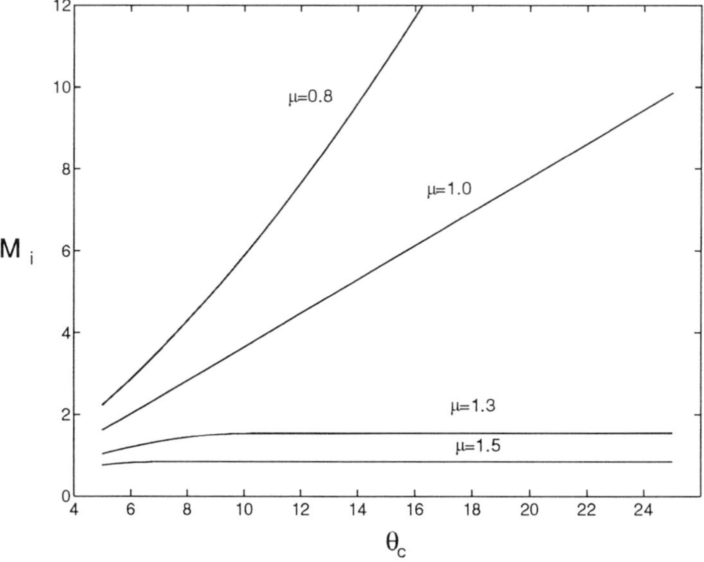

Figure 10: Normalized guaranteed miss distance (M_i) in the game with constraint.

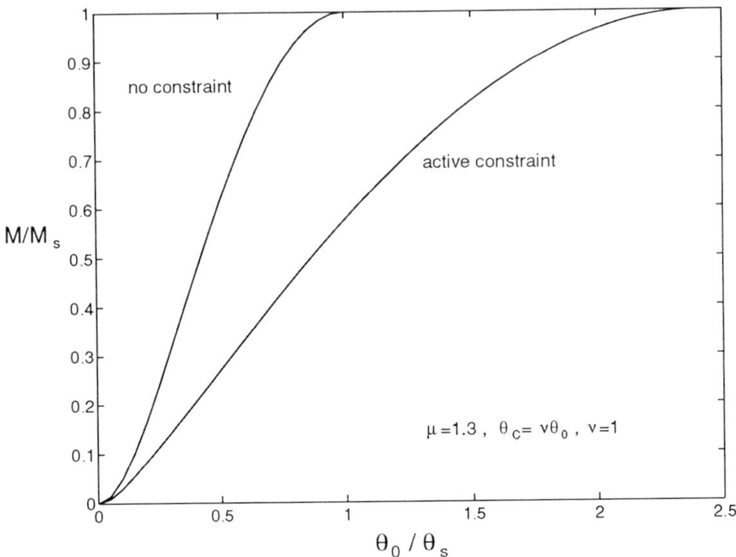

Figure 11: Effect of the constraint on the guaranteed miss distance.

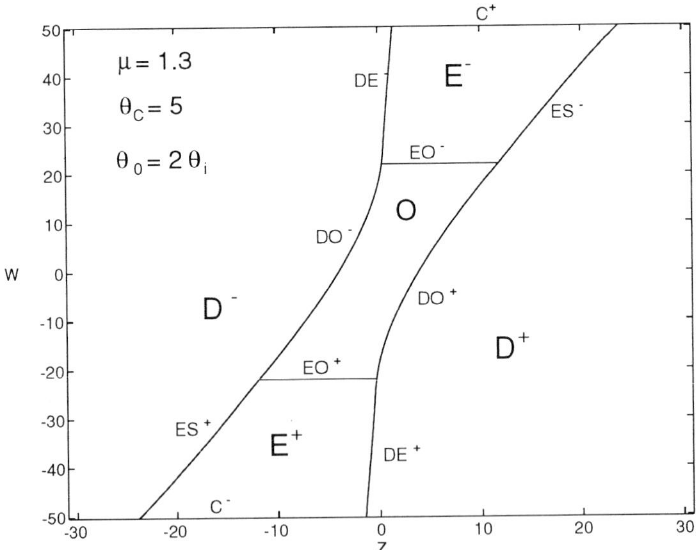

Figure 12: Decomposition of the game space at constant Θ.

the best solution for the defense is to plan the interception at *minimum range* $(x_f = R_{min})$, which leads to

$$\Theta_c = R_{min}/V_A\tau \tag{76}$$

In this case the evader, (assumed to have perfect information) may start to maneuver only at $\Theta = \Theta_i$. The maneuver can be either to the left, or to the right, by a random selection, in order to guarantee the *miss distance* predicted for the *minimum range*. This *guaranteed miss distance* is of course smaller than in the unconstrained game, as shown in Fig. 11 for the example of $\mu = 1.3$ and $\nu = 1$.

If $\Theta_0 < \Theta_i$, a case that may occur only for rather large values of ν, the terminal maneuver of the evader is shortened and the normalized miss distance is less than M_i.

With the definition of the *neutral zone* the decomposition of the game space is completed. Such a decomposition at a constant value of $\Theta_0 = 2\Theta_i$ is shown in Fig.12.

8   Summary

In this paper the complete solution of the perfect information pursuit-evasion game with an active state constraint, modelling future anti-missile defense scenarios, is presented. This solution consists of the partition of the game space into three regions, where the optimal strategies of the players and the corresponding *value* are defined. In Fig. 12 the different regions are shown for a given constant value of $\Theta_0 = 2 \cdot \Theta_i$. As shown, the game space is decomposed into three regions of substantially different optimal strategies.

(i) The primary region D is composed of two fields D^+ and D^- according to the sign of the *miss distance*. The two fields are separated for small values of Θ by a *dispersal surface* and for larger values they become disjointed. The optimal strategies in this region are given by (63)-(65) and the *value* of the game is a function of the initial conditions.

(ii) The region E is divided into two disjoint fields E^+ and E^-, both under the influence of the nearby state constraint. In this region the evader's object is to avoid the constraint and use the optimal strategy given in (73), which is the same as on the nearby constraint. The optimal pursuer strategy in this region is arbitrary. If the game starts in this region the *value* is independent of Z_0.

(iii) The *neutral* region O is the one where most practically significant engagements are likely to start. In this region the optimal strategies of both players are arbitrary and the *value* of the game is constant, depending only on the parameters Θ_c and μ. Such a region exists only if $\Theta_0 > \Theta_i$.

Since arbitrary control strategies (optimal in E and O) are not convenient for implementation, they can be replaced (without changing the outcome of the game) by some linear control strategy, as proposed in [1], or even by using the "bang-bang" strategies of D with the extended interpretation of Z_{ds} beyond the optimal *dispersal surface.*

It can be concluded that the solution has led to the discovery of new phenomena, such as the existence of an *aggregation line* on the *dispersal surface* and an *equivocal surface*, not encountered in the unconstrained game. This perfect information game solution (the worst case from the point of view of the defense) provides a new insight and serves as the basis for an ongoing analysis of the more realistic imperfect information scenario.

REFERENCES

[1] Gutman, S.: "On Optimal Guidance for Homing Missiles", *Journal of Guidance and Control*, Vol. 3, No. 4, 1979, 296-300.

[2] Shinar, J. and Gutman, S.: "Three-Dimensional Optimal Pursuit and Evasion with Bounded Control", *IEEE Trans. on Automatic Control*, Vol. AC-25, No. 3, 1980, 492-496.

[3] Shinar , J., Medina, M. and Biton, M.: "Singular Surfaces in a Linear Pursuit-Evasion Game with Elliptical Vectograms", *Journal of Optimization Theory and Applications*, Vol. 43, No. 3, 1984, 431-458.

[4] Forte, I. and Shinar, J.: "Improved Guidance Law Design Based on Mixed Strategy Concept", *Journal of Guidance, Control and Dynamics*, Vol. 12, No. 2, 1989, 739-745.

[5] Shinar, J. and Forte, I.: "On the Optimal Pure Strategy Sets for a Mixed Missile Guidance Law Synthesis", *IEEE Trans. on Automatic Control*, Vol. AC-36, No. 11, 1991, 1296-1300.

[6] Bernhard, P.: "Singular Surfaces in Differential Games: An Introduction". In: *Differential Games and Applications*. Edited by Hagedorn, P., Knobloch, H. W. and Olsder, G. J., Springer-Verlag, Berlin, 1977, 1-33.

[7] Lipman, Y.: "Analysis of Anti-Missile Defence Scenarios by Using a Linear Model". D.Sc. Thesis, Faculty of Aerospace Engineering, Technion, Haifa, 1995 (in Hebrew).

Three-Dimensional Air Combat:
Numerical Solution of Complex Differential Games

R. Lachner,[*] M. H. Breitner[†]and H. J. Pesch[‡]

Institute of Mathematics, Clausthal University of Technology,

D-38678 Clausthal-Zellerfeld, Germany

Abstract

Complex pursuit-evasion games with complete information under
state variable inequality constraints are investigated. By exploitation
of Isaacs' minimax principle, necessary conditions of first and second
order are derived for the optimal trajectories. These conditions give
rise to multipoint boundary-value problems, which yield open-loop representations of the optimal strategies along the optimal trajectories.
The multipoint boundary-value problems are accurately solved by the
multiple shooting method. The computed open-loop representations
can thereafter be used to synthesize the optimal strategies globally.

As an illustrative example, the evasion of an aircraft from a pursuing missile is investigated. The flight of the aircraft is restricted by
various control variable inequality contraints and by a state variable inequality constraint for the dynamic pressure. The optimal trajectories
exhibit boundary arcs with regular and singular constrained controls.
The influence of various singular surfaces in the state space including
a low-dimensional universal surface is discussed.

Keywords: Differential games, pursuit-evasion games, singular surfaces,
multipoint boundary-value problems, multiple shooting method.

1 Introduction

Differential games are useful to describe many aspects of human interactions. Unlike optimal control problems, they deal with the competitive
character of those interactions. Even unpredictable disturbances can be investigated, if "nature" is taken into account as an additional player. The

---

[*]Assistant Professor. E-mail: lachner@math.tu-clausthal.de.

[†]Assistant Professor. E-mail: breitner@math.tu-clausthal.de.

[‡]Professor of Mathematics. E-mail: pesch@math.tu-clausthal.de.

All authors' address: Institute of Mathematics, Clausthal University of Technology,
Erzstr. 1, D–38678 Clausthal-Zellerfeld, Germany.

solution of differential games is much more involved than the solution of optimal control problems. This is due to the fact that only optimal feedback-type strategies are suitable to describe the global behavior of the players, especially if one of them acts nonoptimally. A global computation of these optimal strategies is generally impossible, since the state space is split up by singular surfaces even for simple differential games. Another reason is that direct optimization methods, which are easy to deal with, do not exist so far for differential games. Direct optimization methods for optimal control problems (see, e.g., von Stryk [30]) are incompatible with the information structures usually associated with differential games. Moreover, feedback control schemes that can be applied in real time have been developed even for complicated optimal control problems (see the survey in [25]) whereas only a few real-time feasible and generally applicable methods can be found which have already been applied to differential games; see [3], [4], [5], and [27].

In the present paper, optimal evasive maneuvers of an aircraft against an optimally guided missile are investigated. The results of [21], where the missile-versus-aircraft scenario has been restricted to a common vertical plane, are extended here to pursuit-evasion maneuvers in three space dimensions. Thereby, the maneuverability of the aircraft is restricted by several control and state variable inequality constraints. Various initial conditions are investigated including the head-on shoot.

The application of Isaacs' minimax principle yields complicated optimal control histories including constrained and singular subarcs. These control histories coincide with the optimal strategies along the optimal trajectories. A combination of all the necessary conditions from the minimax principle gives rise to multipoint boundary-value problems with interior point conditions and jump conditions for the state and the adjoint variables. These multipoint boundary-value problems are then solved by means of the multiple shooting method; see Bulirsch [9], Stoer and Bulirsch [29], and the references cited in [25]. For the most recent code used in this paper, see Hiltmann et al. [15].

By means of homotopy techniques, the optimal trajectories and the associated open-loop representations of the optimal strategies can be computed in the entire capture zone. These open-loop representations provide the information by which optimal strategies can be approximated globally, e. g., by a successive solution of neighboring multipoint boundary-value problems (see [2]), by Taylor series expansions around many optimal trajectories (see [3], [4]), or by neural networks (see [27]). Therefore, this work can be considered as a first step towards the real-time computation of optimal strategies for realistic differential games of pursuit-evasion type.

2 Pursuit-Evasion Game

The following pursuit-evasion game arises as a subproblem for an air combat scenario between two aircraft. Similar problems have been treated recently with various solution methods; see, e. g., Fink [10], Greenwood [11], Guelman, Shinar, Green [13], Gutman, Katz [14], and the papers [6], [7], [19], [20], and [26] of the authors. At time $t = 0$ the pursuing aircraft launches a medium-range air-to-air missile P, which carries on pursuing the evading aircraft E. The missile P is much more maneuverable then the aircraft E, but can accelerate only a short time due to its limited fuel supply. As a result, E can always escape if the initial distance between the opponents is large enough. Therefore, only initial constellations for which P can enforce capture against all possible maneuvers of the aircraft are considered here. This aerial scenario can be formulated as a pursuit-evasion game with the missile P identified with the pursuer and the aircraft E with the evader.

Note that the restriction of the flight paths to a common vertical plane as in [6], [7], [19], [20], [21], and [26] is discarded here. Therefore, the computation of optimal flight paths for almost arbitrary initial constellations including the so-called head-on shoot can be performed. As a by-product, the two-dimensional optimal trajectories computed in the above-mentioned papers will turn out to be optimal in the three-dimensional space, too, if the initial velocity vectors lie in a common vertical plane.

The state of P and E is described by their position vectors and their veloctity vectors. The equations of motion for the missile P and the aircraft E then read as follows (see, e. g., Miele [24]),

$$
\begin{aligned}
\dot{x}_P &= f_{x_P} = v_P \cos\gamma_P \cos\chi_P \,, \\
\dot{x}_E &= f_{x_E} = v_E \cos\gamma_E \cos\chi_E \,, \\
\dot{y}_P &= f_{y_P} = v_P \cos\gamma_P \sin\chi_P \,, \\
\dot{y}_E &= f_{y_E} = v_E \cos\gamma_E \sin\chi_E \,, \\
\dot{h}_P &= f_{h_P} = v_P \sin\gamma_P \,, \\
\dot{h}_E &= f_{h_E} = v_E \sin\gamma_E \,, \\
\dot{v}_P &= f_{v_P} = \frac{T_{P,\mathrm{max}}(t) - D_{0,P}(h_P, v_P) - n_P{}^2\, D_{I,P}(t, h_P, v_P)}{m_P(t)} \\
&\qquad\qquad -g \sin\gamma_P \,, \\
\dot{v}_E &= f_{v_E} = \frac{\eta_E\, T_{E,\mathrm{max}}(h_E, v_E) - D_{0,E}(h_E, v_E) - n_E{}^2\, D_{I,E}(h_E, v_E)}{m_E} \\
&\qquad\qquad -g \sin\gamma_E \,, \\
\dot{\gamma}_P &= f_{\gamma_P} = \frac{g}{v_P}\left(n_P \cos\mu_P - \cos\gamma_P\right) \,,
\end{aligned}
$$

(1)

$$\dot{\gamma}_E = f_{\gamma_E} = \frac{g}{v_E}\left(n_E \cos\mu_E - \cos\gamma_E\right),$$

$$\dot{\chi}_P = f_{\chi_P} = \frac{g}{v_P}\frac{n_P \sin\mu_P}{\cos\gamma_P},$$

$$\dot{\chi}_E = f_{\chi_E} = \frac{g}{v_E}\frac{n_E \sin\mu_E}{\cos\gamma_E}.$$

Subscripts P and E refer here to the missile P and the aircraft E. The state variables x, y, h, v, γ, and χ denote the two coordinates of the horizontal position, the altitude, the velocity, and the vertical and horizontal flight path angle of P and E, respectively. The thrust $T_E = \eta_E T_{E,\max}$ of the aircraft is controlled by the throttle $\eta_E \in [\eta_{E,\min}, \eta_{E,\max}]$. The thrust $T_P = T_{P,\max}$ of the missile is not controllable. It is assumed to be piecewise constant with two discontinuities at $t = 3\,[\mathrm{s}]$ and $t = 15\,[\mathrm{s}]$ indicating the transitions from the boost phase to the sustain phase and from the sustain phase to the coast phase. The control variables n_P and n_E denote the load factors, which govern the flight path angles. The bank angles μ_P and μ_E denote control variables peculiar to the three-dimensionality of the flight scenario. They determine the radii of curvature. Since only flight times less then 90 seconds occur, the mass m_E of the aircraft is assumed to be constant neglecting the effects of fuel consumption ($\Delta m_E < 0.1\,m_E$). The mass m_P of the missile decreases linearly during the boost and the sustain phase and is constant during the coast phase. The constant g denotes the gravitational acceleration. The complicated functions $T_{E,\max}$, $D_{0,P}$, $D_{0,E}$, $D_{I,P}$, and $D_{I,E}$ are based on data for the *F15E-Strike Eagle* multi-role combat aircraft and the medium-range air-to-air missile *AMRAAM AIM-120A*. These data provide realistic approximations for the maximum thrust of the aircraft, the zero-lift drag, and the induced drag. The coefficients for the approximations have been determined by the method of least squares.

At time $t = 0$ the game starts with the initial conditions

$$
\begin{aligned}
x_P(0) &= 0\,[\mathrm{m}], & x_E(0) &= x_{E,0}\,[\mathrm{m}], \\
y_P(0) &= 0\,[\mathrm{m}], & y_E(0) &= y_{E,0}\,[\mathrm{m}], \\
h_P(0) &= h_{P,0}\,[\mathrm{m}], & h_E(0) &= h_{P,0}\,[\mathrm{m}], \\
v_P(0) &= v_{P,0}\,[\mathrm{m/sec}], & v_E(0) &= v_{E,0}\,[\mathrm{m/sec}], \\
\gamma_P(0) &= 0\,[\mathrm{deg}], & \gamma_E(0) &= \gamma_{E_0}\,[\mathrm{deg}], \\
\chi_P(0) &= 0\,[\mathrm{deg}], & \chi_E(0) &= \chi_{E,0}\,[\mathrm{deg}].
\end{aligned}
\tag{2}
$$

These initial conditions are abbreviated by $z(0) = z_0$ with

$$z = (x_P,\ x_E,\ y_P,\ y_E,\ h_P,\ h_E,\ v_P,\ v_E,\ \gamma_P,\ \gamma_E,\ \chi_P,\ \chi_E)^{\mathsf{T}}.$$

Note that $x_P(0) = 0$, $y_P(0) = 0$ and $\chi_P(0) = 0$ can be prescribed without loss of generality, since these initial conditions can always be obtained by a translation and rotation of the coordinate system. The dimension of the state space can be reduced from 12 to 9 defining the new variables $x_E - x_P$, $y_E - y_P$, and $\chi_E - \chi_P$. The condition $\gamma_P(0) = 0$ is due to the technical reason that the launch of a missile is dangerous for the carrying aircraft when $\gamma \neq 0$.

By choosing admissible values for the control variables of both players in Eqs. (1), a trajectory is determined uniquely in the state space. The terminal time t_f of the game is determined by the capture condition

$$\left(x_P(t_f) - x_E(t_f)\right)^2 + \left(y_P(t_f) - y_E(t_f)\right)^2 + \left(h_P(t_f) - h_E(t_f)\right)^2 - d^2 = 0, \quad (3)$$

where the capture radius d is chosen as $50\,[\mathrm{m}]$. Equation (3) defines a hypermanifold in the state space called the terminal manifold. If P is not able to enforce capture against all possible maneuvers of E, the outcome of the game is "E not captured" and t_f is set to infinity.

The players choose their strategies Γ_P and Γ_E under the aspect of minimaximizing the objective functional

$$J(\Gamma_P, \Gamma_E, t = 0, z = z_0) = t_f. \quad (4)$$

The pursuer P engages in driving the state z from the initial state z_0 to the terminal manifold in minimal time, whereas the evader E tries to avoid capture or, if this is impossible, tries to maximize the capture time t_f.

For a realistic modelling, several constraints have to be taken into account. The most important constraint for the evader, which shall particularly be discussed here, is the limit of the dynamic pressure q_E,

$$Q(h_E, v_E) = q_E(h_E, v_E) - q_{E,\max} \leq 0 \quad (5)$$

with $q_E = \frac{1}{2}\rho(h_E)\,v_E{}^2$, $q_{E,\max} = 80\,[\mathrm{kPa}]$, and ρ denoting the air density. This first-order state variable inequality constraint keeps the aircraft away from the flutter boundary and limits the static load. For certain initial conditions of the state, it is advantageous for the aircraft to descend, in order to maximize thrust, to gain additional kinetic energy, and to force the missile to follow into regions of high drag. In this case the altitude constraint

$$h_E \geq 0, \quad (6)$$

a second-order state variable inequality constraint, may become active. However, optimal trajectories with altitude-constrained subarcs will not be presented here; see [19] instead. Finally, the control variables are bounded by

$$\begin{aligned}
n_P &\in [n_{P,\min}, n_{P,\max}] &&= [0, 20], \\
n_E &\in [n_{E,\min}, n_{E,\max}] &&= [0, 7], \\
\eta_E &\in [\eta_{E,\min}, \eta_{E,\max}] &&= [0.23, 1].
\end{aligned} \quad (7)$$

Note that, unlike the case in the two-dimensional version of this game in [6], [7], [19], [20], [21], and [26], the $n_{E,\max}$-constraint is more restrictive, since sharp turns of E are optimal for many initial constellations; see also [10]. The Mach limit at high altitudes and the maximum lift coefficient for low speeds are of secondary importance and are not taken into account. These constraints can, however, be included in a similar way. Also a visibility condition for the missile to keep the aircraft in its radar cone can be included analogously.

3   Necessary Conditions for Optimality

3.1   Differential Game Approach

In the sequel it is assumed that the game is started with an arbitrary initial state z_0 for which capture is guaranteed against all admissible strategies of E, provided that P plays optimally. The set of all optimal trajectories emanating from these initial states span a subset in the state space, the capture zone. It is separated from the remainder of the state space, the escape zone, by a submanifold, the barrier. Note that the derivation of necessary conditions in this section is valid only for initial states in the capture zone.

Only admissible strategies

$$u_P(t,z) = \begin{pmatrix} n_P(t,z) \\ \mu_P(t,z) \end{pmatrix} = \Gamma_P(t,z),$$

$$u_E(t,z) = \begin{pmatrix} n_E(t,z) \\ \mu_E(t,z) \\ \eta_E(t,z) \end{pmatrix} = \Gamma_E(t,z) \tag{8}$$

are considered; see, e.g., Başar and Olsder [1]. The players P and E have perfect information about the actual state, but no information about the actual or future control of the opposite player. A strategy $\Gamma_P{}^*$ is called optimal for P if

$$J(\Gamma_P{}^*, \Gamma_E, t, z) \le J(\Gamma_P, \Gamma_E, t, z) \tag{9}$$

holds for all admissible strategies Γ_E, for all t, and for all z within the capture zone. Conversely, a strategy $\Gamma_E{}^*$ is called optimal for E if

$$J(\Gamma_P, \Gamma_E, t, z) \le J(\Gamma_P, \Gamma_E{}^*, t, z) \tag{10}$$

holds for all admissible strategies Γ_P, for all t, and for all z within the capture zone.

The optimal value V^* of the objective function as a function of the actual time t and the actual state z is defined by

$$V^*(t,z) = J(\Gamma_P{}^*, \Gamma_E{}^*, t, z). \tag{11}$$

A trajectory generated by optimal strategies will be called an optimal trajectory. Along optimal trajectories, the optimal strategies yield optimal controls $u_P^*(t) := u_P^*(t, z(t))$ and $u_E^*(t) := u_E^*(t, z(t))$. Henceforth, we will speak about optimal controls if we have the values of the optimal strategies along optimal trajectories in mind.

The minimax principle of Isaacs [16] and [17] yields local necessary conditions which must be satisfied by the optimal controls u_P^* and u_E^*, i. e., for all t along the optimal trajectories. Defining auxiliary functions by

$$
\begin{aligned}
\tilde{u}_P(t, z, V_t^*(t, z), V_z^*(t, z), u_E) &= \arg \min_{u_P} \frac{\mathrm{d}}{\mathrm{d}t} V^*(t, z) \\
&= \arg \min_{u_P} \left(V_t^*(t, z) + V_z^*(t, z) \, f(t, z, u_P, u_E) \right), \\
\tilde{u}_E(t, z, V_t^*(t, z), V_z^*(t, z), u_P) &= \arg \max_{u_E} \frac{\mathrm{d}}{\mathrm{d}t} V^*(t, z) \\
&= \arg \min_{u_E} \left(V_t^*(t, z) + V_z^*(t, z) \, f(t, z, u_P, u_E) \right),
\end{aligned}
\tag{12}
$$

the optimal controls must satisfy

$$
\begin{aligned}
& u_P^*(t, z, V_t^*(t, z), V_z^*(t, z)) \\
& = \arg \min_{u_P} \left(V_t^*(t, z) + V_z^*(t, z) \, f(t, z, u_P, \tilde{u}_E(t, z, V_t^*, V_z^*, u_P))) \right), \\
& u_E^*(t, z, V_t^*(t, z), V_z^*(t, z)) \\
& = \arg \max_{u_E} \left(V_t^*(t, z) + V_z^*(t, z) \, f(t, z, \tilde{u}_P(t, z, V_t^*, V_z^*, u_E), u_E) \right),
\end{aligned}
\tag{13}
$$

for all t.

In Eqs. (12) and (13), the minimization and maximization must be performed only for controls u_P and u_E which are admissible in t. Because of the separability of $\mathrm{d}V^*/\mathrm{d}t$, the auxiliary functions $\tilde{u}_P$ and $\tilde{u}_E$ coincide with the optimal controls u_P^* and u_E^*. Therefore, one need not distinguish upper and lower values of the game. Since the game has a terminal payoff [see Eq. (4)], the optimal value V^* satisfies the nonlinear partial differential equation of first order

$$
V_t^* + V_z^* \, f\left(t, z, u_P^*\left(t, z, V_t^*, V_z^*\right), u_E^*\left(t, z, V_t^*, V_z^*\right)\right) = 0
\tag{14}
$$

for all (t, z) in the capture zone. This equation is known as Isaacs' equation (see [16] and [17]). Unfortunately, Eq. (14) cannot be used directly for computing a pair of optimal strategies, since it is a nonlinear partial differential equation for 13 independent variables here. A state space reduction to the essential variables can decrease the dimension only to 10. Moreover, the function V^* is not continuously differentiable in the entire capture zone. However, an equivalent system of ordinary differential equations along a characteristic curve can be formulated identifying the gradient $(V_t^*, V_{x_P}^*, \ldots)^\mathsf{T}$ at (t, z) with the merely time-dependent adjoint functions $(\lambda_\theta, \lambda_{x_P}, \ldots)^\mathsf{T}$. For this purpose, the system (1) is first rewritten in autonomous form using $\theta = t$ as a new state variable,

$$\frac{\mathrm{d}}{\mathrm{d}t}\begin{pmatrix} \theta \\ z \end{pmatrix} = \begin{pmatrix} 1 \\ f(\theta, z, u_P, u_E) \end{pmatrix}. \tag{15}$$

The new initial condition for θ is $\theta(0) = 0$. Secondly, the system for the adjoint variables is given, according to [17], by

$$\begin{aligned}
\frac{\mathrm{d}}{\mathrm{d}t}\begin{pmatrix} \lambda_\theta \\ \lambda \end{pmatrix} &= -\left(\frac{\partial H}{\partial(\theta, z)}\right)^\mathsf{T} \\
&= -\left(\lambda^\mathsf{T} \frac{\partial f(\theta, z, u_P^*, u_E^*)}{\partial(\theta, z)} \right. \\
&\qquad \left. + \frac{\partial H}{\partial(u_P, u_E)} \frac{\partial}{\partial(\theta, z)} \begin{pmatrix} u_P^* \\ u_E^* \end{pmatrix}\right)^\mathsf{T},
\end{aligned} \tag{16}$$

where the Hamiltonian is defined by

$$H(\theta, z, \lambda_\theta, \lambda, u_P, u_E) = \lambda_\theta + \lambda^\mathsf{T} f(\theta, z, u_P, u_E). \tag{17}$$

Note that there holds

$$H(\theta, z, \lambda_\theta, \lambda, u_P^*, u_E^*) = \frac{\mathrm{d}}{\mathrm{d}t} V^*(\theta, z). \tag{18}$$

Following this procedure, the optimal controls u_P^* and u_E^* in Eq. (13) can now be identified along optimal trajectories through

$$\begin{aligned}
u_P^*(t) &= u_P^*\big(t, z(t)\big) = u_P^*\big(t, z(t), V_t^*(t, z(t)), V_z^*(t, z(t))\big) \\
&= u_P^*\big(t, z(t), \lambda_\theta(t), \lambda(t)\big), \\
u_E^*(t) &= u_E^*\big(t, z(t)\big) = u_E^*\big(t, z(t), V_t^*(t, z(t)), V_z^*(t, z(t))\big) \\
&= u_E^*\big(t, z(t), \lambda_\theta(t), \lambda(t)\big),
\end{aligned} \tag{19}$$

and the optimal strategies Γ_P^* and Γ_E^* are obtained along optimal trajectories from the optimal control histories u_P^* and u_E^* via

$$\Gamma_P^*(t, z(t)) = u_P^*(t), \qquad \Gamma_E^*(t, z(t)) = u_E^*(t). \tag{20}$$

These equations provide an open-loop representation of the optimal strategies; for details see [1].

3.2 Open-Loop Representations of Optimal Strategies for the Missile

In this subsection, it is assumed that the $n_{P,\mathrm{max}}$-constraint is not active during the entire maneuver. This assumption is valid for all numerically computed optimal trajectories presented in Section 4. Furthermore, it can be shown that $n_P^*(t) = n_{P,\mathrm{min}}$ holds only for isolated t; see Grimm [12]. From

$$H(\theta, z, \lambda_\theta, \lambda, n_P, \mu_P, n_E, \mu_E, \eta_E) = H_1(\theta, z, \lambda)\, n_P^2$$
$$+ H_2(z, \lambda)\, n_P\, \cos\mu_P + H_3(z, \lambda)\, n_P\, \sin\mu_P$$
$$+ H_4(\theta, z, \lambda_\theta, \lambda, n_E, \mu_E, \eta_E)\,, \tag{21}$$

the optimal controls of the missile P are computed using the minimax principle,

$$\frac{\partial}{\partial n_P} H(\theta, z, \lambda_\theta, \lambda, n_P^*, \mu_P^*, n_E, \mu_E, \eta_E) \;=\; 0\,,$$

$$\frac{\partial}{\partial \mu_P} H(\theta, z, \lambda_\theta, \lambda, n_P^*, \mu_P^*, n_E, \mu_E, \eta_E) \;=\; 0\,,$$

$$\frac{\partial^2}{\partial n_P^2} H(\theta, z, \lambda_\theta, \lambda, n_P^*, \mu_P^*, n_E, \mu_E, \eta_E) \;\geq\; 0\,, \tag{22}$$

$$\frac{\partial^2}{\partial \mu_P^2} H(\theta, z, \lambda_\theta, \lambda, n_P^*, \mu_P^*, n_E, \mu_E, \eta_E) \;\geq\; 0\,.$$

This yields

$$n_P^* = \frac{m_P\, g\, U_P}{2\, D_{I,P}\, v_P\, \lambda_{v_P}}\,, \tag{23}$$

$$\cos\mu_P^* = -\frac{\lambda_{\gamma_P}}{U_P}\,, \tag{24}$$

$$\sin\mu_P^* = -\frac{\lambda_{\chi_P}}{U_P\, \cos\gamma_P} \tag{25}$$

with

$$U_P = \sqrt{\lambda_{\gamma_P}{}^2 + (\lambda_{\chi_P}/\cos\gamma_P)^2}$$

and the necessary sign condition $\lambda_{v_P} \leq 0$.

3.3 Open-Loop Representations of Optimal Strategies for the Aircraft

The different optimal control laws for E depend on the function

$$U_E = \sqrt{\lambda_{\gamma_E}{}^2 + (\lambda_{\chi_E}/\cos\gamma_E)^2}\,. \tag{26}$$

During the numerical computations it turned out that the singular case $U_E \equiv 0$ can occur only on dynamic-pressure-constrained subarcs. Hence, we will have to distinguish between regular and singular constrained subarcs. However, we first consider state-unconstrained subarcs.

3.3.1 Dynamic Pressure Constraint Inactive

The optimal controls of E can be obtained by maximizing the Hamiltonian

$$H(\theta, z, \lambda_\theta, \lambda, n_P, \mu_P, n_E, \mu_E, \eta_E) = \bar{H}_1(z, \lambda)\, n_E{}^2$$
$$+ \bar{H}_2(z, \lambda)\, n_E \, \cos \mu_E + \bar{H}_3(z, \lambda)\, n_E \, sin\mu_E$$
$$+ \bar{H}_4(z, \lambda)\, \eta_E + \bar{H}_5(\theta, z, \lambda_\theta, \lambda, n_P, \mu_P) \tag{27}$$

over the set of admissible controls $[n_{E,\min}, n_{E,\max}] \times\,]-\pi, \pi] \times [\eta_{E,\min}, \eta_{E,\max}]$. This yields

$$n_E{}^* = \begin{cases} n_{E,\max}\,, & \text{if } \lambda_{v_E} \leq 0, \\[2ex] \min\left\{ \dfrac{m_E\, g\, U_E}{2\, D_{I,E}\, v_E\, \lambda_{v_E}}, n_{E,\max} \right\}\,, & \text{if } \lambda_{v_E} > 0, \end{cases} \tag{28}$$

$$\cos \mu_E{}^* = \frac{\lambda_{\gamma_E}}{U_E}\,, \tag{29}$$

$$\sin \mu_E{}^* = \frac{\lambda_{\chi_E}}{U_E \, \cos \gamma_E}\,, \tag{30}$$

$$\eta_E{}^* = \begin{cases} \eta_{E,\min}\,, & \text{if } \lambda_{v_E} < 0, \\ \text{undefined}\,, & \text{if } \lambda_{v_E} = 0, \\ \eta_{E,\max}\,, & \text{if } \lambda_{v_E} > 0. \end{cases} \tag{31}$$

3.3.2 Dynamic Pressure Constraint Active

Unlike the situation in the case of optimal control problems, state variable inequality constraints, or more precisely, state constraints of order greater than zero, are not well understood in differential games. This obstacle can generally be remedied by strengthening the state constraint by a mixed control-state constraint; see, e. g., [2], and [21]. Their theoretical treatment is well understood. Unfortunately, this procedure leads to optimal solutions with a chattering control here due to the nonconvexity of the hodograph. An extended class of strategies would have then been permitted which, however, cannot be implemented for aircraft. Therefore, the well-known necessary conditions of optimal control theory for state constraints of higher order (see, e. g., Bryson, Denham, and Dreyfus [8], Jacobson, Lele, and Speyer [18], and Maurer [23]) are applied to the present differential game despite the lack of a rigorous proof in the context of differential games. This approach can be justified according to [21] as follows. The state-constrained differential game problem can be imbedded into a one-parameter family of more stringent mixed-state-control-constrained problems. By weakening these control-state-constraints, numerical convergence of the associated one-parameter family of solutions has been observed in [21] towards the solution associated with those necessary conditions. For more details, see [21].

If the dynamic pressure constraint holds, the set of admissible controls is restricted by the condition

$$\rho(h_E)\, v_E \left(\frac{1}{2} \frac{\partial \ln(\rho(h_E))}{\partial h_E} v_E{}^2 \sin \gamma_E - g \sin \gamma_E \right.$$

$$\left. + \frac{1}{m_E} \left(\eta_E\, T_{E,\max} - D_{0,E} - n_E{}^2 D_{I,E} \right) \right) \leq 0. \qquad (32)$$

This is implied by the first derivative of the dynamic pressure constraint (5).

The aircraft E again attains its optimal controls by maximizing the Hamiltonian. Since the set of admissible controls is restricted by Eq. (32), this condition is, according to [8], adjoined to H by a Lagrange multiplier ν_{q_E} which necessarily must be nonpositive. The optimal controls are then given by

$$n_E{}^* = \left[\frac{1}{D_{I,E}} \left(\frac{1}{2} \frac{\partial \ln(\rho(h_E))}{\partial h_E} m_E\, v_E{}^2 \sin \gamma_E \right. \right.$$

$$\left. \left. - m_E\, g \sin \gamma_E + T_{E,\max} - D_{0,E} \right) \right]^{1/2}, \qquad (33)$$

$$\cos \mu_E{}^* = \frac{\lambda_{\gamma_E}}{U_E}, \qquad (34)$$

$$\sin \mu_E{}^* = \frac{\lambda_{\chi_E}}{U_E \cos \gamma_E}, \qquad (35)$$

$$\eta_E{}^* = \eta_{E,\max}, \qquad (36)$$

provided that $\lambda_{v_E} > 0$ and $U_E > 0$. From a discussion of the Hamiltonian, there follows that $\lambda_{v_E} < 0$ cannot occur on dynamic-pressure-constrained subarcs; compare [21]. Note that the adjoint equations differ from those on unconstrained subarcs, denoted by the superscript "free",

$$\dot{\lambda} = \dot{\lambda}^{\text{free}} - \nu_{q_E} \left(\frac{\partial}{\partial z} \frac{\mathrm{d}}{\mathrm{d}t} Q(z) \right)^{\mathsf{T}}. \qquad (37)$$

Furthermore, the interior point condition $q_E(t_{\text{entry}}) = q_{E,\max}$ at the junction point t_{entry} induces discontinuities in the adjoint variables, namely

$$\lambda(t_{\text{entry}}^{+}) = \lambda(t_{\text{entry}}^{-}) - \sigma \left(\frac{\partial}{\partial z} Q(z) \right)^{\mathsf{T}} \tag{38}$$

with a necessarily nonpositive parameter σ,

$$\sigma = \nu_{q_E}(t_{\text{entry}}^{+}) . \tag{39}$$

If $U_E \equiv 0$ holds on a certain subinterval, the Hamiltonian has no unique maximum on the set of admissible controls. From $\lambda_{\chi_E} \equiv 0$ and $\dot{\lambda}_{\chi_E} \equiv 0$, it can be derived $\chi_E = \text{const}$ yielding

$$\sin \mu_E^{*} \;=\; 0 , \tag{40}$$
$$\cos \mu_E^{*} \;\in\; \{-1, 1\} . \tag{41}$$

Note that on singular subarcs, the optimal flight path of E is in a vertical plane. The control that keeps $\lambda_{\gamma_E} \equiv 0$ can be obtained by differentiating this identity twice with respect to time. Then there holds

$$n_E^{*} \cos \mu_E^{*} = \cos \gamma_E \left(1 + \frac{1}{2\,g} \frac{\partial \ln(\rho(h_E))}{\partial h_E} v_E^2 \right) . \tag{42}$$

Since the activity of the dynamic pressure constraint must be maintained, Eq. (32) with the equality sign yields

$$\eta_E^{*} = \frac{1}{T_{E,\text{max}}} \left(D_{0,E} + m_E\, g \sin \gamma_E \right.$$
$$\left. - \frac{1}{2} \frac{\partial \ln(\rho(h_E))}{\partial h_E} m_E\, v_E^2 \sin \gamma_E + D_{I,E}\, n_E^{*2} \right) \tag{43}$$

An additional necessary condition on singular subarcs is

$$\frac{\partial}{\partial(n_E \cos \mu_E)} \ddot{\lambda}_{\gamma_E} \leq 0 . \tag{44}$$

As long as this condition holds, the singular subarc cannot be left without contradiction to the minimax principle. Indeed, one obtains two local maxima of the Hamiltonian for all $n_E \cos \mu_E$ differing from the respective value (42) on the singular subarc. None of these maxima can be chosen as candidate optimal, since the other immediately yields a larger value of the Hamiltonian; see [21] for more details. Equation (44) is always satisfied here on singular subarcs. Hence, optimal trajectories always terminate on singular subarcs, if they include a singular subarc and if no other constraint is violated by the singular control. If the game terminates on a singular subarc, it can be easily shown from $\dot{\lambda}\gamma_E(t_f) = 0$ that the velocity vector of E at the final time t_f is parallel to the line of sight.

Finally, it should be mentioned that singular subarcs are located in a singular surface of universal type. Entering the universal surface, the adjoint variables are continuous.

3.4 Jump Conditions and Interior Point Conditions

The following conditions are related to the transition points **(i)** between the different thrust phases of the missile, **(ii)** between $n_{E,\max}$-constrained subarcs with $\eta_E^* = \eta_{E,min}$ and $n_{E,\max}$-constrained subarcs with $\eta_E^* = \eta_{E,max}$, **(iii)** between $n_{E,\max}$-constrained subarcs and free subarcs, **(iv)** between free subarcs and $q_{E,\max}$-regular-constrained subarcs, **(v)** between $q_{E,\max}$-regular-constrained subarcs and free subarcs, and **(vi)** between $q_{E,\max}$-regular-constrained subarcs and $q_{E,\max}$-singular-constrained subarcs.

(i) The discontinuities in the right-hand side of the dynamic system (1) induced by the different thrust phases of P, create singular surfaces in the state space which lead to jumps in the adjoint vector when the optimal trajectory penetrates these surfaces. The two transition points $t_1 = 3\,[s]$, and $t_2 = 15\,[s]$ between the different thrust phases must be treated as interior point conditions. The continuity of the Hamiltonian implies that λ_θ has discontinuities at t_i, $i = 1, 2$,

$$\lambda_\theta(t_i^-) - \lambda_\theta(t_i^+) = \frac{\lambda_{v_P}(t_i)}{m_P(t_i)} \left(T_{P,\max}(t_i^+) - T_{P,\max}(t_i^-)\right) ; \qquad (45)$$

see Leitmann [22].

(ii) The switching point t_{η_E} associated with that junction is determined by the interior point condition

$$\lambda_{v_E}(t_{\eta_E}) = 0 . \qquad (46)$$

At this point, a bang of η_E^* from the minimal admissible value to the maximal admissible value takes place.

(iii) The optimal load factor n_E^* maximizing the Hamiltonian H moves from the $n_{E,\max}$-boundary into the interior of $[n_{E,\min}, n_{E,\max}]$. Therefore, the associated switching point t_{n_E} is determined by [compare (28)]

$$\left. \frac{m_E\, g\, U_E}{2\, D_{I,E}\, v_E\, \lambda_{v_E}} \right|_{t=t_{n_E}} = n_{E,\max} . \qquad (47)$$

(iv) It can be shown that the control variables of the aircraft E are continuous at the transition point t_{entry} into a dynamic-pressure-constrained subarc. The adjoint vector is discontinuous, the jump is given by Eq. (38).

Note that the adjoint equations must be modified according to Eq. (37) on those subarcs. The switching point t_{entry} is determined by the interior point condition

$$\frac{1}{2}\rho(h_E(t_{\text{entry}}))\,v_E(t_{\text{entry}})^2 = q_{E,\max}\,. \tag{48}$$

(**v**) The optimal load factor $n_E{}^*$ maximizing the Hamiltonian no longer satisfies the condition $(d/dt)Q = 0$. Therefore, the optimal trajectory leaves the dynamic pressure constraint. The exit point t_{exit} is determined by [compare (28)]

$$\left.\frac{m_E\,g\,U_E}{2\,D_{I,E}\,v_E\,\lambda_{v_E}}\right|_{t=t_{\text{exit}}} = n_E{}^*(t_{\text{exit}}{}^-)\,. \tag{49}$$

Note that this condition is equivalent to $\nu_{q_E}(t_{\text{exit}}{}^-) = 0$.

(**vi**) The optimal control variables $n_E{}^*$ and $\eta_E{}^*$ are discontinuous at the transition point t_{sing} into a singular subarc. The following conditions can be shown,

$$U_E(t_{\text{sing}}{}^-) = 0\,, \tag{50}$$

$$\dot{\chi}_E(t_{\text{sing}}{}^-) = 0 \quad \Rightarrow \quad \begin{cases} \sin\mu_E{}^*(t_{\text{sing}}{}^-) = 0\,, \\[2mm] \cos\mu_E{}^*(t_{\text{sing}}{}^-) \in \{-1, 1\}\,, \end{cases} \tag{51}$$

$$\dot{\lambda}_{\gamma_E}(t_{\text{sing}}{}^-) = 0\,, \tag{52}$$

$$\dot{\lambda}_{\chi_E}(t_{\text{sing}}{}^-) = 0\,. \tag{53}$$

3.5 Two-Point Boundary Conditions

The initial conditions are given by Eqs. (2). In order to obtain conditions at $t = t_f$ for the adjoint variables, the terminal manifold described by Eq. (3) is parametrized; see [17]. Partial differentiation of the value $V^*(\theta(t_f), z(t_f))$ with respect to the parameters yields a system of linear equations for $\lambda(t_f)$. One obtains, for trajectories not ending on the dynamic pressure limit,

$$\lambda_{x_E}(t_f)\,[y_E(t_f) - y_P(t_f)] - \lambda_{y_E}(t_f)\,[x_E(t_f) - x_P(t_f)] = 0\,,$$

$$\lambda_{x_E}(t_f)\,[h_E(t_f) - h_P(t_f)] - \lambda_{h_E}(t_f)\,[x_E(t_f) - x_P(t_f)] = 0\,,$$

$$\lambda_{x_P}(t_f) + \lambda_{x_E}(t_f) = 0\,, \qquad \lambda_{y_P}(t_f) + \lambda_{y_E}(t_f) = 0\,,$$

$$\lambda_{h_P}(t_f) + \lambda_{h_E}(t_f) = 0\,, \tag{54}$$

$$\lambda_{v_P}(t_f) = 0\,, \qquad\qquad \lambda_{v_E}(t_f) = 0\,,$$

$$\lambda_{\gamma_P}(t_f) = 0\,, \qquad\qquad \lambda_{\gamma_E}(t_f) = 0\,,$$

$$\lambda_{\chi_P}(t_f) = 0\,, \qquad\qquad \lambda_{\chi_E}(t_f) = 0\,,$$

$$\lambda_\theta(t_f) = 1\,,$$

and for those which,

$$\lambda_{x_E}(t_f)\left[y_E(t_f) - y_P(t_f)\right] - \lambda_{y_E}(t_f)\left[x_E(t_f) - x_P(t_f)\right] = 0,$$

$$\lambda_{x_E}(t_f)\left[h_E(t_f) - h_P(t_f)\right] - \lambda_{h_E}(t_f)\left[x_E(t_f) - x_P(t_f)\right] = 0,$$

$$\lambda_{x_P}(t_f) + \lambda_{x_E}(t_f) = 0, \qquad \lambda_{y_P}(t_f) + \lambda_{y_E}(t_f) = 0,$$

$$\lambda_{h_P}(t_f) + \lambda_{h_E}(t_f) - \frac{1}{2}\,\lambda_{v_E}(t_f)\,\frac{\partial \ln(\rho(h_E(t_f)))}{\partial h_E(t_f)}\,v_E(t_f) = 0, \qquad (55)$$

$$\lambda_{v_P}(t_f) = 0, \qquad\qquad \lambda_{v_E}(t_f) = \text{arbitrarily nonnegative},$$

$$\lambda_{\gamma_P}(t_f) = 0, \qquad\qquad \lambda_{\gamma_E}(t_f) = 0,$$

$$\lambda_{\chi_P}(t_f) = 0, \qquad\qquad \lambda_{\chi_E}(t_f) = 0,$$

$$\lambda_\theta(t_f) = 1.$$

Together with the Isaacs equation

$$H(\theta, z, \lambda_\theta, \lambda, n_P{}^*, \mu_P{}^*, n_E{}^*, \mu_E{}^*, \eta_E{}^*)\big|_{t=t_f} = 0, \qquad (56)$$

the adjoint variables are determined at $t = t_f$. An additional terminal condition which determines the final time t_f is given by the capture condition (3).

3.6 Singular Subarcs

On state-constrained subarcs

$$q_E(h_E, v_E) \equiv q_{E,\max}, \qquad (57)$$

the optimal controls may become singular indicated by $U_E \equiv 0$. This equation is equivalent to

$$\lambda_{\gamma_E} \;\equiv\; 0, \qquad\qquad (58)$$

$$\lambda_{\chi_E} \;\equiv\; 0. \qquad\qquad (59)$$

These three identities define a universal-type singular surface of codimension three.

For the 12 state variables, the 13 adjoint variables, and the parameters t_{entry}, t_{sing}, and t_f, there are 28 interior and boundary conditions to be satisfied, namely the 12 initial conditions (2), the capture condition (3), the 13 terminal conditions (55) and (56), the entry condition (48) into the state-constrained subarc, and the entry condition (50) into the singular subarc.

It should be mentioned that Eq. (50) itself cannot be chosen for numerical reasons. Instead, one of the conditions

$$\lambda_{\gamma_E}(t_{\text{sing}}) \;=\; 0, \qquad\qquad (60)$$

$$\lambda_{\chi_E}(t_{\text{sing}}) \;=\; 0. \qquad\qquad (61)$$

is prescribed. The other one is checked afterwards and can always be found to be fulfilled with an absolute error of at most 10^{-6}.

If additional constraints become active here, e.g., the altitude constraint (6), the switching structure might become even more complicated; see [12] and Seywald, Cliff, and Well [28] for similar problems in optimal control where more complicated switching structures occur due to prescribed terminal conditions.

This section now completes the derivation of well-defined multipoint boundary-value problems for all switching structures that have been obtained during the numerical computations by means of the multiple shooting method.

4 Numerical Results

Although superior with respect to accuracy and reliability, the multiple shooting method suffers from the fact that an initial estimate of all variables must be provided. This difficulty is surmounted by using a trajectory from [21] as an initial guess where flight scenarios in a vertical plane have been investigated. For if the initial positions and the velocity vectors of P and E are placed in a common vertical plane, the optimal trajectories remain in this plane. The procedure in [21] is based on the following idea: First, an optimal control problem is established which is related to a family of differential games including the differential game to be solved. Secondly, the optimal control problem is solved by means of a robust direct collocation method due to [30] which is not so exacting with respect to an appropriate initial guess. This method additionally yields an approximation for the adjoint variables and a hypothesis of the switching structure. Thereafter, the multiple shooting method is employed in connection with homotopy techniques to solve the differential games. Finally, the trajectories for arbitrary initial conditions in the capture zone are computed by means of homotopy techniques as well.

To illustrate this, the following set of initial conditions is chosen:

$$
\begin{aligned}
x_{P,0} &= 0 \,[\mathrm{m}], & x_{E,0} &= 12500 \,[\mathrm{m}], \\
y_{P,0} &= 0 \,[\mathrm{m}], & y_{E,0} &= 1000 \,[\mathrm{m}], \\
h_{P,0} &= 5000 \,[\mathrm{m}], & h_{E,0} &= 6000 \,[\mathrm{m}], \\
v_{P,0} &= 250 \,[\mathrm{m/s}], & v_{E,0} &= 400 \,[\mathrm{m/s}], \\
\gamma_{P,0} &= 0 \,[\mathrm{deg}], & \gamma_{E,0} &= 0 \,[\mathrm{deg}], \\
\chi_{P,0} &= 0 \,[\mathrm{deg}], & \chi_{E,0} &= 120 \,[\mathrm{deg}].
\end{aligned}
\tag{62}
$$

Figure 1 shows the optimal flight paths of P and E in a (x, y, h)-coordinate system. Additionally, the projections of the flight paths into the coordinate

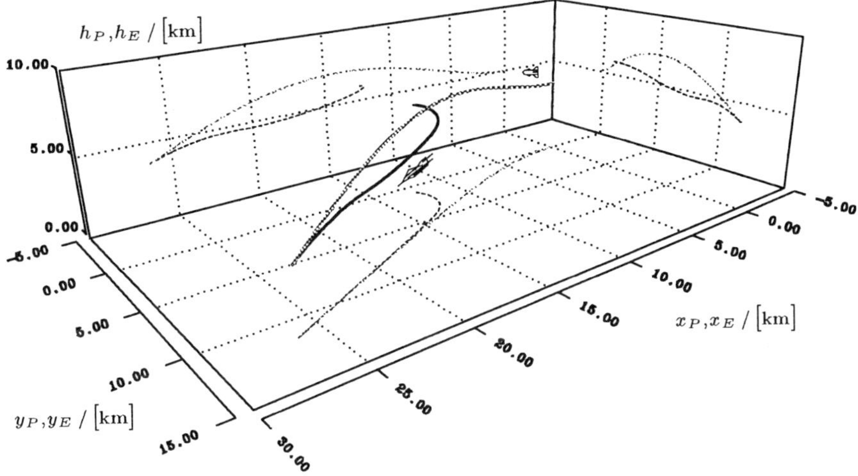

Figure 1: Optimal flight path of P (gray) and E (black) and projections into the coordinate planes.

planes are drawn (thin dashed lines). It can be seen that the optimal flight paths of P and E enter a common vertical plane. The missile P ascends immediately after the launch in order to exploit the low drag in the higher regions of the atmosphere. Since the initial angle $\chi_{E,0} = 120$ [deg] is quite unfavourable, the aircraft E is forced to turn right as sharply as possible. Hence, a $n_{E,max}$-constrained subarc occurs during the first 7 seconds of the maneuver; see Fig 3.

Figure 2 shows the history of the optimal trajectory in an altitude-velocity diagram. The shaded area marks the set of all pairs (h_E, v_E) for which E can maintain a stationary horizontal flight. The boundary of this set is called the flight envelope of the aircraft E. Note that the lower right part of the flight envelope is determined by the dynamic pressure constraint. It should be mentioned that only the mid-course guidance of the missile is modelled appropriately here. The final approach immediately before the hit

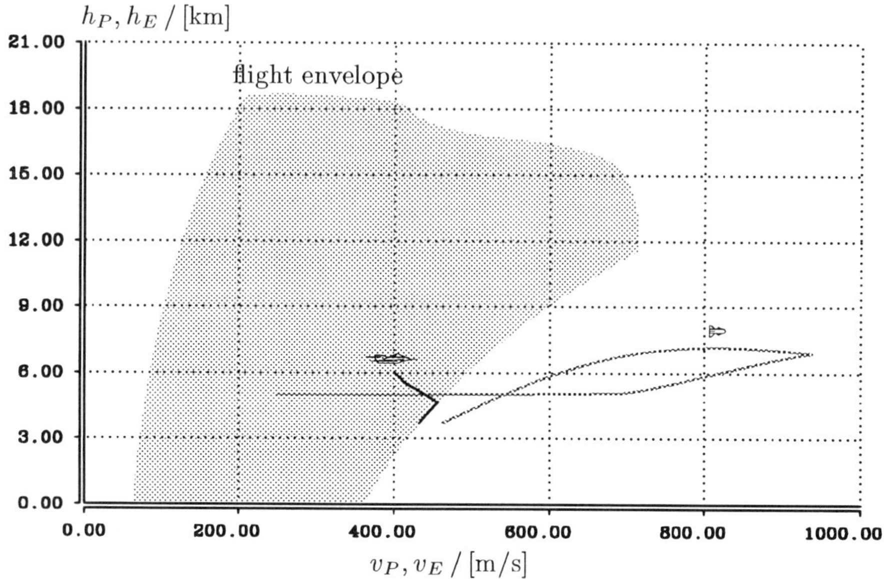

Figure 2: History of h and v for P (gray) and E (black).

must be left to special guidance laws. The corners in the curve of P mark
the two transitions between the different thrust phases of the missile. The
effects of these transition points also govern the history of the optimal load
factor n_P^* in Fig. 3. The corner in the optimal load factor n_E^* at $t = 27$ [s]
marks the junction point between the inactive and active dynamic pressure
constraints. At $t = 29$ [s] the entry point of the singular subarc is reached.
Note that n_E^* as well as μ_E^* are discontinuous at t_{sing}; see Fig. 4. Figure 5
shows the history of the optimal throttle η_E^* which is also discontinuous
at t_{sing}. However, the jump in η_E^* is below drawing accuracy.

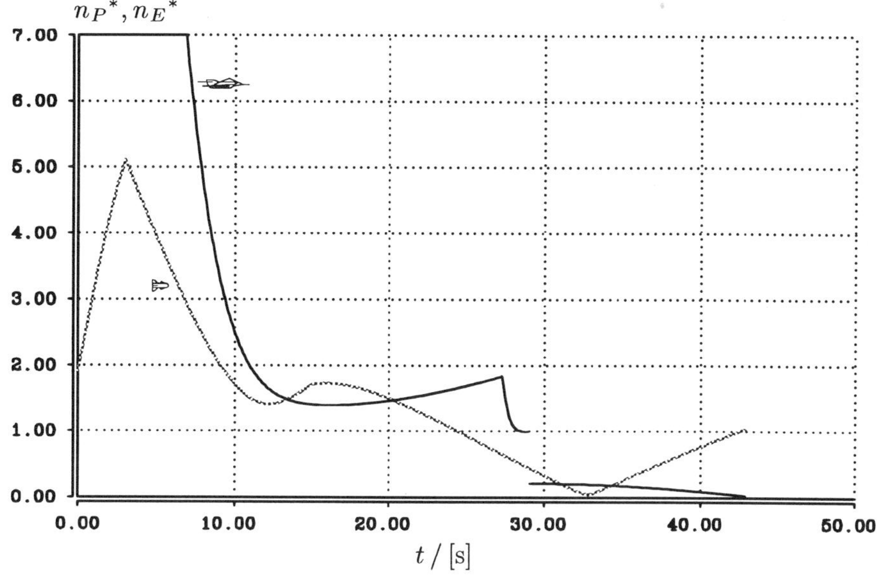

Figure 3: Optimal load factors for P (gray) and E (black).

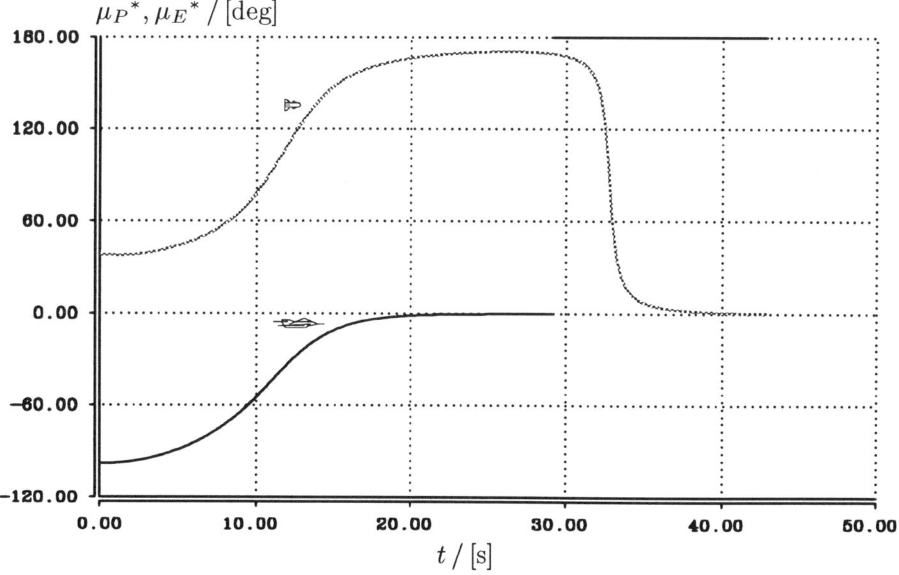

Figure 4: Optimal bank angles for P (gray) and E (black).

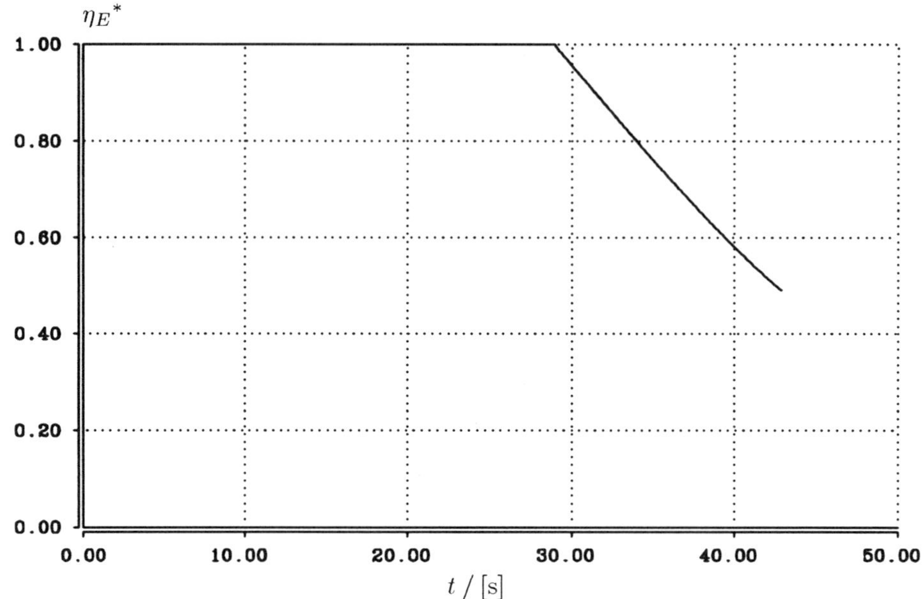

Figure 5: Optimal throttle for E.

In Fig. 6 eight optimal flight paths are depicted in the (x, y, h)-space. For the sake of visibility, only the projections into the (x, y)-plane are drawn. These flight paths belong to the initial conditions

$$
\begin{aligned}
x_{P,0} &= 0\,[\text{m}]\,, \\
y_{P,0} &= 0\,[\text{m}]\,, & y_{E,0} &= 0\,[\text{m}]\,, \\
h_{P,0} &= 5000\,[\text{m}]\,, & h_{E,0} &= 10500\,[\text{m}]\,, \\
v_{P,0} &= 250\,[\text{m/s}]\,, & v_{E,0} &= 300\,[\text{m/s}]\,, \\
\gamma_{P,0} &= 0\,[\text{deg}]\,, & \gamma_{E,0} &= 0\,[\text{deg}]\,, \\
\chi_{P,0} &= 0\,[\text{deg}]\,, & \chi_{E,0} &= 90\,[\text{deg}]\,,
\end{aligned}
\tag{63}
$$

$$
x_{E,0}\,/\,[\text{m}] \in \{3000, 4500, 6000, 7500, 9000, 10500, 12000, 13500\}\,.
$$

The dynamic pressure constraint becomes active only for the trajectory with the largest initial distance. The two optimal trajectories related to the smallest and the largest value of $x_{E,0}$ lie close to the barrier. The differences of the optimal flight paths, depending on the initial distance between P and E, are considerable.

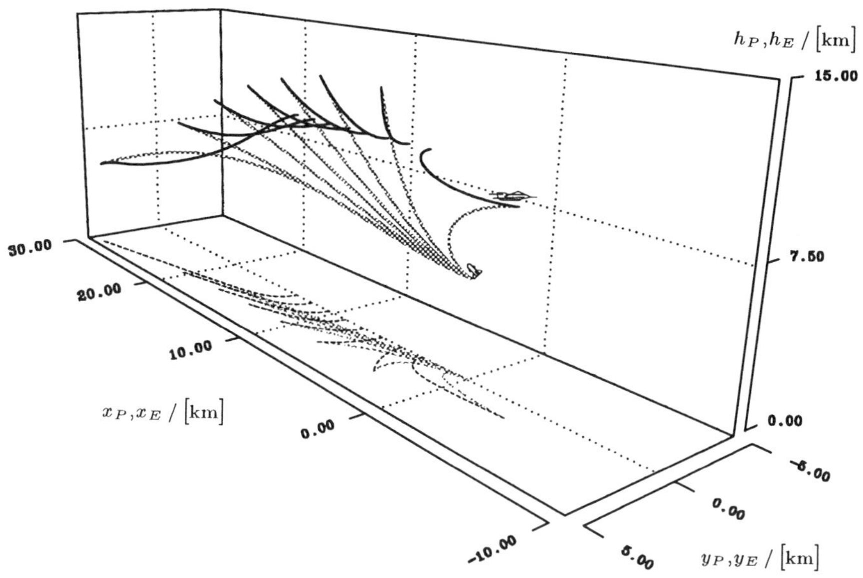

Figure 6: Optimal flight paths for different initial constellations. Flight paths of P are gray and those of E are black. Initial conditions differ only in $x_{E,0}$.

Finally, Fig. 7 shows seven optimal flight paths related to the following set of initial conditions

$$
\begin{aligned}
x_{P,0} &= 0\,[\text{m}]\,, & x_{E,0} &= 15000\,[\text{m}]\,, \\
y_{P,0} &= 0\,[\text{m}]\,, & y_{E,0} &= 0\,[\text{m}]\,, \\
h_{P,0} &= 10000\,[\text{m}]\,, & h_{E,0} &= 10000\,[\text{m}]\,, \\
v_{P,0} &= 250\,[\text{m/s}]\,, & v_{E,0} &= 500\,[\text{m/s}]\,, \\
\gamma_{P,0} &= 0\,[\text{deg}]\,, & \gamma_{E,0} &= 0\,[\text{deg}]\,, \\
\chi_{P,0} &= 0\,[\text{deg}]\,,
\end{aligned}
\tag{64}
$$

$$
\chi_{E,0}\,/\,[\text{deg}] \in \{0, 30, 60, 90, 120, 150, 177\}\,.
$$

As in Fig. 6, the turns of the optimal paths into common vertical planes can be seen. Each trajectory terminates on a singular subarc. Note that the almost head-on initial constellation for $\chi_{E,0} = 177$ [deg] is close to a dispersal surface in the state space beyond which a symmetric situation can be found. For the tail-chase initial constellations, the missile P again ascends into the regions of low drag. This behavior is optimal for P if the duration of the game is not too short.

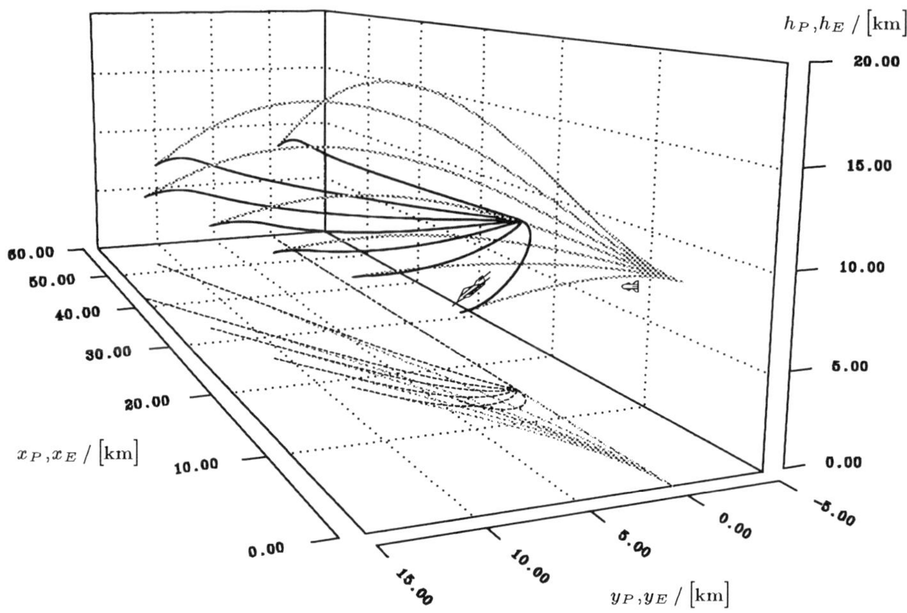

Figure 7: Optimal flight paths for different initial constellations. Flight paths of P are gray, those of E are black. Initial conditions differ only in $\chi_{E,0}$.

5　Conclusions

Optimal trajectories and their associated open-loop representations of the optimal strategies can be computed for complicated pursuit-evasion games in the entire capture zone. These open-loop representations provide information by which optimal strategies can be approximated globally, e. g., by the subsequent solution of neighboring boundary-value problems [2], by Taylor series expansions around many optimal trajectories ([3], and [4]), or by neural networks [27]. The last two of these methods are real-time feasible. This idea is generally applicable and not confined to the air-combat problem investigated in this paper.

This air-combat problem includes a first-order state variable inequality constraint for the dynamic pressure of the aircraft. The numerical solutions of this differential game have been obtained by applying the known necessary conditions of optimal control theory. In addition, the controls may become singular on state-constrained subarcs. The associated singular surface is of universal type and has codimension three. All the necessary conditions have been examined for not contradicting the minimax principle. More detailed and rigorous investigations concerning differential games with state constraints of order greater than zero are still pending.

Acknowledgements

The authors would like to thank Professor R. Bulirsch of Munich University of Technology, Department of Mathematics, for his encouragement. The authors are also indebted to the Reviewers for their helpful comments. This work has been partly funded by the German National Science Foundation within the Project "Applied Optimization and Control", and by FORTWIHR, the Bavarian Consortium on High Performance Scientific Computing.

REFERENCES

[1] BAŞAR, T. , and OLSDER, G. J., *Dynamic Noncooperative Game Theory*, Academic Press, London, 1982, 2nd ed., 1994.

[2] BREITNER, M. H., *Construction of the Optimal Feedback Controller for Constrained Optimal Control Problems with Unknown Disturbances*, in: R. Bulirsch and D. Kraft (Eds.), Computational Optimal Control, International Series of Numerical Mathematics **115**, Birkhäuser, Basel, 1994, 147–162.

[3] BREITNER, M. H., *Real-Time Capable Approximation of Optimal Strategies in Complex Differential Games*, in: M. Breton, G. Zaccour (Eds.), Proceedings of the Sixth International Symposium on Dynamic Games and Applications, St-Jovite, Québec, GERAD, École des Hautes Études Commerciales, Montreal, 1994 370–374.

[4] BREITNER, M. H., *Robust-optimale Rückkopplungssteuerungen gegen unvorhersehbare Einflüsse: Differentialspielansatz, numerische Berechnung und Echtzeitapproximation*, PhD thesis, in preparation.

[5] BREITNER, M. H., and PESCH H. J., *Reentry Trajectory Optimization under Atmospheric Uncertainty as a Differential Game*, in: T. Başar, and A. Haurie (Eds.), Advances in Dynamic Games and Applications, Annals of the International Society of Dynamic Games **1**, 1994, Birkhäuser, Basel, 70–88.

[6] BREITNER, M. H., GRIMM, W., and PESCH, H. J., *Barrier Trajectories of a Realistic Missile/Target Pursuit-Evasion Game*, in: R. P. Hämäläinen, and H. K. Ehtamo (Eds.), Differential Games — Developments in Modelling and Computation, Lecture Notes in Control and Information Sciences **156**,1991, Springer, Berlin, 48–57.

[7] BREITNER, M. H., PESCH, H. J., and GRIMM, W., *Complex Differential Games of Pursuit-Evasion Type with State Constraints, Part 1: Necessary Conditions for Optimal Open-Loop Strategies, Part 2: Numerical Computation of Optimal Open-Loop Strategies*, Journal of Optimization Theory and Applications **78**, 1993, 419–441, 443–463.

[8] BRYSON, A. E., DENHAM, W. F., and DREYFUS, S. E., *Optimal Programming Problems with Inequality Constraints, I: Necessary Conditions for Extremal Solutions*, AIAA Journal **1**, 1963, 2544–2550.

[9] BULIRSCH, R., *Die Mehrzielmethode zur numerischen Lösung von Randwertproblemen und Aufgaben der optimalen Steuerung*, Report of the Carl-Cranz-Gesellschaft, DLR, Oberpfaffenhofen, Germany, 1971; Reprint, Department of Mathematics, Munich University of Technology, Munich, 1985 and 1993.

[10] FINK, G., *Numerische Berechnung optimaler Strategien für ein realitätsnahes Differentialspiel einer Verfolgung im dreidimensionalen Raum*, Diploma thesis, Department of Mathematics, Munich University of Technology, Munich, 1994.

[11] GREENWOOD, N. J. C., *Applied Differential Games in the Aerial Scenario*, PhD thesis, Department of Mathematics, University of Queensland, Brisbane, 1994.

[12] GRIMM, W., *Lenkung von Hochleistungsflugzeugen. Vergleich von optimaler Steuerung und fastoptimaler Echtzeitsteuerung*, PhD thesis, Department of Mathematics, Munich University of Technology, Munich, 1992.

[13] GUELMAN, M., SHINAR, J., and GREEN, J., *Qualitative Study of a Planar Pursuit Evasion Game in the Atmosphere*, in: Proceedings of the AIAA Guidance, Navigation and Control Conference, Minneapolis, 1988.

[14] GUTMAN, S., and KATZ, D., *On Guaranteed-Cost, Closed-Form Guidance via Simple Linear Differential Games*, in: Proceedings of the 27th IEEE Conference on Decision and Control, Austin, 1988.

[15] HILTMANN, P., CHUDEJ, K., and BREITNER, M. H., *Eine modifizierte Mehrzielmethode zur Lösung von Mehrpunkt-Randwertproblemen — Benutzeranleitung*, Report No. 14, Deutsche Forschungsgemeinschaft, Sonderforschungsbereich 255 "Transatmosphärische Flugsysteme", Department of Mathematics, Munich University of Technology, Munich, Germany, 1993.

[16] ISAACS, R., *Games of Pursuit*, Paper No. P-257, RAND Corporation, Santa Monica, California, 1951.

[17] ISAACS, R., *Differential Games*, Wiley, New York, New York, 1965; Krieger, New York, New York, 1975.

[18] JACOBSON, D. H., LELE, M. M., and SPEYER, J. L., *New Necessary Conditions of Optimality for Control Problems with State Variable Inequality Constraints*, Journal of Mathematical Analysis and Applications **35**, 255–284, 1971.

[19] LACHNER, R., *Realitätsnahe Modellierung eines Differentialspieles mit Zustandsbeschränkungen und numerische Berechnung optimaler Trajektorien*, Diploma thesis, Department of Mathematics, Munich University of Technology, Munich, Germany, 1992.

[20] LACHNER, R., BREITNER, M. H., and PESCH, H. J., *Optimal Strategies of a Complex Pursuit-Evasion Game*, in: R. Fritsch, and M. Toepell (Eds.), Proceedings of the 2nd Gauss Symposium, Munich, Germany, 1993; to appear in Journal of Computing and Information, 1995.

[21] LACHNER, R., BREITNER, M. H., and PESCH, H. J., *Efficient Numerical Solution of Differential Games with Application to Air Combat*, Report No. 466, Deutsche Forschungsgemeinschaft, Schwerpunkt "Anwendungsbezogene Optimierung und Steuerung", Department of Mathematics, Munich University of Technology, Munich, 1993.

[22] LEITMANN, G., *Einführung in die Theorie optimaler Steuerung und der Differentialspiele*, R. Oldenburg, Munich, 1974.

[23] MAURER, H., *Optimale Steuerprozesse mit Zustandsbeschränkungen*, Habilitationsschrift, University of Würzburg, Würzburg, 1976.

[24] MIELE, A., *Flight Mechanics I, Theory of Flight Paths*, Addison Wesley, Reading, 1962.

[25] PESCH, H. J., *Offline and Online Computation of Optimal Trajectories in the Aerospace Field*, in: A. Miele and A. Salvetti (Eds.), Applied Mathematics in Aerospace Science and Engineering, Plenum, New York, 1994, 165–219.

[26] PESCH, H. J., *Solving Optimal Control and Pursuit-Evasion Game Problems of High Complexity*, in: R. Bulirsch, and D. Kraft (Eds.), Computational Optimal Control, International Series of Numerical Mathematics **115**, Birkhäuser, Basel, 1994, 43–64.

[27] PESCH, H. J., GABLER, I., MIESBACH, S., and BREITNER, M. H., *Synthesis of Optimal Strategies for Differential Games by Neural Networks*, in: G. J. Olsder (Ed.), Advances in Dynamic Games and Applications, Annals of the International Society of Dynamic Games, this volume, Birkhäuser, Boston, 1995.

[28] SEYWALD, H., CLIFF, E. M., and WELL, K. H., *Range Optimal Trajectories for an Aircraft Flying in the Vertical Plane*, Journal of Guidance, Control, and Dynamics **17**, 1994, 389-398.

[29] STOER, J., BULIRSCH, R., *Introduction to Numerical Analysis*, Springer, New York, 2nd ed., 1993.

[30] VON STRYK, O., *Numerische Lösung optimaler Steuerungsprobleme: Diskretisierung, Parameteroptimierung und Berechnung der adjungierten Variablen*, Fortschritt-Berichte VDI, Reihe 8, Nr. 441, VDI-Verlag, Düsseldorf, 1995

Control of Informational Sets in a Pursuit Problem*

S.I. Kumkov and V.S. Patsko

Institute of Mathematics & Mechanics

S.Kovalevskaya str., 16, Ekaterinburg, 620219, Russia

Abstract

The paper deals with one model problem of pursuit with incomplete information. The pursuer makes an error in measuring the angular velocity of the pursuer-evader line of sight and tries to minimize the final miss. Corresponding to the original problem, the auxiliary differential game is formulated, where the informational set and the residual resource of impulse control constitute the position of the game. The first player's strategy is the base for constructing pursuer's control in the original problem, the second player's strategy gives the "worst" method for producing disturbed measurements.

1   Introduction

In "usual" differential games, the game position consists of a time moment and a phase state at that moment. In games with incomplete information and geometric restrictions on a measurement error, the game position consists of a time moment and an informational set. The last describes a totality of the system phase states, coordinated with the history of the process.

From the theoretical viewpoint, the differential games with incomplete information under geometric restrictions on a measurement disturbance were intensively investigated in the mid 70s [1, 2, 3, 4, 5]. At the same time, pithy examples of such problems, solved analytically or with computer realized algorithms, were absent. This paper deals with one such problem. A similar problem had been investigated in [6, 7] under the assumption of statistical distribution of the measurement error.

The paper is devoted to a pursuit problem with incomplete information. As prototypes, problems of pursuit and approach in the space can be used when the pursuer-vehicle (PV) has a restricted resource for the impulse changing of its velocity with a determined value for every control impulse. The case when PV's control operates orthogonally to the vector of initial

---

*Research supported by the Russian Foundation of Fundamental Investigations (94-01-00350).

nominal relative velocity is typical. In the homing process, PV measures an angular velocity of the pursuer-evader line of sight (LOS) with an error and tries to minimize the final miss. The error obeys a geometric restriction. The evader's vector control has dimension (meaning) of acceleration and is also restricted geometrically. The additional assumption is as follows: the relative velocity variations appearing in consequence of the pursuer and evader controls are small. It is supposed that initial uncertainties of location and velocity are relatively small also. It is necessary to construct the pursuer's feedback control law providing satisfactory decision in the game problem of miss minimization under incomplete information.

In the mathematical investigation of the problem, the case of pursuit in the plane is considered. Our approach is connected with construction of the informational sets and control of them. The formalized statement of the problem is given in frames of an auxiliary antagonistic differential game of two persons with incomplete information. The first player governs by impulse control, the second player forms the measurements. The evader's acceleration is taken into account in the dynamics of the informational set development. The game position includes the time moment, the informational set and the residual store of the impulse control. The definitions of the best guaranteed results for the first and the second players are given. The players' strategies are suggested. The strategies become optimal in one particular and natural case. So, the considered problem is one of a small number [3, 8, 9], for which one has succeeded in finding optimal (or close to optimal) solutions.

The strategies of the auxiliary differential game are applied in the original problem. The results of the numerical computer simulation of the homing process are represented.

2   Description of the pursuit problem

We consider the pursuit problem of one material point to another in the plane when pursuer P tries to catch up with the evader E under incomplete information about the latter. Put up zero of the relative coordinate system (RCS) at the P point. Suppose that at the initial moment t_0, both the nominal (precalculated) location of the evader E_{NOM} in RCS and the nominal vector V_{NOM} of the relative velocity are given. Let us consider also that the vector V_{NOM}, applied at the point E_{NOM}, is directed to the point P, i.e. to the beginning of the RCS. The axis x of the RCS system is directed opposite to the vector V_{NOM} (Fig. 1), and the axis z is orthogonal to the axis x. Directions of the axes are constant on t. The true location of E at the initial moment can differ from the nominal one, and the true initial vector of relative velocity can differ from the preliminary given one. Modulus of the vector V_{NOM} is denoted as e. The control of P is impulse-like. The pursuer

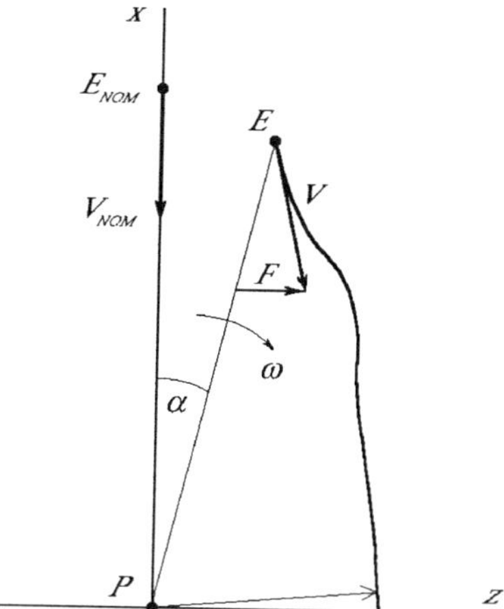

Figure 1: Motion of the relative point.

can change its velocity (and consequently the relative velocity) by a jump. It is supposed that control impulses are orthogonal to the axis x, i.e., they are oriented along the axis z. Such direction of impulses provides the economic expenditure of the control store in the miss minimization problem. Both the value σ of one impulse and the initial store N of impulses are given a priori. Hence $N\sigma$ is the pursuer's store for changing the velocity. The vector control v of the evader with components v_x, v_z has dimension of an acceleration and is restricted as $v \in Q$, where Q is a convex closed bounded set, symmetrical relative to zero.

The pursuer's control is realized in a discrete scheme. Let $k(t_i)$ be some number of impulses with prescribed sign, given at a moment t_i ($k(t_i)$ is either negative integer, or a positive integer, or zero). Corresponding to the problem formulation, the restriction $\sum_i \mid k(t_i) \mid \le N$ must be satisfied. At every current discrete moment t_i, the pursuer evaluates the LOS angular velocity $\omega_M(t_i)$. The measured value and true magnitude $\omega(t_i)$ of the angular velocity are connected by the relation

$$\omega_M(t_i) = \omega(t_i) + \xi(t_i). \tag{1}$$

Here, $\xi(t_i)$ is a measurement error. We assume that the error obeys the restriction

$$\mid \xi(t_i) \mid \le c_1 \mid \omega(t_i) \mid + c_2, \quad 1 > c_1 \ge 0, \quad c_2 \ge 0. \tag{2}$$

If $c_1 = 0$, then the restriction does not depend on the true value of the angular velocity. If a constant c_2 is equal to zero, then relation (2) gives the restriction on relative error in measurement of the angular velocity.

We suppose that the coordinates of the initial state $(x(t_0), z(t_0))^T$ of the relative point are given inexactly and bounded by the geometric restriction

$$(x(t_0), z(t_0))^T \in B. \tag{3}$$

Similarly, the initial value of the relative velocity is

$$(V_x(t_0), V_z(t_0))^T \in D. \tag{4}$$

Here B and D are convex closed bounded sets. Relations $(1)-(4)$ are known to the pursuer.

The goal of pursuit is the minimization of the miss. We mean the miss is the minimal distance $\sqrt{x^2(t) + z^2(t)}$ in the process of moving P and E. The aim of this study is to construct the feedback pursuer control, which gives satisfactory result in the game problem of miss minimization under incomplete information.

The problem investigation is carried out under the following simplifying assumption of weak controllability. We shall suppose that the relative velocity variations, which appear in consequence of the pursuer and evader controls, are small. Suppose, that the size of the set D and the swing of the set B along the axis z are small also.

3   Original dynamic equations. Transition to equivalent coordinates

The dynamic equations in relative coordinates x, z are:

$$\begin{aligned}
\ddot{z}(t) &= v_z - \sigma \sum_i k(t_i)\delta(t - t_i), \\
\ddot{x}(t) &= v_x,
\end{aligned} \tag{5}$$

$$\sum_i \mid k(t_i) \mid \leq N, \quad (v_x, v_z)^T \in Q.$$

Here δ is the delta-function symbol. We rewrite the system (5) in coordinates α, ω, x, V_x. To carry this out, let us twice differentiate on t the relation $z(t) = x(t)\mathrm{tg}\alpha(t)$. We obtain

$$\ddot{z}(t) = \ddot{x}(t)\mathrm{tg}\alpha(t) + \frac{2\dot{x}(t)\dot{\alpha}(t)}{\cos^2\alpha(t)} + \frac{2x(t)\mathrm{tg}\alpha(t)\dot{\alpha}^2(t)}{\cos^2\alpha(t)} + \frac{x(t)\ddot{\alpha}(t)}{\cos^2\alpha(t)}.$$

From here

$$\ddot{\alpha}(t) = \frac{\ddot{z}(t)\cos^2\alpha(t)}{x(t)} - \frac{\ddot{x}(t)\sin\alpha(t)\cos\alpha(t)}{x(t)} - \frac{2\dot{x}(t)\dot{\alpha}(t)}{x(t)} - 2\mathrm{tg}\alpha(t)\dot{\alpha}^2(t).$$

We have as a result:

$$\dot{\alpha}(t) = \omega(t),$$
$$\dot{\omega}(t) = -2V_x(t)\omega(t)/x(t) - 2\mathrm{tg}\,\alpha(t)\omega^2(t) - \sin\alpha(t)\cos\alpha(t)v_x/x(t) +$$
$$+ \cos^2\alpha(t)v_z/x(t) - (\sigma\cos^2\alpha(t)/x(t))\sum_i k(t_i)\delta(t - t_i),$$
$$\dot{x}(t) = V_x(t), \tag{6}$$
$$\dot{V}_x(t) = v_x,$$

$$\sum_i \mid k(t_i)\mid\, \le N, \quad (v_x, v_z)^T \in Q.$$

4 Dynamic equations for the auxiliary differential game

The system in equivalent coordinates is very handy because the measurable value ω is the phase variable. More than that, an opportunity of passing on to auxiliary problems with different degrees of simplification appears.

Indeed, the assumption about weak controllability permits us to account the angle α as small on a respectively long interval of the motion, which begins from the initial moment. Let us assume that at some moment t_* the angle α is large. Note that the cone of possible future motions is small (the cone axis is close to vertical). So the final miss will be large a fortiori, and the miss will be the greater when the time interval increases from the moment t_* up to the final moment. Hence, if we are interested in smaller misses, then we can consider the angle α to be small. So, in simplified formulation of the auxiliary game, let us change $\sin\alpha$ and $\mathrm{tg}\,\alpha$ to 0 and $\cos\alpha$ to 1. Because the measurable value is ω and the angle α now does not enter into relation for $\dot{\omega}$, the equation $\dot{\alpha}(t) = \omega(t)$ can be omitted. In doing so, we reduce the phase vector dimension on one and obtain the three-dimensional auxiliary problem. In this investigation, we make additional simplifications and pass to the two-dimensional problem.

The assumptions about weak controllability and orientation of the vector V_{NOM} along the axis x lead to the fact that calculation of the miss for every motion can be approximately substituted by evaluation of the coordinate z modulus at the moment when the motion crosses the axis z (Fig. 1). The variation of the velocity along the axis z affects the miss more strongly than its variation along the axis x. Simplifying the system (6), we suppose $V_x(t_0)$ to be known exactly and equal $V_{NOM x} = -e$, and control v_x to be equal to zero identically. The corresponding variation of the coordinate x is described by relation $x(t) = x(t_0) - e(t - t_0)$. Suppose at last that in simplified formulation, the value v_z is chosen from the interval $[-\nu, \nu]$, which is the projection of the set Q on the axis z. As a result, we obtain:

$$\dot{\omega}(t) \;=\; 2e\omega(t)/x(t) + v_z/x(t) - (\sigma/x(t))\sum_i k(t_i)\delta(t - t_i),$$

$$x(t) \;=\; x(t_0) - e(t - t_0),$$

$$\sum_i \mid k(t_i) \mid \le N, \quad \mid v_z \mid \le \nu. \tag{7}$$

When impulse control is applied at a moment t_i, the coordinate ω changes step-wise on a value $-k(t_i)\sigma/x(t_i)$. So the auxiliary differential game dynamics has been described. Formulas (1), (2) relate the value $\omega(t_i)$ and its measured value $\omega_M(t_i)$. We take the initial uncertainties as

$$x(t_0) \in [x_0, x^0], \quad \omega(t_0) \in A(x(t_0)). \tag{8}$$

Here, $[x_0, x^0]$ is the projection of the set B on the axis x; $A(x(t_0))$ is an interval (upper evaluation), which contains for every $x(t_0)$ all initial values $\omega(t_0)$ of the angular velocity admissible with the restrictions (3), (4).

We give now a formula for coordinate ω changing on an interval $[\hat{t}, t]$:

$$\omega(t) = \omega(\hat{t})\frac{x^2(\hat{t})}{x^2(t)} + \frac{1}{x^2(t)} \int_{\hat{t}}^{t} x(\tau)v_z(\tau)d\tau - \frac{\sigma}{x^2(t)} \sum_{t_i \in [\hat{t},\, t]} k(t_i)x(t_i). \tag{9}$$

A maximal (minimal) value of the second addend is achieved at $v_z \equiv \nu$ $(v_z \equiv -\nu)$. We let

$$\zeta(\hat{t}, t) = \frac{1}{x^2(t)} \int_{\hat{t}}^{t} x(\tau)\nu d\tau = \frac{\nu(t - \hat{t})}{x^2(t)} \left(x(\hat{t}) - \frac{e}{2}(t - \hat{t}) \right). \tag{10}$$

5   Formalization of the auxiliary differential game with incomplete information

Now we shall formulate an auxiliary differential game of two players, in which the game position at a moment t_i is a pair: an informational set in the plane ω, x and a residual quantity of control impulses. Consider the axis ω to be horizontal, and the axis x to be vertical. Let us fix a parameter $\varepsilon > 0$.

5.1   Informational sets

As an initial informational set $I_\ominus(t_0)$, let us take an arbitrary closed bounded set in the plane ω, x $(x \ge \varepsilon)$, whose section at any x is an interval.

Let now a bounded closed set $I_\ominus(t_i)$, $i \ge 0$, be given in the space ω, x $(x \ge \varepsilon)$ so that its section for any x is an interval. So $I_\ominus(t_i)$ is a union of intervals, parallel to the axis ω. At a moment t_i, a measurement ω_M appears. Denote as $H(t_i)$ the uncertainty set, which corresponds to the measurement. This set is a totality of all points in the plane ω, x such that, for every point, the received measurement ω_M is admissible in accordance with formulas $\omega_M = \omega + \xi$, $|\xi| \leq c_1 |\omega| + c_2$. The set H is a vertical band. Its projection $\mathcal{H}$ on the axis ω is described by the relations

$$\mathcal{H} = \begin{cases} [(\omega_M - c_2)/(1+c_1), (\omega_M + c_2)/(1-c_1)], & \text{if} \quad \omega_M \geq c_2, \\ [(\omega_M - c_2)/(1-c_1), (\omega_M + c_2)/(1-c_1)], & \text{if} \quad -c_2 < \omega_M < c_2, \\ [(\omega_M - c_2)/(1-c_1), (\omega_M + c_2)/(1+c_1)], & \text{if} \quad \omega_M \leq -c_2. \end{cases}$$

We put $I(t_i) = I_\ominus(t_i) \bigcap H(t_i)$. Suppose such an intersection is not empty.

At the moment t_i (after receiving a measurement), the first player, which governs the impulse choice, can apply his control. The set $I(t_i)$ comes into the set $I_\oplus(t_i)$. Transition $I(t_i) \to I_\oplus(t_i)$ is a shift of every interval (under constant x) on a value $-k(t_i)\sigma/x$ along the axis ω. Let us extract the part $J_\varepsilon(t_i)$ from the set $I_\oplus(t_i)$, which lies strictly below the level $x = \varepsilon + e(t_{i+1} - t_i)$. At the moment t_{i+1}, this part will go down under the level ε. Put $I_{\oplus\varepsilon}(t_i) = I_\oplus(t_i) \setminus J_\varepsilon(t_i)$.

We define $I_\ominus(t_{i+1})$ as a prognosis of the system (7) state at the moment t_{i+1} if the previous state at the moment t_i was $I_{\oplus\varepsilon}(t_i)$ and the first player's control on the interval $(t_i, t_{i+1}]$ was zero. When the set $I_\ominus(t_{i+1})$ is constructed, every interval from $I_{\oplus\varepsilon}(t_i)$ goes down along the axis x on a value $e(t_{i+1} - t_i)$. Its left edge coordinate on the axis ω takes a value (in accordance with formulas (9), (10)) $\omega_{*\ominus}(t_{i+1}) = \omega_{*\oplus}(t_i)x^2(t_i)/x^2(t_{i+1}) - \zeta(t_i, t_{i+1})$. Here, $\omega_{*\oplus}(t_i)$ is the left edge coordinate of the considered interval at the moment t_i. The state $\omega^*_\oplus(t_i)$ of the right edge changes to $\omega^*_\ominus(t_{i+1}) = \omega^*_\oplus(t_i)x^2(t_i)/x^2(t_{i+1}) + \zeta(t_i, t_{i+1})$.

Every set from the collection $I_\ominus(t_i), I(t_i), I_\oplus(t_i), I_{\oplus\varepsilon}(t_i)$ is called the informational set (before a measurement, after a measurement, after an impulse operation, after ε-cut-off). Let us call the set $I_\ominus(t_i)$ the prognosis set also. So the sequence of informational sets is defined recurrently.

In the differential game with incomplete information, a temporal changing of the informational set and a residual quantity of the control impulses is called a motion. The first player governs the impulse control, the second produces the measurement. We consider that the impulse control at a moment t_i is applied after the measurement received at this moment. The influence of the v_z is taken into account when the set $I_\ominus(t_i)$ is constructed. The parameter ε is introduced with the stipulation that, in the considered problem, its termination is connected with the passing of the level $x = 0$, but the system (7) has a singularity at $x = 0$.

5.2 Admissible strategies

A totality (t_i, n, I), where t_i is a time moment, n is the residual number of the control impulses, and I is the informational set after a measurement, is called a game position. As an admissible strategy of the first player, we denote the rule $U : (t_i, n, I) \to k$, which, for every position of the game, gives a corresponding number of impulses with necessary sign, and $\mid k \mid \leq n$.

Let consider the totality $(t_i, n, I_\ominus)$ as the game position of the second player. Here, $I_\ominus$ is an informational set before a measurement (the prognosis set). As an admissible strategy of the second player, we call the rule $\Omega :$ $(t_i, n, I_\ominus) \to \omega_M$, which, for every position of the game, gives a corresponding measurement ω_M. We stipulate that $I_\ominus \bigcap H \neq \emptyset$, where H is the uncertainty set, constructed by means of ω_M.

Having a pair of admissible strategies U, Ω, a step Δ (which connects adjacent discrete time moments t_i and t_{i+1}), a parameter ε, and an initial position $(t_0, n(t_0), I_\ominus(t_0))$, we have defined the motion of the system in time.

5.3 Payoff functional

Determine a payoff value on the motion in the auxiliary differential game. For an arbitrary pair ω, x $(x > 0)$, we let $\Pi(\omega, x) = \mid \omega \mid x^2/e$. The value $\Pi(\omega, x)$ is approximately a passive prognosis miss from the state ω, x, i.e. it is a miss corresponding to the moment when the axis z is crossed by free motion of the system (7). For exact calculation of the passive prognosis miss, it would be necessary to give not only ω, x, but the angle α also: $\mid \omega \mid x^2/(e \cos^2 \alpha)$. The meaning of this formula is clear from the fact that $\omega x / \cos^2 \alpha$ is a value of noncompensated linear velocity F (Fig. 1), and x/e is a residual time till the axis z is crossed. Let $\widehat{\Pi}(\omega, x) = \Pi(\omega, x) + \nu(x/e)^2/2$. The addition $\nu(x/e)^2/2$ is the maximum possible increasing of the miss, stipulated by an acceleration $v_z, \mid v_z \mid \leq \nu$. Put a number

$$\check{\Pi}(M) = \sup_{(\omega, x) \in M} \widehat{\Pi}(\omega, x)$$

in correspondence with any arbitrary set M in the space ω, x $(x > 0)$.

For every certain motion of the informational set, we note by symbol T_ε the totality of moments t_i such that $J_\varepsilon(t_i) \neq \emptyset$. Let us call the value

$$\Phi(t_0, n(t_0), I_\ominus(t_0), U, \Omega, \varepsilon, \Delta) = \max_{t_i \in T_\varepsilon} \check{\Pi}(J_\varepsilon(t_i))$$

a miss, which corresponds to the initial position $(t_0, n(t_0), I_\ominus(t_0))$, strategies U, Ω, discrete Δ, and parameter ε.

5.4   The best guaranteed results

The best strategies of the first and second players under fixed $t_0, n(t_0), I_\ominus(t_0)$ are determined by the relations

$$\Gamma^{(1)}(t_0, n(t_0), I_\ominus(t_0)) = \inf_U \varlimsup_{\varepsilon \to 0} \varlimsup_{\Delta \to 0} \sup_\Omega \Phi(t_0, n(t_0), I_\ominus(t_0), U, \Omega, \varepsilon, \Delta),$$

$$\Gamma^{(2)}(t_0, n(t_0), I_\ominus(t_0)) = \sup_\Omega \varliminf_{\varepsilon \to 0} \varliminf_{\Delta \to 0} \inf_U \Phi(t_0, n(t_0), I_\ominus(t_0), U, \Omega, \varepsilon, \Delta).$$

Strategies on which the best guaranteed results are achieved are called optimal.

Let us note the simplifications which appear in one particular case when $x(t_0)$ is supposed to be known exactly, i.e. when $x_0 = x^0$. In this case, the informational set consists of one interval only; the moment of the game termination becomes fixed and equal to $t_0 + x(t_0)/e$; for given ε and Δ, there is the unique moment t_i when $J_\varepsilon(t_i) \neq \emptyset$.

6   The first player's strategy

We denote as $\partial^* I$ the totality of the right edges of the intervals, which constitute the set I. Analogously, let $\partial_* I$ be the totality of the left edges. Suppose

$$\pi^*(I) = \max_{(\omega, x) \in \partial^* I} \left\{ \frac{\omega\, x^2}{e} + \frac{\nu}{2} \left(\frac{x}{e} \right)^2 \right\}, \tag{11}$$

$$\pi_*(I) = \min_{(\omega, x) \in \partial_* I} \left\{ \frac{\omega\, x^2}{e} - \frac{\nu}{2} \left(\frac{x}{e} \right)^2 \right\}. \tag{12}$$

Let $(\omega^*, x^*), (\omega_*, x_*)$ be such points, on which the maximum in (11) and minimum in (12) are achieved. Let us call the interval with edges $\pi_*(I), \pi^*(I)$ an interval of prognosis miss, which corresponds to the set I.

If at the current moment the first player applies the control with k impulses, then the miss for the former maximizing point in (11) is changed instantly on the value $\Delta \omega x^{*\,2}/e = -k\sigma x^{*\,2}/x^* e = -k\sigma x^*/e$, and for the minimizing point in (12), the changing is on the value $\Delta \omega x_*^2/e = -k\sigma x_*/e$. Compose the symmetry relation $\pi^* - k\sigma x^*/e = -(\pi_* - k\sigma x_*/e)$. Solving for the unknown integer k, we obtain

$$k_S = \left[\frac{(\pi^* + \pi_*)e}{(x^* + x_*)\sigma} \right]. \tag{13}$$

Here square brackets mean the integer part.

The strategy U_S for keeping up the symmetry of the prognosis miss is determined as a function, which for the position (t_i, n, I) yields the number

k_S, calculated by the formula (13) if $\mid k_S \mid \leq n$, and the number $n \mathrm{sign} k_S$ if $\mid k_S \mid > n$.

The strategy U_S is optimal for the case $x_0 = x^0$. The proof is mentioned in [10]. If $x^0 - x_0$ is small, then the strategy U_S can be supposed quasi-optimal.

7   The second player's strategy

Let $(\widetilde{\omega}, \widetilde{x})$ be the point from $I_\ominus$, at which the maximum of the following expression is achieved

$$\widetilde{\pi}(n, I_\ominus) = \max_{(\omega, x) \in I_\ominus} \left\{ \frac{\mid \omega \mid x^2}{e} + \frac{\nu}{2}\left(\frac{x}{e}\right)^2 - \frac{n \sigma x}{e} \right\}.$$

The first two terms in the curly brackets give the miss prognosis (in modulus) under zero control of the first player. The third term gives the maximum possible compensation of the miss via the first player's control.

Select in the set $I_\ominus$ the interval, which contains the point $(\widetilde{\omega}, \widetilde{x})$, and project it on the axis ω. If $\widetilde{\omega}$ is the right edge of this projection, then choose $\omega_M \leq \widetilde{\omega}$ so that the projection $\mathcal{H}$ of the uncertainty set $H(\omega_M)$ on the axis ω keeps $\widetilde{\omega}$ on its right edge. If $\widetilde{\omega}$ is the left edge, then choose $\omega_M \geq \widetilde{\omega}$ so that $\widetilde{\omega}$ will be situated on the left edge of the projection $\mathcal{H}$. Such a measurement provides the location of the point $(\widetilde{\omega}, \widetilde{x})$ in the informational set $I = I_\ominus \bigcap H$. Denote by the symbol $\widetilde{\Omega}$ the second player's strategy, based on the choice of the point $(\widetilde{\omega}, \widetilde{x})$.

The strategy $\widetilde{\Omega}$ is optimal for the case $x_0 = x^0$ [10].

8   Simulation results

We simulate the motion of the system (1). The incompleteness of information is given by relations (1)–(4).

The list of the input data (in relative coordinates):

− the nominal initial range on the vertical axis $x_{NOM} = 80000$ m,

− the set, which restricts possible initial locations, is

$$B = \{(x, z) : \mid x - 80000 \mid \leq 10000 \text{ m}, \mid z \mid \leq 2000 \text{ m}\},$$

− the nominal value of the initial velocity on the vertical axis is $V_{NOMx} = -5000$ m/s,

− the set, which restricts the initial velocity vector, is

$$D = \{(V_x, V_z) : \mid V_x + 5000 \mid \leq 100 \text{ m/s}, \mid V_z \mid \leq 100 \text{ m/s}\},$$

   – the value of one control impulse is $\sigma = 5$ m/s,
   – the quantity of impulses is $N = 70$,
   – the set, which restricts the evader's acceleration, is

$$Q = \{(v_x, v_z) : \mid v_x \mid \le \mu, \ \mid v_z \mid \le \nu\}, \quad \mu = 2 \text{ m/s}^2, \ \nu = 2 \text{ m/s}^2,$$

   – the constants in the restriction on the maximal value of the error in the LOS angular velocity measurement are $c_1 = 0.3$, $c_2 = 0.0008$ rad/s.

The circumscribed input data is known to pursuer. Emphasize that B, D are the preliminary given sets, in which the location and velocity of the relative point must lie. The pursuer does not possess any more exact information before the beginning of the pursuit process.

We shall apply for the pursuer a method of control based on the strategy U_S. Denote this by SM. Choose the step Δ of the discrete scheme of observation and control equal to 0.1 s.

The initial informational set $I_\Theta(t_0)$ is given in accordance with (8): $[x_0, x^0] = [70000, 90000]$, its section for all $x \in [x_0, x^0]$ is an interval with boundary points (in projection on the axis ω)

$$\omega_0(x) = \frac{-100x - 5100 \cdot 2000}{x^2}, \quad \omega^0(x) = \frac{100x + 5100 \cdot 2000}{x^2}.$$

The true value of the angular velocity at a moment t is calculated by

$$\omega(t) = \frac{\dot{z}(t)x(t) - \dot{x}(t)z(t)}{x^2(t) + z^2(t)}.$$

Therefore the set $I_\Theta(t_0)$ involves all such possible pairs $\omega(t_0), x(t_0)$ in the plane ω, x, which are consistent with restrictions (3), (4) for sets B and D, determined above.

The informational set for numerical realization is given by a finite number of intervals. For the results, presented below, the number of intervals in the initial set $I_\Theta(t_0)$ will be equal to 21. During the process, the number of intervals can decrease. If at some moment the quantity is smaller than the given number (in our case, this number is eleven), then the additional intervals are introduced so that their general number is duplicated. In the algorithm for constructing the informational sets, we use the value of restriction on the evader's acceleration along the axis z with a little larger magnitude than its given value, namely 2.5 m/s^2 instead of 2 m/s^2. It is performed to exclude degeneration of the informational sets in numerical realization. The threshold is taken as $\varepsilon = 500$ m. The pursuer's control ceases when the informational set, calculated by the pursuer via formulas from section 5, is completely below the level ε on coordinate x or when the distance from the true relative point $(x(t), z(t))^T$ to zero becomes less than ε. The last condition has not been mentioned earlier. It reflects the fact that the pursuer's control system cannot provide information about the angular velocity of the LOS when the distance to the evader is too small.

With regard to the measurements $\omega_M(t_i), i = 0, 1, 2, \ldots$, and the controls v_x and v_z, we consider two variants. Denote them as RN, GM.

1) The value $\omega_M(t_i)$ at any moment t_i is produced by means of the random number generator with uniform distribution in interval $[\omega(t_i)-\chi, \omega(t_i)+\chi]$, where $\chi = c_1 \mid \omega(t_i) \mid +c_2$, and $\omega(t_i)$ is the true angular velocity. The controls v_x and v_z are constant through time interval of pursuit; at the initial moment t_0, they are produced from the set Q by means of the random number generator with uniform distribution.

2) The second variant of producing ω_M, v_x, v_z is called "gaming". In this variant, ω_M is produced as in the construction of the strategy $\widetilde{\Omega}$, described in section 7, but also taking into account the additional demand that the measurement $\omega_M(t_i)$ must lie in interval $[\omega(t_i) - \chi, \omega(t_i) + \chi]$. The choice of ω_M: if the point $(\widetilde{\omega}, \widetilde{x})$ lies on the right (left) edge of the set $I_\ominus(t_i)$, then we define $\omega_M(t_i)$ so that the right (left) edge of the uncertainty set $H(\omega_M(t_i))$ is maximally close to this point. Determine the choice of the controls v_x, v_z as follows: if $x(t_0) \geq (x_0+x^0)/2$, then $v_x \equiv \mu$; in the case of $x(t_0) < (x_0+x^0)/2$, let $v_x \equiv -\mu$. Thus, the control v_x is constant on the whole interval of motion. The control v_z is chosen anew at every moment t_i and holds constant on the interval $[t_i, t_{i+1})$: if the point $(\widetilde{\omega}, \widetilde{x})$, calculated at the moment t_{i+1}, lies on the right edge of the set $I_\ominus(t_{i+1})$, we assume that $v_z(t_i) = \nu$; if the point lies on the left edge, then $v_z(t_i) = -\nu$.

The current value $\omega(t_i)$ of the true LOS angular velocity is calculated using coordinates $x(t_i), z(t_i), \dot{x}(t_i), \dot{z}(t_i)$ of the system (1).

Graphs in Fig. 2 show variations of parameters along the unit realizations for the combinations SM$-$RN and SM$-$GM. We let $t_0 = 0$. The initial coordinates are: $x(t_0) = 80000$ m, $z(t_0) = 100$ m, $\dot{x}(t_0) = -5000$ m/s, $\dot{z}(t_0) = 10$ m/s. The acceleration in the method RN is: $v_x = v_z \equiv 2$ m/s^2. The curves of the measured ω_M, true ω angular velocities, and graphs of the current impulse control are represented depending on the current time t. In contrast to the method RN, if the gaming method is applied, then the graph $\omega_M(t)$ does not have a character of a random function, but is rather the same type as the $\omega(t)$. The miss for the combination SM$-$RN was 0.17 m, and it was 1.88 m for the combination SM$-$GM. In the first variant, the expenditure of the control impulses was 43; in the second variant, 64 impulses were spent.

Variations of the informational sets for the case of the gaming disturbance are shown in Figs. 3, 4. Fig. 3 corresponds to the transition from the initial moment $t_0 = 0$ to the next discrete moment $t_0 + \Delta = 0.1$. The pass from the moment $t = 13.9$ s to the moment $t = 14$ s is represented in Fig. 4. The dotted line marks the uncertainty set, calculated for the current measurement.

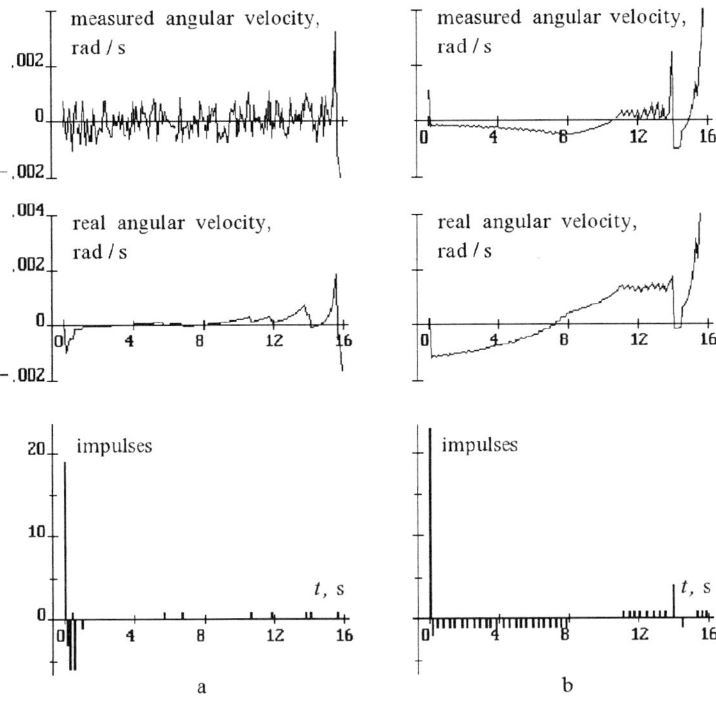

Figure 2: Measurement of the angular velocity, real angular velocity, impulse control; a) random disturbance, b) gaming disturbance.

The cross marks the true relative point (in coordinates ω, x). Close to the symbol of the moment, the quantity of expended impulses is shown. It can be seen that, at the end of the approaching process (Fig. 4, approximately 2 s till the end), the variation of the informational set is more active than at the beginning. Particularly, the vertical size of the informational set strongly decreases. The greatest lower interval of the set $I_{\oplus}(13.9)$ would be far to the left, and out of the graph field. This interval, moving from the set $I_{\oplus}(13.9)$ to the set $I_{\ominus}(14)$, went under the threshold $\varepsilon = 500$ m and disappeared. The greatest lower interval of the set $I_{\ominus}(14)$ is also out of the graph field.

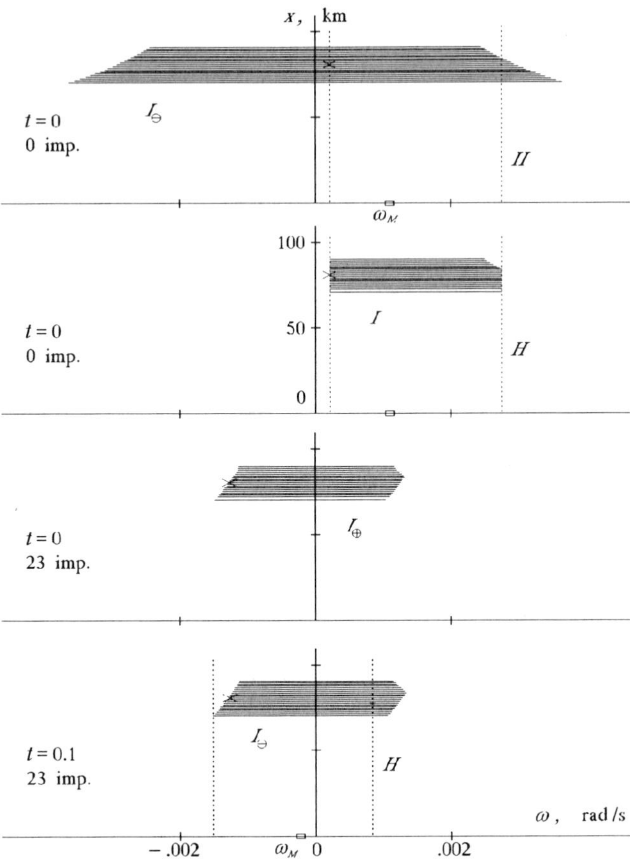

Figure 3: Informational sets on the initial stage of pursuit.

The results of simulation, based on on a large number of realizations, are represented in [10, 11]. Significant misses and essentially greater expenditure of impulses were obtained in the case of gaming disturbance in comparison with the case of random disturbance. Of course, the gaming method for measurement constructing is speculative. Nevertheless, it can be used to test different variants of the pursuer's control laws. In [10], the results of comparison of the SM method with the method based on using the Kalman filtration are represented. When the distribution of the random disturbance is close to Gaussian, the results on the miss and impulses expenditure are close to each other. For the random disturbance with uniform distribution and to a greater extent, for the gaming disturbance, the SM method gives essentially better results.

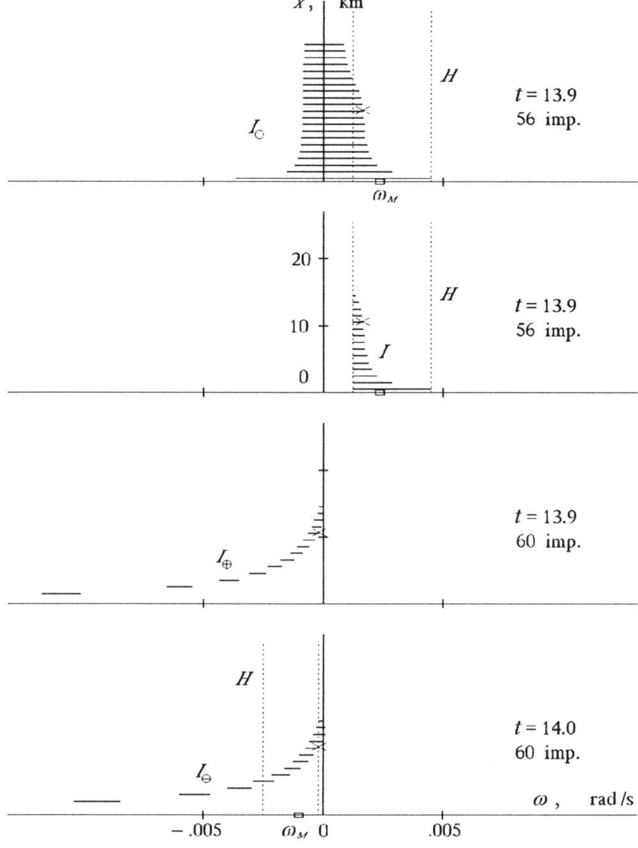

Figure 4: Informational sets on the final stage of pursuit.

REFERENCES

[1] Krasovskii N. N. and Subbotin A. I., *Positional Differential Games,* Moscow, Nauka, 1974 (in Russian); French transl., *Jeux différentiels,* Moscow, Mir, 1977.

[2] Kurzhanskii A. B., *Control and Observation in Uncertainty Conditions,* Moscow, Nauka, 1977 (in Russian).

[3] Chernous'ko F. L. and Melikyan A. A., *Game Problems of Control and Searching,* Moscow, Nauka, 1978 (in Russian).

[4] Krasovskii N. N. and Osipov U. S., *On the Theory of Differential Games with Incomplete Information,* Soviet Math. Dokl., Vol. 15, No. 2, 1974, 587–591.

[5] Subbotina N. N. and Subbotin A. I., *A Game Problem of Control in the Case of Incomplete Information,* Engrg. Cybernet., Vol. 15, No. 5,1977, 1–10.

[6] Merz A. W., *Stochastic Guidance Laws in Satellite Pursuit-Evasion,* Comput. Math. Applic., Vol. 13, No. 1–3, 1987, 151–156.

[7] Merz A. W., *Noisy Satellite Pursuit-Evasion Guidance,* J. Guidance., Vol. 12, No. 6, 1989, 901–905.

[8] Patsko V. S., *A Model Example of Game Pursuit Problem with Incomplete Information,* Part I, Part II, Differents. Uravneniya, Vol. 7, No. 3, 1971, 424–435; Vol. 8, No. 8, 1972, 1423–1434 (in Russian).

[9] Melikyan A. A. and Chernous'ko F. L., *Certain Minimax Control Problems with Incomplete Information,* J. Appl. Math. Mech., Vol. 35, No. 6, 1971, 907–916 (in Russian).

[10] Kumkov S. I. and Patsko V. S., *Pursuit Problem with Incomplete Information,* Preprint, Institute of Mathematics and Mechanics, Urals Branch of Acad. Sci. of Russia, 1993 (in Russian).

[11] Kumkov S. I. and Patsko V. S., *Model Problem of Impulse Control with Incomplete Information,* Transactions of the Institute of Mathematics and Mechanics, Vol. 1, Ural Branch of Russian Acad. of Sci., Ekaterinburg, 1992, 106–121 (in Russian).

Decision Support System for Medium Range Aerial Duels Combining Elements of Pursuit-Evasion Game Solutions with AI Techniques

Stéphane Le Ménec
MATRA-DÉFENSE - BP 1
78146 Vélizy-Villacoublay Cedex
France[*]

Pierre Bernhard
INRIA - BP 93
06902 Sophia-Antipolis Cedex
France[†]

Abstract

The improvement of guidance possibilities of medium range missiles with new missiles like the Mica/Amraam[1] increases the number of phases in aerial duels and implies more complex firing and escape strategies. Therefore we are interested in developing algorithmic methods to study these new duels, which are difficult to study merely with the classical techniques of game theory.

The paper describes a decision support system for a fighter pilot in medium-range game combat. The design of the system is based on combining pursuit-evasion game solutions with AI techniques, such as decision trees, by taking advantage of an existing expert system shell called SMECI[2]. This system improves on a previous study about a Pilot Advisory System outlined in [7] and develops new concepts for further support systems, optimizing pilot decisions in air combat.

The article describes firstly what aerial medium range duels are, before studying parts of them as differential subgames. Then we explain how to design a decision support system with several simulations,

---

[*]tél : (33 1) 34 88 32 22
fax : (33 1) 34 88 44 55
lemenec@sophia.inria.fr
[†]bernhard@sophia.inria.fr
[1]Missile d'Interception de Combat et d'Auto-défense / Advanced Medium Range Air to Air Missile
[2]Système Multi-Expert de Conception en Ingénierie / Multi Expert System for Engineering Design

using barriers of differential subgames. At the end of the paper, we give some examples of this decision support system called ADAM[3].

This study has been supported by DRET[4], which is interested in new methodologies for pilot decision support systems, contract n^0 90/532 : "Decision Support System for Aerial Duels".

1   Introduction

The object of game theory is the mathematical study of situations containing a conflict of interests [3]. In the case of the pursuit of an aircraft by a self-guided short range missile, we consider the aircraft and the missile as two players in order to calculate a capture zone and an escape zone (or non-capture zone) separated by a barrier giving the configurations leading to the destruction of the aircraft or the loss of the missile. We calculate the initial conditions of the pursuit, characterized by state variables of the game allowing the aircraft to evade any guidance law of the missile and allowing the missile to destroy any maneuvering target.

This paper considers a medium range duel opposing two identical aircraft (figure 1). We call the blue aircraft (BA) and the red aircraft (RA). Each aircraft has a Mica/Amraam, called the blue missile (BM) for BA and red missile (RM) for RA. We restrict this study to a co-planar game, because in medium range duels the altitude parameter does not have the same importance as in dogfight duels. The aircraft begin the duel with a pre-launch phase at a range of about twice their firing range.

These medium range missiles use several guidance modes. After firing a Mica/Amraam flies uplink[5] as long as the aircraft can forward information to the missile. When the uplink is broken, the missile is self-guided. The missile self-guidance law uses past information to extrapolate the target position until the missile can lock its active radar seeker on the target[6]. The firing and all phases, except the phase with the missile active radar seeker locked on the target, are undetectable.

The outcome of the game is given for BA. A victory of BA corresponds to a defeat of RA (*win* outcome) and a defeat of BA to a victory of RA (*lose* outcome). A *win* outcome corresponds to the destruction of RA with a successful evasion of BA, while in a *lose* outcome, BA is the destroyed aircraft. A duel can end with other outcomes too. We speak about a *draw* outcome when no missile reaches its goal and about a *mutual kill* outcome if both aircraft are destroyed.

---

[3] Aide au Duel Aérien Moderne / Decision Support System in Modern Aerial Duel

[4] Direction des Recherches Etudes et Techniques / French Defence Advanced Research Agency

[5] Lam (Liaison Avion Missile) mode

[6] Ad (Auto-directeur) mode

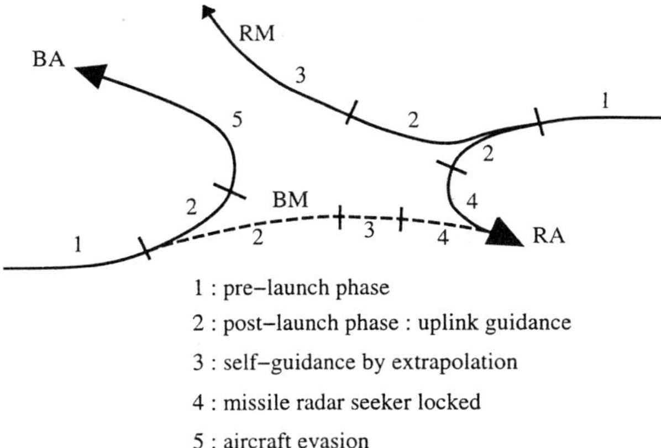

Figure 1: Two aircraft in medium range duel

The theory of games also solves air combat games between two aircraft with fire and forget missiles. But the aerial duels set new problems such as that of role determination of the aircraft, since an aircraft plays both as a pursuer and as an evader [5].

At the beginning of realistic medium range duels each pilot has generally the same chance to win, and to say that the optimal solution of the game is a *draw* or a *mutual kill* is a quite poor operational result. Aircraft engage in an air-to-air combat only if they believe they have some chance to win. An aircraft can win only by taking advantage of the other player's errors; that is why we develop a decision support system like ADAM to advise BA of reprisal strategies during the duel.

AI techniques are required to study medium range duels for other reasons as well. The number of possible outcomes and of missile guidance phases make aerial duels with Mica/Amraam complex to study. We have also to decide whether a player prefers to end a duel by a *draw* outcome or by a *mutual kill* outcome, if it cannot win. Moreover, both pilots can play cooperative strategies to obtain a *draw* outcome instead of ending by a *mutual kill* outcome, if both players prefer a *draw* outcome to a *mutual kill* outcome. The theory of differential games is principally interested in non-cooperative games.

We have studied the different phases of a medium range duel as subgames to realize simulations of modern aerial duels using information of subgame barriers. These improved simulations allow us to define a guaranteed evasion strategy for BA and shorten the decision support system simulations developed by ADAM during the real duel to test several BA reprisal strategies. This new method allows us to solve a complex game, generally studied up to now by heuristic methods.

A previous article by J. Shinar [6] introduces firing envelopes for aircraft in terms of differential game barriers. These envelopes, named using barrier vocabulary, were computed on-line by forward simulations according to different assumptions of aircraft behavior. This work gave us the idea to introduce real differential game barriers to calculate firing envelopes for aircraft and capture zones for fired missiles. All our firing domains are constructed off-line by differential game techniques with backward integration as explained in sections 5 and 6.

2   Hypotheses

In the study of Mica/Amraam duels, we have made the following assumptions :

1. An aircraft executes only one evasion which is definitive.

2. An aircraft cannot fire during its evasion maneuver.

3. An aircraft cannot stay uplink with its missile during the evasion phase.

4. An aircraft evades systematically when it is locked by the enemy active radar seeker.

5. An aircraft does not fire after the opponent's evasion.

6. In uplink guidance, a missile has the same information as in autonomous guidance with its radar seeker locked.

7. An aircraft does not detect the opponent uplink.

These assumptions simplify the complexity of simulations and the decision tree processing by decreasing the number of BA alternatives to test for BA reprisal strategies. Though this analysis is limited to horizontal "head-on" duel type (1 X 1) encounters only, these hypotheses are reasonable.

This study is the first step towards a more complex realistic decision support system dealing with multiple aircraft. According to this goal, the four subgames described below represent a reasonable description of the scenario.

3   Subgames

Since there exist different guidance modes for medium range missiles, we define several subgames to study some parts of the complete duel. One of these subgames corresponds to the final short distance game when the missile has the radar seeker locked on the target. Other subgames describe the initial phase with uplink guidance. Hypotheses on the duel allow one to define, from the final phase with radar seeker locked, the end conditions of previous subgames. We study the pursuit subgames between RM and BA. These subgames have to be seen also as pursuits between BM and RA since the missiles and the aircraft are identical:

- **SR**: Short Range optimal pursuit subgame between RM and BA with the radar seeker locked on BA.

- **MR**: Medium Range subgame with perfect information (each player knows the state of the game) dealing with the RM uplink guidance phase. We consider BA detecting RM during the post-firing phase. This situation does not correspond to the real medium range duel situation, but as explained in section 4, this subgame is useful to define a guaranteed BA evasion considering hypothetic RM firings.

- **CMR**: Constrained Medium Range subgame identical to the previous one with a restricted evasion of BA. This subgame describes the evasion of an aircraft staying in uplink guidance with its missile (realistic firing domain)

- **MRWE**: Medium Range subgame Without Evasion identical to the subgame **MR** with no BA escape. The barrier of this subgame gives the maximum firing range of a missile.

4 Evasion strategy of BA

We want to help BA make its decisions in the duel and in particular to choose its evasion time [4]. The aircraft does not detect the opponent firing; it sees the enemy missile only when the enemy active radar seeker locks on and then it may be too late to evade. To be sure of escaping from RM, BA must not enter into the capture zone of RM of the subgame **MR**, when BA has not yet detected the RM firing. A secure evasion of BA corresponds to an evasion started before entering this capture zone.

Since BA does not see RM, BA protects itself against an RM that RA can fire at the present time and against all RMs that RA may have shot in the past. That is why we introduce in all ADAM simulations some hypothetic RMs to perform the BA evasion strategy:

- At each time step in the simulation, RA fires a hypothetic missile (hypothetic RM) as soon as BA crosses the barrier of the subgame **MRWE** of an RM supposed not yet to be shot.

- BA evades as soon as it reaches the capture zone of the subgame **MR** of an hypothetic RM fired or not.

On the one hand, because of the duel hypotheses and in particular of hypothesis 5, this evasion prevents BA from losing and assures BA of an outcome at least equal to a *draw* outcome, and the BA evasion does not depend on the real RM trajectory. On the other hand, with such an evasion strategy, BA takes no risk and can miss a possible *win* outcome.

The BA evasion strategy looks like "the principle of min-max certainty equivalence" of P. Bernhard [1] which says that one must look for what is actually the worst possible state with the available information and play the strategy which would be optimal if we were certain that the state is effectively this state. Of course, we prove no optimality of this strategy in the present context.

RA does not use hypothetic BM for its escape maneuver and does not know where BM is. Therefore RA cannot choose its evasion instant to make an evasion as efficient as BA, even if RA uses the barriers of subgames that it can manipulate because of the imperfect knowledge of the position of BM. RA uses subgame barriers only in considering BM not fired, i.e. BM with initial energy at BA position. If both RA and BA were to evade considering hypothetic enemy missiles, ADAM would be without interest, since the duel would always lead to a *draw* outcome.

Figure 2 represents a duel in ADAM. This decision support system simulates in SMECI complex kinematics for the aircraft and the missiles, with Proportional Navigation guidance law for the missiles. The aircraft use an evasion strategy designed by us considering the barriers calculated previously. The barriers determine the BA time to evade and the direction of its turn. During the evasion phase, BA executes a sharp turn before going back in a straight line. Subgame capture zones give a good approximation of realistic firing domains, even if the differential games consider simplified kinematics and optimal strategies, which are not implemented in ADAM.

In figure 2, RA fires several hypothetic missiles. Some drawings on the trajectories explain the aircraft positions in the state spaces of the subgames. Other drawings explain the guidance mode of the missiles.

The BA evasion strategy considering hypothetic RM firings is the first use of subgame barriers in ADAM. As soon as the BA evasion strategy is defined, the combinations of ADAM decision trees to propose an efficient BA reprisal strategy decreases.

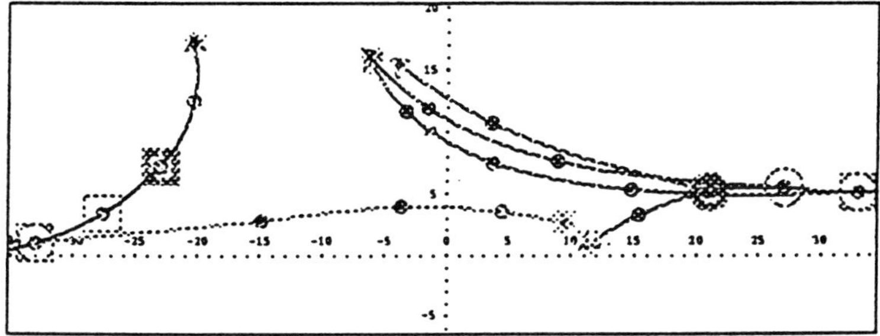

Figure 2: Duel in ADAM with BA at left and RA at right

5 Short range subgame

The geometry of planar pursuit defining the state variables of the game is depicted in figure 3. The missile P possessing a velocity V_P and a minimum admissible turning radius r_c is pursuing in planar motion an aircraft E, assumed to be flying with a constant velocity V_E and without constraint on its turning rate. The two constants a and b describe the missile drag. u (with $-1 \leq u \leq 1$) and γ_E are respectively the control of the pursuer and the control of the evader. The game terminates by capture when the pursuer approaches the evader within the distance $R = R_f$.

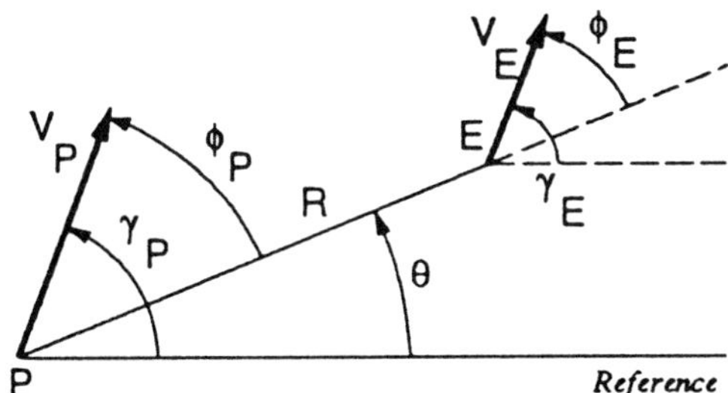

Figure 3: Geometry of missile (P) - aircraft (E) pursuit games centered on P

The kinematic equations are :

$$\dot{V}_P = -V_P^2\left(a + bu^2\right), \tag{1}$$

$$\dot{R} = V_E \cos\left(\gamma_E - \theta\right) - V_P \cos\left(\gamma_P - \theta\right), \tag{2}$$

$$\dot{\theta} = \frac{[V_E \sin\left(\gamma_E - \theta\right) - V_P \sin\left(\gamma_P - \theta\right)]}{R}, \tag{3}$$

$$\dot{\gamma}_P \;=\; \frac{V_P}{r_c} u. \tag{4}$$

The number of independent variables can be reduced to three, which is the minimum representation of the system, if use is made of the pursuer and evader relative angles with respect to the line of sight : ($\phi_P = \gamma_P - \theta, \phi_E = \gamma_E - \theta$). The use of the reduced system complicates the analysis, but allows the representation of the capture zone in a 3D state space. In the reduced system, E uses the control ϕ_E, and the game target set is defined as a plane of equation $R = R_f$, because no additional conditions are imposed on V_P and ϕ_P.

The short-range subgame is a new model and has not yet been published in the literature, but we can compare our investigation to a previous version of such a dynamic model given in [2]. This other dynamic model looks like ours except that the authors consider an additional state variable to constrain the minimum turning radius of the aircraft.

Fortunately, as in [2], the adjoint equations of our game can be analytically integrated in terms of state variables and their final values. When λ exists, λ is the gradient of the barrier. The final value of the adjoint vector λ on the game target is: $\lambda_f = (0, 1, 0)$. Without losing any generality, the final line of sight is used as the angular reference: $\theta_f = 0$. The adjoint vector of optimal trajectories on the natural barrier is:

$$\lambda_{V_P} = -\frac{V_E}{V_P}(t_f - t), \quad \lambda_R = \cos\theta, \quad \lambda_{\phi_P} = R\sin\theta. \tag{5}$$

The capture of the evader only occurs in the usable part of the game target. To capture an optimal evader, the pursuer must satisfy the compromise between its final speed and its final angle of attack given by the following condition:

$$V_{P_f} \geq \frac{V_E}{\cos\phi_{P_f}}.$$

The limit of the usable part: $V_{P_f} = \frac{V_E}{\cos\phi_{P_f}}$ defines the final conditions of optimal trajectories of the natural barrier. Since E has no constraint on its turning rate, the optimal control strategy of the evader on the natural barrier is to take the final line of sight direction. We note the optimal controls of the evader and the pursuer respectively γ_E^* (ϕ_E^* in the 3D state space) and u^*.

$$\gamma_E^* = 0 \qquad\qquad \phi_E^* = -\theta$$

The analysis of the Hamiltonian of the system (equations 1 to 4) with the analytic solution of the adjoint vector (equation 5) gives u^* on the natural barrier.

$$u^* \;=\; \max\left[-1, \min\left(1, u_0\right)\right] \tag{6}$$

$$\forall\, t < t_f \qquad u_0 \;=\; \frac{-R\sin\theta}{2V_E b r_c(t_f - t)} \tag{7}$$

This expression is not available on the game target, where the expression

$$u_0 = \frac{-\tan \phi_{P_f}}{2br_c}$$

must be used.

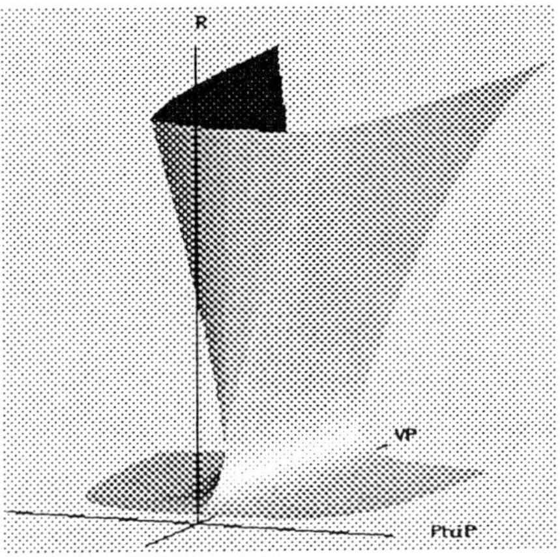

Figure 4: Barrier of the short range subgame in the 3D state space V_P, R, ϕ_P

The natural barrier separates the capture zone and the non capture zone in the closeness of the game target, but the natural barrier is not sufficient to close the capture zone for ϕ_P small and R superior to a value R_1. To close the barrier of this pursuit game, we have built a focal line in the plan $\phi_P = 0$ starting at $R = R_1$ with R growing in backward time. On the focal line, the evader maximizes the capture time while the pursuer plays the control:

$$u_{focal} = \frac{V_E r_c}{V_P R} \sin \phi_E$$

to keep its velocity $\vec{V}_P$ in the direction of the line of sight, i.e., to keep $\phi_P = 0$. The barrier of the short range subgame is closed with optimal trajectories tangentially reaching the focal line in forward time.

Figure 4 represents the barrier of this pursuit game in the 3D state space (V_P, R, ϕ_P). The focal line and the trajectories reaching it appear on Figure 4 at the front of the barrier. In the reduced state space (V_P, R, ϕ_P), the focal line is unique, but this focal line summarizes two different behaviors of E and P. If the evader turns left optimally, then the pursuer turns left with the control u_{focal}. E can also turn right optimally and then P turns right as explained in the equation of u_{focal}. Figure 5 shows the focal line with E turning right in the earth reference frame (x, y). Optimal pursuits reaching

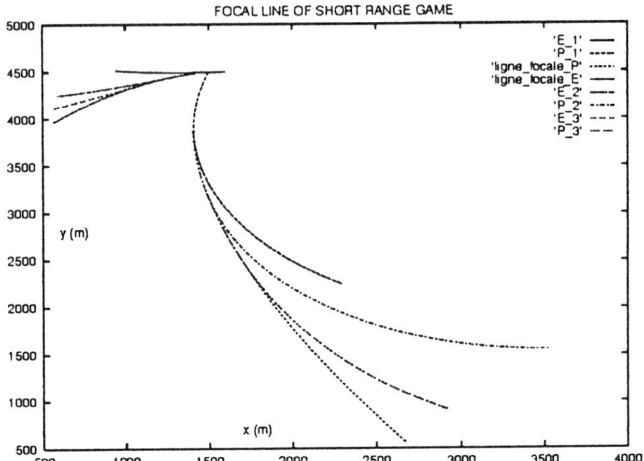

Figure 5: Trajectories of P and E in the earth reference frame reaching the focal line of the short range subgame

the focal line of the short range subgame barrier are also drawn in the figure 5 (P_1 pursuing E_1, P_2 pursuing E_2 and P_3 pursuing E_3).

We use the part of the barrier of the short range subgame corresponding to $R = R_{lock}$ when the missile locks on the aircraft to design the game target of medium range subgames and to calculate in a similar way the medium range subgames. We call this new game target a "surcible". This way to define a differential game target is new, which is why we introduce the concept of "surcible".

6   Medium range subgames

To simplify the analysis of the pursuit, we modify the kinematics of medium range subgames. We change equation (1) into equation (8). During the post-firing phase, the importance of the drag factor b decreases according to the smallness of u. This modification reduces the number of state variables, because equation (8) is now time integrable. In the medium range subgames, we constrain u to vary between -0.1 and 0.1 to be consistent with the assumption of u being small. The constant r_c of medium range subgames is set equal to $10 * r_c$ of the short range subgame to keep the same missile characteristics in both phases.

$$\dot{V}_P = -aV_P^2 \tag{8}$$

The "surcible" gives the state initial conditions, but we have to calculate the new adjoint initial values since the dynamics are different between the short and the medium range phase. The solution of the medium range subgame consists of an optimal singular arc in the plan $\phi_P = 0$ of the 3D state space. This singular arc corresponds to a straight line pursuit in the earth reference frame (x, y). This behavior of E and P is reasonable since the missile drag factor b is equal to 0, and since P does not lose energy in turning. When $\phi_P < 0$, $u^* = 0.1$ and when $\phi_P > 0$, $u^* = -0.1$. Along each optimal trajectory γ_E^* is constant.

Figure 6 represents a part of the optimal medium range subgame barrier corresponding to the missile firing time when $V_P = V_{P_{max}}$.

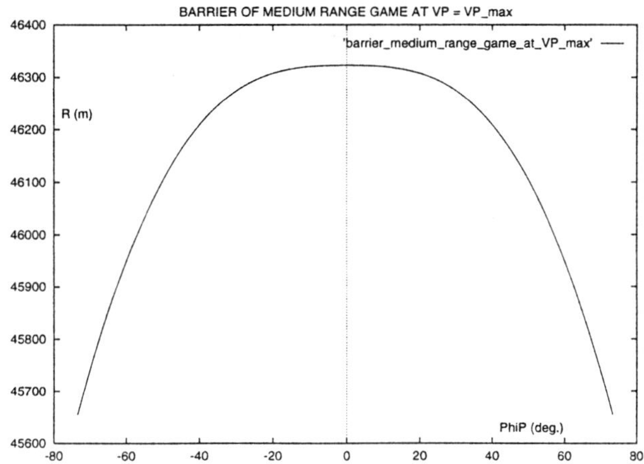

Figure 6: Capture domain of the subgame **MR** at the missile shooting instant

The **CMR** subgame is an optimal differential game with perfect information similar to the **MR** subgame except that we put a constraint on the domain of the evader control ϕ_E (figure 7):

$$|\pi - \phi_E| \leq \alpha. \tag{9}$$

When the missile of the evader is in uplink guidance, E must fly in the direction of the other aircraft, which corresponds approximately to the direction of the missile P.

The **MRWE** subgame is identical to the **CMR** subgame with $\alpha = 0$, i.e.:

$$\phi_E = \pi. \tag{10}$$

Figure 8 presents the part of the barriers of **CMR** and **MRWE** subgames corresponding to $V_P = V_{P_{max}}$.

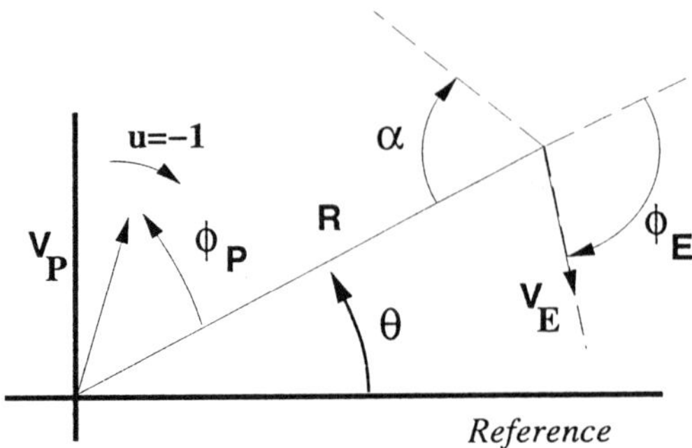

Figure 7: Geometry of medium range subgame with constraint on evader direction

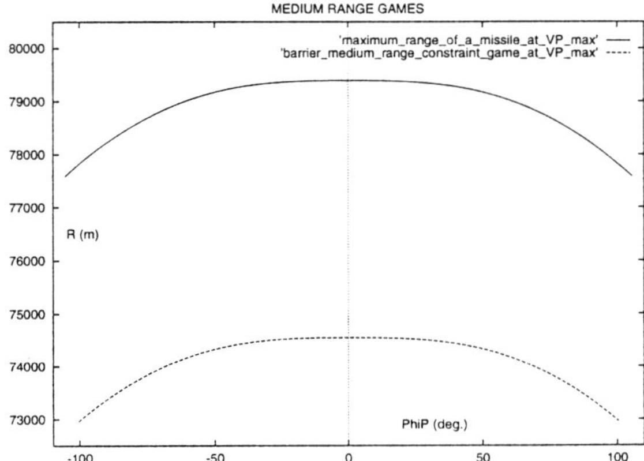

Figure 8: Capture domain of **CMR** and **MRWE** subgames at the missile firing instant

7   Firing window

With the barriers of the studied subgames, we cannot give exactly the optimal strategies of the players, but we have results to help BA not lose (BA guaranteed evasion strategy) and not to consider strategies without victory possibilities (firing window). With capture domains considered at $V_P = V_{P_{max}}$, we define a firing window (figure 9). To fire BM before RA reaches the barrier of the **MRWE** subgame makes no sense, because BM is necessarily lost. In the same way, to fire after RA crosses the barrier of the **SR** subgame corresponds to RA taking a too large risk, because as soon as RA is in the capture zone of the **SR** subgame, BM catches any maneuvering RA.

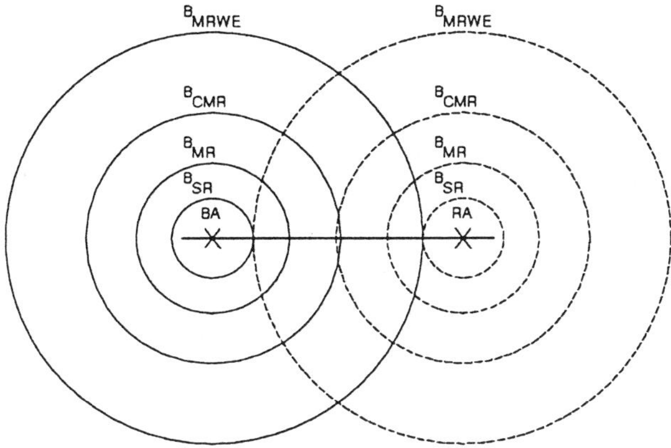

Figure 9: Simplified description of aircraft firing windows

8 Adam

A pilot does not exactly know when to fire. The later he fires, the more efficient his missile is, but then he has to fly closer to his opponent and the opponent's missile. Each player wants to guide his missile (to maintain the uplink) as long as possible to increase the performance of his missile, but he wants to begin his evasion as early as possible too, to evade with success. Moreover, as the firings are undetectable, a player can bluff. He has an influence on the opponent's decisions by executing maneuvers which mimic firing.

That is why this system optimizes the firing instant and the maneuvers of the pre-launch and post-launch phases of BA according to RA choices. We do not study different BA evasion maneuvers. An RA behavior (feedback control) or a user interface specifies the RA strategy during the duel. The most powerful RA behaviors use RA and BA firing windows to choose the RA firing and evasion maneuvers.

This decision support system building and analysing decision trees is composed of several expert systems (figure 10) :

- **ES duel**[7] : simulation of the real duel.

- **ES observer**: analysis of RA aggressiveness in the **ES duel** according to its previous maneuvers to put weights on **ES simulator** decision trees.

- **ES simulator**: construction of a game tree corresponding to BA alternatives against a particular RA behavior.

---

[7]Expert System duel

- **ES analyser**: analysis of game trees to propose a BA reprisal strategy.

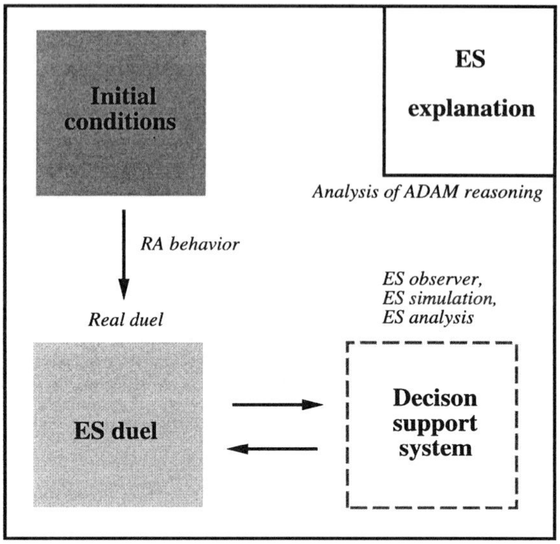

Figure 10: ADAM: a decision support system

The **ES duel** simulates the real duel between any RA and a supported BA. During the real duel, BA chooses its decisions asking an **ES observer**. The **ES duel** calls regularly on the **ES observer** to refine the BA reprisal strategy according to duel improvements. The **ES observer** assumes the RA aggressiveness level in the **ES duel** to put weights on **ES simulator** decision trees. Actually, the **ES simulator** builds three BA decision trees (one with an aggressive RA behavior, one with a normal RA behavior and one with a prudent RA behavior). In the first mode, the **ES observer** deductions provide AI techniques as expert system rules. In the second mode, transition matrices using an RA state model based on Markov chains describe the RA transition probabilities from one phase to another for the three RA aggressiveness levels. After the **ES simulator** sessions, an **ES analyser** uses the values of the different BA alternatives in the three decision trees to propose a BA reprisal strategy (figure 11).

9 Smeci

To develop our decision trees, we use the facility of programming with expert systems techniques. An expert system lets one easily manipulate way the concepts of trees and heuristics. Take the example of the expert system shell SMECI. SMECI offers a formalism of knowledge representation by frames and rules clustered in tasks. Objects called categories or classes define hierarchies of structures, which we instantiate to create the objects describing the domain.

Its reasoning consists in creating a state tree, called a reasoning tree obtained by the application of rules. A state includes objects characterized by the values of their slots and the relations linking them. A state can lead to several concurrent reasoning lines. Depending on the rule bases, a SMECI state can easily correspond to decision nodes similar to the nodes of ADAM decision trees.

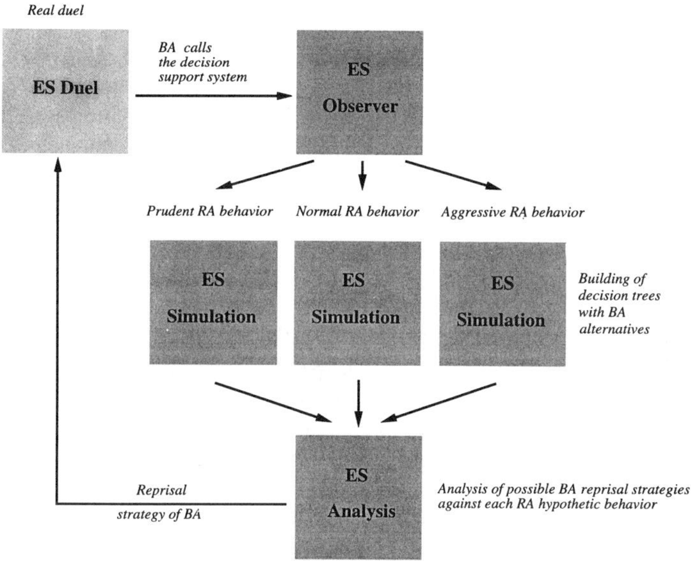

Figure 11: Organization of ADAM

SMECI can handle a big decision tree and has some functionalities to manipulate it. For instance, we can navigate from one state to another, define an order of preference to expand the state tree, cut branches and visualize rapidly the information in this tree. Writing heuristics is easy thanks to the rule formalism. SMECI includes a complete environment for design, with tools to create the graphic applications we use to draw the trajectories of the simulated games with all necessary information.

We have implemented each ADAM expert system with the expert system shell SMECI and an additional module, called GAMES, which is a tool to expand simultaneously several state trees in Smeci. GAMES allows the manipulation in the same SMECI session of all expert systems described above. In fact, without this additional module, SMECI is a multi expert systems shell in terms of multi knowledge bases and not in terms of multi expert systems.

10   Player strategies in Smeci

We simulate with SMECI in discrete time many duels between BA and RA (the real duel in the ES duel and the simulations of ES simulator to know the influence of players' decisions). We discretize the time using a clock object. We have discretized the set of admissible strategies of each player too. In ADAM, a strategy is composed of decisions described by the instant at which the decision is taken (change of the turn control, shooting time ...) and, when it applies, the value of this decision (turning rate ...).

The form of the player strategies is very important, because the realistic character of our simulations depends on these strategies. The more complex the SMECI simulation strategies are, the more probable the simulated duels are, but the bigger the ES simulator decision trees are. The player strategy description makes a trade-off between realistic strategies, speed of decision tree construction and size of these trees.

11   Marks of decision tree nodes

The ES simulator simulates methodically, for the three RA behaviors, all possible BA reprisal strategies called $S_{B_{plan}}$ and grades them according to their efficiency. In SMECI the best marks are the smallest.

At the end of the three ES simulator sessions, an ES analyser compares the $S_{B_{plan}}$ strategy marks built by the ES simulator. When we have enough time to expand the complete choice tree corresponding to a RA behavior, the $S_{B_{plan}}$ marks are in accordance with the outcomes of the games developed with these reprisal strategies.

In order to avoid slowing down the real duel with too many long calls to the decision support system, we set a limit on the execution time of an ES simulator session. We also grade alternatives, which have not been simulated to their end. The mark of an incomplete simulation is given according to aircraft and missile positions relative to the barriers. The mark of a decision tree leaf is small if the chance for BM to reach RA is big, if BA must not evade early, etc.

A decision tree is composed of nodes with two or five sons. The alternatives with five sons correspond to different turns and the binary alternatives correspond to a decision to fire or not to fire. The ES simulator grades the decision tree states to expand them with a strategy in "best first". The most interesting nodes become the minimum mark. SMECI uses the same heuristic to grade a node or a leaf in a decision tree according to the following evaluation function:

- The mark decreases when RA is near BM capture zones. This heuristic is more precise than a comparable heuristic not using the subgame barriers as: "We modulate the mark of a state in accordance with the energy of BM and the range between BM and RA".

- The mark increases if BA is close to hypothetic RM capture zones.

- In a simulation, when BM is not yet fired, the mark is better if the direction of BA is far from the line of sight. We search for a *win* strategy for BA and not only a strategy leading to a *draw* outcome.

- When a simulation is finished, the mark is according to the game outcome :

  - The mark corresponds to a high constant if BA evades without firing (outcome: early evasion).

  - We assign to a leaf of a state tree a very low mark in the case of a *win* outcome...

The **ES analyser** manipulates some lists like:

$$(\text{prudent}_{RA}, \quad (S_{B_{plan_1}}, \text{mark}_{11}), \ ..., \ (S_{B_{plan_i}}, \text{mark}_{1i}), \ ..., \ (S_{B_{plan_n}}, \text{mark}_{1n}))$$
$$(\text{regular}_{RA}, \quad (S_{B_{plan_1}}, \text{mark}_{21}), \ ...,(S_{B_{plan_i}}, \text{mark}_{2i}), \ ...,(S_{B_{plan_n}}, \text{mark}_{2n}))$$
$$(\text{aggressive}_{RA},(S_{B_{plan_1}}, \text{mark}_{31}), \ ..., \ (S_{B_{plan_i}}, \text{mark}_{3i}), \ ..., \ (S_{B_{plan_n}}, \text{mark}_{3n}))$$

The **ES analyser** proposes to BA the strategy $S_{B_{plan_i}}$ with the best mark ($\min_i [\text{mark of } S_{B_{plan_i}}]$), $i \in \{1..n\}$ and with $[\text{mark of } S_{B_{plan_i}}]$ equal to the sum of marks of $S_{B_{plan_i}}$ on the three RA behaviors tested:

$$[\text{mark of } S_{B_{plan_i}}] = \sum_{j=1}^{3} mark_{ji}$$

We can be pessimistic and imagine that RA uses the worst supposed behavior for the chosen BA reprisal strategy. BA also plays the strategy insuring it the minimum mark against any supposed behavior of RA:

$$[\text{mark of } S_{B_{plan_i}}] = \max_{j\in\{1,2,3\}} mark_{ji}$$

In the third mode (mostly used mode), we use the weighted sum associated to the three RA behaviors considered by the **ES simulator**; e.g., according to the probability distribution (p_j) established by the **ES observer** on the decision trees:

$$[\text{mark of } S_{B_{plan_i}}] = \sum_{j=1}^{3} p_j \, mark_{ji}$$

12   Example

Figures 12 and 13 give two medium range aerial duel examples with BA on the left side and RA at right. Theses drawings picture BM, RM and all hypothetic red missiles.

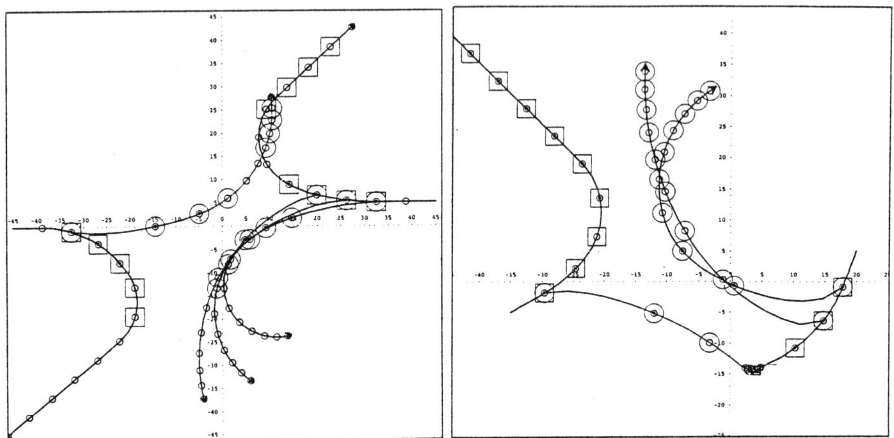

Figure 12: Draw outcome   Figure 13: Win outcome

13   Conclusion

This study illustrates how to combine pursuit-evasion game solutions, simulations and AI techniques to model and solve a complex aerial duel with modern missiles. As an example of this approach, we have chosen the Mica/Amraam duel and we have used advanced programming to build a decision support system. This example shows how game theory proposes, in particular, a secure evasion to an aircraft against a missile such as a Mica/Amraam with only a few hypotheses.

In this study, we have designed carefully the aircraft and missile kinematic models and the player strategies in SMECI. We have chosen realistic dynamics for the aircraft and the missiles in the SMECI simulations as in the differential games.

With other differential games and especially with a more complex simulator, it would be possible to do the same process for 3D duels. The assumption of only one opponent is not a restriction for future interesting prospects because BA already plays against several hypothetic RMs.

As the `ES simulator` does not have to try several BA evasion strategies (the `ES duel` uses hypothetic RM firings to calculate a BA guaranteed evasion), the number and the size of decision trees to build decreases. This last remark still shows well how it is possible to optimize a decision taking process combining classical decision trees and other techniques.

REFERENCES

[1]   Pierre Bernhard. Application of the min-max certainty equivalence principle to sampled data h-infinity optimal control. *Systems & Control Letters*, July 1990.

[2]   M. Guelman, J. Shinar, and A. Green. Qualitative study of a planar pursuit evasion game in the atmosphere. In *Proceedings of the AIAA Guidance, Navigation and Control Conference*, Minneapolis, August 15-17 1988.

[3]   Rufus Isaacs. *Differential Games, a Mathematical Theory with Applications to Warfare and Pursuit, Control and Optimization*. Applied Mathematics. SIAM, New York, 1965.

[4]   Stéphane Le Ménec. Differential games and symbolic programming to calculate a guaranteed aircraft evasion in modern aerial duels. In *Proceedings of the 33rd IEEE Conference on Decision and Control*, pp FP7 – Nonlinear Aircraft Control, Lake Buena Vista Florida, December 14–16, 1994.

[5]   G.J. Olsder and J.V. Breakwell. Role determination in an aerial dogfight. *International Journal of Game Theory*, 1974, 3:47 – 66.

[6]   J. Shinar, A.W. Siegel, and Y.I. Gold. On the analysis of a complex differential game using artificial intelligence techniques. In *Proceedings of the 27th IEEE Conference on Decision and Control*, p. TP5, Austin, December 1988.

[7]   J. Shinar, A.W. Siegel, and Y.I. Gold. A medium-range air combat game solution by a pilot advisory system. In *Proceedings of the AIAA Guidance Navigation and Control Conference*, Boston, August 1989.

Optimal Selection of Observation Times
in a Costly Information Game

Geert Jan Olsder
Delft University of Technology
The Netherlands

Odile Pourtallier
INRIA Sophia-Antipolis
France

Abstract

Pursuit evasion games with costly and asymmetric information
are studied. It is supposed that the evader has perfect information
about the position of the pursuer (and himself) at each instant of
time whereas the pursuer gets information about the position of the
evader only at discrete instants. Furthermore we suppose that the
pursuer cannot move during a given period of time while he gathers
this information. We investigate both the case in which the instants
of observation are chosen by the pursuer in an open loop way (at the
beginning of the game) and the case in which he chooses these instants according to the last information obtained (i.e., he chooses in a
feedback way).

1   Introduction

We study a pursuit-evasion game with asymmetric and costly information. We suppose that the pursuer does not see the evader if he (the pursuer)
is moving. Whenever he wants to know where the evader is, he must stop
during a fixed period of time δ. We suppose that at the end of this period the
pursuer knows the state of the evader perfectly at that instant of time. Thus
the information is costly in the sense that pursuer cannot move towards the
evader during the information period, and consequently the evader can move
away from him. The criterion which the pursuer wants to minimize is the
distance between himself and the evader at a given, a priori known instant
of time. We are mainly interested in the pursuer's behavior, and want to
find a strategy for him in the worst possible situation. In other words we are
interested by a lower value, and thereby assume that the evader has perfect
information about the pursuer's state at each instant of time.

Up to now it seems that there are two streams of works which consider this particular asymetric information stucture (full information, continuous time, for one player; information at discrete time instants only for the other player) in generalized pursuit-evasion games (target problems). The first one uses techniques introduced by Pontryagin in [7]; the latter is used in [8] and [9] to solve pursuit-evasion games with piecewise open loop strategies. They are based on the use of morphological operations, namely erosion and dilatation, and the aim is to determine the zone of initial points that lead to capture in a given number of stages. In [6] the link between this method of capture zones and dynamic programming has been shown in the setup of a discrete game. Algorithms with appropriate discretization of the state space have been discussed there. The papers [4], [5] investigate the problems of costly information. The choice of the different stages is not a result of an arbitrary discretization of time but is a consequence of the control exerted by the pursuer, and has a cost associated with it. Analytical results are presented for a linear first order model. The extension of these results to higher order models does not seem tractable if the aim is to obtain analytical results. Nevertheless, numerical algorithms are given to obtain numerical results. Other results related to asymetric and discrete information have been obtained by Melikian (e.g. [2], [3]). In [2] (directly related to the game under study), a problem of encounter where only a limited number of observation instants is available has been studied. The information instants are chosen at the beginning of the game, i.e. in an open loop fashion.

In the present paper we extend the ideas developed in [2] . In the first part we use a more general dynamic and add the aspect of information cost. In the second part we allow the moments at which information is gathered to depend on the current state (these instants are chosen in a state feedback fashion).

2   Problem Statement

The Dynamics

We are given two players: $P1$, called the pursuer, and $P2$, called the evader. The state of $P1$ at time t is indicated by $x(t)$, with $x \in \mathbb{R}^m$, and its time evolution is governed by

$$\dot{x}(t) = A(t)x(t) + B(t)u(t) + f(t). \tag{1}$$

In this equation, $u(t)$ refers to $P1$'s control at time t with $u \in \mathbb{R}^r$; $A(t)$ and $B(t)$ are matrices of size $m \times m$ and $m \times r$ respectively. The m-vector function $f(t)$ denotes a known inhomogeneous term. The symbol $\dot{}$ refers to the time derivative. It is assumed that the entries of A, B and f are all continuous with respect to time, though generalizations are possible. In the same vein, the state of $P2$ at time t is indicated by $y(t)$, with $y \in \mathbb{R}^n$, and its time evolution is governed by

$$\dot{y}(t) = F(t)y(t) + G(t)v(t) + g(t). \tag{2}$$

In this equation, $v(t)$ refers to $P2$'s control at time t with $v \in \mathbb{R}^s$; $F(t)$ and $G(t)$ denote matrices of size $n \times n$ and $n \times s$ respectively. The n-vector function $g(t)$ denotes a known inhomogeneous term. Just as with A, B and f, it is assumed that the entries of F, G and g are all continuous with respect to time.

Player $P1$ must choose his control in such a way that $||u(t)||_{p_u} \leq 1$ for each time instant t. The subscript refers to the standard p-norm with $1 \leq p \leq \infty$. Similarly, $v(t)$ must satisfy $||v(t)||_{p_v} \leq 1$ for each t.

The Information

Player $P1$ knows, at time t, his own past states $x(s)$, $s \leq t$. Moreover he observes the state of his opponent, $y(t)$, at time instants a_i, $i = 0, 1. \ldots, N$. Player $P2$ knows, at each time t, both his own position and that of his opponent. Hence $P2$ continuously knows what the state of his opponent is, whereas $P1$ receives this information (state of $P2$) only at the time instants a_i. $P1$ can choose the time instants a_i at which he wants to receive knowledge about his opponent's state. The price to be paid for this information is that during a period of δ time units prior to a_i, $P1$ cannot choose $u(t)$ freely: during $a_i - \delta \leq t \leq a_i$, $u(t)$ must be chosen such that $\dot{x}(t) = 0$, provided of course that such a $u(t)$ exists. At time a_i $P1$ gets to know $y(t)$ with $t = a_i$. Because of these restrictions, it will be assumed that

$$0 = a_0 \leq a_1 - \delta < a_1 \leq a_2 - \delta < a_2 \leq \cdots < a_{N-1} \leq a_N - \delta < T = a_N. \tag{3}$$

At the beginning of the game, $t = 0$, it is assumed that $P1$ obtains a measurement. It is assumed that the N-th measurement is obtained at a given time T, the final time of the game, which is supposed to be known to both players.

The Criterion

We will be concerned with the criterion $\min_{u,\{a_i\}} \max_v J$, where J has one of the following two forms (to distinguish between these forms, we will call them respectively J_1 and J_2):

- $J_1 = |c'x(T) - h'y(T)|$, where c, h are vectors with m, n components respectively. The symbol $'$ denotes transposition.

- $J_2 = ||x(T) - y(T)||_2.$

In both criteria it is assumed that T is fixed and known to the players. The last criterion only makes sense if $m = n$. The emphasis in this paper will be on the first criterion; brief remarks about the other one will be made.

Change of variables

For future purposes it turns out to be useful to perform the coordinate changes $\hat{x}(t) = \Phi_A(T,t)x(t)$ and $\hat{y}(t) = \Phi_F(T,t)y(t)$. The matrix Φ_A is the transition matrix of A, i.e. it is uniquely determined by $(d/dt)\Phi_A(t,t_0) = A(t)\Phi_A(t,t_0)$ and $\Phi_A(t_0,t_0)$ being the identity matrix. The matrix Φ_F is likewise defined with respect to the matrix F. In the new coordinates, the equations of motion are

$$\dot{\hat{x}} = \Phi_A(Bu(t) + f(t)), \tag{4}$$
$$\dot{\hat{y}} = \Phi_F(Gv(t) + g(t)). \tag{5}$$

The form of the criteria does not change with the coordinate transformation since $\hat{x}(T) = x(T)$ and $\hat{y}(T) = y(T)$ and the criteria only depend on $x(T)$ and $y(T)$. The solutions of the new equations of motion can be written as

$$\hat{x}(a_{k+1}) = \hat{x}(a_k) + \int_{a_k}^{a_{k+1}} \Phi_A[Bu(t) + f(t)]dt, \tag{6}$$

$$\hat{y}(a_{k+1}) = \hat{y}(a_k) + \int_{a_k}^{a_{k+1}} \Phi_F[Gv(t) + g(t)]dt, \quad k = 0, 1, \ldots, N - 1. \tag{7}$$

The first of these equations can be rewritten as

$$\hat{x}(a_{k+1}) = \hat{x}(a_k) + \int_{a_k}^{a_{k+1}-\delta} \Phi_A[Bu(t) + f(t)]dt \tag{8}$$

since it was assumed that during the observation period $a_k - \delta \leq t \leq a_k$ we would have $Bu(t) + f(t) \equiv 0$.

3   Open loop choice of $\{a_i\}$; General Analysis

3.1   Criterion J_1

Assume for the time being that the sequence $\{a_i\}$ is given and it satisfies (3). We will apply the dynamic programming principle with respect to the number of observations still to be made by $P1$. Consider the last step, i.e. the current states are $\hat{x}(a_{N-1})$ and $\hat{y}(a_{N-1})$ and we want to calculate $V_1(\hat{x}(a_{N-1}), \hat{y}(a_{N-1})) \stackrel{\text{def}}{=} \min_u \max_v J_1$. The index 1 of the V(alue) function refers to 1 step to go.

$$
\begin{aligned}
V_1(\hat{x}(a_{N-1}), \hat{y}(a_{N-1})) &= \min_u \max_v |c'\hat{x}(T) - h'\hat{y}(T)| \\
&= \min_{||u||_{p_u} \le 1} \max_{||v||_{p_v} \le 1} |c'\hat{x}(a_{N-1}) - h'\hat{y}(a_{N-1}) \\
&\qquad + \int_{a_{N-1}}^{a_N - \delta} c'\Phi_A(T,t)(Bu(t) + f(t))dt \\
&\qquad - \int_{a_{N-1}}^{a_N} h'\Phi_F(T,t)(Gv(t) + g(t))dt| \\
&= \min_{||u||_{p_u} \le 1} \max_{||v||_{p_v} \le 1} |\Lambda_1 + \int_{a_{N-1}}^{a_N - \delta} c'\Phi_A(T,t)Bu(t)dt \\
&\qquad - \int_{a_{N-1}}^{a_N} h'\Phi_F(T,t)Gv(t)dt|,
\end{aligned}
$$

$$(9)$$

where

$$
\Lambda_1 \overset{\text{def}}{=} c'\hat{x}(a_{N-1}) - h'\hat{y}(a_{N-1}) + \int_{a_{N-1}}^{a_N - \delta} c'\Phi_A(T,t)f(t)dt - \int_{a_{N-1}}^{a_N} h'\Phi_F(T,t)g(t)dt.
$$

Since

$$
\min_{||u||_{p_u} \le 1} \int_{a_{N-1}}^{a_N - \delta} c'\Phi_A(T,t)Bu(t)dt = - \int_{a_{N-1}}^{a_N - \delta} ||c'\Phi_A(T,t)B||_{q_u} dt, \quad (10)
$$

see [1], where q_u is defined through $1/p_u + 1/q_u = 1$, and

$$
\max_{||v||_{p_v} \le 1} \int_{a_{N-1}}^{a_N} h'\Phi_F(T,t)Gv(t)dt = \int_{a_{N-1}}^{a_N} ||h'\Phi_F(T,t)G||_{q_v} dt, \quad (11)
$$

where $1/p_v + 1/q_v = 1$, the value can be written as

$$
V_1 = \begin{cases} |\Lambda_1| - B_{N-1} + C_{N-1}, & \text{if} \quad |\Lambda_1| > B_{N-1}, \\ C_{N-1}, & \text{if} \quad |\Lambda_1| \le B_{N-1}. \end{cases}
$$

Here we have written

$$
B_{N-1} \overset{\text{def}}{=} \int_{a_{N-1}}^{a_N - \delta} ||c'\Phi_A(T,t)B||_{q_u} dt, \quad C_{N-1} \overset{\text{def}}{=} \int_{a_{N-1}}^{a_N} ||h'\Phi_F(T,t)G||_{q_v} dt.
$$

More concisely, we will write

$$
V_1(\hat{x}(a_{N-1}), \hat{y}(a_{N-1})) = \max(C_{N-1}, |\Lambda_1| - B_{N-1} + C_{N-1}). \quad (12)
$$

Now we continue with the next to last step of the dynamic programming algorithm. The value function with 2 steps to go is indicated by V_2;

$$
V_2(\hat{x}(a_{N-2}), \hat{y}(a_{N-2})) =
$$

$$
\begin{aligned}
&= \min_{u} \max_{v} V_1(\hat{x}(a_{N-1}), \hat{y}(a_{N-1})) \\
&= \min_{u} \max_{v} \max(C_{N-1}, |\Lambda_1| - B_{N-1} + C_{N-1}) \\
&= \max(C_{N-1}, \min_{u} \max_{v} |\Lambda_1| - B_{N-1} + C_{N-1}) \\
&= \max(C_{N-1}, \min_{u} \max_{v} | \Lambda_2 + \int_{a_{N-2}}^{a_{N-1}-\delta} c'\Phi_A(T,t)Bu(t)dt \\
&\qquad - \int_{a_{N-2}}^{a_{N-1}} h'\Phi_F(T,t)Gv(t)dt| - B_{N-1} + C_{N-1}),
\end{aligned}
$$

where

$$
\Lambda_2 \stackrel{\text{def}}{=} c'\hat{x}(a_{N-2}) - h'\hat{y}(a_{N-2}) \; + \int_{a_{N-2}}^{a_{N-1}-\delta} c'\Phi_A(T,t)f(t)dt \\
- \int_{a_{N-2}}^{a_{N-1}} h'\Phi_F(T,t)g(t)dt.
$$

If we now define

$$
B_{N-2} \stackrel{\text{def}}{=} \int_{a_{N-2}}^{a_{N-1}-\delta} ||c'\Phi_A(T,t)B||_{q_u} dt,
$$

$$
C_{N-2} \stackrel{\text{def}}{=} \int_{a_{N-2}}^{a_{N-1}} ||h'\Phi_F(T,t)G||_{q_v} dt,
$$

we can write

$$
V_2(\hat{x}(a_{N-2}), \hat{y}(a_{N-2})) \tag{13}
$$
$$
= \max(C_{N-1}, C_{N-1} + C_{N-2} - B_{N-1}, |\Lambda_2| + C_{N-1}
$$
$$
+ C_{N-2} - B_{N-1} - B_{N-2}).
$$

In this way we continue; with k steps to go we get

$$
V_k(\hat{x}(a_{N-k}), \hat{y}(a_{N-k})) = \tag{14}
$$
$$
\max(C_{N-1}, C_{N-1} + C_{N-2} - B_{N-1}, \ldots, C_{N-1} + \cdots + C_{N-k} - B_{N-1}
$$
$$
- \cdots - B_{N-k+1}, |\Lambda_k| + C_{N-1} + \cdots + C_{N-k} - B_{N-1} - \cdots - B_{N-k}),
$$

where

$$
C_{N-k} \stackrel{\text{def}}{=} \int_{a_{N-k}}^{a_{N-k+1}} ||h'\Phi_F(T,t)G||_{q_v} dt, \tag{15}
$$

$$
B_{N-k} \stackrel{\text{def}}{=} \int_{a_{N-k}}^{a_{N-k+1}-\delta} ||c'\Phi_A(T,t)B||_{q_u} dt, \tag{16}
$$

$$\Lambda_k \stackrel{\text{def}}{=} c'\hat{x}(a_{N-k}) - h'\hat{y}(a_{N-k}) + \left(\int_{a_{N-1}}^{a_N-\delta} + \cdots \right. \tag{17}$$

$$\left. + \int_{a_{N-k}}^{a_{N-k+1}-\delta} \right) c'\Phi_A(T,t)f(t)dt - \int_{a_{N-k}}^{a_N} h'\Phi_F(T,t)g(t)dt.$$

Finally, the last step, i.e. the first step measured in forward time, yields

$$V_N(\hat{x}(a_0), \hat{y}(a_0)) = \max(g_{N-1}, g_{N-2}, \ldots, g_0, g_{-1}),$$

where

$$g_{N-1} \stackrel{\text{def}}{=} C_{N-1}, \tag{18}$$

$$g_{N-k} \stackrel{\text{def}}{=} C_{N-1} + C_{N-2} + \cdots + C_{N-k} - B_{N-1} - \cdots - B_{N-k+1}, \tag{19}$$

$$k = 2, 3 \ldots, N,$$

$$g_{-1} \stackrel{\text{def}}{=} |\Lambda_N| + C_{N-1} + \cdots + C_0 - B_{N-1} - \cdots - B_0. \tag{20}$$

So far it was assumed that the a_i were fixed. If now $P1$ can choose these time instants also, subject to (3), then he faces the problem

$$V_N(\hat{x}(a_0), \hat{y}(a_0)) = \min_{\{a_i\}} \max(g_{N-1}, g_{N-2}, \ldots, g_0, g_{-1}). \tag{21}$$

Notice that the initial condition $(\hat{x}(a_0), \hat{y}(a_0)) = (\hat{x}(0), \hat{y}(0))$ only shows up in the term g_{-1}. Since we cannot control the initial condition (which is assumed to be given), the term g_{-1} is not fully 'manipulatable' and therefore there is an independent interest to study

$$\Psi \stackrel{\text{def}}{=} \min_{\{a_i\}} \max(g_{N-1}, g_{N-2}, \ldots, g_0). \tag{22}$$

The term g_{N-k},

$$g_{N-k} = \int_{a_{N-k}}^T ||h'\Phi_F(T,t)G||_{q_v} dt - \int_{a_{N-k+1}}^T ||c'\Phi_A(T,t)B||_{q_u} dt$$

$$+ \sum_{N-k+2}^N \int_{a_j-\delta}^{a_j} ||c'\Phi_A(T,t)B||_{q_u} dt,$$

depends on the time instants a_i, $i = N-k, N-k+1, \ldots, N$, only.

Suppose that the minimization in (22), not taking into account the constraints (3), would lead to optimal a_i values such that $g_{N-k} = g_{N-k+1}$, $k = 2, 3, \ldots, N$. This then would lead to

$$\int_{a_{N-k-1}}^{a_{N-k}} ||h'\Phi_F(T,t)G||_{q_v} dt = \int_{a_{N-k}}^{a_{N-k+1}} ||c'\Phi_A(T,t)B||_{q_u} dt$$

$$- \int_{a_{N-k+1}-\delta}^{a_{N-k+1}} ||c'\Phi_A(T,t)B||_{q_u} dt.$$

Hence, if we would have $a_{N-k} > a_{N-k-1}$, then the left-hand side of this expression is nonnegative which leads to $a_{N-k} \leq a_{N-k+1} - \delta$, equivalently, $a_{N-k+1} \geq a_{N-k} + \delta$. Thus the assumption $a_{N-k} > a_{N-k-1}$ leads to $a_{N-k+1} \geq a_{N-k} + \delta$. Hence the only criterion to be checked whether (3) holds is whether $a_1 \geq a_0 + \delta = \delta$ is satisfied. This property will be formulated as

Theorem 3.1 *Suppose the minimization problem as given by (22), not taking into account the constraints (3), leads to equal values of the g_{N-k}, $k = 1, 2, \ldots, N$, and to $a_1 \geq \delta$, then the constraints (3) are satisfied.*

The assumption in this theorem that all g_{N-k} values are equal after the minimization is not as strange as it might look. We will come back to this in Theorem 4.2. Conversely, if we assume $\Psi \overset{\text{def}}{=} g_{N-1} = \cdots = g_0$, then we have N equations with precisely N unknowns (a_i, $i = 1, \ldots, N-1$ and Ψ). One could apply a standard algorithm to solve these (nonlinear) equations in order to obtain the optimal a_i values.

3.2   Criterion J_2

In this section some remarks will be made about the problem in which criterion J_2 is considered. The last step of the application of the dynamic programming principle yields

$$V_1(x(a_{N-1}), y(a_{N-1})) = \min_u \max_v ||x(T) - y(T)||_2 \tag{23}$$

$$= \min \max ||\mathcal{L}_1 + \int_{a_{N-1}}^{a_N - \delta} \Phi_A(T,t)Bu(t)dt - \int_{a_{N-1}}^{a_N} \Phi_F(T,t)Gv(t)dt||_2,$$

where

$$\mathcal{L}_1 \overset{\text{def}}{=} \hat{x}(a_{N-1}) - \hat{y}(a_{N-1}) + \int_{a_{N-1}}^{a_N - \delta} \Phi_A(T,t)f(t)dt - \int_{a_{N-1}}^{a_N} \Phi_F(T,t)g(t)dt,$$

and which we symbolically write as

$$V_1(x(a_{N-1}), y(a_{N-1})) = \min_\mu \max_\nu ||\mathcal{L}_1 + \mathcal{B}_{N-1}\mu - \mathcal{C}_{N-1}\nu||_2 \tag{24}$$

for suitably defined operators $\mathcal{B}_{N-1}$ and $\mathcal{C}_{N-1}$. Consider Figure 1 in which the circle with the center at the origin and radius $V_1(x(a_{N-1}), y(a_{N-1}))$ separates the two compact sets $\mathcal{S}_v$ and $\mathcal{S}_u$ which are defined as

$$\mathcal{S}_u \overset{\text{def}}{=} \mathcal{L}_1 + \mathcal{B}_{N-1}\mu - \mathcal{C}_{N-1}\nu^* \quad \text{for all admissible } \mu, \tag{25}$$

$$\mathcal{S}_v \overset{\text{def}}{=} \mathcal{L}_1 + \mathcal{B}_{N-1}\mu^* - \mathcal{C}_{N-1}\nu \quad \text{for all admissible } \nu, \tag{26}$$

where μ^* and ν^* are the optimal values of μ and ν respectively in (24).

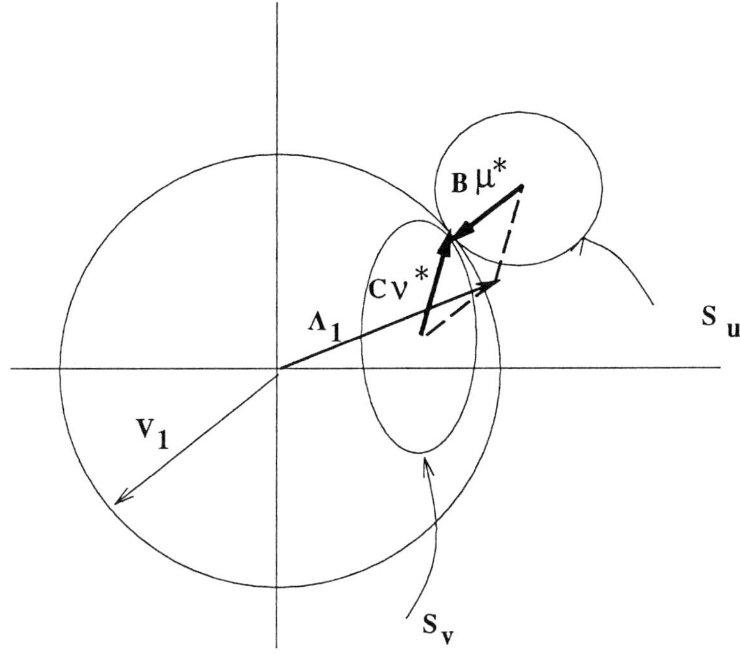

Figure 1: Separation of the two reachable sets

The optimal μ^* and ν^* values satisfy

$$||\mathcal{L}_1 + \mathcal{B}_{N-1}\mu^* - \mathcal{C}_{N-1}\nu^*||_2^2 \leq \tag{27}$$
$$< \mathcal{L}_1 + \mathcal{B}_{N-1}\mu^* - \mathcal{C}_{N-1}\nu^*, \mathcal{L}_1 + \mathcal{B}_{N-1}\mu - \mathcal{C}_{N-1}\nu^* >,$$
$$||\mathcal{L}_1 + \mathcal{B}_{N-1}\mu^* - \mathcal{C}_{N-1}\nu^*||_2^2 \geq \tag{28}$$
$$< \mathcal{L}_1 + \mathcal{B}_{N-1}\mu^* - \mathcal{C}_{N-1}\nu^*, \mathcal{L}_1 + \mathcal{B}_{N-1}\mu^* - \mathcal{C}_{N-1}\nu > .$$

for all admissible μ and ν; the notation $< .,. >$ refers to the inner product. This problem is clearly far more difficult than the problem with the first criterion since the optimal μ^* and ν^* values will in general depend on $\hat{x}(a_{N-1}) - \hat{y}(a_{N-1})$, which was not the case in the J_1-problem. If the sets $\mathcal{S}_u$ and $\mathcal{S}_v$ happen to be balls, as is the case in [2], then the optimal μ^* and ν^* do not depend on $\hat{x}(a_{N-1}) - \hat{y}(a_{N-1})$ and the problem becomes much simpler. A particular case in which this last condition occurs is when $\mathcal{B}_i\mu$ and $\mathcal{C}_i\nu$ can respectively be written $\beta_i\mu$ and $\gamma_i\nu$ with β_i and γ_i scalar.

4 Some Special Cases

4.1 No Time Loss During Observations

In this subsection it is assumed that $\delta = 0$, i.e. the observations do not require a special control $u(t)$ during part of the time. Put differently, $P1$ does not lose time during which he would be forced to keep $\hat{x}$ at rest. We will study (21) and (22) in some more depth for this case. We now have

$$g_{-1} = |\Lambda_N| + \int_0^T ||h'\Phi_F(T,t)G||_{q_v}\, dt - \int_0^T ||c'\Phi_A(T,t)B||_{q_u}\, dt,$$

which is independent of the $\{a_i\}$. The function Λ_N in this expression equals

$$\Lambda_N = c'\hat{x}(0) - h'\hat{y}(0) + \int_0^T c'\Phi_A(T,t)f(t)dt - \int_0^T h'\Phi_F(T,t)g(t)dt.$$

The other g_i-elements also become simpler:

$$g_{N-k} = \int_{a_{N-k}}^T ||h'\Phi_F(T,t)G||_{q_v}\, dt - \int_{a_{N-k+1}}^T ||c'\Phi_A(T,t)B||_{q_u}\, dt,$$

$k = 1, 2, \ldots, N$, and hence the element g_{N-k} only depends on a_{N-k} and a_{N-k+1}. In fact, the function g_{N-k} does not directly depend on the index $N - k$ (which could therefore be omitted), but it does depend on a_{N-k} and a_{N-k+1} and we will also write $g(a_{N-k}, a_{N-k+1})$ rather than g_{N-k}. We can now apply a lemma from [2], which is repeated here for the sake of completeness.

Lemma 4.1 *Let the function $g(a_i, a_{i+1})$ be continuous on the square $0 \leq a_i$, $a_{i+1} \leq T$ and differentiable on the open square $0 < a_i, a_{i+1} < T$, where it also satisfies*

$$\partial g/\partial a_i < 0, \quad \partial g/\partial a_{i+1} > 0.$$

The minimum of $\max(g(a_0, a_1), g(a_1, a_2), \ldots, g(a_{N-1}, a_N))$, *subject to* $0 = a_0 \leq a_1 \leq \cdots \leq a_{N-1} \leq a_N = T$, *is achieved at unique values of the a_i values, denoted by a_i^*. Besides, $a_i^* < a_{i+1}^*$ and $g(a_i^*, a_{i+1}^*) = g(a_{i+1}^*, a_{i+2}^*)$, $i = 0, 1, \ldots, N - 2$.*

As in (22), this latter value will be indicated by Ψ. A direct consequence of this lemma and of the continuity of g_{N-k} with respect to δ is:

Theorem 4.2 *If the optimal values of a_i are substituted and if δ is sufficiently close to zero (or equal to zero), then we have $\Psi = g_{N-k}$, $k = 1, 2, \ldots, N$.*

4.2 Simple Dynamics

In this subsection it is assumed that the dynamics are given by

$$\dot{x} = u, \tag{29}$$

$$\begin{pmatrix} \dot{y}_1 \\ \dot{y}_2 \end{pmatrix} = \begin{pmatrix} 0 & 1 \\ 0 & 0 \end{pmatrix} \begin{pmatrix} y_1 \\ y_2 \end{pmatrix} + \begin{pmatrix} 0 \\ 1 \end{pmatrix} v, \tag{30}$$

where x and u are one dimensional, as is v. The vector y is two dimensional. We will consider the criterion $J_1 = |x(T) - y_1(T)|$, i.e. $c = 1$ and $h = (1\ 0)'$. The value of δ is not necessarily zero. Since

$$\Phi_F(T,t) = \begin{pmatrix} 1 & T-t \\ 0 & 1 \end{pmatrix},$$

we easily obtain $\dot{y}_1(t) = (T-t)v(t)$. Hence it is straightforward to derive

$$C_{N-k} = ((T - a_{N-k})^2 - (T - a_{N-k+1})^2)/2, \quad B_{N-k} = a_{N-k+1} - a_{N-k} - \delta,$$

and hence

$$\begin{aligned}
g_{N-1} &= (T - a_{N-1})^2/2, &(31)\\
g_{N-k} &= (T - a_{N-k})^2/2 - (T - a_{N-k+1} - (k-1)\delta), k = 2, 3, \ldots, N &(32)\\
g_{-1} &= |\hat{x}(0) - \hat{y}_1(0)| + T^2/2 - (T - N\delta). &(33)
\end{aligned}$$

Suppose we want to find the optimal a_i values by solving the equations $g_{N-1} = g_{N-2} = \cdots = g_0(= \Psi)$. Theorem 3.1 can be applied to check whether the inequalities (3) hold. More explicitly, these equations read

$$T - a_i = (T - a_{i-1})^2/2 + (N-i)\delta - \Psi, \quad i = 1, 2, \ldots, N. \tag{34}$$

If we substitute the left-hand sides into the right-hand sides of these equations, then this leads to a polynomial in Ψ:

$$\Psi = \frac{1}{2}(\frac{1}{2}\frac{1}{2}(\cdots(\frac{1}{2}\frac{1}{2}(T-a_0)^2+(N-1)\delta-\Psi)^2+(N-2)\delta-\Psi)^2+\cdots)^2+2\delta-\Psi)^2+\delta-\Psi)^2.$$

Once Ψ has been determined, the a_i values can recursively be determined from (34).

We will end this subsection by briefly suggesting another algorithm to determine the a_i values. The initial values are $a_i = iT/N$, which would be the solution if $\delta = T/N$. Such a δ value is not very interesting since then $P1$ will not move at all. These initial values will be updated in the following way and in the limit approach their optimal values. If the initial a_i values are substituted into the g_j functions, we get

$$g_i = T^2(1 - i/N)^2/2, \quad i = 0, 1, \ldots, N-1.$$

We now assume $\delta = T/N - \epsilon$, where ϵ is a small positive number, and linearize the g_j functions around the initial a_i values. We write $a_i = iT/N + \alpha_i$. The linearized g_i functions are

$$\begin{aligned}
g_0 &= T^2/2 + \alpha_1 - \epsilon(N-1); \\
g_i &= T^2(1 - i/N)^2/2 + \alpha_{i+1} - \alpha_i T(1 - i/N) - \epsilon(N - i - 1); \\
g_{N-1} &= T^2(1/N)^2 - \alpha_{N-1}T/N.
\end{aligned}$$

The linear programming problem $\min \Delta$ with respect to α_i and subject to

$$\begin{aligned}
a_{i+1} &\geq a_i + \delta, \quad i = 0, 1, \ldots, N-1; \\
\Delta &\geq T^2/2 + \alpha_1 - \epsilon(N-1); \\
\Delta &\geq T^2(1 - i/N)^2/2 + \alpha_{i+1} - \alpha_i T(1 - i/N) - \epsilon(N - i - 1), \\
& \qquad\qquad\qquad\qquad\qquad\qquad i = 1, 2, \ldots, N-2; \\
\Delta &\geq T^2(1/N)^2 - \alpha_{N-1}T/N,
\end{aligned}$$

yields the optimal α_i and hence a_i values if $\delta = T/N - \epsilon$. This procedure of linearization and solving an LP problem is repeated until the true δ value is reached.

5   Number of observations fixed; closed loop choice for $\{a_i\}$

General analysis

In the previous sections the instants of observations were chosen at the beginning of the game, i.e. in an open loop way. The wish to have the possibility to select these instants during the evolution of the game as in [4], [5] seems quite natural. One could then decide about the next time instant of observation once the information of the current observation had become known.

In this part we assume that the sequence $\{a_i\}$ of information times are to be chosen by the pursuer. Two slightly different versions are considered. In the first one we suppose that the final time $a_N = T$ together with the initial time a_0 are fixed. In a second one we consider the case that a_0 is fixed also, but that the final time $a_N = T$ is to be chosen by the pursuer. For convenience, (simultaneous descriptions of the two versions, and use of dynamic programming), we will suppose that, in the second version, the initial time is to be chosen by the pursuer, whereas the final time is fixed to an arbitrary value T. This can be done since the problem statement does not depend on time explicitly, and keeping the final time fixed rather then the intial time only leads to a time shift in the solution. As a consequence, we allow the game to start at a negative time a_0.

Let k denote the number of observations done. This number k will be used as the parameter of the following, equivalent discrete description of the game, where an extra state variable a_k, representing the date of the beginning of the $k + 1$st stage, is used, and where x_k and y_k denote respectively $\hat{x}(a_k)$ and $\hat{y}(a_k)$, the state variables at time a_k. One has

$$
\begin{cases}
x_{k+1} = x_k + \displaystyle\int_{a_k}^{a_{k+1}-\delta} \Phi_A[Bu_k(t) + f(t)]dt, \\[2em]
y_{k+1} = y_k + \displaystyle\int_{a_k}^{a_{k+1}} \Phi_F[Gv_k(t) + g(t)]dt, \\[1em]
a_{k+1} = \bar{u}_k, \qquad k = 0, 1, \ldots, N-1
\end{cases}
\tag{35}
$$

The quantity $\bar{u}_k$ is considered as a control of the pursuer $P1$ at stage k. At each stage k, the pursuer chooses the control $\bar{u}_k$ together with the (measurable) control function $u_k(t)$ on the time interval $[a_k, a_{k+1} - \delta]$. The evader chooses the control function $v_k(t)$ on the time interval $[a_k, a_{k+1}]$. Note that it is implicitly assumed that the evader knows the choice $\bar{u}_k$ at the time he decides about $v_k(t)$ on the time interval $[a_k, a_{k+1}]$. Though this may not seem very realistic, it makes sense in our setup where we are interested in the worst case situation from the viewpoint of the pursuer. The first and the last set of $P1$'s controls differ slightly from the intermediate ones just described. At the 'first' step, that by definition will correspond to $k = -1$, the state variables are (x_{-1}, y_{-1} and $a_{-1} = -\infty$). The set of controls is restricted to be ($\bar{u}_{-1} = a_0, u_{-1}(t) = 0$). Consequently x_0 and y_0 will be equal to x_{-1} and y_{-1} respectively. At the last step, corresponding to $k = N-1$, the pursuer is constrained to choose $\bar{u}_{N-1} = T$, since T is the fixed final time of the game.

The criterion we are going to study, if the number of observations is N, and with $(\hat{x}_0, \hat{y}_0)$ given, is

$$
\min_{\{u_k(\cdot)\}} \ \min_{\{\bar{u}_k\}} \ \max_{\{v_k(\cdot)\}} \ J^N
$$

where $J^N = | c'x_N - h'y_N |$, where c and h are vectors with m and n components respectively. For all k the controls are restricted to satisfy $\| u_k(\cdot) \|_{p_u} \leq 1$ and $\| v_k(\cdot) \|_{p_v} \leq 1$.

In order to simplify the above analysis we will introduce the following notations:

$$
z_k = c'x_k - h'y_k
$$

$$
B(a,b) = \int_a^{b-\delta} \| c'\Phi_A(T,t)B \|_{q_u} \, dt,
$$

$$
C(a,b) = \int_a^b \| h'\Phi_F(T,t)G \|_{q_v} \, dt,
\tag{36}
$$

$$
D(a,b) = \int_a^b h'\Phi_F(T,t)g(t)dt - \int_a^{b-\delta} c'\Phi_A(T,t)f(t)dt
$$

where q_u and q_v satisfy $1/p_u + 1/q_u = 1$ and $1/p_v + 1/q_v = 1$, with $k = 1, \ldots, N$. With these notations the dynamics (35) can be rewritten as

$$a_k = \bar{u}_{k-1},$$
$$z_k = z_{k-1} + D(a_{k-1}, \bar{u}_{k-1}) - \int_{a_{k-1}}^{\bar{u}_{k-1}} h' \Phi_F(T, t) G v_k(t) dt + \int_{a_{k-1}}^{\bar{u}_{k-1}-\delta} c' \Phi_A B u_k(t) dt.$$

$$(37)$$

As before, V_k will refer again to the 'k steps to go' value function. One has $V_0(z, T) = |z_N|$. Consider now the last step; x_{N-1}, y_{N-1} and a_{N-1} or, equivalently, z_{N-1} and a_{N-1} are assumed fixed. Recall that at this last step the pursuer does not choose $\bar{u}_{N-1}$, since the next observation time is $a_N = T = \bar{u}_{N-1}$ which is given. We get

$$
\begin{aligned}
V_1(z_{N-1}, a_{N-1}) &= \min_{u_{N-1}(\cdot)} \max_{v_{N-1}(\cdot)} |z_N| \\
&= \min_{u_{N-1}(\cdot)} \max_{v_{N-1}(\cdot)} |z_{N-1} + D(a_{N-1}, T) \\
&\quad - \int_{a_{N-1}}^{T} h' \Phi_F(T, t) G v_{N-1}(t) dt \\
&\quad + \int_{a_{N-1}}^{T-\delta} c' \Phi_A(T, t) B u_{N-1}(t) dt \, | \, .
\end{aligned}
$$

Since $\bar{u}_{N-1}$ is fixed we can repeat the reasoning of section 3.1 (see page 231), and together with the notations (36) we obtain

$$
V_1(z_{N-1}, a_{N-1}) =
\begin{cases}
|z_{N-1} + D(a_{N-1}, T)| - B(a_{N-1}, T) + C(a_{N-1}, T), \\
\qquad \text{if } |z_{N-1} + D(a_{N-1}, T)| > B(a_{N-1}, T) \\
\\
C(a_{N-1}, T), \text{ if } |z_{N-1} + D(a_{N-1}, T)| \le B(a_{n-1}, T)
\end{cases}
$$

hence

$$
V_1(z_{N-1}, a_{N-1}) = \max \left(
\begin{array}{l}
\bullet \, |z_{N-1} + D(a_{N-1}, T)| - B(a_{N-1}, T) + C(a_{N-1}, T) \; ; \\
\bullet \, C(a_{N-1}, T)
\end{array}
\right)
$$

$$(38)$$

(Bullets are here to mark more clearly the items in the max operation). At the next to last step ($k = N - 1$ in (37)) player $P1$ must choose both the next time of observation $\bar{u}_{N-2}$ and an open loop control on the interval $[a_{N-1}, \bar{u}_{N-1}]$. For the sake of economy we will write V_2 instead of $V_2(z_{N-2}, a_{N-2})$. We have

$$
V_2 = \min_{\bar{u}_{N-2}} \min_{u_{N-2}(\cdot)} \max_{v_{N-2}(\cdot)} V_1(z_{N-1}, \bar{u}_{N-2})
$$

that is, using (38),

$$V_2 = \min_{\bar{u}_{N-2}} \min_{u_{N-2}(\cdot)} \max_{v_{N-2}(\cdot)} \max \left(\begin{array}{l} \bullet \mid z_{N-1} + D(\bar{u}_{N-2}, T) \mid -B(\bar{u}_{N-2}, T) \\ \qquad\qquad\qquad\qquad\qquad +C(\bar{u}_{N-2}, T) \;\; ; \\[2mm] \bullet \; C(\bar{u}_{N-2}, T) \end{array} \right)$$

and, ultimately, together with (37),

$$V_2 = \min_{u_{N-2}} \min_{\bar{u}_{N-2}(\cdot)} \max_{v_{N-2}(\cdot)} \max \left(\begin{array}{l} \bullet \mid z_{N-2} + D(a_{N-2}, \bar{u}_{N-2}) + D(\bar{u}_{N-2}, T) \\ \qquad + \int_{a_{N-2}}^{\bar{u}_{N-2}-\delta} c'\Phi_A B u_{N-2}(t)dt \\ \qquad - \int_{a_{N-2}}^{\bar{u}_{N-2}} h'\Phi_F(T, t)_{N-2}(t)dt \mid \\ \qquad -B(\bar{u}_{N-2}, T) + C(\bar{u}_{N-2}, T) \;\; ; \\[2mm] \bullet \; C(\bar{u}_{N-2}, T) \end{array} \right)$$

We note that the "min" with respect to $u_{N-2}(\cdot)$ and the "max" with respect to $v_{N-2}(\cdot)$ apply only to the first member of the maximization. Therefore we can move these minimizations and maximizations inside the outside parentheses. We also make use of equalities (10) and (11) to obtain

$$\begin{aligned} &V_2(z_{N-2}, a_{N-2}) \\ &= \min_{\bar{u}_{N-2}} \max \left(\begin{array}{l} \bullet \mid z_{N-2} + D(a_{N-2}, \bar{u}_{N-2}) + D(\bar{u}_{N-2}, T) \mid \\ \qquad -B(a_{N-2}, \bar{u}_{N-2}) - B(\bar{u}_{N-2}, T) \\ \qquad +C(a_{N-2}, \bar{u}_{N-2}) + C(\bar{u}_{N-2}, T) \\ \bullet \; C(a_{N-2}, \bar{u}_{N-2}) + C(\bar{u}_{N-2}, T) - B(\bar{u}_{N-2}, T) \\ \bullet \; C(\bar{u}_{N-2}, T) \end{array} \right) \end{aligned}$$

Now we cannot interchange the remaining minimization and the maximization since every term of the maximization depends on $\bar{u}_{N-2}$. If we want to continue the analytical treatment, we need to make more assumptions on the dynamics. Let us suppose that one of the following two sets of hypotheses are satisfied:

$$(H_1) \left\{ \begin{array}{l} A = 0 \\ f(t) = f \\ B(t) = B \end{array} \right. \tag{39}$$

or

$$(H_2) \quad \delta = 0 \tag{40}$$

Let us emphasize that these assumptions are needed only because we want to be able to continue the analysis and to obtain tractable formulas. The dynamic programming algorithm applies without any restriction now. With the above notations it is straightforward to see that

$$D(a, b) + D(b, c) = D(a, c) + \delta c'.f, \tag{41}$$

$$B(a,b) + B(b,c) = B(a,c) - \delta \parallel c'.B \parallel, \tag{42}$$

$$C(a,b) + C(b,c) = C(a,c), \tag{43}$$

the latter equality being true even if neither (H_1) nor (H_2) hold. This allows us to interchange minimization and maximization since the first term does not depend on $\bar{u}_{N-2}$ anymore, which results in

$$V_2 = \max \left(\begin{array}{l} \bullet \ \mid z_{N-2} + D(a_{N-2}, T) + \delta c' f \mid \ -B(a_{N-2}, T) \\ \quad + C(a_{N-2}, T) + \delta \parallel c'B \parallel \ ; \\ \bullet \ \min_{\bar{u}_{N-2}} \max \left(\begin{array}{l} \bullet \ C(a_{N-2}, T) - B(\bar{u}_{N-2}, T) \ ; \\ \bullet \ C(\bar{u}_{N-2}, T) \end{array} \right) \end{array} \right)$$

or V_2 is the maximum of

$$\begin{array}{l} \bullet \ \mid z_{N-2} + D(a_{N-2}, T) + \delta c'.f \mid \ -B(a_{N-2}, T) + C(a_{N-2}, T) + \delta \parallel c'B \parallel; \\ \bullet \quad \min_{\{\bar{u}_{N-i}\} \in \mathcal{U}_2(a_{N-2})} \max_{i=1,2} \{ C(\bar{u}_{N-i-1}, T) - B(\bar{u}_{N-i}, T) + (i-1)\delta \parallel c'B \parallel \} \end{array} \tag{44}$$

where the set $\mathcal{U}_2(a_{N-2})$ denotes the sequence $\bar{u}_{N-2}, \bar{u}_{N-1}$ which should satisfy

$$a_{N-2} + \delta \leq \bar{u}_{N-2}; \bar{u}_{N-2} + \delta \leq \bar{u}_{N-1} = T.$$

Let us note that the first term of the maximization depends only on the present state, that is, on z_{N-2} and a_{N-2}, but does not depend on the future choice of observation times.

Recursively, using the same kind of arguments as used previously, we can prove that

$$V_k = \max \left(\begin{array}{l} \bullet \ \mid z_{N-k} + D(a_{N-k}, T) + (k-1)\delta c'f \mid \\ \quad -B(a_{N-k}, T) + C(a_{N-k}, T) + (k-1)\delta \parallel c'B \parallel; \\ \bullet \quad \min_{\{\bar{u}_{N-i}\} \in \mathcal{U}_k(a_{N-k})} \max_{i=1\ldots k} \{ C(\bar{u}_{N-i-1}, T) - B(\bar{u}_{N-i}, T) \\ \hspace{6cm} + (i-1)\delta \parallel c'B \parallel \} \end{array} \right)$$

where $\mathcal{U}_k(a_{N-k})$ is the sequence $\{\bar{u}_i\}, i = N-k, \ldots N-1$, such that

$$a_{N-k} + \delta \leq \bar{u}_{N-k}, \quad \bar{u}_i + \delta \leq \bar{u}_{i+1}, i = N - (k+1), \ldots, N-2; \quad \bar{u}_{N-1} = T.$$

In particular for $k = N$ we have

$$V_N(z_0, a_0) = \max \left(\begin{array}{l} \bullet \ \mid z_0 + D(a_0, T) + (N-1)\delta c'f \mid \\ \quad -B(a_0, T) + C(a_0, T) + (N-1)\delta \parallel c'B \parallel; \\ \bullet \quad \min_{\{\bar{u}_{N-i}\} \in \mathcal{U}_N(a_0)} \max_{i=1\ldots N} \{ C(\bar{u}_{N-(i+1)}, T) - B(\bar{u}_{N-i}, T) \\ \hspace{6cm} + (i-1)\delta \parallel c'B \parallel \} \end{array} \right) \tag{45}$$

with $\mathcal{U}_N(a_0) = \{\{u_i\}; \ \bar{u}_1 = a_0, \bar{u}_i + \delta \leq \bar{u}_{i+1}, i = 1, \ldots, N-2; \ \bar{u}_{N-1} = T\}$. This yields the value function for the first version of this game.

To solve the second version of the game with the final time, or, equivalently, after some rescaling described above, with the initial time not fixed, we need one additional step in order to determine the initial instant of the game, i.e. a_0. This leads to the last minimization problem

$$J_N(z_0) = \min_{a_0} V_N(z_0, a_0),$$

or

$$J_N(z_0) = \min_{a_0} \max \left(\begin{array}{l} \bullet \ | \ z_0 + D(a_0, T) + (N-1)\delta c' f \ | \\ \quad -B(a_0, T) + C(a_0, T) + (N-1)\delta \parallel c'B \parallel; \\ \bullet \quad \min_{\{\bar{u}_{N-i}\} \in \mathcal{U}_N(a_0)} \max_{i=1\ldots N} \{C(\bar{u}_{N-(i+1)}, T) - B(\bar{u}_{N-i}, T) \\ \qquad\qquad\qquad\qquad\qquad\qquad\qquad +(i-1)\delta \parallel c'B \parallel\} \end{array} \right)$$

$$(46)$$

Remarks

For both versions of the game just described, we needed to consider the following problem at each stage

$$\Psi_k(a_{N-k}) \overset{\text{def}}{=} \min_{\{a_i\} \in \mathcal{A}_k(a_{N-k})} \max\{g_{N-1}, \ldots, g_k\} \tag{47}$$

where

$$g_i = g_i(a_{i+1}, a_i) = C(a_{i+1}, T) - B(a_i, T) + (N - i - 1)\delta \parallel c'B \parallel$$

and

$$\begin{aligned} \mathcal{A}_k(a_{N-k}) \ &\overset{\text{def}}{=} \ \{(a_N, a_{N-1}, \ldots, a_{N-k+1}), \\ a_N \ &= \ T; \ a_i + \delta \leq a_{i+1}, i = N-1, \ldots, N-k\}. \end{aligned}$$

We now make a few remarks about this latter problem. First let us state a rather straightforward generalisation of Lemma 4.1.

Lemma 5.1 *Let the functions $g_i(x, y)$ be continuous and differentiable on $\mathbb{R}^2$, where they also satisfy*

$$\partial g_i/\partial x < 0, \quad \partial g_i/\partial y > 0,$$

and suppose furthermore that $g_{i+1}(x, y) = g_i(x, y) + c$ where c is a positive scalar. If there exists $(a_1^*, \ldots, a_{N-1}^*) \in \mathcal{A}_N(a_0)$ such that

$$g_i(a_i^*, a_{i+1}^*) = g_{i+1}(a_{i+1}^*, a_{i+2}^*), \quad i = 0, 1, \ldots, N-2, \tag{48}$$

then Ψ_N ($= g_i(a_i^*, a_{i+1}^*)$, $i = 0, 1, \ldots, N-2$) is the minimum of

$$\max(g_0(a_0, a_1), g_1(a_1, a_2), \ldots, g_{N-1}(a_{N-1}, a_N))$$

We omit the proof since it is not difficult and very similar to the one of Lemma 4.1. A direct consequence of this lemma is that if a_k is such that there exists a solution of equalities (48) confined to $i = k-1, k, \ldots, N-2$ (i.e. we only need to consider the set $\mathcal{A}_{N-k}(a_k)$), then $\Psi_{k+1} = \Psi_{k+2} = \ldots = \Psi_{N-1}$, and the sequence of optimal observation times up to the end of the game will be the sequence computed at step k whatever the evader does. As a matter of fact, if equality (48) is satisfied for $(\bar{u}_{k+1}^*, \ldots, \bar{u}_{N-1}^*, \bar{u}_N^* = T)$ in $\mathcal{U}_{N-k}(a_k)$, it will also be satisfied for $(\bar{u}_{k+2}^*, \ldots, \bar{u}_{N-1}^*, \bar{u}_N^* = T)$ which belongs to $\mathcal{U}_{N-k-1}(\bar{u}_k^*)$.

Note that for $k = N$, we obtain exactly (22) (with the simplifications introduced by hypothesis (39) or (40)). We obtain the same sequence of observation time as in the case where the pursuer chooses these times in open loop.

The direct consequence of these remarks is that if the initial time a_0 is such that (48) has a solution in $\mathcal{A}_N(a_0)$, then for this problem, i.e. with hypothesis (39) and/or (40), the optimal observation instants are the same for the open loop strategy and the closed loop strategy case. Note furthermore that these observation instants do not depend on the initial state z_0, but only on the duration of the game and the number N of observations imposed.

Lower bound for the infinite horizon problem

We now give a lemma that will provide us with a lower bound for the value function of the second version of the game (total duration of the game not fixed).

Lemma 5.2 *Consider the minimization problem*

$$\min_{\{a_i\}} \max_{i=0,\ldots,N-1} \{g_i(a_i, a_{i+1})\}$$

under the condition

$$a_i + \delta \leq a_{i+1}, \quad i = 0, \ldots, N-1, \quad and \quad a_N = T \tag{49}$$

where the g_i satisfy the same hypotheses as previously. The minimum of this problem is obtained for a sequence $\{a_0^*, \ldots, a_{N-1}^*\}$ such that the following equalities hold

$$g_i(a_i^*, a_{i+1}^*) = \psi, \quad i = 1, \ldots, N-1, \tag{50}$$

for some number ψ, and furthermore a_o^* is the smallest scalar such that the equalities (50) hold under the constraint (49).

Note that this lemma provides us with a condition to compute the lower bound of the value function. As a matter of fact, in (46) we minimize the maximum of two terms that depend on a_0 with respect to this same quantity a_0. The minimum with respect to a_0 of only the second term gives us a lower value for $J_N(z_0)$. Notice that this lower value is independent of the initial state z_0. It can be interpreted as a critical value. The pursuer can never be sure to come closer to the evader whatever the control of the evader. These kinds of results were already found in [4] and in [5]. Another consequence of the first term in the maximization being dependent on a_0, is that the seqence of observation times will depend on the initial state z_0.

Proof If a_0 is such that all equalities of (50) are possible then a sequence that satisfies these equalities attains the minimum (Lemma 5.1). Let $\{a_0^*, \ldots, a_{N-1}^*,$

$a_N\}$ be such a sequence. It satisfies (48) and (49) and assume furthermore its first element is the smallest one such that these equalities are satisfied. We will prove that it is not possible to perturb a_0^* and to obtain a new sequence that decreases the minimum.

Define a new sequence by $a_0^{*'} = a_0^* - h$ with $h > 0$, $a_i^{*'} = a_i^*$, for $i = 1, \ldots, N$. Since g_0 is decreasing in its first variable, $g_0(a_0^{*'}, a_1^{*'})$ is greater than $g_0(a_0^*, a_1^*)$ and subsequently $\max_i\{g_i(a_i^{*'}, a_{i+1}^{*'})\}$ is equal to $g_0(a_0^{*'}, a_1^{*'})$ and is greater than $\phi \overset{\text{def}}{=} \max_i\{g_i(a_i^*, a_{i+1}^*)\}$. We see that we must perturb a_1^* if we want the first term not to be greater than ϕ. But then, by the same reasoning, the second term will be greater than ϕ. Recursively we prove that we cannot obtain a new sequence $a_i^{*'}$, such that $a_0^{*'} = a_0^*$, $a_N^{*'} = T$ and $\max_i\{g_i(\bar{a}_i, \bar{a}_{i+1})\} \leq \phi$.

Let us now increase a_0^* and look at the sequence $a_i^{*'}$ such that $a_0^{*'} = a_0^* + h$, $h > 0$, and $a_i^{*'} = a_i^*$ for $i = 1, \ldots, N$. Two possibilities can be distinguished. The first one is $a_0^{*'} + h + \delta \leq a_1^*$, and $g_0(a_o^{*'}, a_1^*) \leq g_0(a_0^*, a_1^*)$, which does not modify the maximum $\max_i\{g_i(a_i^*, a_{i+1}^*)\}$. Any other perturbation will increase the maximum since two consecutive terms will be affected. The other possibility deals with $a_0^{*'} + h + \delta > a_1^*$. Other perturbations in the sequence must be introduced in order to satisfy the condition (49) and then two consecutive terms will be affected which increases the maximum. $\blacksquare$

6 Number and sequence of observation instants not fixed

The analysis of the previous section gave us a solution for the game where the players strive to maximinimize the criterion for a given number of observation instants. As a byproduct of this analysis we can easily obtain the solution of the game where the players are not given a fixed number of observations but are given a maximal duration for the pursuit. Denote this maximal duration by Δ. Δ can obviously correspond to a maximal autonomy of the pursuer. Again, let T be an arbitrary fixed final time. For any initial state $(\hat{x}, \hat{y})$ or z we have computed, for all N, the value $V_N(z, a)$, a referring to the initial state. The solution to the game just formulated is then given by the minimization

$$\min_{N} \min_{a \geq (T-\Delta)} V_N(z, a).$$

Note that we just add a minimization over a finite bounded discrete variable, which does not present any mathematical difficulty.

Acknowledgement

This research was performed during a sabbatical leave of the first author at INRIA, Sophia-Antipolis, France, which he gratefully acknowledges.

REFERENCES

[1] D.G. Luenberger, *Optimization by Vector Space Methods*. John Wiley, New York, 1969.

[2] A.A. Melikian, *On minimal observations in a game of encounter*. PMM, 37(3), 1972, 426–433.

[3] A.A Melikian, *On optimal selection of noise intervals in differential games of encounter*. PMM, 37(2), 1973, 195–203.

[4] P. Bernhard, O. Pourtallier, *Pursuit Evasion Game with Costly Information*. Dynamics and Control.

[5] P. Bernhard, J. M. Nicolas, O. Pourtallier, *Pursuit games with costly information, two approaches*. Fifth International Symposium on Dynamic Games and Applications, Grimentz, Switzerland July 1992.

[6] V. Laporte, J.M. Nicolas, P. Bernhard, *About the resolution of discrete pursuit games and its applications to naval warfare*. Differential Games-Developments in Modelling and Computation. Springer Verlag, 1991.

[7] N. S. Pontryagin, *Linear Differential Games, I and II*. Soviet Math. Doklady 8, 1967.

[8] G.V. Tomski, *Jeux dynamiques qualitatifs*, Cahier du CEREMADE n°7934, Université Paris 9 Dauphine, 1979.

[9] P. Bernhard, G. Tomski, *Une construction rétrograde dans les jeux différentiels qualitatifs, et application à la régulation*, RAIRO, J 16:1, 1982, 71–84.

Pursuit Games with Costly Information: Application to the ASW Helicopter Versus Submarine Game *

D. Neveu, J.P. Pignon, A. Raimondo
Thomson Sintra Activités Sous Marines,
1, avenue Aristide Briand - 94117 Arcueil-Cedex, FRANCE

J.M. Nicolas
Thomson-CSF, Laboratoire Central de Recherches,
Domaine de Corbeville - 94404 Orsay-Cedex, FRANCE

O. Pourtallier
Institut National de Recherche en Informatique et Automatique,
2004, route des lucioles - BP 93 - 06902 Sophia-Antipolis, FRANCE

Abstract

This paper deals with an application of game theory to a problem of an ASW (Anti Submarine Warfare) helicopter versus a submarine pursuit game: the helicopter (the pursuer) tries to reach, with respect to duration constraints (autonomy), a relative location to the submarine close enough to deliver a weapon; the submarine (the evader) maneuvers in order to escape to a secure position. Under realistic assumptions, the game therefore consists of a two-player pursuit game, in which each player gets information about the other one only at discrete dates (corresponding to helicopter dipping stations), these dates being chosen by the pursuer. Furthermore, information access is costly for the helicopter due to the intrinsic duration of a dipping station. Then the pursuit game is characterized by the fact that the pursuer suffers some penalty when he wants to obtain information about the position of the evader, and the evader remains blind except when the pursuer takes information.

Theoretical aspects of this problem have been previously treated. Assuming order one dynamics for both players, and a limited capture zone, the Pontrjagin approach yields an explicit formula for the sets of initial states which the pursuer can capture in at most (i.e. in the worst case) a given number of stages. An explicit formula can also

---

*This research was sponsored by Direction des Recherches, Etudes et Techniques, FRANCE.

be deduced when the pursuer aims at minimizing the duration of the games.

Performance evaluations of different dynamic games strategies are presented, achieved within realistic environment (via a high level of realism in computer simulation), which illustrate the interest of games techniques for various applications in the tactical domain, and suggest some deeper exploration.

1   Introduction

The problem addressed here is the definition of optimal strategies in the helicopter versus submarine warfare context using differential game theory. We essentially focus on the optimal strategy of the helicopter during a submarine prosecution fight phase.

Operational missions given to submarine-hunters helicopters can be roughly divided into two classes: on the one hand, the ASW helicopter is launched, either from a land base or a ship, to detect the possible presence of a submarine in a given (fixed or mobile) area without any previous detection or information on the target; on the other hand, the helicopter is launched towards a contact (i.e. a previously detected, and eventually identified, target), with the objective of approaching and then attacking it (firing a torpedo for instance). The first class of mission refers to search game theory; the second one (called prosecution) mixes phases associated respectively to search games and to pursuit games. We only deal here with this second class of missions.

As a first step, we briefly present the operational problem, the actors and their equipment. As a second step, we describe the associated model (in terms of game theory) and some theoretical results previously obtained in the domain. A third section is dedicated to the presentation of performance evaluation of these strategies in Monte-Carlo simulation: after a presentation of the proposed evaluation methodology and procedure, we describe the used simulation software testbed; experimental results are given for different tactical and environmental situations. We finally give some conclusions on the interest of the differential game theory approach for treating such problems.

2   Submarine prosecution from an ASW helicopter

In helicopter anti-submarine warfare, a prosecution phase begins when the ASW helicopter is launched on the basis of a predetected and, at least roughly, localized and identified target (submarine).

In the presented case, the helicopter is only equipped with a dipping active sonar (i.e. no onboard acoustic sonobuoys processing). The mission of the helicopter consists of making and keeping contact with the submarine, and simultaneously getting closer and closer to it with respect to duration constraints (flight autonomy), in order to send a weapon with an acceptable probability of hitting the adversary ship. Thus, the procedure is the following: to get information concerning the submarine position and behavior, the helicopter has to choose, and then to reach, a dip position; it then stops and uses its dipping active sonar to detect and localize (in relative polar position and eventually in radial velocity) the submarine; according to the estimated target position and kinematics parameters, it then chooses, if possible in an optimal way (referring to operational criteria), the next dip position, the next sonar dip depth and ping parameters. No information is available between the two dips.

During the same time, the submarine tries to optimize its behavior regarding both operational constraints (assigned ongoing mission and goals) and its surrounding situation assessment. Thus, the submarine looks at identifying the presence of the helicopter (using its acoustic and eventually optronic onboard equipment), estimating if it is detected or not, and, according to these decisions, breaking contact and escaping to a secure position or not, while pursuing its assigned mission (transit, reaching a precise location, etc.). The submarine may detect the helicopter from its passive sonar detection system, either on its radiated noise when the helicopter is close enough to the detection submarine arrays, or on the active emissions of the dipping sonar (via its passive sonar interceptor). In the present case, we suppose that only this latter capability is used by the helicopter; so the submarine only gets information on its pursuer when the helicopter uses its active sonar. After each intercepted ping, the submarine makes assumptions on the probability of its being detected, and deduces whether it should change its depth, remain quiet or maintain optimized course parameters.

3 Theoretical aspects

3.1 Introduction

The situation is modeled by a pursuit evasion game played in the following way: at the beginning of the game both the evader and the pursuer know the state. The pursuer chooses the next place and time it will stop in order to get information about the state. It then suffers some penalty when it wants to obtain information about the evader. During each stage, the two players do not have any information and then play using open loop controls.

This game has been studied, using "erosion-dilatation" techniques. The main idea, initially introduced by Pontrjagin [8], is to find successive sets C_n

where C_0 is the given target and C_n is the set of initial states such that the capture is possible within n stages. The equivalence between these techniques and the dynamic programming equation for a classical discrete time pursuit-evasion game has been shown in [6]. In [2] and [3], these techniques have been applied to the specific game under studies, that is the pursuit-evasion game with costly information. This game has also been studied with more classical dynamic programming techniques [7], following ideas of Melikian [1] who studied a related game. In the first part, we recall the mathematical modeling of the game and mention the results obtained for the computation of sets C_n. The second part is devoted to the implementation in a simulator of the previous results, together with an analysis of their contribution to the operational situation. In a last section we make brief comparison remarks.

3.2   Game rules

Let us note y_P and y_E respectively the pursuer's (P) and evader's (E) positions at each time, and note $x(t)$ the evader's position in the pursuer's coordinate system, that is $x(t) = y_E(t) - y_P(t)$. We suppose that the state variable $x(t)$ follows the dynamics equations:

$$\dot{x}(t) = g(v(t)) - f(u(t)), \text{ with } u(t) \in U, \quad v(t) \in V,$$

where U and V are the set of pursuer's and evader's controls. The implementation of this theory has been made with simple motions for both players, that is

$$f(u(t)) = u(t), \quad \| u(t) \| \le a$$
$$g(v(t)) = v(t), \quad \| v(t) \| \le b.$$

Note that the first reason to choose such simple dynamics lies in the fact that results are really easy to implement; thus it is possible to have a first measure of the impact of game theory approach to this problem. Then a first order dynamic is optimistic for the evader but not for the pursuer (at least if we consider that each stage is long enough, which will be the case). The fact of being optimistic for the evader and not for the pursuer goes in the same direction as the wish to find a "safe" strategy for the pursuer. The pursuer overestimates the evader's strategy, so the pursuer's strategy will be adapted to a less serious threat.

The game is played with stages. During each stage both the pursuer and the evader play with open loop controls, and obtain information on the state x only at the end of each stage. We consider that at the end of each stage the pursuer loses some time, let us say δ, in order to obtain some information about the state. We suppose that this duration is constant, although we could consider that the information duration is dependent on the previous information (this does not alter the theoretic analysis, but leads to cumbersome notations). At the beginning of each stage (let us say at time

t_i), the pursuer's control consists of the choice of the duration τ of the stage and the choice of a speed function $u(t), t \in [t_i, t_i + \tau - \delta]$, while the evader's control only consists of a speed function $v(t)$.

Let x_0 be the initial state of the system, and τ the length of the first stage. At the end of this stage, we have:

$$x(\tau) = x_0 + \int_0^\tau v(s)ds - \int_0^{\tau-\delta} u(s)ds.$$

Or again, if we note $Q_\tau = B(0, b\tau)$ and $P_\tau = B(0, a(\tau - \delta))$, the sets of E's and P's respective possible movements during a period of duration τ, we have

$$x(\tau) = x_0 + q_\tau - p_\tau, \qquad q_\tau \in Q_\tau, \quad p_\tau \in P_\tau.$$

3.3   The sets C_n

We suppose that the capture occurs whenever the distance between helicopter and submarine is smaller than a given radius R_0. The initial target is then $C_0 = B(0, R_0)$.

Let the set C_n be given and compute the set C_{n+1}. C_{n+1} is the set of initial states such that there exists a pursuer's control (duration of the stage and motion) such that, whatever the evader does, the state will be in set C_n at the end of the stage. $\bigcup_n C_n$ will be the set of initial states that lead to capture in at most n stages.

We obtain, see [3] for more details,[1]

$$C_{n+1} = \bigcup_{\tau \in I_n} C_{n+1}(\tau), \quad I_n = \{\tau \text{ such that } \tau > \delta \text{ and } C_n \overset{*}{-} Q_\tau \neq \emptyset\}.$$

where

$$C_{n+1}(\tau) = (C_n \overset{*}{-} Q_\tau) + P_\tau.$$

Using the simple motion dynamics we obtain

$$C_n = \bigcup_{t \in J_n} C_n(t),$$

with

$$C_n(t) = \begin{cases} B(0, (a - b)t - na\delta + R_0) & \text{if } t \in J_n \\ \emptyset & \text{otherwise,} \end{cases}$$

---

[1] $A \overset{*}{-} B = C \Rightarrow B + C \subset A$ and $B + C \subset A \Rightarrow C \subset A \overset{*}{-} B$ where $B + C = \{b + c \mid b \in B, \ c \in C\}$, or similarly, $A \overset{*}{-} B = \{a \in A \text{ such that } \forall b \in B, \quad a + b \in A\}$.

and

$$
J_n = \begin{cases}
\emptyset \text{ if } R_{n-1}(t^s_{n-1}) \le b\delta \\
\quad t^m_n = \sup\left(t^m_{n-1} + \delta \,,\, \dfrac{na\delta}{a-b} \right) \;;\; t^m_0 = 0 \\
[t^m_n, t^s_n] \\
\quad t^s_n = \dfrac{R_{n-1}(t^s_{n-1})}{b} + t^s_{n-1} \;;\; t^s_0 = 0.
\end{cases}
$$

The control of the pursuer at each stage is the following: he goes at maximal speed towards the last position of the evader during the duration:

$$
\tau = \frac{\| x \| - R^s_{n-1} + a\delta}{a - b}
$$

and then gets another measurement of the state.

3.4   The dual problem

Up to now we have investigated the problem where the target is fixed and the pursuer has no limitation on the game duration. Another interesting problem from an operational point of view is the case where the pursuer has a limited autonomy. We suppose that the pursuer wants to know how close he can be (in the worst situation) at the end of a given period of duration T. We call this problem the "dual problem," and we can derive its solution from the initial problem. Indeed, supposing that the initial state x together with T are given, the problem is then to solve the minimization problem:

$$
\min_n R_0(x, T, n),
$$

with :

$$
R_0(x, T, n) = \min R_0,
$$

such that R_0 satisfies :

$$
\begin{cases}
\exists T' \le T \\
T' \in T_n(T') \\
x \in C_n(T', R_0).
\end{cases}
$$

4   Game theory strategies performance evaluation

4.1   Performance evaluation methodology

In order to estimate and quantify as much as possible the interest of this approach for an operational application, a performance evaluation methodology has been used based on intensive use of a realistic environmental simulation software testbed. An analysis of a real environment has been done, from which significant parameters have been extracted. Realistic influence of these parameters on mobile dynamics has been analyzed and modeled, and sets of realistic values of these parameters have been defined. Realistic behaviors of the two players in realistic scenarios have been analyzed, and operational behavior rules have been synthetised, in order to be able to build reference situations for comparisons. The main idea is to compare the performance of the expert rule controls and the game theory controls in the "real" environment. Let us emphasize the fact that the expert rule controls take into account other parameters that those used in the game theory approach, and are supposed to be very close to those that would be used in a real conflict.

4.2   Simulation software testbed

The software tactical simulator [5] used for evaluation is mainly dedicated:

- to assess performances of a dipping sonar system during a prosecution mission,

- to test specific tactical rules either for the helicopter or the submarine.

Built-in traditional simulation capabilities are proposed to the user either to make a one shot simulation (with pause/go or step by step options) or to estimate statistical performance on specified parameter sets via Monte-Carlo procedures. A single scenario begins when the ASW helicopter has detected the submarine, and then has initial information about its target (position, velocity, etc.); it ends either when the submarine is within a possible torpedo firing zone (capture zone), or when helicopter flight autonomy is reached.

This simulator is developed using an Object Oriented Language, and realistic mobile behaviors (i.e. kinematics and sensor management strategies) are implemented using an expert system rules based approach. These technologies allow us to simulate event-driven behaviors for the mobiles, and then with realistic and complex unpredictable scenarios. Fine modeling of the environment has been emphasised in order to guarantee validity of the results: these realistic models impact, for instance, the navigation of day/night visual environments and of meteorological parameters (such as wind), and underwater acoustic detection and on sonar settings management of sea velocity profile and acoustic detection probability laws.

This simulator has been enriched with behavioral rules given by dynamics game theory in order to evaluate the impact of application of the obtained theoretical results to the helicopter pursuit strategy: these rules implement the Pontrjagin construction, (and the mixed strategies when contact is lost by the helicopter); they run on the same scenarios as the kinematics "expert" rules in order to get comparisons on the same basis. As game theory does not give information on questions such as optimal sonar settings, the game rules use the same equipment management rules in both cases.

4.3 Experimental results

Different scenarios have been tested in the Monte-Carlo mode for different seasons and geographic zones (and then different bathycelerimetry profiles), for different initial pursuer-evader initial distances, and for different initial speeds and headings of the evader. The output estimated parameters are the helicopter mission success ratio and the number of dipping sonar stations made by the helicopter during the prosecution phase to reach mission success.

The following figures illustrate results for different significant scenarios.

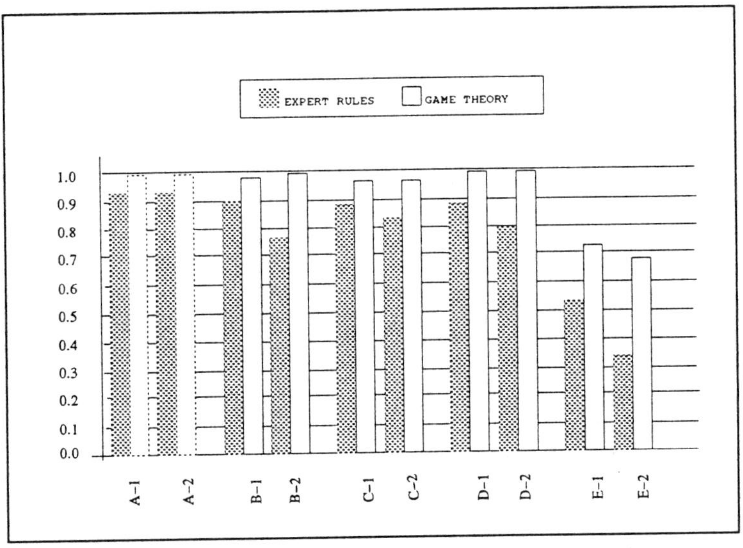

Figure 1: Helicopter mission success ratio vs. different scenarios

Notations:

— A,B,C,D and E correspond to 5 different sets of scenarios, seasons and bathycelerimetry profiles,

— 1 corresponds to the case where the initial speeds of the two players are in the same direction, 2 to the case where they have opposite directions.

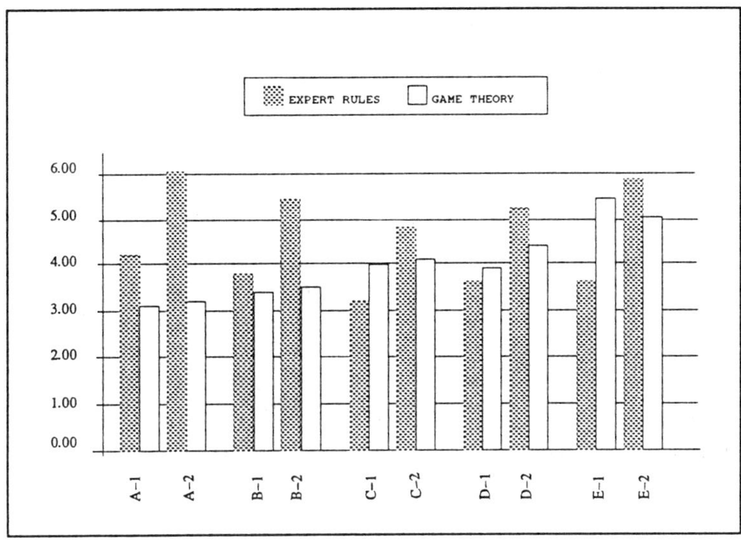

Figure 2: mean number of helicopter dipping stations for different scenarios

4.4 Comments

Study of the obtained results, part of which has been reported here, concludes with a global identity of results obtained with the "expert" rules and with the game theory ones, with a trend to superior interest for dynamics games rules.

This basic result has to be analyzed by taking different factors into account:

- reference expert rules used in this evaluation (a) are not optimized for the simulated helicopter equipment (new generation active dipping sonar), and (b) are built on an operational doctrine which is less "offensive" than the one underlying dynamic game theory; this deals with pessimistic results obtained using the expert rules,

- parameter value settings in the prospective scenarios and fine tuning of free parameters in dynamic game equations deal with the intrinsic advantage of game theory results,

- as said previously, optimized expert equipment management rules are used in both cases, which add also some advantage to the games results.

Nevertheless, the most important thing to notice is that the use of dynamic game theory results allows us to reach very interesting success probability ratios. That means that hypotheses which have been introduced to solve the problem with the dynamic game theory only have a reduced impact on the quality of the results, and then that this theory can be of great help for defining tactical behaviors and rules, even if modeling of all the parameters is mathematically difficult.

5   Conclusion

The reported work deals with the application of dynamic game theory for solving real operational problems. The chosen example is the prosecution mission of a submarine by an ASW helicopter. A mathematical model of this problem has been built in order to take into account its specificity (incomplete information, costly information for the pursuer, etc.). The theoretical strategies obtained from the game theory approach have been added to a realistic tactical software testbed dealing with this problem. The software has been intensively used for evaluation of the proposed strategies, and comparisons with expert rules (via statistical evaluation of significant parameters) have been performed. Evaluation results (after correction of artifacts external to this problem, due to the difference of approach underlying the two models) lead to quasi-identical performances in both cases, and then illustrate the interest of dynamic games for solving this type of problem. This work allowed us to identify some interesting fields of research in the tactical domain, for the future, relative to the increasing complexity of mathematical models (higher order dynamics, fine modelization of new parameters, etc.) in order to make this approach more efficient, and to design software tools of real operational interest.

REFERENCES

[1]   A.A. Melikian, "On minimal observations in a game of encounter", *PMM, vol. 37(3), 1973, 426-433.*

[2]   P. Bernhard, J.M. Nicolas, O. Pourtallier, *Pursuit evasion game with costly information, two approaches,* Proceeding of the 5th symposium on differential games and applications, Grimentz, July 1992.

[3]   P. Bernhard, O. Pourtallier, *Pursuit evasion game with costly information,* Dynamics and Control, Vol 4, N 4, October 1994, 365–382. .

[4]   P. Bernhard, G. Tomski, *Une construction rétrograde dans les jeux différentiels qualitatifs et application à la régulation,* RAIRO, (J 16), 1982, 71–84.

[5] T. Gach, Y. Lagoude, D. Neveu and M.Revol, *A tool for dipping sonar system evaluation*, Undersea Defence Technology 1991 Proceedings (UDTO91), Paris, 1991.

[6] V. Laporte, J.M. Nicolas, P. Bernhard, *About the resolution of discrete pursuit games and its applications to naval warfare*, Proceeding of the 4th symposium on differential games and applications, Helsinki, July 1990.

[7] G.J Olsder, O. Pourtallier, *Optimal selection of observation times in a costly information game*, Annals of the International Society of Dynamic Games, N2, 1995.

[8] N.L. Pontrjagin, *Linear differential games, I and II*, Soviet Math. Doklady, 8, 1968.

Linear Avoidance in the Case of Interaction of Controlled Objects Groups

Arkadij A. Chikrii and Pavel V. Prokopovich
Glushkov Institute of Cybernetics
252187 Kiev, Ukraine

Abstract

In this paper we study a conflict interaction of n pursuers and m evaders in Euclidean space R^k. All objects are linear and of the same type. The qualitative conclusion about avoidance problem solvability is developed depending on the ratio of numbers n, m and k. The avoidance problem from given initial states is considered as auxiliary. The results have a bearing on the research reported in [1-7].

1 Basic assumptions and preliminary results

Let R^k be a k – dimensional real euclidian space, let (x, y) denote an inner product of two vectors x and y from R^k and let $\|x\| = \sqrt{(x, x)}$. We denote by $\mathrm{int}X$, ∂X, $\mathrm{co}X$, $\mathrm{con}X$ interior, boundary, convex and conical hulls of an arbitrary set $X \subseteq R^k$ respectively. Let $\Omega(R^k)$ $(\mathrm{co}(\Omega(R^k)))$ be a space of all non-empty compact sets (non-empty convex compact sets) in R^k with a Hausdorff distance. Let $S = \{\, x \in R^k : \|x\| \leq 1 \,\}$ be a closed unit ball in R^k and let $N = \{\, 1, 2, \ldots, q \,\}$. If the set X consists of a finite number of elements, then $|X|$ denotes this number.

Let $F \in \Omega(R^k)$. We denote by $c(F, \cdot) : R^k \to R$ its support function defined by $c(F, w) = \max_{f \in F}(f, w)$, $w \in R^k$. If $w_0 \in R^k$, $\|w_0\| \neq 0$, then

$$U(F, w_0) = \{\, f \in F : (f, w_0) = c(F, w_0) \,\}$$

is called a support set to F in the direction w_0. If $U(F, w_0)$ consists of a unique point, then it is said that $F \in \Omega(R^k)$ is strictly convex in the direction $w_0 \in R^k$. It is said that $F \in \Omega(R^k)$ is strictly convex if it is strictly convex in any nonzero direction $w_0 \in R^k$.

We also say that $F \in \Omega(R^k)$ is a compact with smooth boundary if

$$U(F, w) \cap U(F, \hat{w}) = \emptyset \qquad \forall w, \hat{w} \in \partial S, \ w \neq \hat{w}.$$

We recall that both compact with smooth boundary and strictly convex compact can be nonconvex sets. For example, the unit sphere ∂S in R^k is a strictly convex compact with smooth boundary.

Consider the control system

$$\dot{y} = Ay + v, \qquad v \in V, \ V \in \Omega\,(\mathrm{R}^k), \tag{1}$$

where y is a k–dimensional vector of the system phase state, v is a k–dimensional vector of control and A is a square matrix of order k. Admissible control on $I = [\,0,\, t_1\,]$ is any measurable function $v : I \to V$.

We denote by $X(\,t_1;\ G,\ V\,)$ the reachable set of the control system (1) at time $t_1 \geq 0$ from the set of initial conditions $G \in \Omega\,(\mathrm{R}^k)$

$$X(\,t_1;\ G,\ V\,) = \exp(t_1 A)G + \int_0^{t_1} \exp((t_1 - s)A)V ds.$$

Let y be a solution of (1) corresponding to the control v and the initial condition $y(0) \in G,\ G \in \Omega\,(\mathrm{R}^k)$. It is said that the control-trajectory pair $(\,v,\, y\,)$ satisfies the maximum principle on $[\,0,\, t_1\,]$ and the transversality condition on G if the solution ψ of the auxiliary adjoint system

$$\dot{\psi} = -A^* \psi \tag{2}$$

with the initial condition $\psi(0) \in \partial S$ exists such that

1. $(\,v(t),\, \psi(t)\,) = c\,(\,V,\, \psi(t))$ for almost all $t \in [\,0,\, t_1]$;

2. $(\,y(0), \psi(0)) = c\,(\,G, \psi(0))$.

Lemma 1.1 *Let $G \in \mathrm{co}\,\Omega\,(\mathrm{R}^k)$. A solution of (1) at time $t_1 > 0$ satisfies $y(t_1) \in \partial X(\,t_1;\ G,\ V\,)$ if and only if the pair $(\,v,\, y\,)$ satisfies the maximum principle on $[\,0,\, t_1\,]$ and the transversality condition on G.*

Lemma 1.2 *Let y_j be a solution of (1) corresponding to the control v_j and the initial condition $y_j(0) \in G,\ G \in \Omega\,(\mathrm{R}^k)$. Let $(\,v_j,\, y_j\,)$ satisfy the maximum principle on $I = [\,0,\, t_1\,]$, $t_1 > 0$ and the transversality condition on $G \in \Omega\,(\mathrm{R}^k)$; ψ_j is a corresponding solution of the adjoint system (2), $j = 1, 2$. If $y_1(0) \neq y_2(0)$ and at least for one $j \in N_2$ the support function $c\,(\,V,\, \cdot\,)$ is differentiable at $\psi_j(t)$ for almost all $t \in I$, then $y_1(t_1) \neq y_2(t_1)$.*

Lemma 1.3 *If $V \in \Omega\,(\mathrm{R}^k)$ has a smooth boundary and $G \in \mathrm{co}\,\Omega\,(\mathrm{R}^k)$, then for any $t_1 > 0, X(\,t_1;\ G,\ V\,)$ is a convex compact with smooth boundary.*

Consider the following differential game. The motions of objects in R^k $(k \geq 2)$ are described by equations:

$$P_i : \dot{x}_i = Ax_i + u_i, u_i \in U_i; \ E_j : \dot{y}_j = Ay_j + v_j, v_j \in V; \tag{3}$$

$$U_i, V \in \Omega\,(\mathrm{R}^k), \ U_i \subseteq \mathrm{co}\,V, \ i = 1, \ldots, n, \ j = 1, \ldots, m,$$

with initial states $x_i(0) = x_i^0$, $y_j(0) = y_j^0$,

$$x_i^0 \neq y_j^0, \qquad i = 1, \ldots, n, \ j = 1, \ldots, m. \tag{4}$$

Here x_i, y_j are phase coordinates of the pursuer P_i and evader E_j respectively; A is a given square matrix of order k. Controls of players are supposed to be measurable functions $u_i : [0, +\infty) \to U_i$, $v_j : [0, +\infty) \to V$, $i = 1, \ldots, n, \ j = 1, \ldots, m$.

We consider evader E_j to be caught at the moment $t \geq 0$ if $y_j(t) = x_i(t)$ for at least one $i \in N_n$. Escape from the given initial state $z^0 = (x(0), y(0)) = (x_1^0, \ldots, x_n^0, y_1^0, \ldots, y_m^0)$ is possible, if there exist controls of players E_j, $j = 1, \ldots, m$, such that for at least one $s \in \{1, \ldots, m\}$: $x_i(t) \neq y_s(t)$ for all $i \in N_n$, $t \in [0, +\infty)$ and all controls of players P_i, $i = 1, \ldots, n$. At any moment $t \geq 0$ pursuers choose their controls using information about game state

$$z(t) = (x_1(t), \ldots, x_n(t), \ y_1(t), \ldots, y_m(t))$$

and evaders may use any additional information.

The global avoidance problem in the game (3) is solvable, when the escape is possible from any initial state satisfying (4).

2 Escape in the class of programme strategies

Let G be a non-empty subset of R^k. For the initial game state z^0 we define the following sets

$$I(x(0), G) = \{i \in \{1, \ldots, n\} : x_i^0 \in G\},$$
$$J(y(0), G) = \{j \in \{1, \ldots, m\} : y_j^0 \in G\},$$
$$J_0(y(0), \partial G) = \{j \in \{1, \ldots, m\} : y_j^0 \in \partial G\}.$$

Here if $y_{j_1}^0 = y_{j_2}^0 = \ldots = y_{j_s}^0$ for some $j_l \in J_0(y(0), \partial G)$, $l = 1, \ldots, s$, $s > 1$, $j_1 < j_2 < \ldots < j_s$, then we assume that $j_l \notin J_0(y(0), \partial G)$, $l = 2, \ldots, s$.

Let $G \in \mathrm{co}\,\Omega\,(\mathrm{R}^k)$. For all $y \in \mathrm{R}^k \setminus \mathrm{int}\,G$ we define

$$P(G, y) = \partial S \cap [\mathrm{con}\,(y - G)]^*.$$

Here $[\mathrm{con}\,(y - G)]^*$ is a positive polar cone to $\mathrm{con}\,(y - G)$. We denote by $\psi_j(\cdot, r_j)$ the solution of (2) corresponding to the initial condition $\psi(0) = r_j$, $r_j \in P(G, y_j^0)$, $j \in J_0(y(0), \partial G)$.

Theorem 2.1 *If there exists $G \in \mathrm{co}\,\Omega\,(\mathrm{R}^k)$ such that*

$$|J_0\,(\,y(0),\,\partial G\,)| > |I\,(\,x(0),\,\mathrm{R}^k\backslash G\,)|$$

and for any $j \in J_0\,(\,y(0),\,\partial G\,)$ there exists $r_j \in P(\,G,\,y_j^0\,)$ such that the support function $c\,(\,V,\,\cdot\,)$ is differentiable at $\psi_j(\,t\,,\,r_j\,)$ for almost all $t \in [0,\,+\infty)$, then in the game (3) the escape is possible from the initial state z^0.

Proof. For any $j \in J_0\,(\,y(0),\,\partial G\,)$ we choose $r_j \in P(\,G,\,y_j^0\,)$ such that the support function $c\,(\,V,\,\cdot\,)$ is differentiable at $\psi_j(\,t\,,\,r_j\,)$ for almost all $t \in [0,\,+\infty)$. We define the control v_j on $[0,\,+\infty)$, $j \in J_0\,(\,y(0),\,\partial G\,)$, as a measurable function satisfying

$$(\,v_j(t),\,\psi_j(\,t\,,\,r_j\,)\,) = c\,(\,V,\,\psi_j(\,t\,,\,r_j\,)\,). \tag{5}$$

Since $c\,(\,V,\,\cdot\,)$ is differentiable at $\psi(\,t\,,\,r_j\,)$ for almost all $t \in [0,\,+\infty)$, then control v_j on $[0,\,+\infty)$ is uniquely defined. The uniqueness is meant in the following sense: two measurable functions are equal if their values coincide almost everywhere. The controls of evaders E_j, $j \in N_m\backslash J_0\,(\,y(0),\,\partial G\,)$, on $[0,\,+\infty)$ may be chosen arbitrarily.

By Lemmas 1.1, 1.2, we deduce that pursuer P_i, $i \in I\,(\,x(0),\,G\,)$, can not catch any evader E_j if $j \in J_0\,(\,y(0),\,\partial G\,)$, and pursuer P_i, $i \in I\,(\,x(0),\,\mathrm{R}^k\backslash G\,)$, can catch no more than one evader E_j, $j \in J_0\,(\,y(0),\,\partial G\,)$, on the semi-infinite time interval. Since $|J_0\,(\,y(0),\,\partial G\,)| > |I\,(\,x(0),\,\mathrm{R}^k\backslash G\,)|$, the proof is complete. $\square$

We note that function $c\,(\,V,\,\cdot\,)$ is differentiable at $\psi_0 \in \mathrm{R}^k$, $\|\psi_0\| \neq 0$, if and only if the compact V is strictly convex in the direction ψ_0.

Corollary 2.1 *In the game (3), let V be a strictly convex compact and there exists $j \in N_m$ such that $y_j^0 \notin \mathrm{intco}\{\,x_1^0,\,\ldots,\,x_n^0\,\}$. Then the escape is possible from the initial state z^0.*

This corollary extends the well-known result in the simple motion case, which can be found in [6].

Corollary 2.2 *In the game (3), let V be a strictly convex compact, $n = k+1$ and $m = 2$. Then the global avoidance problem is solvable.*

Proof. We may assume, without loss of generality, $y_1^0 \neq y_2^0$. We denote by H the hyperplane passing through points x_i^0, $i = 1,\,\ldots,\,k-2$, y_1^0, y_2^0. If such a hyperplane is not unique, then we choose any of them. It is clear that there exists an open half-space defined by H and containing no more than one point of x_{k-1}^0, x_k^0, x_{k+1}^0. Therefore, there exists a convex compact G such that y_1^0, $y_2^0 \in \partial G$ and $|I\,(\,x(0),\,\mathrm{R}^k\backslash G\,)| \leq 1$. $\square$

From this proposition, in particular, it follows that in the planar game of three pursuers and two evaders the global avoidance problem is solvable.

Corollary 2.3 *In the game (3), let V be a strictly convex compact, $n = 2k - 1$ and $m = k$. Then the global avoidance problem is solvable.*

Corollary 2.4 *In the game (3), let V be a strictly convex compact, $n = 2k$ and $m = k$, the initial state $z^0 = (x_1^0, \ldots, x_n^0, y_1^0, \ldots, y_m^0)$ such that $y_s^0 \neq y_q^0$ for all $s, q \in N_k$, $s \neq q$, and initial positions of some $k + 1$ players lie in the same hyperplane. Then the escape is possible from the initial state z^0.*

3 Avoidance problem in the general case

Now we shall formulate more delicate sufficient conditions for solvability of the avoidance problem from a given initial state.

Theorem 3.1 *In the game (3), let V be a strictly convex compact with smooth boundary. If there exist $G_1, G_2 \in \mathrm{co}\,\Omega\,(\mathrm{R}^k)$ such that $x_i^0 \in G_1 \cup G_2$ for any $i \in N_n$ and*

$$|I\,(\,x(0),\, G_1 \backslash G_2\,)| < |J\,(\,y(0),\, \mathrm{R}^k \backslash (\, G_1 \cup G_2\,)\,)| + |J_0\,(\,y(0),\, \partial G_2\,)|.$$

Then the escape is possible from the initial state z^0.

Proof. Since $X(\,\delta;\; G_2,\, V\,)$ is a convex compact with smooth boundary for arbitrarily small $\delta > 0$, then we can assume that G_2 has smooth boundary. In this case, $P(\,G_2,\, y_j^0\,)$ for any $j \in J_0\,(y(0),\, \partial G_2)$ consists of the unique point r_j. We define the control v_j on $[\,0,\, +\infty)$ for any $j \in J_0\,(\,y(0),\, \partial G_2\,)$ from (5), where $\psi_j(\,t\,,\, r_j\,)$ is the solution of (2) corresponding to the initial condition $\psi_j(0) = r_j$. Controls of evaders

$$E_j,\; j \in N_m \backslash [J\,(\,y(0),\, \mathrm{R}^k \backslash (\, G_1 \cup G_2\,)\,) \cup J_0\,(\,y(0),\, \partial G_2\,)],$$

may be chosen arbitrarily.

We shall prove the Theorem by induction on number of evaders, whose initial states belong to $\mathrm{R}^k \backslash (\, G_1 \cup G_2\,)$. Let $l = |J\,(\,y(0),\, \mathrm{R}^k \backslash (\, G_1 \cup G_2\,)\,)|$.

Consider the case $l = 1$. For all $i \in I\,(\,x(0),\, \partial G_2\,)$ we define the trajectory $\bar{x}_i$ on $[\,0,\, +\infty)$ starting at x_i^0 and corresponding to the control $\bar{u}_i$ chosen from

$$(\,\bar{u}_i(t),\, \psi_i(\,t\,,\, \bar{r}_i\,)\,) = c\,(\,V,\, \psi_i(\,t\,,\, \bar{r}_i\,)\,),$$

where $\psi_i(\,\cdot\,,\, \bar{r}_i\,)$ is the solution of (2) for $\psi_i(0) = \bar{r}_i$, $\bar{r}_i \in P(\,G_2,\, x_i^0\,)$. Since V is a strictly convex compact and G_2 is a compact with smooth boundary, then trajectories y_j, $j \in J_0\,(\,y(0),\, \partial G_2\,)$, $\bar{x}_i$, $i \in I\,(\,x(0),\, \partial G_2\,)$ on $[\,0,\, +\infty)$ are uniquely defined.

We define the control v_j on $[0, t(r_j))$, $j \in J(y(0), \mathrm{R}^k \backslash (G_1 \cup G_2))$, from (5), where $\psi_j(\cdot, r_j)$ is the solution of (2) starting at $r_j \in P(G_1, y_j^0)$ such that

$$y_j(t(r_j), r_j) \neq x_i(t(r_j)) \qquad \forall\, i \in I(x(0), \partial G_2) \tag{6}$$

$$y_j(t(r_j), r_j) \neq y_s(t(r_j)) \qquad \forall\, s \in J_0(y(0), \partial G_2) \tag{7}$$

Here $y_j(\cdot, r_j)$ is the corresponding trajectory of E_j, $j \in J(y(0), \mathrm{R}^k \backslash (G_1 \cup G_2))$; $t(r_j)$ is the first moment when $y_j(t, r_j) \in X(t; G_2, V)$. If $y_j(t, r_j) \notin X(t; G_2, V)$ for all $t \geq 0$, then any player E_j, $j \in J(y(0), \mathrm{R}^k \backslash (G_1 \cup G_2))$ cannot be caught by Lemmas 1.1, 1.2. Therefore, we can assume that $t(r_j) < +\infty$ for any $j \in J(y(0), \mathrm{R}^k \backslash (G_1 \cup G_2))$.

We shall prove that for every $j \in J(y(0), \mathrm{R}^k \backslash (G_1 \cup G_2))$ there exists $r_j \in P(G_1, y_j^0)$ such that inequalities (6), (7) hold. Indeed, since V is a compact with smooth boundary, then for all $t > 0$, $r_j^1, r_j^2 \in P(G_1, y_j^0)$, $r_j^1 \neq r_j^2$,

$$y_j(t, r_j^1) \neq y_j(t, r_j^2). \tag{8}$$

We note that if for some $r_j^1 \in P(G_1, y_j^0)$ there exists $s \in I(x(0), \partial G_2)$ such that

$$y_j(t(r_j^1), r_j^1) = \bar{x}_s(t(r_j^1)), \tag{9}$$

then for any $r_j^2 \in P(G_1, y_j^0)$, $r_j^1 \neq r_j^2$, (we assume that $t(r_j^2) < +\infty$)

$$y_j(t(r_j^2), r_j^2) \neq \bar{x}_s(t(r_j^2)). \tag{10}$$

To show this, we suppose the opposite: there exists $r_j^2 \in P(G_1, y_j^0)$, $r_j^1 \neq r_j^2$, such that

$$y_j(t(r_j^2), r_j^2) = \bar{x}_s(t(r_j^2)). \tag{11}$$

From (8), (9), (11) it follows that $t(r_j^1) \neq t(r_j^2)$. Without loss of generality, $t(r_j^1) < t(r_j^2)$. Inasmuch as

$$\bar{x}_s(t(r_j^1)) \in \partial X(t(r_j^1); y_j^0, V),$$
$$\bar{x}_s(t(r_j^2)) \in \partial X(t(r_j^2); y_j^0, V),$$

and (9) holds, then, by Lemmas 1.1-1.3, $\bar{x}_s(t(r_j^2)) = y_j(t(r_j^2), r_j^1)$. It is contrary to (8). The inequality (10) has been proved.

Thus, evader E_j, $j \in J_0(y(0), \mathrm{R}^k \backslash (G_1 \cup G_2))$ using the information about initial positions of all players P_i, $i \in I(x(0), \partial G_2)$, E_q, $q \in J_0(y(0), \partial G_2)$, chooses its control from (5), where $\psi_j(\cdot, r_j)$ is the solution of (2) starting at $r_j \in P(G_1, y_j^0)$ such that (6), (7) are fair.

It is clear that for all controls of players P_i, $i = 1, \ldots, n$, on $[0, t(r_j)]$ the state $z(t(r_j))$ meets conditions of Theorem 2.1. We take $X(t(r_j); G_2, V)$ as G mentioned in Theorem 2.1.

Suppose that conditions of Theorem 2.1 hold and for $l \le r$ in the game (3) the escape is possible from initial state z^0. Let us show that the theorem is fair for $l = r + 1$. Fix $F \in \mathrm{co}\,\Omega\,(\mathrm{R}^k)$ such that $0 \in \mathrm{int}\,F$. We can assume that

$$J\,(y(0),\,\mathrm{R}^k\backslash(\,G_1 \cup G_2\,)) = N_{r+1},$$
$$y_s^0 \in \partial(\,G_1 + \epsilon_s F\,), \qquad \epsilon_s > 0, \quad s = 1, \ldots, r+1,$$

and $\epsilon_1 > \epsilon_2 > \ldots > \epsilon_{r+1} > \epsilon_{r+2} = 0$. If there exists $q \in \{\,2, \ldots, r+1\,\}$ such that $\epsilon_{q-1} = \epsilon_q$, $\epsilon_q > \epsilon_{q+1}$, then we can choose controls of evaders E_j, $j = 1, \ldots, r+1$, on the semi-interval $[\,0, \delta\,)$ ($\delta > 0$ is arbitrary small) when the following inclusions are satisfied

$$y_s(\delta) \in \partial(\,X(\,\delta;\,G_1,\,V\,) + \epsilon_s \exp(\delta A)F\,) \qquad \forall s \in N_{r+1}\backslash\{q\},$$
$$y_q(\delta) \in \partial(X(\,\delta;\,G_1,\,V\,) + \epsilon_q' \exp(\delta A)F\,), \qquad \epsilon_{q-1} > \epsilon_q' > \epsilon_{q+1}.$$

We define the control v_j for any $j \in N_{r+1}$ from (5), where $\psi_j(\,\cdot\,,\,r_j\,)$ is the solution of (2) starting to $r_j \in P(\,G_1 + \epsilon_{j+1}\,F,\,y_j^0\,)$ satisfying (6), (7). Suppose that for any $j \in N_{r+1}$ there exists the first moment $t = t\,(\,r_j\,)$ such that $y_j(\,t,\,r_j\,) \in X(\,t;\,G_2,\,V\,)$.

Let $t^* = \min_{j \in N_{r+1}} t\,(r_j)$. Taking into account the described method of evader's controls construction, Lemmas 1.1, 1.2 and the induction hypothesis, we conclude that in the game (3) the escape is possible from the state $z\,(t^*)$.

$\square$

Corollary 3.1 *In the game (3) let V be a strictly convex compact with smooth boundary, $n = 2k$, $m = k$; the initial state z^0 is such that $y_s^0 \ne y_q^0$ for all s, $q \in N_k$, $s \ne q$, and there exist different in pairs $i_1, \ldots, i_{k+1} \in N_{2k}$ such that*

$$y_j^0 \notin \mathrm{intco}\,\{\,x_{i_1}^0, \ldots, x_{i_{k+1}}^0\,\}, \qquad j = 1, \ldots, k.$$

then the escape is possible from the initial state z^0.

Proof. Suppose that $y_j^0 \notin \mathrm{intco}\,\{\,x_1^0, \ldots, x_{k+1}^0\,\}$, $j = 1, \ldots, k$. Consider sets $G_1 = \mathrm{co}\,\{\,x_{k+2}^0, \ldots, x_{2k}^0\,\}$, $G_2 = \mathrm{co}\,\{\,x_1^0, \ldots, x_{k+1}^0\,\}$. If for some $j \in N_k$ $y_j^0 \in G_1$, then escape is possible from z^0 by Corollary 2.4. If, conversely, $y_j^0 \notin G_1$, $j = 1, \ldots, k$, then for the initial state z^0 all conditions of Theorem 3.1 are satisfied. $\square$

Remark 1. Let x_i, $i = 1, \ldots, n$, $n \ge k+2$, y_j, $j = 1, \ldots, m$, $m \ge k$, are given points in R^k and for all different pairs $i_1, \ldots, i_{k+1} \in N_n$ there exists $j \in N_m$ such that

$$y_j \in \mathrm{intco}\,\{\,x_{i_1}, \ldots, x_{i_{k+1}}\,\}.$$

Then for all different pairs in $i_1, \ldots, i_{k+2} \in N_n$

$$x_{i_{k+2}} \notin \mathrm{co}\,\{\,x_{i_1}, \ldots, x_{i_{k+1}}\,\}.$$

Remark 2. Let $x_i, i = 1, \ldots, k+2, y_1, y_2$ be given points in R^k ($k \geq 3$). Then there exist different pairs of points $i_1, \ldots, i_{k+1} \in N_{k+2}$ such that

$$y_j \notin \mathrm{intco}\,\{\,x_{i_1}, \ldots, x_{i_{k+1}}\,\},\; j = 1, 2. \tag{12}$$

Proof. We shall prove our result by contradiction. Suppose that for all different pairs $i_1, \ldots, i_{k+1} \in N_{k+2}$ there exists $j \in N_2$ such that

$$y_j \in \mathrm{intco}\,\{\,x_{i_1}, \ldots, x_{i_{k+1}}\,\}.$$

Without loss of generality, we have

$$y_1 \in \mathrm{intco}\,\{\,x_1, \ldots, x_{k+1}\,\},$$
$$y_2 \in \mathrm{intco}\,\{\,x_1, \ldots, x_{k+2}\,\} \setminus \mathrm{co}\,\{\,x_1, \ldots, x_{k+1}\,\}.$$

The simplex $B = \mathrm{co}\,\{\,x_1, \ldots, x_{k+1}\,\}$ can be presented as the intersection of $k+1$ closed half-spaces $\bar{H}_i^+$. Since $x_{k+2} \notin B$, then x_{k+2} does not belong to at least one half-space $\bar{H}_i^+$. Suppose

$$x_{k+2} \in \bigcup_{r=l}^{k+1} \bar{H}_r^+ \setminus \left(\bigcup_{r=1}^{l} \bar{H}_r^+\right),\; l \geq 1.$$

If $l > 1$, then we consider l k-dimensional simplexes, each of them a convex hull of union of $\{x_{k+2}\}$ and $(k-1)$–dimensional major surface, which lies in hyperplane H_r, $r \in N_l$. Here H_r is a hyperplane restricting $\bar{H}_i^+$. It is clear that interiors of these l simplexes do not intersect in pairs and initial positions of at least $l+1$ evaders belong to interior of $\mathrm{co}\,\{\,x_1, \ldots, x_{k+2}\,\}$.

Let $l = 1$. We may assume that hyperplane H_1 crosses points $x_1, \ldots, x_k$ and $x_{k+1} \in \bar{H}_i^+$. Consider k – dimensional simplexes

$$A_1 = \mathrm{co}\,\{\,x_2, x_3, \ldots, x_k, x_{k+1}, x_{k+2}\,\},$$
$$A_2 = \mathrm{co}\,\{\,x_1, x_3, \ldots, x_k, x_{k+1}, x_{k+2}\,\},$$
$$\cdots$$
$$A_k = \mathrm{co}\,\{\,x_1, x_2, \ldots, x_{k-1}, x_{k+1}, x_{k+2}\,\}.$$

We shall prove that interiors of each pair of them do not intersect. For example, take A_1, A_2 and find a hyperplane separating them. We consider cone W being a union of all rays starting at x_{k+1} and crossing $\mathrm{co}\,\{\,x_1, \ldots, x_k\,\}$. Its interior contains point x_{k+2}. Therefore, hyperplane H crossing $x_3, x_4, \ldots, x_{k+1}, x_{k+2}$ has non-empty intersection with interior of W. So hyperplane H separates sets A_1 and A_2. In this case, initial positions of at least k evaders belong to $\mathrm{intco}\,\{\,x_1, \ldots, x_{k+2}\,\}$. We have obtained a contradiction. $\square$

4   Model examples

Let us use obtained results for studying the global avoidance problem in some special examples.

Theorem 4.1 *In the game (3), let V be a strictly convex compact with smooth boundary, $n = k + 2$ and $m = 2$. Then the avoidance problem is solvable.*

Proof. Let $z^0 = (x_1^0, \ldots, x_{k+2}^0, y_1^0, y_2^0)$ be an arbitrary initial state. Without loss of generality, we assume $y_1^0 \neq y_2^0$. Suppose $k = 2$ and

$$y_j^0 \in \mathrm{intco}\,\{\, x_1^0, \ldots, x_4^0 \,\},\ j = 1, 2.$$

If there exist different pairs in $i_1, i_2, i_3 \in N_4$ such that

$$y_j^0 \notin \mathrm{intco}\,\{\, x_{i_1}^0, x_{i_2}^0, x_{i_3}^0 \,\},\ j = 1, 2$$

then the avoidance problem is solvable by Corollary 3.1. In particular we have the following result: if initial positions of any three players lie on the same line then the escape is possible from the corresponding state. This proposition follows also from the Corollary 2.4.

Let us suppose that for all different pairs in $i_1, i_2, i_3 \in N_4$ there exists $j \in N_2$ such that $y_j^0 \in \mathrm{intco}\,\{\, x_{i_1}^0, x_{i_2}^0, x_{i_3}^0 \,\}$. If

$$y_1^0 \in \mathrm{intco}\,\{\, y_2^0, x_{i_1}^0, x_{i_2}^0 \,\},\ i_1, i_2 \in N_4, \tag{13}$$

then it is easy to see that

$$y_2^0 \in \mathrm{intco}\,\{\, y_1^0, x_{i_3}^0, x_{i_4}^0 \,\}, \qquad i_3, i_4 \in N_4 \setminus \{\, i_1, i_2 \,\}. \tag{14}$$

So set N_4 can be divided into two nonintersecting sets $I_1 = \{\, i_1, i_2 \,\}$, $I_2 = \{\, i_3, i_4 \,\}$ such that (13), (14) hold.

At the initial moment we take two sets $F_1, F_2 \in \mathrm{co}\,\Omega\,(\mathrm{R}^k)$ such that

$$x_{i_l}^0,\ x_{i_{l+1}}^0 \in \mathrm{int}\,F_j,\ l = 2j - 1,\quad y_j^0 \in \partial F_j,\ j = 1,\, 2,$$

and there exist vectors $r_j \in P\,(F_j, y_j^0),\ j = 1,\, 2$, for which

$$y_1(t, r_1) \notin y_2(t, r_2) \qquad \text{for all } t \geq 0. \tag{15}$$

Here $y_j(t, r_j)$ is a trajectory of player E_j corresponding to the control v_j chosen from (5), where $\psi_j(\cdot, r_j)$ is a solution of (2) starting at r_j. Let us fix $r_j \in P(F_j, y_j^0),\ j = 1,\, 2$, satisfying (15).

We denote by $t(r_j)$ the first moment when $y_j(t, r_j) \in X(t, F_l, V)$, $l \in N_2 \setminus \{\, j \,\},\ j = 1,\, 2$. If for some $q \in N_2,\ y_q(t, r_q) \notin X(t, F_l, V)$,

$l \in N_2 \setminus \{q\}$ for all $t \geq 0$, then E_q can avoid the capture. That is why we assume that $t(r_j) < +\infty$, $j = 1, 2$. Without loss of generality, $t(r_1) \geq t(r_2)$.

Up to the first moment $t' \in (0, t(r_1))$, when three players lie on the same line, controls v_j, $j = 1, 2$, are defined by (5). Such a moment t' exists because

$$y_j(t(r_1), r_j) \in \partial X(t(r_1); F_2, V), \quad j = 1, 2,$$
$$x_{i_3}(t(r_1)), \ x_{i_4}(t(r_1)) \in \operatorname{int} X(t(r_1); F_2, V)$$

for all controls u_{i_3}, u_{i_4} on $[0, t(r_1)]$ and, thus,

$$y_2(t(r_1), r_2) \notin \operatorname{co}\{y_1(t(r_1), r_1), \ x_{i_3}(t(r_1)), \ x_{i_4}(t(r_1))\}.$$

By Corollary 2.4 we conclude that escape is possible from the state $z(t') = (x_1(t'), \ldots, x_4(t'), y_1(t'), y_2(t'))$. If $k > 2$, the solvability of the global escape problem with $n = k + 2$, $m = 2$ follows from Remark 2. $\qquad \square$

Theorem 4.2 *In the game (3), let V be a strictly convex compact with smooth boundary, $n = 2k$ and $m = k$. Then the global avoidance problem is solvable.*

Proof. For $k = 2$ this proposition has already been proved. Let $k \geq 3$ and $z^0 = (x_1^0, \ldots, x_{2k}^0, y_1^0, \ldots, y_k^0)$ be an arbitrary initial state. If there exist different pairs in $i_1, \ldots, i_{k+1} \in N_{2k}$ such that

$$y_j^0 \notin \operatorname{intco}\{x_{i_1}^0, \ldots, x_{i_{k+1}}^0\}, \quad j = 1, \ldots, k. \tag{16}$$

then, by Corollary 3.1, the evasion problem is solvable.

Suppose, for all different pairs in $i_1, \ldots, i_{k+1} \in N_{2k}$ there exists $j \in N_k$ such that

$$y_j^0 \in \operatorname{intco}\{x_{i_1}^0, \ldots, x_{i_{k+1}}^0\}. \tag{17}$$

It follows from Remark 1 that

$$x_{i_{k+2}}^0 \notin \operatorname{co}\{x_{i_1}^0, \ldots, x_{i_{k+1}}^0\}$$

for all different pairs in $i_1, \ldots, i_{k+2} \in N_{2k}$. Then for any $l \in N_k$ $x_{k+l}^0 \notin \operatorname{co}\{x_1^0, \ldots, x_{k+l-1}^0\}$ and there exists $j_l \in N_k$ such that

$$y_{i_l}^0 \in \operatorname{intco}\{x_1^0, \ldots, x_{k+l}^0\} \setminus \operatorname{co}\{x_1^0, \ldots, x_{k+l-1}^0\}.$$

Since there are k evaders in the game, then initial positions of exactly l evaders belong to $\operatorname{intco}\{x_1^0, \ldots, x_{k+l}^0\}$ and, moreover, for all different pairs in $i_1, \ldots, i_{k+1} \in N_{k+l}$ there exists $j \in \{j_1, \ldots, j_l\}$ such that (17) holds.

However, even for $l = 2$, by Remark 2, there exist different pairs in $i_1, \ldots, i_{k+1} \in N_{k+2}$ such that $y_{j_l}^0 \notin \text{intco}\{x_{i_1}^0, \ldots, x_{i_{k+1}}^0\}$, $l = 1, 2$. We have got a contradiction. Therefore for any initial state z^0 there exist different pairs in $i_1, \ldots, i_{k+2} \in N_{2k}$, such that (16) holds. $\square$

Upper estimation of the minimal number of evaders providing the solvability of avoidance problem in the simple motion game with n pursuers was obtained in [2]. Let $[a]$ stand for the integer part of $a \in R$.

Theorem 4.3 *In the game (3), let V be a strictly convex compact with smooth boundary and $n \geq 2$, $m \geq (p+1)2^{p+1} + 2$, $p = [\log_2(n-1)]$. Then the global escape problem is solvable.*

REFERENCES

[1] L. S. PONTRYAGIN AND E. F. MISCHENKO, Avoidance Problem in Linear Differential Games, *Dif. Uravn.* 7 (1971), 436-445. [in Russian]

[2] N. N. PETROV AND N. N. PETROV, On a Differential Game of "Cossacks-Robbers", *Dif. Uravn.* 19 (1983), 1366–1374. [in Russian]

[3] P. V. PROKOPOVICH AND A. A. CHIKRII, A Problem of Interaction of Groups of Controlled Objects, in Theory of Optimal Solutions, Institute of Cybernetics, Kiev, 1987, 71–75. [in Russian]

[4] A. A. CHIKRII AND P. V. PROKOPOVICH, Pursuit and Evasion Problem for Interacting Groups of Moving Objects, *Cybernetics*, 25 (1989), 634–640.

[5] P. V. PROKOPOVICH AND A. A. CHIKRII, Quasi-linear Conflict-controlled Processes with Non-fixed Time, *Prikladnaya Matematika i Mekhanika* , 55 (1991), 63–71. [in Russian]

[6] A. A. CHIKRII AND P.V.PROKOPOVICH, Simple Pursuit of One Evader by a Group, *Cybernetics and Systems Analysis*, 28 (1992), 438-444.

[7] A. A. CHIKRII AND P. V. PROKOPOVICH, Evasion Problem for Interacting Groups of Linear Objects, *Doklady Academii Nauk SSSR*, 333 (1993), 735–739. [in Russian]

PART III
Solution methods

Convergence of Discrete Schemes
for Discontinuous Value Functions
of Pursuit-Evasion Games

Martino Bardi,* Sandra Bottacin
Dipartimento di Matematica P. e A., Università di Padova,
via Belzoni 7, I-35131 Padova, Italy.

Maurizio Falcone
Dipartimento di Matematica, Università di Roma "La Sapienza",
P.Aldo Moro 2, I-00185 Roma, Italy.

Abstract

We describe an approximation scheme for the value function of general pursuit-evasion games and prove its convergence, in a suitable sense. The result works for problems with discontinuous value function as well, and it is new even for the case of a single player. We use some very recent results on generalized (viscosity) solutions of the Dirichlet boundary value problem associated to the Isaacs equation, and a suitable variant of Fleming's notion of value. We test the algorithm on some examples of games in the plane.

Introduction

In the last ten years the theory of viscosity solutions for first and second order partial differential equations initiated by M. Crandall and P.L. Lions has provided a rigorous framework for studying the Isaacs equations of zero sum deterministic and stochastic differential games, see e.g. the survey paper [18], and [22], [21], [6], [7], [26], [34], [35], the introductory paper [1] and the references therein. A typical result of this theory is the characterization of the value function of a game as the unique solution of a suitable boundary value problem for the Isaacs equation, no matter what definition of value is adopted. Another successful application is the convergence of approximation schemes to the value function, sometimes with explicit estimates of the error, see [37], [17], [8], [13], [5] and the references therein.

---

*Partially supported by M.U.R.S.T., project "Problemi nonlineari nell'analisi e nelle applicazioni fisiche, chimiche e biologiche".

However, most of this theory works as long as there exists a continuous solution to the boundary value problem under investigation. Unfortunately it is well known that the value functions of many differential games are not continuous, for instance in the classical (deterministic) pursuit-evasion problem as soon as barriers occur. The few papers proving uniqueness of some notion of non-continuous viscosity solution do not include true Isaacs equations, that is, equations with Hamiltonian $H(x,p)$ not convex with respect to p, but just Bellman equations, see [12], [10], [14], [27], [11], [36]. Some of them study the convergence of discrete schemes to discontinuous solutions, for convex $H(x,.)$ [10], [5], [36].

During the 1980s A. Subbotin and his school developed another theory of weak solutions to first order Hamilton-Jacobi-Isaacs equations, see [38], [39], [40] and the references therein. These solutions were eventually named minimax solutions, and they were shown to be equivalent to viscosity solutions [41]. Also this theory is mostly developed for solutions which are at least continuous.

More recently, however, Rozyev and Subbotin [33] proposed a notion of non-continuous minimax solution, and Subbotin [42] proved an existence and uniqueness theorem applicable to the Dirichlet problem associated with pursuit-evasion games.

His result has been reformulated in the forthcoming book of Bardi and Capuzzo Dolcetta [4] within the theory of viscosity solutions. They introduce the equivalent notion of envelope (viscosity) solution, briefly e-solution, and prove existence and uniqueness by different methods. The notion of e-solution extends naturally to 2nd order fully nonlinear degenerate elliptic equations, including the Isaacs equation for stochastic games. In this general context Bardi and Bottacin [16], [2] have very recently proved an existence and uniqueness theorem for Dirichlet problems, as well as some results on the stability of e-solutions with respect to perturbations of the data.

In the present paper we apply this rather general PDE theory to pursuit-evasion games. Our main results are (i) the existence of a value under general assumptions, without any request of continuity, and its characterization as the unique e-solution of the Dirichlet problem for the Isaacs equation; (ii) the convergence to the value function of a suitable variant of the approximation scheme studied by Bardi, Falcone and Soravia [5]; (iii) the numerical tests of our algorithm on some pursuit-evasion games in the plane with discontinuous value function.

To our knowledge the theorem in Section 3 is the first convergence result for approximations of Hamilton-Jacobi-Isaacs equations holding in case of discontinuous solutions. It reflects a general stability property of e-solutions which is studied in [2]. Therefore it works for other approximation schemes, essentially for any monotone, stable and consistent scheme, in the terminology of [13]. Moreover it can be extended to the 2nd order degenerate elliptic equations arising in stochastic games. This will be done in a future paper.

The convergence theorem is new even for Bellman equations, because the result in [5] assumes some regularity of the boundary (the target) which is not always satisfied in applications, while [36] proves just the convergence of a lower weak limit. A rather different approximation result has been given by Cardaliaguet, Quincampoix and Saint-Pierre [19], [20] using the methods of viability theory.

Here is a more detailed description of the contents of the paper. In Section 1 we define e-solutions and non-continuous minimax solutions, we prove their equivalence and state the existence and uniqueness theorem.

In Section 2 we define a variant of Fleming's value for pursuit-evasion games [25]. We prove that the upper and lower values exist and are the e-solutions of the Dirichlet problem for the corresponding upper and lower Isaacs equations. As a consequence they coincide, and so the game has a value, if the Isaacs' condition on the Hamiltonians holds. We also compare this notion of value with those of Krasovskii and Subbotin [30] and of Varaiya, Roxin, and Elliott-Kalton [22], [21], [6].

In Section 3 we prove that if we take any "good approximation scheme", e.g. the one in [5], we apply it to the game where the true target $\mathcal{T}$ is replaced by its ϵ-neighbourhood $\mathcal{T}_\epsilon$, we send to 0 the approximation step h and ϵ with "h linked to ϵ" in a suitable sense, then we get the value function in the limit. This result is a bit abstract, especially because we cannot give a formula linking h to ϵ. However it gives the first rigorous justification for using the algorithm of [5] for games with barriers, and we already observed in some simple examples in [5] that the numerical performances are good for such games as well.

In Section 4 we present some numerical results for pursuit-evasion game tests in $I\!\!R^2$.

1 Envelope viscosity solutions of the Isaacs equation

In this section we present a notion of weak solution for the boundary value problem

$$\begin{cases} u + F(x, Du) = 0, & \text{in } \Omega, \\ u = 0, & \text{on } \partial\Omega, \end{cases} \tag{1}$$

where $\Omega \subset I\!\!R^N$ is an open set and $F : I\!\!R^{2N} \to I\!\!R$ is continuous and such that

$$|F(x,p) - F(y,q)| \le L|x - y|(1 + |p|) + K(1 + |x|)|p - q|. \tag{2}$$

We are interested in the Hamiltonians arising in the Isaacs equations, namely

$$H(x,p) := \min_{b \in B} \max_{a \in A} \{-f(x,a,b) \cdot p - 1\}, \tag{3}$$

$$\hat{H}(x,p) := \max_{a \in A} \min_{b \in B} \{-f(x,a,b) \cdot p - 1\},$$

which satisfy (2) if

$$\begin{cases} f : I\!\!R^N \times A \times B \to I\!\!R^N \text{ is continuous, } A \text{ and } B \text{ are compact,} \\ |f(x,a,b) - f(y,a,b)| \leq L|x-y|, \text{ for all } x,y,a,b. \end{cases} \tag{4}$$

We recall that a function $u : \overline{\Omega} \to I\!\!R$ bounded and upper semicontinuous (respectively lower semicontinuous) is a (viscosity) subsolution (respectively supersolution) of (1) if for all $\phi \in C^1(\Omega)$ such that $u - \phi$ attains a local maximum point at x (respectively minimum) we have $u(x) + F(x, D\phi(x)) \leq 0$ (respectively $u(x) + F(x, D\phi(x)) \geq 0$), and $u(x) \leq 0$ for all $x \in \partial\Omega$ (respectively $u(x) \geq 0$). A (viscosity) solution of (1) is a subsolution which is also a supersolution.

We recall the comparison principle

Theorem 1.1 *Assume (2). Then any subsolution w and supersolution W of (1) satisfy $w \leq W$.*

For the proof see [9].

Definition 1.1 *Let S, Z be respectively the sets of all subsolutions and all supersolutions of (1). If $u : \overline{\Omega} \to I\!\!R$ is locally bounded, we will say that u is an* envelope viscosity solution *or e-solution of (1) if there exist two nonempty subsets $S(u) \subset S$ and $Z(u) \subset Z$ such that for all $x \in \overline{\Omega}$*

$$u(x) = \sup_{w \in S(u)} w(x) = \inf_{W \in Z(u)} W(x).$$

Note that by Theorem 1.1 there is at most one e-solution. In fact if u and v are e-solutions, then

$$u = \sup_{w \in S(u)} w(x) \leq \inf_{W \in Z(v)} W(x) = v$$

and the opposite inequality is obtained by exchanging the roles of u and v.

The existence of the e-solution is much less trivial and it is given by the following result, which is a special case of a much more general existence theorem for a fully nonlinear degenerate elliptic second order equation due to Bardi and Bottacin [16], [2]. It is also equivalent to a special case of the existence result for minimax solutions of first order equations by Subbotin [42].

Theorem 1.2 *Assume (2), and*

$$-1 \le F(x,0) \le 0, \ \text{for all } x \in \Omega.$$

Then S, Z are nonempty, and

$$u(x) := \sup_{w \in S} w(x) = \min_{W \in Z} W(x) \tag{5}$$

is the e-solution of (1). Moreover

$$u(x) = \sup_{\epsilon > 0} u_\epsilon(x), \ \text{for all } x \in \overline{\Omega}, \tag{6}$$

where

$$u_\epsilon(x) := \sup\{w(x) : \ w \in S, \ w(x) = 0 \text{ if } \text{dist}(x, \partial\Omega) \le \epsilon\}. \tag{7}$$

For the proof see [16], [2].

Next we compare the notion of e-solution with that of (generalized) minimax solution of (1) introduced by Rozyev and Subbotin [33]. This definition is based on Subbotin's theory of minimax solutions of the Hamilton-Jacobi-Isaacs equation [39], [40]. Since Subbotin proved that lower (respectively upper) minimax solutions of the PDE appearing in (1) are equivalent to viscosity subsolutions (respectively supersolutions) [41], his definition of solutions of the boundary value problem (1) can be formulated as follows.

Definition 1.2 *[42] A function $u : \overline{\Omega} \to I\!\!R$ is a* minimax solution *of (1) if there exist two sequences $w_n \in S$, $W_n \in Z$, such that $w_n = W_n = 0$ on $\partial\Omega$, w_n is continuous at each point of $\partial\Omega$ and*

$$\lim_n w_n(x) = u(x) = \lim_n W_n(x) \text{ for all } x \in \overline{\Omega}. \tag{8}$$

Proposition 1.1 *Under the assumptions of Theorem 1.2 u is the e-solution of (1) if and only if it is its minimax solution.*

Proof. A minimax solution is the e-solution by formula (5). If, on the other hand, u is the e-solution of (1), we choose $W_n = u$ for all n and $w_n = (u_{1/n})^* :=$ the minimal upper semicontinuous function above $u_{1/n}$. By definition w_n is null in a neigbourhood of $\partial\Omega$, and it is a subsolution because $u_{1/n}$ is the sup of a set of subsolutions, see [29]. $\Box$

We end this section by recalling another notion of weak solution of the boundary condition in (1). It is useful for proving the convergence results of sections 2 and 3 by virtue of its stability with respect to the weak limits in the viscosity sense, for which the definition is recalled next. We refer to [16], [2] for a detailed comparison between e-solutions and this boundary condition.

Definition 1.3 *A lower semicontinuous (respectively upper semicontinuous) function $u : \overline{\Omega} \to I\!\!R$ is a (viscosity) subsolution (respectively supersolution) of the boundary condition*

$$u = 0 \ or \ u + F(x, Du) = 0 \ on \ \partial\Omega$$

if for all $x \in \partial\Omega$ and for all functions $\phi \in C^1(\overline{\Omega})$ such that $u - \phi$ attains a local maximum point at x (respectively minimum)

$$u(x) \le 0 \ (resp. \ge 0), \ or \ u(x) + F(x, D\phi(x)) \le 0 \ (resp. \ge 0).$$

Definition 1.4 *If the functions $u_h : \overline{\Omega} \to I\!\!R$, $h > 0$, are locally bounded uniformly with respect to h, we define their upper and lower weak limits, respectively, as follows*

$$\limsup_{h \searrow 0}{}^* u_h(x) := \inf_{\delta > 0} \sup\{u_h(y) : \ |x - y| < \delta, \ 0 < h < \delta\},$$

$$\liminf_{h \searrow 0}{}_* u_h(x) := \sup_{\delta > 0} \inf\{u_h(y) : \ |x - y| < \delta, \ 0 < h < \delta\}.$$

2   Existence of a value

In this section we recall the definition of value for pursuit-evasion games following Fleming, we give a new notion of value by modifying Fleming's definition, and prove the existence of the value function under very weak hypotheses. Finally we compare it with other notions of value.

We consider the dynamical system controlled by two players

$$\begin{cases} y' = f(y, a, b), & t > 0, \\ y(0) = x, \end{cases} \tag{9}$$

where $y(t) \in I\!\!R^N$, $a \in \mathcal{A}$, $b \in \mathcal{B}$ and

$$\mathcal{A} := \{a : [0, +\infty) \to A, \ \text{measurable}\},$$

$$\mathcal{B} := \{b : [0, +\infty) \to B, \ \text{measurable}\},$$

are the sets of admissible *controls*. We assume (4).

Given the closed *target* $\mathcal{T} \subset I\!\!R^N$, we define the first time the trajectory hits the target

$$t_x(a, b) := \inf\{t : \ y_x(t, a, b) \in \mathcal{T}\} \le +\infty,$$

where $t_x = +\infty$ if the trajectory does not hit the target. The goal of the game for the first player "a" is to minimize this time, for the second one "b" to maximize it. As in [6], [7] we consider the game with discounted payoff $\Psi(t_x(a,b))$ where

$$\Psi(r) := \begin{cases} 1 - e^{-r}, & \text{if } r < +\infty, \\ 1, & \text{if } r = +\infty. \end{cases}$$

From the solution of this game one recovers easily the solution of the original one.

To define Fleming's value, we need to introduce the discrete-time games with step $h > 0$

$$\begin{cases} y_{n+1} = y_n + hf(y_n, a_n, b_n), \\ y_0 = x. \end{cases}$$

The set of nonanticipating strategies for the first player is

$$\Lambda := \{\alpha : B^{I\!N} \to A^{I\!N} : b_j = \hat{b}_j \text{ for all } j \leq \hat{j} \Rightarrow \alpha[b]_j = \alpha[\hat{b}]_j \text{ for all } j \leq \hat{j}\},$$

where $X^{I\!N}$ denotes the set of sequences taking values in X.

Likewise we define the set Θ of nonanticipating strategies for the second player. Denote by n_h the number of steps taken by a trajectory to reach the target, that is,

$$n_h(x, a, b) := \inf\{j \in I\!N : y_j \in T\} \leq +\infty.$$

Then the lower value and the upper value of the discrete-time game are respectively

$$N_h(x) := \inf_{\alpha \in \Lambda} \sup_{b \in B^{I\!N}} n_h(x, \alpha[b], b),$$

$$\hat{N}_h(x) := \sup_{\beta \in \Theta} \inf_{a \in A^{I\!N}} n_h(x, a, \beta[a]),$$

whereas

$$v_h(x) := \Psi(hN_h(x)), \quad \hat{v}_h(x) = \Psi(h\hat{N}_h(x))$$

are, respectively, the approximated (discounted) lower and upper value.

Definition 2.1 *[25] For given x, the* lower *and* upper values *of the game are, respectively,* $\lim_{h \searrow 0} v_h(x)$ *and* $\lim_{h \searrow 0} \hat{v}_h(x)$ *if they exist. If in addition they coincide, then* Fleming's value *is*

$$v(x) := \lim_{h \searrow 0} v_h(x) = \lim_{h \searrow 0} \hat{v}_h(x).$$

It was proved in [8] that Fleming's value exists everywhere if
i) f is bounded on $\partial T \times A \times B$ and the following Isaacs' condition holds

$$H(x,p) = \hat{H}(x,p), \quad \text{for all } x, p, \tag{10}$$

ii) the target is the closure of an open set whose boundary is a Lipschitz manifold ,

iii) the boundary value problem (1) with $F = H$ has a continuous viscosity solution.

Simple explicit examples show that Fleming's value does not exist for all x if either ii) or iii) is violated, see [16], [3]. This motivates the following modification of Fleming's notion of value.

Consider, for all $\epsilon > 0$ the discrete-time game with target

$$\mathcal{T}_\epsilon = \{x \in I\!R^N : \ \mathrm{dist}(x, \partial \mathcal{T}) \leq \epsilon\},$$

and let v_h^ϵ, $\hat{v}_h^\epsilon$ be, respectively, the lower and upper approximated value function for that game.

We first give a precise meaning to the convergence of v_h^ϵ to v as $h, \epsilon \searrow 0$ "with h linked to ϵ". The same notion of convergence will be used in the next section for fully discrete approximation schemes.

Definition 2.2 *Let v_h^ϵ, $v : Y \to I\!R$, for $\epsilon > 0$, $h > 0$, $Y \subset I\!R^N$. We say that v_h^ϵ converges to v as $(\epsilon, h) \searrow (0,0)$ with h linked to ϵ at the point x, and write*

$$\lim_{\substack{(\epsilon,h)\searrow(0,0) \\ h \leq h(\epsilon)}} v_h^\epsilon(x) = v(x),$$

if for all $\gamma > 0$, there exists a function $\tilde{h} :]0, +\infty[\to]0, +\infty[$ and $\bar{\epsilon} > 0$ such that

$$|v_h^\epsilon(y) - v(x)| \leq \gamma, \ \text{for all } y : \ |x - y| \leq \tilde{h}(\epsilon), \tag{11}$$

for all $\epsilon \leq \bar{\epsilon}$, $h \leq \tilde{h}(\epsilon)$.

Similarly, we define the limit as $(\epsilon, n) \to (0^+, \infty)$ with n linked to ϵ, and write

$$\lim_{\substack{(\epsilon,n)\to(0^+,\infty) \\ n \geq n(\epsilon)}} v_n^\epsilon(x) = v(x),$$

by replacing $h \leq \tilde{h}(\epsilon)$ with $n \geq \tilde{h}(\epsilon)$.

To justify this definition we note that:

i) it implies that for any $\epsilon_n \searrow 0$ there is a sequence $h_n \searrow 0$ such that $v_{h_n}^{\epsilon_n}(x_n) \to v(x)$ for any sequence x_n such that $|x - x_n| \leq h_n$, e.g. $x_n = x$ for all n, and the same holds for any sequence $h'_n \geq h_n$;

ii) if $\lim_{h \searrow 0} v_h^\epsilon(x)$ exists for all small ϵ and its limit as $\epsilon \searrow 0$ exists, then it coincides with the limit of Definition 2.2;

iii) if the convergence of Definition 2.2 occurs on a compact set K where the limit v is continuous, then (11) can be replaced, for all $x \in K$ and redefining $\tilde{h}$ if necessary, with

$$|v_h^\epsilon(y) - v(y)| \leq 2\gamma, \ \ \text{for all } y : |x - y| \leq \tilde{h}(\epsilon),$$

and by a standard compactness argument we obtain the uniform convergence in the following sense:

Definition 2.3 *Let K be a subset of $\mathbb{R}^N$ and v_h^ϵ, $v : K \to \mathbb{R}$, for all $\epsilon, h > 0$ and small, we say that v_h^ϵ converge uniformly on K to v as $(\epsilon, h) \searrow (0,0)$ (respectively $(\epsilon, n) \to (0^+, \infty)$) with h linked to ϵ (respectively n linked to ϵ) if for any $\gamma > 0$ there are $\bar{\epsilon} > 0$ and $\tilde{h} :]0, +\infty[\to]0, +\infty[$ such that*

$$\sup_K |v_h^\epsilon - v| \leq \gamma$$

for all $\epsilon \leq \bar{\epsilon}$, $h \leq \tilde{h}(\epsilon)$ (respectively $n \geq \tilde{h}(\epsilon)$).

Now we can give our definition of values.

Definition 2.4 *Assume that for given x there exist the limits of v_h^ϵ and $\hat{v}_h^\epsilon$ as $(\epsilon, h) \searrow (0,0)$ with h linked to ϵ. We define the* lower value function *at x*

$$V(x) := \lim_{\substack{(\epsilon,h) \searrow (0,0) \\ h \leq h(\epsilon)}} v_h^\epsilon(x).$$

Similarly the upper value function *at x is*

$$\hat{V}(x) := \lim_{\substack{(\epsilon,h) \searrow (0,0) \\ h \leq h(\epsilon)}} \hat{v}_h^\epsilon(x).$$

If in addition the limits coincide, we define the value function *at x as*

$$v(x) := V(x) = \hat{V}(x).$$

The next theorem is the main result of this section: it shows that the value function exists and it is the e-solution of the Dirichlet problem associated with the game.

Theorem 2.1 *Assume (4). Then*
   i) there exists the lower value and it is the e-solution of (1) where $F = H$;
   ii) there exists the upper value and it is the e-solution of (1) where $F = \hat{H}$;
   iii) if in addition Isaacs' condition (10) holds, then the game has value.

Proof. We give the proof only for i) and iii), the same proof works for ii). From now on we will denote by (1) the Dirichlet problem where $F = H$. Note that the hypotheses of Theorem 1.2 are satisfied, so there exists the e-solution u.

It was proved in [8] that v_h^ϵ is solution in $\Omega_\epsilon := \mathbb{R}^N \setminus \mathcal{T}_\epsilon$ of the problem

$$\begin{cases} v_h(x) - \sup_{b \in B} \inf_{a \in A} e^{-h} v_h(x + h f(x, a, b)) - 1 + e^{-h} = 0, & \text{in } \Omega_\epsilon, \\ v_h = 0, & \text{on } \partial \Omega_\epsilon. \end{cases}$$

Consider, now, the weak limits

$$\underline{v}_\epsilon(x) := \liminf_{h\searrow 0}{}_* v_h^\epsilon(x), \ \overline{v}_\epsilon(x) := \limsup_{h\searrow 0}{}^* v_h^\epsilon(x).$$

We know (see [8]) that $\underline{v}_\epsilon$ and $\overline{v}_\epsilon$ are respectively a supersolution and subsolution of

$$\begin{cases} v + H(x, Dv) = 0, & \text{in } \Omega_\epsilon, \\ v = 0 \text{ or } v + H(x, Dv) = 0, & \text{on } \partial\Omega_\epsilon. \end{cases} \tag{12}$$

We claim that $\overline{v}_\epsilon$ is a subsolution of (1). In fact $v_h^\epsilon \equiv 0$ in $\Omega \setminus \Omega_\epsilon$, so $\overline{v}_\epsilon \equiv 0$ in the interior of $\Omega \setminus \Omega_\epsilon$ and then in this set it is a subsolution.

In Ω_ϵ we have already seen that $\overline{v}_\epsilon$ is upper semicontinuous and it is a subsolution. It remains to check what happens on $\partial\Omega_\epsilon$. Given $\hat{x} \in \partial\Omega_\epsilon$, we must prove that for all $\phi \in C^1(\Omega)$ such that $\overline{v}_\epsilon - \phi$ attains a local maximum at $\hat{x}$, we have

$$\overline{v}_\epsilon(\hat{x}) + H(\hat{x}, D\phi(\hat{x})) \leq 0. \tag{13}$$

1^0 CASE $\overline{v}_\epsilon(\hat{x}) > 0$.

Since $\overline{v}_\epsilon$ satisfies the boundary condition on $\partial\Omega_\epsilon$ of problem (12), then for all $\phi \in C^1(\overline{\Omega}_\epsilon)$ such that $\overline{v}_\epsilon - \phi$ attains a local maximum at $\hat{x}$ (13) holds. Then the same inequality holds for all $\phi \in C^1(\Omega)$ as well, because $C^1(\Omega) \subset C^1(\overline{\Omega}_\epsilon)$.

2^0 CASE $\overline{v}_\epsilon(\hat{x}) = 0$.

Let $\phi \in C^1(\Omega)$ be such that $\overline{v}_\epsilon - \phi$ attains a local maximum at $\hat{x}$. We claim that $D\phi(\hat{x}) = 0$. Indeed for all x near $\hat{x}$ we have

$$\overline{v}_\epsilon(x) - \overline{v}_\epsilon(\hat{x}) \leq \phi(x) - \phi(\hat{x}),$$

and using Taylor's formula for ϕ at $\hat{x}$, and the fact that $\overline{v}_\epsilon(x) \geq 0$, we get

$$-D\phi(\hat{x}) \cdot (x - \hat{x}) \leq o(|x - \hat{x}|).$$

If we assume $D\phi(\hat{x}) \neq 0$, we get a contradiction by choosing $x = \hat{x} - (D\phi(\hat{x})/|D\phi(\hat{x})|)\,|x - \hat{x}|$. Then

$$\overline{v}_\epsilon(\hat{x}) + H(\hat{x}, D\phi(\hat{x})) = H(\hat{x}, 0) \leq 0$$

and $\overline{v}_\epsilon$ is a subsolution of (1).

We now claim that

$$u_\epsilon \leq \underline{v}_\epsilon \leq \overline{v}_\epsilon \leq u \tag{14}$$

where u_ϵ is defined by (7). Indeed, since $\underline{v}_\epsilon$ is a supersolution in Ω_ϵ and $\underline{v}_\epsilon \geq 0$, by Theorem 1.1 $\underline{v}_\epsilon \geq w$ in Ω_ϵ for any $w \in S$ such that $w = 0$ on $\partial\Omega_\epsilon$. Moreover $\underline{v}_\epsilon \equiv 0$ on $\Omega \setminus \Omega_\epsilon$, so we get $\underline{v}_\epsilon \geq u_\epsilon$ in $\overline{\Omega}$. To prove the last inequality we note that u is a supersolution of (1) by (5), which implies $\overline{v}_\epsilon \leq u$ by Theorem 1.1.

Now we fix x, $\epsilon > 0$, $\gamma > 0$ and note that, by definition of the lower weak limit, there exists $\overline{h} = \overline{h}(x, \epsilon, \gamma) > 0$ such that

$$\underline{v}_\epsilon(x) - \gamma \leq v_h^\epsilon(y)$$

for all $h \leq \overline{h}$ and $y \in B(x, \overline{h})$. Similarly there exists $\overline{k} = \overline{k}(x, \epsilon, \gamma) > 0$ such that

$$v_h^\epsilon(y) \leq \overline{v}_\epsilon(x) + \gamma$$

for all $h \leq \overline{k}$ and $y \in B(x, \overline{k})$. By (6) there exists $\overline{\epsilon}$ such that

$$u(x) - \gamma \leq u_\epsilon(x), \text{ for all } \epsilon \leq \overline{\epsilon}.$$

Then, using (14), we get

$$u(x) - 2\gamma \leq v_h^\epsilon(y) \leq u(x) + \gamma$$

for all $\epsilon \leq \overline{\epsilon}$, $h \leq \tilde{h} := \min\{\overline{h}, \overline{k}\}$ and $y \in B(x, \tilde{h})$, and this completes the proof of (i).

To prove (iii) it is enough to observe that the Isaacs'condition (10) and the uniqueness of e-solutions imply $V = \hat{V}$. $\square$

Next we compare the modified Fleming values defined in Definition 2.4 with the notions of value of Krasovskii and Subbotin [30], briefly KS value, and of Varaiya-Roxin-Elliot-Kalton [21], [22], [7], briefly VREK value.

Let θ^0 be the KS value function of the game with payoff t_x, i.e. the first time the trajectory hits the target. In his paper [42] on discontinuous minimax solutions (see Definition 2.1), Subbotin gives the discounted value $w^0 := \Psi(\theta^0)$ as an example of the minimax solution of (1) with $F = H$. In view of Proposition 1.1 on the equivalence between minimax and e-solutions, and on Theorems 1.2 and 2.1, assuming (4) and Isaacs'condition (10) we conclude that the discounted KS value w^0 coincides with the unique e-solution of (1) and therefore with the value v of Definition 2.4.

The connection with VREK values is given by the following results.

Proposition 2.1 *Let v_ϵ and $\hat{v}_\epsilon$ be, respectively, the lower and upper VREK values of the game with payoff $\Psi(t_x^\epsilon)$, where t_x^ϵ is the first time the trajectory hits the target $\mathcal{T}_\epsilon$. If (4) holds, then*

$$v_\epsilon \nearrow V, \ \hat{v}_\epsilon \nearrow \hat{V}, \ as \ \epsilon \searrow 0.$$

If in addition Isaacs'condition (10) holds, then

$$v_\epsilon = \hat{v}_\epsilon \nearrow V = \hat{V} = v.$$

Proof. By Theorem 2.1 V is the e-solution of (1), and (6) holds. We claim that

$$u_\epsilon \le v_{\epsilon*} \le v_\epsilon^* \le V = u,$$

where $v_{\epsilon*}$ (respectively v_ϵ^*) denotes the minimal lower semicontinuous function below v_ϵ (respectively the maximal upper semicontinuous function above v_ϵ). In fact v_ϵ is a solution of the Dirichlet problem (12) (see [9]) and by definition $v_\epsilon \ge 0$ in Ω_ϵ and $v_\epsilon \equiv 0$ in $\mathcal{T}_\epsilon$. Thus $v_{\epsilon*}$ is supersolution in Ω_ϵ, $v_{\epsilon*} \ge 0$ in $\partial\Omega_\epsilon$, and $v_{\epsilon*} \equiv 0$ in the interior of $\mathcal{T}_\epsilon$. Therefore we get $u_\epsilon \le v_{\epsilon*}$ by means of Theorem 1.1. Moreover v_ϵ^* is a subsolution of (1), by the same argument we used in Theorem 2.1 to prove that the upper weak limit $\overline{v}_\epsilon$ was a subsolution of (1). Then Theorem 1.1 yields the claim.

Now (6) forces $v_\epsilon \nearrow V$ as $\epsilon \searrow 0$.

Some slight changes in the proof show that $\hat{v}_\epsilon \nearrow \hat{V}$. The last assertion is an easy consequence of the fact that if Isaacs' condition (10) holds, then $v_\epsilon = \hat{v}_\epsilon$ and $V = \hat{V}$. $\square$

Proposition 2.2 *Assume (4). If the lower (respectively upper) VREK value is continuous, then it coincides with the lower (respectively upper) value of Definition 2.4.*

Proof. If the VREK value is continuous then it is the solution of (1) (see [6]) so it is also the e-solution. By Theorem 2.1 and the uniqueness of the e-solution we get the conclusion. $\square$

3   Convergence of approximation schemes

In this section we construct an approximation scheme converging to the e-solution of (1) merely under assumption (4) and without any continuity assumption on the solution, starting from any "good" scheme which is, roughly speaking, a convergent scheme in the case of continuous value function. Here is the precise definition.

Definition 3.1 *We say an approximation scheme is good for the equation $u + F(x, Du) = 0$ if, for any closed target $\mathcal{S}$, it generates a sequence of functions $v_n : \mathbb{R}^N \to \mathbb{R}$ which are locally bounded uniformly with respect to n, and such that their weak limits $\limsup_n^* v_n$ and $\liminf_{n*} v_n$ are, respectively, subsolution and supersolution of*

$$\begin{cases} u + F(x, Du) = 0, & in \ \mathbb{R}^N \setminus \mathcal{S}, \\ u = 0 \ or \ u + F(x, Du) = 0, & on \ \partial\mathcal{S}, \end{cases}$$

see Definitions 1.3 and 1.4.

This definition is justified by recalling what happens when problem (1) with $\Omega = \mathbb{R}^N \setminus \mathcal{S}$ has a continuous solution u, that is, if the (lower) value of the game with target $\mathcal{S}$ is continuous. In this case it is easy to see that the above property of the weak limits of v_n is necessary for the convergence of v_n to u, locally uniformly. Vice-versa, if $\mathcal{S}$ is the closure of an open set with Lipschitz boundary and f is bounded on $\partial \mathcal{S} \times A \times B$, then such a property is also sufficient for the convergence of v_n to u, locally uniformly. This follows from a comparison theorem in [9], see e.g. the proof of Theorem 3.1 in [8] or Theorem 1 in [5].

The main example of good scheme we have in mind is the discretization in space of the Dynamic Programming equation for discrete-time games described in [5], see the Lemma in [5].

In practice one may want to use variants to that scheme, for instance discretizing (9) by higher order methods, as in [23]. The property of Definition 3.1 is rather general, since it is verified by any monotone, stable and consistent scheme, according to the terminology of [13]. Another explicit example of a good scheme is described in the very recent paper [32] where the game is approximated by stochastic discrete-state games, see Lemma 4.1 in [32].

Here is the main result of this section.

Theorem 3.1 *Assume we are given a good approximation scheme for* $u + H(x, Du) = 0$ *with* H *given by (3) and* f *satisfying (4). Let* v_n^ϵ *be the sequence generated by the scheme corresponding to the target* $\mathcal{T}_\epsilon = \{x : \mathrm{dist}(x, \mathcal{T}) \le \epsilon\}$*. Then the limit*

$$v_{lim}(x) = \lim_{\substack{(\epsilon,n) \to (0^+, \infty) \\ n \ge n(\epsilon)}} v_n^\epsilon(x), \tag{15}$$

(see Definition 2.2) exists for all x *and coincides with the e-solution of (1) with* $F = H$*, and so*

$$v_{lim} = V$$

where V *is the lower value of the game with target* $\mathcal{T}$ *(see Definition 2.4). Moreover the convergence in (15) is uniform (see Definition 2.3) on every compact set where* V *is continuous.*

Proof. By the definition of a good scheme the weak limits

$$\underline{v}_\epsilon(x) := \liminf_n {}_* v_n^\epsilon(x), \quad \overline{v}_\epsilon(x) := \limsup_n {}^* v_n^\epsilon(x)$$

are, respectively, supersolution and subsolution of (12). The rest of the proof is the same as the proofs of Theorem 2.1 and of the remark (iii) after Definition 2.2. $\square$

4 Numerical experiments

In this section we present the numerical results related to three tests. In all the tests the pursuit-evasion game is set in a subdomain of $I\!R^2$. Then, working in natural coordinates, the corresponding value function will be defined in a subdomain of $I\!R^4$. The algorithm is essentially based on the approximation scheme described in [5]. We just recall its main features for the reader's convenience. The approximate value function w is computed by the scheme

$$\begin{cases} w(x) = \sum_j \lambda_j w(x_j) & \text{if } x = \sum_j \lambda_j x_j, \\ w(x_i) = \gamma \max_b \min_a w(x_i + hf(x_i,a,b)) + 1 - \gamma & \text{if } x_i \in Q\backslash\mathcal{T}, \end{cases} \qquad (16)$$

on a domain Q having nonempty intersection with the target, where x_i are the nodes of the grid, $\gamma \equiv e^{-h}$ and $h = \Delta t$. Of course, in order to have a unique solution one should complement (16) with some Dirichlet boundary conditions on ∂Q (see [5] for details). In the three tests we have adopted the boundary condition $w = 1$, which modifies the game giving an advantage to the evader (he can win also reaching ∂Q). This choice is completely arbitrary but has the advantage of making clear which kind of error we should expect in the computations. In fact, let w_1 represent the lower value of the pursuit-evasion game in Q with that boundary condition and w be the restriction to Q of the lower value in the whole space, we will always have $w_1 \geq w$ in Q. Then, if the evader is captured according to w_1 he will also be captured in the real game. If not, we can expect that our solution will converge to the exact solution as the influence of the boundary conditions vanishes, i.e. for the diameter of Q going to infinity (see Theorem 1 in [5] for a precise result).

In order to treat problems in $I\!R^4$ we simplified the algorithm to reduce the global number of variables and to speed up the computations.

The first modification is that we use a simple "square" grid so that there is no need to keep it in memory. In fact a "cell" is a hypercube in dimension 4 and the problem of recovering the cell which contains the points $y_i \equiv x_i + hf(x_i,a,b)$ can be solved very easily. The second shortcut is that we do not compute the local coordinates of the points y_i with respect to the vertices of the cell, we just assign to a cell a single value which corresponds to the value in its baricenter (i.e. to each point in the cell we assign uniform local coordinates $\lambda_{ij} = 1/2^4$). Finally, the code has been written in Fortran 90 and it has been completely parametrized with respect to the space dimension. The numerical results have been obtained by a code running on the parallel computer IBM-SP2 at the IBM-ECSEC in Rome (see [31] and [24] for details on the implementation).

Although the following tests are rather simple they give some interesting information about the way the algorithm is working. Since Theorem 3.1

does not give precise indications on the choices of the parameters of the discretization, we made several tests to get a deeper understanding on how the parameters (in particular $k \equiv \Delta x$, Δt, ϵ) influence the convergence. This is of course just an experimental insight and we hope to obtain precise a priori estimates in the near future. Moreover, the three tests we present are simple enough that one has a clear idea of what the optimal trajectories should look like.

We start observing that the algorithm needs some compatibility assumptions on the parameters to be fulfilled in order to produce reasonable results. First of all the target should be "visible" in the grid, i.e. there should be nodes of the grid belonging to the target $\mathcal{T}$. A reasonable choice for ϵ (the parameter controlling the radius of the tube around $\mathcal{T}$) is to guarantee that the tube $\mathcal{T}$ contains at least the cells having a non empty intersection with $\mathcal{T}$. If not, the algorithm (in particular the one using the values of w at the baricenters of the cells instead of the linear interpolation) will not "see" the difference between $\mathcal{T}$ and $\mathcal{T}_\epsilon$.

The computations have been made in a hypercube Q of $\mathbb{R}^4$. A point $x = (x_1, x_2, x_3, x_4)$ in Q will represent the positions of player-1 (x_1, x_2 coordinates) and player-2 (x_3, x_4 coordinates). We will also use a simplified notation denoting by z_1 the coordinates of player-1 (the pursuer) and by z_2 those of player-2 (the evader).

Test 1
We consider the following dynamics for the players,

$$\begin{cases} \dot{z}_1 &= v_1\, a \\ \dot{z}_2 &= v_2\, b \end{cases} \tag{17}$$

where v_1, v_2 are two positive real parameters representing their relative velocities. We choose $A \equiv B(0,1)$ and $B \equiv B(0,1)$.

We study this game in $Q \equiv [-1,1]^4$. Since Q is not invariant with respect to the trajectories we add the boundary condition $w = 1$. We set $\mathcal{T} \equiv \{(z_1, z_2) : z_1 = z_2\}$ and we use in the algorithm the "fat target" $\mathcal{T}_\epsilon \equiv \{(z_1, z_2) \in \mathbb{R}^4 : \text{dist}(z_1, z_2) \leq \epsilon\}$. Starting the game from (z_1^0, z_2^0), the optimal strategy for the evader will be to move as fast as he can along the line $\overline{z_1 z_2}$ in the opposite direction with respect to the position of the pursuer. The optimal strategy for the pursuer will be to follow the evader on the same line moving at his maximum speed. Then the computation of the optimal capture time is rather easy and we get

$$T(z_1^0, z_2^0) = [-\epsilon + |z_1^0 - z_2^0|]^+ / (v_1 - v_2)$$

Figures 1, 2, 3 show the numerical approximation of the value function $w(0, 0, x_3, x_4)$ (remember that w is the discounted capture time function), i.e. we fix the pursuer at the origin and represent the value as a function of the initial position of the evader. The result in Figure 1 has been obtained by applying the algorithm on a grid of $23^4 = 279841$ nodes ($k \equiv \Delta x = 0.09$) and for $h \equiv \Delta t = 0.1$. The control sets A and B have been discretized using 41 controls in each of them (1 for the origin plus 40 directions).

Looking at Figure 1 we can see that it matches with the real solution in that it is radially symmetric and it grows with the distance from the origin (i.e. from the position of the pursuer). However, w is almost flat near the boundary due to the effect of the boundary condition.

If we compare Fig. 2 with Fig. 1 and with the graph of the exact solution, we see that decreasing the time step h does not lead to an improvement if the space step k is kept fixed. This agrees with Theorem 1 of [5], saying that the algorithm converges as the ratio k/h goes to 0. In fact here $h = 0.1$ and $k/h = 0.9$ in Fig.1, whereas $h = 0.03$ and $k/h = 3$ in Fig.2.

Comparing Figure 1 and Figure 3 we can see the influence of ϵ on the solution. For ϵ increasing from 0.2 to 0.25 the graph of the solution becomes more "fat" which corresponds to the intuition since the time of capture decreases if we enlarge the target. Also notice that the effect of the boundary condition is less important in Figure 3 (look also at the contour lines). Figures 4 and 5 represent the approximate optimal trajectories computed integrating numerically along the trajectory by applying the feedback controls given by w. The letters E and P indicate the initial position of the evader and the pursuer, respectively, and each move is represented on the trajectory either by a small star or by a small circle. The qualitative behaviour is very close to the exact solution. However, we can observe some changes of direction mainly determined by the error in the approximation of the value function and by the discretization of the control sets A and B. The circle centered at the last position of the pursuer has radius ϵ so that one can visually check that the evader has been captured.

Test 2
Let the dynamics of the pursuer and the evader be given by

$$\begin{cases} \dot{z}_1 & = v_1\, a \\ \dot{z}_2 & = v_2\, b \end{cases} \tag{18}$$

where a, b, v_1, v_2 have the same meaning as in Test 1 and $B \equiv B(0,1)$. The set of admissible control for the pursuer has now been restricted to $A \equiv \{a \in B(0,1) : a \cdot (0,1) \geq -\sqrt{2}/2\}$. We study the game in the square $Q \equiv [-1,1]^4$ and set $\mathcal{T}$ and $\mathcal{T}_\epsilon$ as in Test 1. It should be noted that the dynamics of the pursuer now have a "forbidden sector" pointing south with angular width $\theta = \pi/2$. This does not imply that the pursuer cannot move south, since he can go south zig-zagging from south-west to south-east. Of course, this means that he will make a longer path with respect to the evader if he has to go south. For example, if the pursuer has to move south for 1 unit his path will be made by two segments of length $\sqrt{2}/2$ so that his optimal time corresponding to that path will be $t^* = \sqrt{2}/v_1$. It should also be noted that the pursuer can reach any point in the plane and that he has several equivalent optimal paths to reach any single point in the southern hemisphere (they depend on the number of switchings in his direction). We can also compute the exact solution, e.g. when $\epsilon = 0$.

Figure 6 shows the approximate value function $w(0, 0, x_3, x_4)$ and Figures 7 and 8 show the approximate optimal trajectories corresponding to various initial positions. In particular, Figures 7a and 8a show how the pursuer can follow the evader in a southerly direction still respecting the constraints imposed on his dynamics. Figures 7b shows the effect of the boundary condition $w = 1$: the evader wins just leaving the domain Q. Finally, Figure 8b shows a case where the behaviour of the two players is the same as in Test 1.

Test 3

Let the dynamics of the pursuer and the evader be given by

$$\begin{cases} \dot{z}_1 &= v_1\, a \\ \dot{z}_2 &= v_2\, b \end{cases} \tag{19}$$

where a, b, v_1, v_2 have the same meaning as in Test 1 and $B \equiv B(0, 1)$. The set of admissible control for the pursuer has now been restricted to $A \equiv \{a \in B(0, 1) : a \cdot (0, 1) \geq 0\}$. We study the game in the square $Q \equiv [-1, 1]^4$ and set $\mathcal{T}$ and $\mathcal{T}_\epsilon$ as in Test 1. It should be noted that the dynamics of the pursuer now have a "forbidden sector" pointing south with angular width $\theta = \pi$. In this test the pursuer cannot reach the points in his southern hemisphere. The value function $w(0, 0, x_3, x_4)$ is certainly equal to 1 below the line $x_4 = -\epsilon$ and it is 0 on this line at $x_3 = 0$, so it is discontinuous. It should be noted that the value function is equal to 1 also at some points of the northern hemisphere. In fact, if the evader starts far enough from the origin, he can reach the southern hemisphere before being captured.

Figure 9 shows the value function for the pursuer fixed at the origin which has been computed over a grid of $29^4 = 707281$ nodes for $h = 0.1$. Figure 10a shows the approximate optimal trajectories in the interesting situation mentioned above: the evader start in the northern hemisphere and he wins because he can reach the southern hemisphere before being captured.

Finally, Figure 10b shows what happens when the initial position of the evader is in the souther hemisphere. The pursuer tries to follow him but he cannot move south. In this situation there is a big ambiguity since the evader can win by simply remaining in the southern hemisphere (e.g. he can even decide to stand still in the initial position). The algorithm compares all the different possibilities and, if some of them are equivalent, it chooses the last one. This is why the evader starts moving to the right. Note that the approximated strategy of the pursuer is coherent with the strategy of the evader and keeps them at the minimal distance.

Acknowledgments. We wish to thank Monica Marinucci and Piero Lanucara (CASPUR) for their contribution to the development of the code which produced the numerical results and the IBM-ECSEC for its technical support.

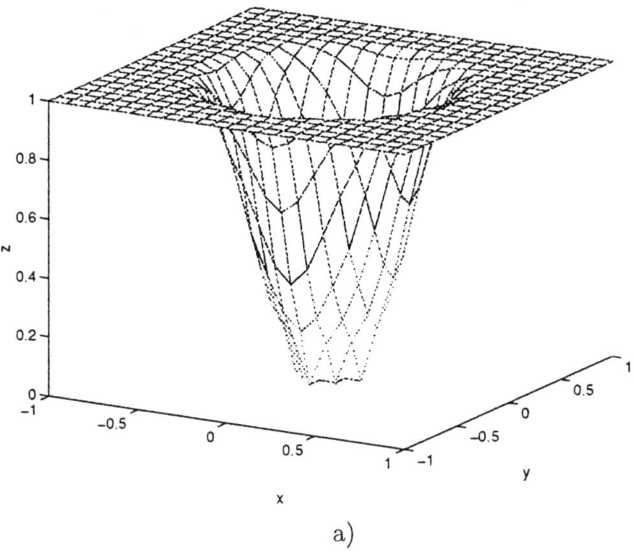

a)

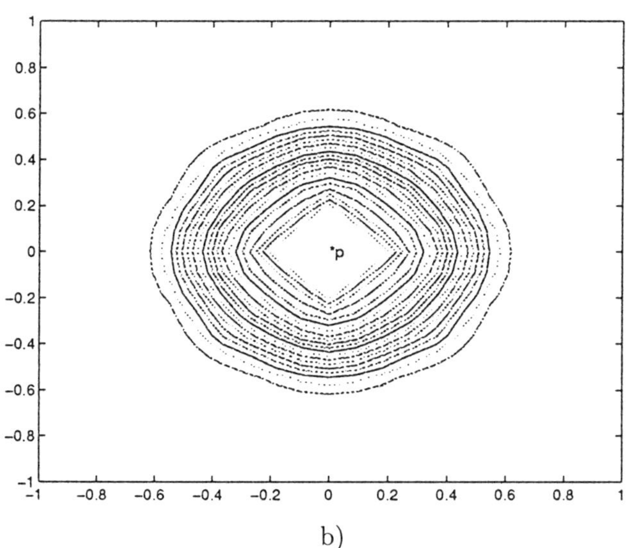

b)

Figure 1. Test 1: velocities $v_1 = 2$, $v_2 = 1$.
Nodes=23^4, $k = 0.09$, $h = 0.1$, $k/h = 0.9$, $\epsilon = 0.2$, controls=(41,41).
a) the approximate value function $v(0, 0, x_3, x_4)$
b) its level curves.

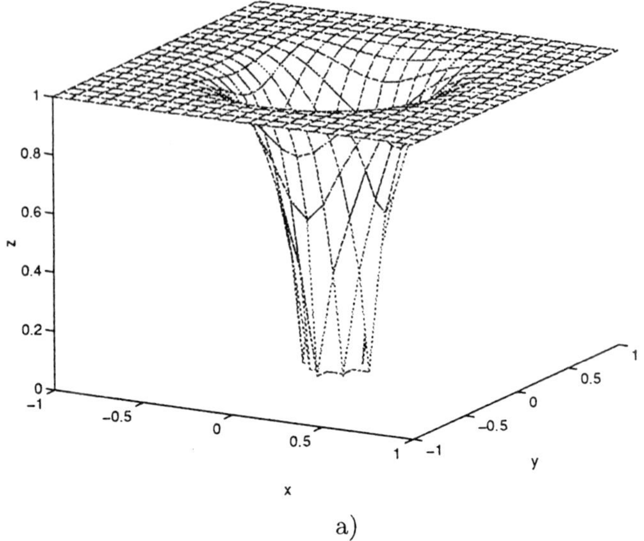

a)

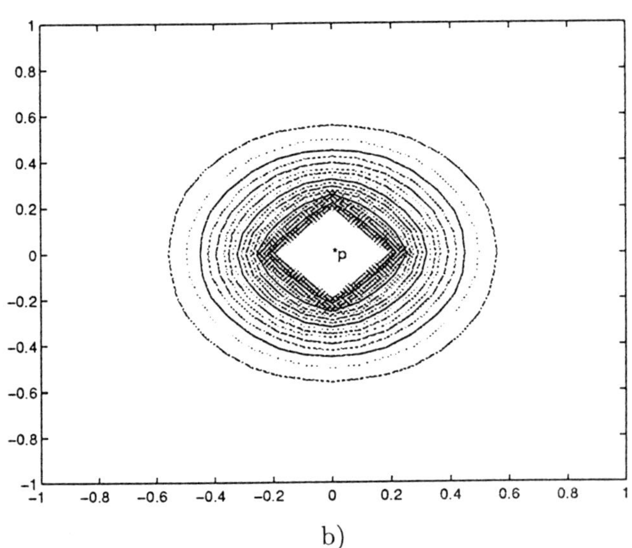

b)

Figure 2. Test 1: velocities $v_1 = 2, v_2 = 1$
Nodes=23^4, $k = 0.09$, $h = 0.03$, $k/h = 3$, $\epsilon = 0.2$, controls=(41,41).
a) the approximate value function $v(0, 0, x_3, x_4)$
b) its level curves.

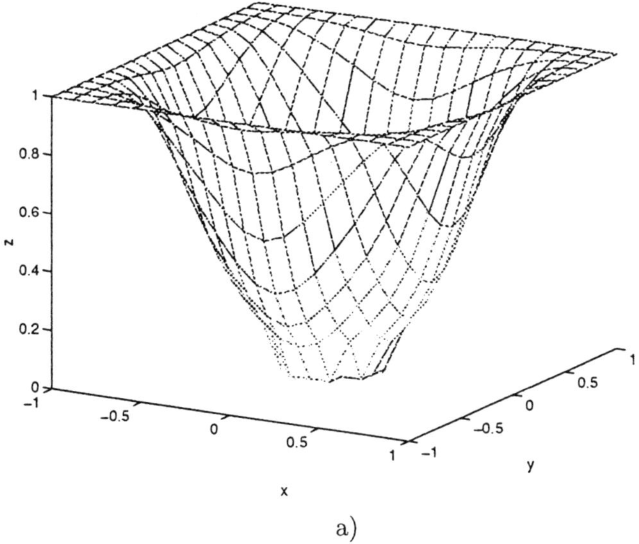

a)

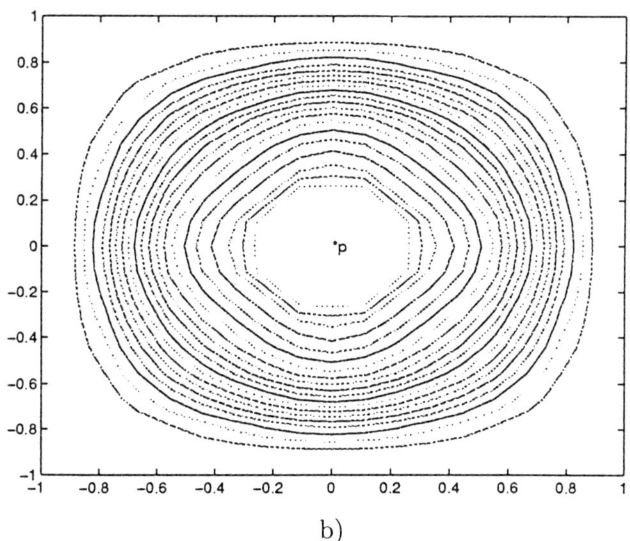

b)

Figure 3. Test 1: velocities $v_1 = 2, v_2 = 1$.
Nodes=19^4, $k = 0.11$, $h = 0.1$, $k/h = 1.1$, $\epsilon = 0.25$, controls=(65,65).
a) the approximate value function $v(0, 0, x_3, x_4)$
b) its level curves.

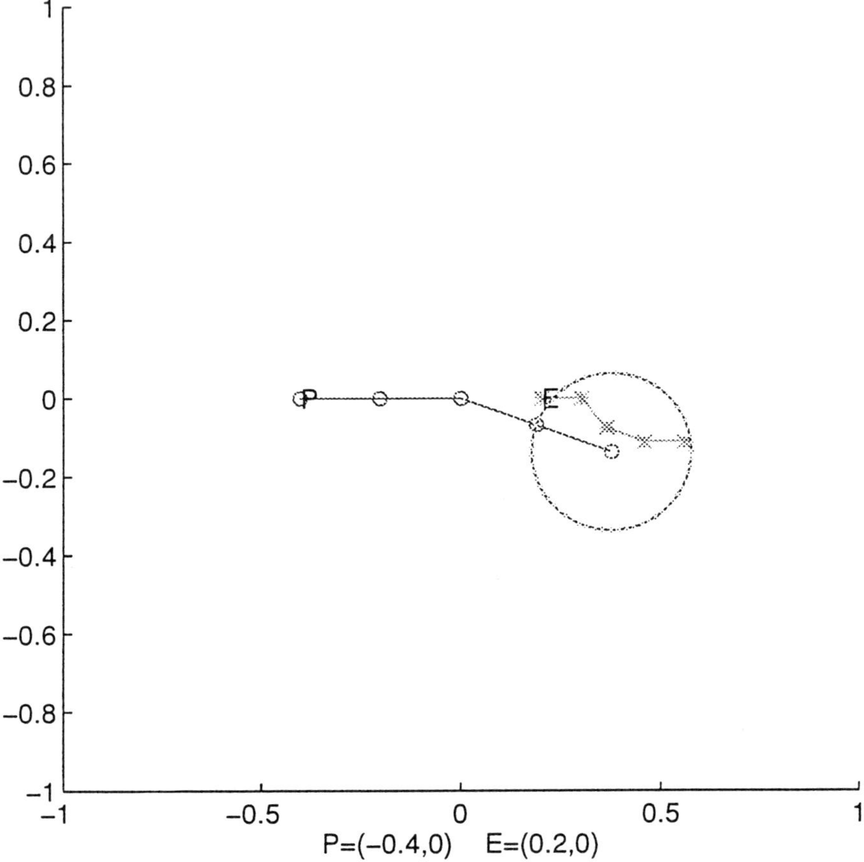

Figure 4. Test 1: velocities $v_1 = 2, v_2 = 1$.
Nodes=23^4, $k = 0.09$, $h = 0.1$, $\epsilon = 0.2$, controls=(41,41).
Approximate optimal trajectories: P captures E in 5 moves

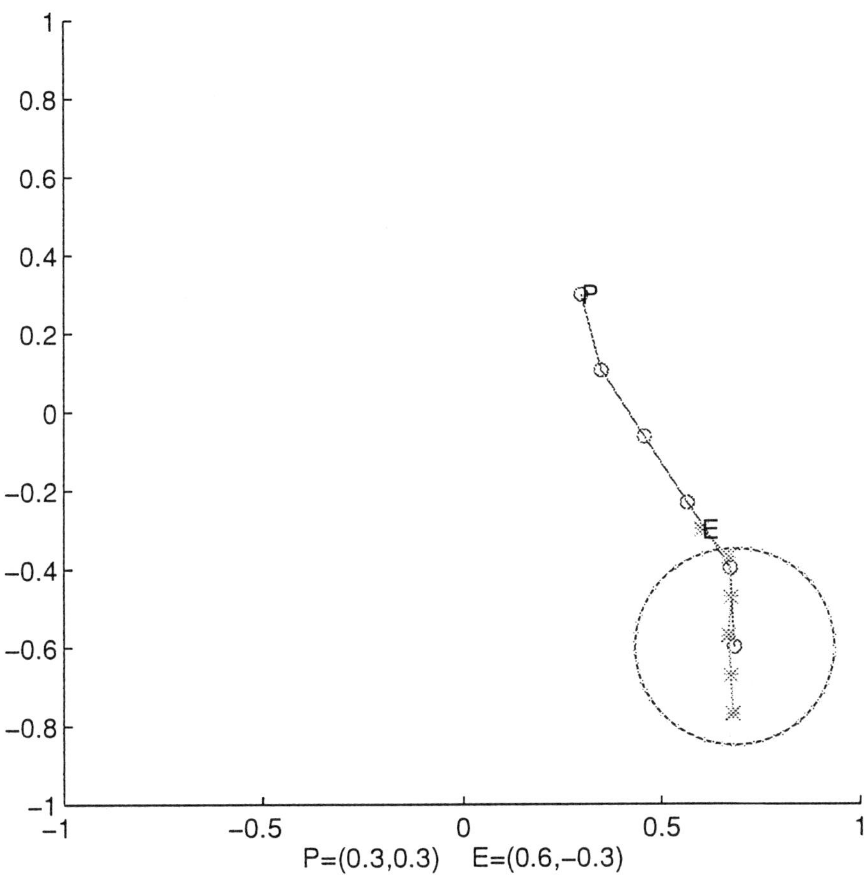

Figure 5. Test 1: velocities $v_1 = 2, v_2 = 1$.
Nodes=19^4, $k = 0.11$, $h = 0.1$, $\epsilon = 0.25$, controls=(65,65).
Approximate optimal trajectories: P captures E in 6 moves

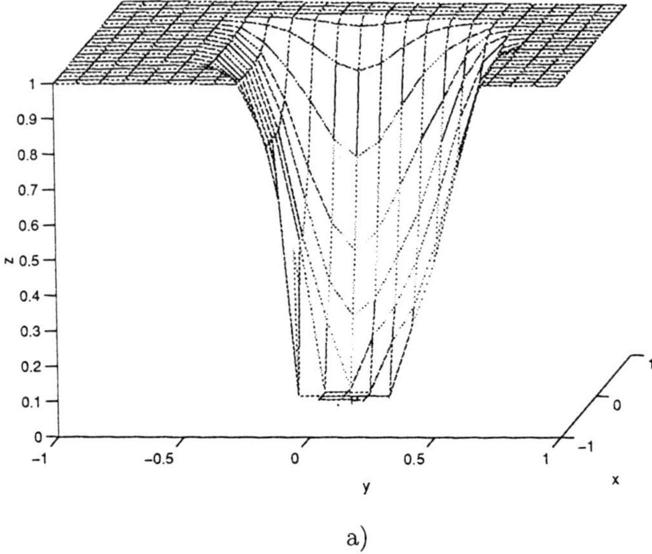

a)

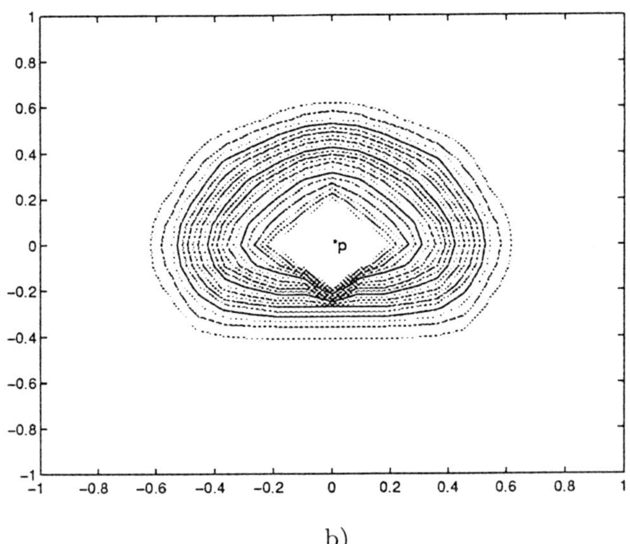

b)

Figure 6. Test 2: velocities $v_1 = 2, v_2 = 1$.
Nodes=23^4, $k = 0.09$, $h = 0.1$, $\epsilon = 0.2$, controls=(28,41).
a) the approximate value function $v(0, 0, x_3, x_4)$
b) its level curves

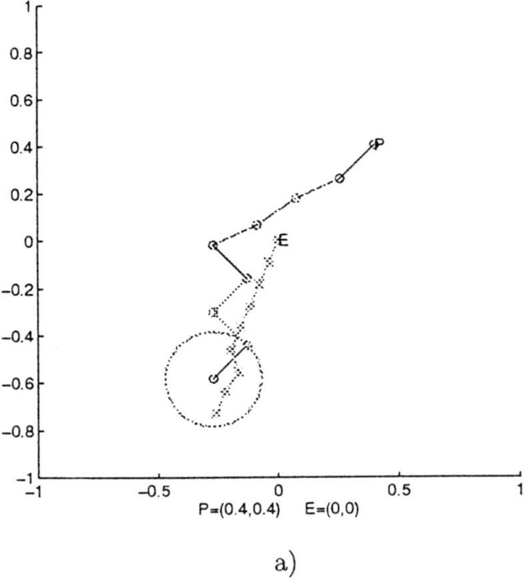

a)

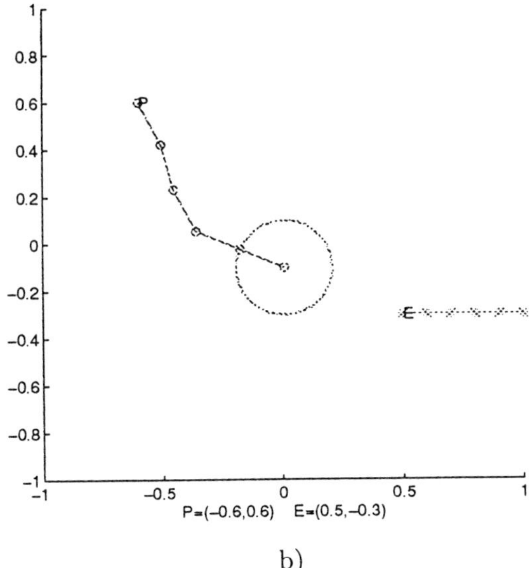

b)

Figure 7. Test 2: velocities $v_1 = 2, v_2 = 1$.
Nodes$=23^4$, $k = 0.09$, $h = 0.1$, $\epsilon = 0.2$, controls$=(28,41)$.
Approximate optimal trajectories
a) P captures E in 9 moves.
b) E wins going out of Q

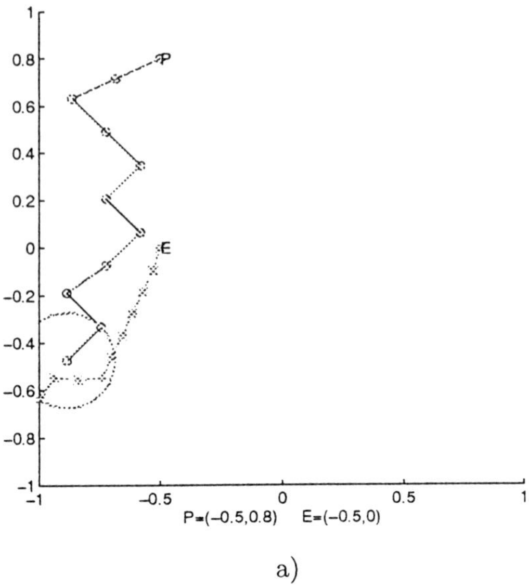

a)

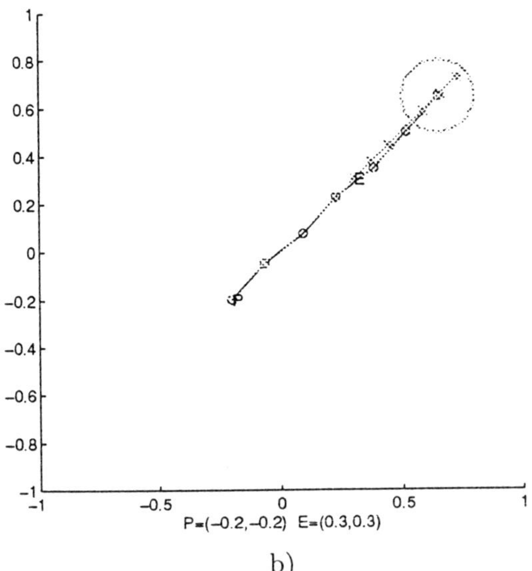

b)

Figure 8. Test 2: velocities $v_1 = 2, v_2 = 1$.
Nodes=23^4, $k = 0.09$, $h = 0.1$, $\epsilon = 0.2$, controls=(28,41).
Approximate optimal trajectories
a) P captures E in 11 moves
b) P captures E in 7 moves

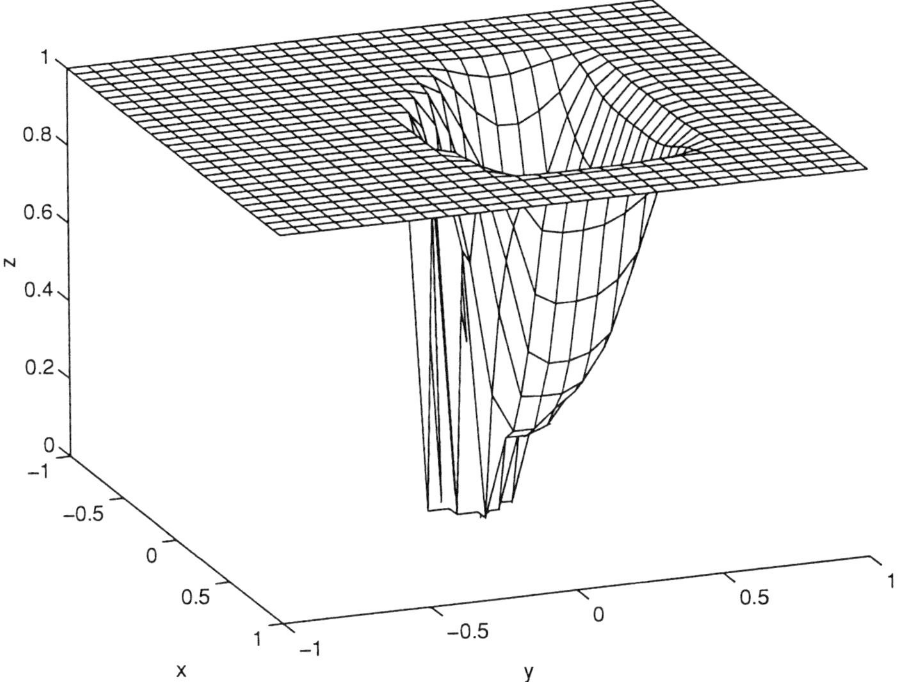

Figure 9. Test 3: velocities $v_1 = 2, v_2 = 1$.
Nodes=29^4, $k = 0.07$, $h = 0.1$, $\epsilon = 0.15$, controls=(20,41).
The approximate value function $v(0, 0, x_3, x_4)$

                                                        M. Bardi, S. Bottacin and M. Falcone

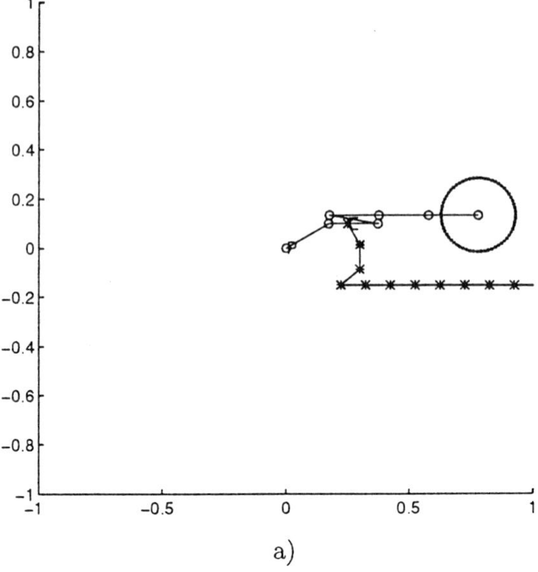

a)

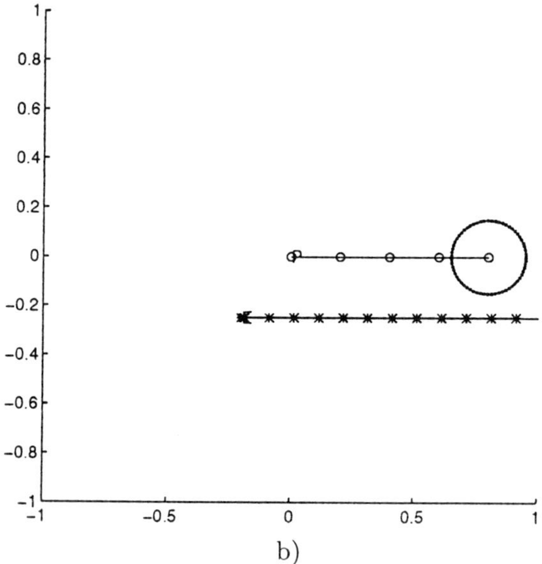

b)

Figure 10. Test 3: velocities $v_1 = 2, v_2 = 1$.
Nodes=29^4, $k = 0.07$, $h = 0.1$, $\epsilon = 0.15$, controls=(20,41).
a)$P = (0,0)$, $E = (0.25, 0.1)$: E enters the southern hemisphere
b)$P = (0,0)$, $E = (-0.2, -0.25)$: E starts in the southern hemisphere
and wins

REFERENCES

[1]   M. Bardi: *Viscosity solutions of Isaacs'equation and existence of a value*, in "Lectures on Games", G.Ricci and C.Torricelli eds., Springer, to appear.

[2]   M. Bardi, S. Bottacin: *Discontinuous solutions of degenerate elliptic boundary value problems*, preprint Dipartimento di Matematica, Università di Padova, 22 (1995).

[3]   M. Bardi, S. Bottacin: work in preparation.

[4]   M. Bardi, I. Capuzzo Dolcetta: *Optimal control and viscosity solutions of Hamilton-Jacobi-Bellman equations*, Birkhäuser, to appear.

[5]   M. Bardi, M. Falcone, P. Soravia: *Fully discrete schemes for the value function of pursuit-evasion games*, Advances in dynamic games and applications, T.Basar and A.Haurie eds., Birkhäuser, (1994), 89–105.

[6]   M. Bardi, P. Soravia: *A PDE framework for games of pursuit-evasion type*, in "Differential Games and Applications," T.Başar and P.Bernhard, eds., Lecture Notes in Control and Information Sciences, Springer-Verlag, (1989), 62–71.

[7]   M. Bardi, P. Soravia: *Hamilton-Jacobi equations with singular boundary conditions on a free boundary and applications to differential games*, Trans. Amer. Math. Soc. 325, (1991), 205–229.

[8]   M. Bardi, P. Soravia: *Approximation of differential games of pursuit-evasion by discrete-time games*, Differential games-developments in modelling and computation, R.P. Hamalainen and H.K.Ethamo eds., Lecture Notes in Control and Information Sciences 156, Springer-Verlag, (1991), 131–143.

[9]   M. Bardi, P. Soravia: *A comparison result for Hamilton-Jacobi equations and applications to some differential games lacking controllability*, Funkcial. Ekvac. 37, (1994), 19–43.

[10]   M. Bardi, V. Staicu: *The Bellman equation for time-optimal control of noncontrollable, nonlinear systems*, Acta Appl. Math. 31, (1993), 201–223.

[11]   G. Barles: *Discontinuous viscosity solutions of first order Hamilton-Jacobi equations: a guided visit*, Nonlinear Anal. T.M.A, 20, (1993), 1123–1134.

[12]   G. Barles, B. Perthame: *Discontinuous solutions of deterministic optimal stopping time problems*, RAIRO Model. Math. Anal. Num. 21, (1987), 557–579.

[13]   G. Barles, P.E. Souganidis: *Convergence of approximation schemes for fully nonlinear second order equations*, Asymptotic Anal. 4 (1991), 271–283.

[14]   E.N. Barron, R. Jensen: *Semicontinuous viscosity solutions of Hamilton-Isaacs equation with convex Hamiltonians*, Comm. Partial Differential Equations 15, (1990), 1713–1742.

[15]   T. Başar, G.J. Olsder: *Dynamic noncooperative game theory*, Academic Press, 1982.

[16]   S. Bottacin: *Soluzioni discontinue del problema di Dirichlet per equazioni ellittiche degeneri non lineari*, Dissertation, Università di Padova, March 1994.

[17]   I. Capuzzo Dolcetta, M. Falcone: *Viscosity solutions and discrete dynamic programming*, Ann. Inst. H. Poincaré Anal. Non Lin. 6, (Supplement), (1989), 161–183.

[18]   M. Crandall, H. Ishii, P. Lions: *User's guide to viscosity solutions of second order partial differential equations*, Bull. of the Amer. Math. Soc. 27, (1992), 1-67.

[19]   P. Cardaliaguet, M. Quincampoix, P. Saint-Pierre: *Optimal times for constrained controlled systems without local controllability*, C. R. Acad. Sci. Paris, 318, Serie I,(1994), 607-612.

[20]   P. Cardaliaguet, M. Quincampoix, P. Saint-Pierre: *Optimal times for constrained non-linear control problems without local controllability*, preprint CEREMADE 9364.

[21]   R. Elliott: *Viscosity solutions and optimal control*, Longman, (1987).

[22]   L.C. Evans, P.E. Souganidis: *Differential Games and Representation Formulas for Solutions of Hamilton-Jacobi-Isaacs Equations*, Indiana Univ. Math. J. 33, (1984), 773–797.

[23]   M. Falcone, R. Ferretti: *Discrete time high-order schemes for viscosity solutions of Hamilton-Jacobi-Bellman equations*, Numerische Mathematik, 67 (1994), 315–344.

[24]   M. Falcone, P. Lanucara: *Vector and parallel algorithms for Hamilton–Jacobi–Bellman equations*, in preparation.

[25]   W. Fleming: *The convergence problem for differential games*, Math. Anal. and Appl. J. 3, (1961), 102-116.

[26]   W.H. Fleming, P.E. Souganidis: *On the existence of value functions of two-players, zero-sum stochastic differential games*, Indiana Univ. Math. J. 38, (1989), 293–314.

[27]   H. Frankowska: *Lower semicontinuous solutions of Hamilton-Jacobi-Bellman equations*, SIAM J. Control Optim. 31, (1993), 257–272.

[28]   R. Isaacs: *Differential games*, John Wiley & Sons, New York, 1965.

[29]   H. Ishii: *Perron's method for Hamilton-Jacobi equations*, Duke Math. J. 55, (1987), 369–384.

[30]   N. Krasovskii, A.I. Subbotin: *Game-theoretical control problems*, Springer-Verlag, New York, 1988.

[31]   M. Marinucci: *Giochi differenziali: metodi numerici recenti e loro parallelizzazione*, Tesi di Laurea, Rome, 1995.

[32]   O. Pourtallier, M. Tidball: *Approximation of the value function for a class of differential games with target*, in Preprint Volume of the 6th International Symposium on Dynamic Games and Applications, M. Breton and G. Zaccour, eds., (1994), 263–272.

[33]   I. Rozyev, A.I. Subbotin: *Semicontinuous solutions of Hamilton-Jacobi equations*, Prikl. Matem. Mekhan. U.S.S.R. 52, (1988), 141-146.

[34]   P. Soravia: *The concept of value in differential games of survival and viscosity solutions of Hamilton-Jacobi equations*, Differential Integral Equations 5, (1992), 1049–1068.

[35]   P. Soravia: *Pursuit-evasion problems and viscosity solutions of Isaacs equation*, SIAM J. Control Optim. 31, (1993), 604–623.

[36]   P. Soravia: *Discontinuous viscosity solutions to Dirichlet problems for Hamilton-Jacobi equations with convex Hamiltonians*, Comm. Partial Diff. Equations 18, (1993), 1493-1514.

[37]   P.E. Souganidis: *Max-min representations and product formulas for the viscosity solutions of Hamilton-Jacobi equations with applications to differential games*, Nonlinear Anal. T.M.A. 9, (1985), 217–257.

[38]   A.I. Subbotin: *Generalization of the main equation of differential game theory*, Optim. Th. Appl.J. 43, (1984), 103–133.

[39]   A.I. Subbotin: *Existence and uniqueness results for Hamilton-Jacobi equations*, Nonlinear Anal. T.M.A. 16, (1991), 683–699.

[40]   A.I. Subbotin: *Minimax and viscosity solutions of Hamilton-Jacobi equations*, Nauka, (1991).(In Russian)

[41]   A.I. Subbotin: *On a property of the subdifferential*, Math. USSR Sbornik 74, (1993), 63–78.

[42]   A.I. Subbotin: *Discontinuous solutions of a Dirichlet type boundary value problem for first order P.D.E.*, Russian Numer. Anal. Math. Modelling J. 8, (1993), 145–164.

Undiscounted Zero Sum Differential Games with Stopping Times

Mabel M. Tidball*

Facultad de Ciencias Exactas, Ingeniería y Agrimensura
Universidad Nacional de Rosario
Pellegrini 250, (2000) Rosario
Argentina

Abstract

We propose a discretization scheme for an undiscounted zero sum differential game with stopping times. The value function of the original problem satisfies an integral inequality of Isaacs type that we can discretize using finite difference or finite element techniques.

The fully discrete problem defines a stochastic game problem associated with the process, which may have, in general, multiple solutions. Among these solutions there exists one which is naturally associated with the value function of the original problem.

The main contribution of this paper is the complete characterization of the set of solutions and the description of a procedure to identify the desired solution. We also present accelerated algorithms in order to efficiently compute the discrete solution.

1   Introduction

We consider an undiscounted zero sum differential game problem with stopping times. The same problem with discount rate ($\lambda > 0$) was studied in [20] and a discretization procedure was described in [22].

For the case $\lambda > 0$ it is known, see [20], that the value function of the game is a Hölder continuous function; but in the present case ($\lambda = 0$), we lose, in general, this regularity property. In spite of this discouraging result we can prove that the value function is always well defined and unique when the stopping costs verify a natural condition which correspond to a non trivial game; but even though we have regular data, the value function can be discontinuous. Works related with this type of phenomenon can be found in [1]. Problems without discounting have been studied in [18] for ergodic control problems.

---

*This work was partially done during a visit of the author at INRIA - Sophia Antipolis, France

The aim of this work is to obtain a numerically computable discrete solution of this game and to indicate precisely when this solution represents the value function of our problem.

By the dynamic programming principle, we can prove that the value function satisfies an integral inequality system of Isaacs type (see [10], [11], [20], [22]). Since an exact solution of this type of problem is usually impossible, we must discretize the problem and solve it numerically. In order to obtain a discrete problem we approximate, using finite element techniques (see [21]), these inequalities. The discrete problem that we obtain may have in general multiple solutions. It defines a stochastic game problem with stopping time associated with the underlying Markov process.

Among these solutions, there exists one that is naturally associated with the value function of the original problem (recall that the solution of our problem is unique). In this work we completely characterize the set of solutions of the discrete problem and we describe a procedure to identify this particular discrete solution. We present, in order to compute it, accelerated algorithms based on the techniques given in [9] and [22].

We also present special cases where we can prove regularity properties of the value function and the existence of a unique solution of the discrete problem. In these cases we can prove the convergence, in the viscosity sense (see [8]), of the discrete solution to the value function of the original problem.

2 Description of the problem

2.1 The differential game problem

In this differential game problem, the decision variables used by the players are the stopping times for the uncontrolled system. The evolution of the system is described by an ordinary differential equation of the form:

$$\frac{dy(t)}{dt} = g(y(t)); \qquad y(0) = x \tag{1}$$

with $x \in \Omega \subset \mathbb{R}^n$, an open bounded domain. We denote by τ_1 and τ_2 the stopping times chosen by players 1 and 2 respectively. To those stopping times is associated a payoff J which takes the form:

$$J(x, \tau_1, \tau_2) = \tag{2}$$

$$\overline{\lim_{\tau \to \infty}} \int_0^{\tau_1 \wedge \tau_2 \wedge \tau \wedge \theta} f(y(s))ds + \phi_1(y(\tau_1))\chi_{\tau_1 < \tau_2 \wedge \theta} + \phi_2(y(\tau_2))\chi_{\tau_2 < \tau_1 \wedge \theta}$$

$$+ \phi_0(y(\theta))\chi_{\tau_2 \wedge \tau_1 > \theta}.$$

In (2), χ_E is the characteristic function of set E, $\theta(x)$ is the first time when the trajectory, solution of (1) with initial condition x reaches $\partial\Omega$ ($\theta(x) = \infty$ when the trajectory always remains in Ω, that is:

$$\theta(x) = \inf\{t : y(t) \notin \Omega\}.$$

ϕ_0 is the final cost to pay if the game stops because the system leaves Ω , f is the instantaneous cost, ϕ_i, $i = 1, 2$ is the final cost if the game stops because player i decides first to stop the game, with

$$\phi_2(x) \geq \phi_1(x) \qquad \forall x \in \Omega. \tag{3}$$

As we have a zero sum game, the goal of the first player is to maximize the payoff J and the goal of the second player is to minimize it. Our objective is to find (if they exist) $\bar{\tau}_1$ and $\bar{\tau}_2$ such that:

$$V(x) = \sup_{\tau_1} \inf_{\tau_2} J(x, \tau_1, \tau_2) = J(x, \bar{\tau}_1, \bar{\tau}_2)$$

where V is the value function of the game and $\bar{\tau}_1$ and $\bar{\tau}_2$ are optimal strategies or saddle points of the functional J.

Inequality (3) indicates that the game is not a trivial game because no player can obtain advantages of priority of actions. It plays a key role in the proof of the existence of the value function but it does not represent a restriction as it is shown in [20].

We assume that:

$$(H) \begin{cases} g, & f, & \phi_i, & i = 0, 1, 2 \text{ are bounded functions in } \Omega \\ g, & f, & \phi_i, & i = 0, 1, 2 \text{ are Lipschitz continuous functions in } \Omega. \end{cases}$$

Remark 2.1 : *In (2) we do not consider $\tau_1 = \tau_2$ because finite strategies $\tau_1 = \tau_2$ are not realistic strategies for both players.*

Remark 2.2 : *The limsup in (2) means that player 2 must minimize his worst situation.*

It is necessary to consider the limsup (or the liminf, in case we study the worst situation for player 1) in order to have a well defined payoff functional J. The fact that a limit need not exist can be easily seen in the following example. Let $x = (x_1, x_2)$; $\phi_2(x) = M$, $\phi_1(x) = -M$, $f(x) = x_1$, $\Omega = (-2, 2)$ and

$$\frac{dx_1}{dt}(t) = x_2(t); \qquad x_1(0) = 1, \qquad \frac{dx_2}{dt}(t) = -x_1(t); \qquad x_2(0) = 0.$$

It is possible to choose M large enough such that $\tau_1 = \tau_2 = +\infty$. In this case $\int_0^\infty f(y(s))ds$ does not exist.

2.2   Properties of the value function

We can define the lower and upper values of the game $\underline{V}$ and $\overline{V}$ respectively, in the following way:

$$\underline{V}(x) = \sup_{\tau_1} \inf_{\tau_2} J(x, \tau_1, \tau_2) \quad \overline{V}(x) = \inf_{\tau_2} \sup_{\tau_1} J(x, \tau_1, \tau_2).$$

If (H) and (3) are valid we could prove, following the techniques used in [20], that $\underline{V} = \overline{V} = V$. V will be the unique value function of the game. We omit this proof for the sake of brevity.

Remark 2.3 : *In terms of the value function V, ϵ-optimal stopping times $\bar{\tau}_1^\epsilon$, $\bar{\tau}_2^\epsilon$ can be found, (see [10]), i.e. $\bar{\tau}_1^\epsilon$, $\bar{\tau}_2^\epsilon$, such that:*

$$J(x, \tau_1, \bar{\tau}_2^\epsilon) - \epsilon \leq J(x, \bar{\tau}_1^\epsilon, \bar{\tau}_2^\epsilon) \leq J(x, \bar{\tau}_1^\epsilon, \tau_2) + \epsilon \quad \forall \tau_1, \tau_2.$$

They are defined in the following way (see [11]):

$$\bar{\tau}_1^\epsilon = \inf\{s / V(y(s)) \leq \phi_1(y(s)) + \epsilon\} \qquad \bar{\tau}_2^\epsilon = \inf\{s / V(y(s)) \geq \phi_2(y(s)) - \epsilon\}.$$

Remark 2.4 : *V can be a discontinuous function as we can see in the following example.*

Let $\Omega = [0, \pi]$ and the dynamics of the system be given by: $\dfrac{dy}{dt}(t) = y(t)$, $y(0) = x$, $f(x) = 10\ \sin x$, $\phi_0 = 0$, $\phi_1(x) = 2x - 1$, $\phi_2(x) = 2x + 1$; then:

$$V(x) = \begin{cases} \phi_2(x) & \text{if} \quad x \neq 0 \\ 0 & \text{if} \quad x = 0. \end{cases}$$

Isaacs inequality. Multiplicity of solutions.
By the dynamic progamming principle we obtain that the function V satisfies the following Isaacs inequality:

$$\phi_1(x) \leq V(x) \leq \phi_2(x) \quad \forall x \in \Omega,$$

$$V(x) < \phi_2(x) \Rightarrow \exists \ \delta_x > 0 \ / \ \forall \delta < \delta_x, \quad V(x) \geq \int_0^\delta f(y(s))ds + V(y(\delta)),$$

$$V(x) > \phi_1(x) \Rightarrow \exists \ \delta_x > 0 \ / \ \forall \delta < \delta_x, \quad V(x) \leq \int_0^\delta f(y(s))ds + V(y(\delta)).$$

Then for δ_x small enough we can insure that $V(x)$ is the fixed point of the following operator:

$$V(x) = (M_{\delta_x} V)(x) = A_I \left\{ \int_0^{\delta_x} f(y(s))ds + V(y(\delta)) \right\} \tag{4}$$

where A_I denotes the projection on interval $I=[\phi_1(x),\,\phi_2(x)]$ i.e:

$$A_I(a) = \begin{cases} \phi_1(x) & \text{if } a \leq \phi_1(x) \\ \phi_2(x) & \text{if } a \geq \phi_2(x) \\ a & \text{everywhere else.} \end{cases}$$

If we consider x fixed and $y_x(t)$ the solution of (1) with initial condition x, we have that (4) has infinitely many solutions forming a manifold with one-dimension. Indeed, we define the following equivalence relation:

$$x \sim x' \text{ if there exists } t_1 \geq 0 \text{ such that } y_x(t_1) = x' \text{ or } y_{x'}(t_1) = x.$$

Let C_x be the equivalence classes; we have that if $V(.)$ is a solution of (4), and φ is a function such that $\varphi : C_x \to \mathbb{R}$, then $\bar{V}(x) = V(x) + \varphi(C_x)$ is also a solution if $\bar{V}(x) \in [\phi_1(x), \phi_2(x)]$.

Remark 2.5 : *Characterization of the set where V is a discontinuous function.*
Define, $\tau(x) = \bar{\tau}_1 \wedge \bar{\tau}_2 \wedge \theta$ if there exist finite optimal strategies; $\tau(x) = +\infty \wedge \theta$ if there do not exist optimal strategies; $D = \{x \in \Omega : V \text{ is discontinuous at } x\}$ and $\Omega_\tau = \{x \in \Omega : \tau(x) \leq \tau\}$. Then it is easy to see that:

$$D \subset \left(\overline{\bigcup_{\tau>0} \Omega_\tau} \right)^c .$$

Remark 2.6 : *Examples where $D = \emptyset$*
a) Problems with boundary.
Consider the case where the following condition holds (see [11] and [19]):

$$\theta(x) < +\infty \qquad \forall x$$

$$\exists a > 0, \gamma > 0 \quad / \quad d(y(s), \partial\Omega) \leq a \Rightarrow \frac{d}{ds}d(y(s), \partial\Omega) \leq -\gamma. \tag{5}$$

If (5) holds it is proved in [11] that there exists a unique Lipschitz continuous value function V and that there exists optimal strategies. This property arises because the exit time is a Lipschitz continuous function of x.
b) Some ergodic processes.
It is possible to give a precise characterization of the solution of our problem for a process such that the system remains in a compact manifold. In this case the solution of (1) defines in Ω a semigroup of translations $S(t)$. If we assume that this semigroup has a unique invariant measure, μ, that satisfies:

$$\lim_{\tau \to \infty} \frac{1}{\tau} \int_0^\tau f(y_x(s))\,ds = \int_\Omega f(z)d(\mu(z)) =: <\mu, f> \tag{6}$$

we say that the process is ergodic. Moreover

$$< \mu, f > \neq 0 \implies \lim_{\tau \to \infty} \int_0^\tau f(y_x(s))ds = \infty \quad sign\left\{\int_\Omega f(z)d(\mu(z))\right\}.$$

We can easily prove that this condition implies that there exist finite optimal strategies.

In order to obtain a numerical solution of the original problem we analyze, in the following two sections, a time discretization and a space discretization which uses finite element techniques (see [7], [13], [14], [15], [16] and [21]).

3   Time discretization

To obtain a time discretization (see [7] for control problems and [1] for games problems), we consider $h > 0$, a uniform partition of $[0, +\infty)$ in intervals of same length h and the Euler approximation for the dynamics. Then, the Isaacs equation for time discretization becomes in this case:

$$V^h(x) = (M^h V^h)(x) = \begin{cases} A_I\left[hf(x) + V^h(x + hg(x))\right] & \text{if} \quad x + hg(x) \in \Omega \\ A_I\left[\phi_0(x)\right] & \text{if} \quad x + hg(x) \notin \Omega. \end{cases} \tag{7}$$

It is easy to see that $V^h \in B(\Omega)$, the set of bounded functions in Ω.

We can define a game problem with discrete stopping times associated with our real problem. The discrete time value function has the following form:

$$J^h(x, \tau_1, \tau_2) =$$

$$\overline{\lim_{\nu \to \infty}} \sum_{\nu=1}^{\tau_1 \wedge \tau_2 \wedge \nu h \wedge \theta} f(y_x^h(\nu)) + \phi_1(y(\tau_1))\chi_{\tau_1 < \tau_2 \wedge \theta} + \phi_2(y(\tau_2))\chi_{\tau_2 < \tau_1 \wedge \theta}$$

$$+ \phi_0(y(\theta))\chi_{\tau_2 \wedge \tau_1 > \theta}$$

where $\theta, \tau_1, \tau_2 \in hZ^+$ and the discrete evolution is:

$$y_x^h(0) = x; \qquad y_x^h(\nu + 1) = y_x^h(\nu) + hg(y_x^h(\nu)). \tag{8}$$

For the sake of simplicity we denote:

$$y^\nu(t) = y_x^h(t) = \left(1 - \frac{t - \nu h}{h}\right) y_x^h(\nu) + \frac{t - \nu h}{h} y_x^h(\nu+1) \quad \text{if} \quad t \in [\nu h, (\nu+1)h]. \tag{9}$$

The time discrete value function is:

$$V^h(x) = \inf_{\tau_1} \sup_{\tau_2} J^h(x, \tau_1, \tau_2). \tag{10}$$

The following theorem characterizes the set of all solutions of (7).

Theorem 3.1 *Equation* (7) *has at least one solution. Moreover:*
i) V^h *given by* (10) *is one of them.*
ii) The following iteration

$$w^{n+1} = M^h w^n, \qquad w^0 \geq \phi_2$$

converges to the maximal solution that we call $\bar{w}$.
iii) $w^{n+1} = M^h w^n$, $\quad w^0 \leq \phi_1$ *converges to the minimal solution,* $\underline{w}$.
iv) If $w(.)$ *is a solution,* $C_x^h = \left\{ x' = y_x^h(t), t \geq 0 \right\}$, $\varphi : C_x^h \to \mathbb{R}$, *then* $\tilde{w}(x) = w(x) + \varphi(C_x^h)$ *is also a solution if* $\tilde{w}(x) \in [\phi_1(x), \phi_2(x)]$.

Proof: We omit the proof of i), which is technical and does not contribute to the main purpose of this paper. We are going to prove item ii), and the other proofs are similar. With condition $w^0 \geq \phi_2$ the sequence $M^h w$ is monotonously decreasing. In effect $w^1 = M^h w^0 \leq \phi_2 \leq w^0$ and as A_I is a monotonous operator, we have $w^{n+1} = M^h w^n \leq M^h w^{n-1} = w^n$. Moreover the sequence is bounded from below by $- \parallel \phi_1 \parallel$. Then w^n converges to $\bar{w}$ when $n \to \infty$ and it is a solution of (7). Moreover it is the maximal solution because if $\hat{w}$ is a solution of (7) $\hat{w}(x) \leq \phi_2(x) \quad \forall x$ then $\hat{w} = M^h \hat{w} \leq \phi_2 \leq w_0$ and $\hat{w} \leq (M^h)^n w_0 = w^n, \quad \forall n$, that implies $\hat{w} \leq \bar{w}$. ■

4 Space discretization

4.1 Elements of the discrete problem

To define a fully discrete problem it is necessary to discretize the set Ω, the space $B(\Omega)$ and the operator defined in (7).
We denote by k, $k \downarrow 0$, the space parameter of the discretization approximating Ω. We consider a polyhedron $\Omega_k = \bigcup_j S_j^k$, where S_j^k is a simplex; $int S_i^k \cap int S_j^k = \emptyset$ for $i \neq j$, $\max_j (diam \ S_j^k) = k$. We denote $X = \{x_i, i = 1, \ldots, N\}$ the set of vertices of Ω_k (nodes), N the cardinality of X and $\partial \Omega_k^+ = \{x_i \in X / x_i + hg(x_i) \notin \Omega_k\}$.
We consider as an approximation of $B(\Omega)$ the set W_k of functions $w : \Omega_k \to \mathbb{R}$, w continuous in Ω_k, with constant partial derivatives in the interior of each simplex of Ω_k.

By definition of W_k, to obtain a solution of the discrete problem it is only necessary to obtain a solution in X.

As the elements of W^k are piecewise affine functions we can write $V(x + hg(x)) = \sum_j P_{i,j} V(x_i)$, $x_i \in X$, where $P_{i,j}$ are defined such that if $x + hg(x) \in S_j^k$, then, $x + hg(x) = \sum_j P_{i,j} x_i$, $P_{i,j} \geq 0$, $\sum_j P_{i,j} = 1$ and $P_{i,j} = 0$ if $x_i \notin S_j^k$.

We obtain the formulation of the **fully discrete problem**; (problem $Prob_k^h$) that is: Find a fixed point of the operator $M^k : W_k \to W_k$ where $(M^k V)(x_i) = A_I \left[hf(x_i) + \sum_j P_{i,j} V(x_j) \right]$.

Through the natural isomorphism $W_k \leftrightarrow \mathbb{R}^N$ we can rewrite Problem $Prob_k^h$ as: Find the fixed point of operator $M^k : \mathbb{R}^N \to \mathbb{R}^N$, with
$$(M^k w)(i) = A_I \left(Pw + F \right)(i).$$
If we call V_k^h a solution of the fully discrete problem we have:

$$V_k^h(i) = A_I \left(PV_k^h + F \right)(i) \tag{11}$$

where A_I denotes the projection on interval $[\phi_1(i), \phi_2(i)]$; P is a matrix with components $P_{i,j}$ given by:

$$\begin{cases} x_i + hg(x_i) = \sum_{j=1}^N P_{i,j} x_j, & \text{if } x_i \notin \partial\Omega_k^+ ; x_i \in \Omega_k, \ \sum_{j=1}^N P_{i,j} = 1 \\ P_{i,j} = 0 & \text{if } x_i \in \partial\Omega_k^+ \end{cases} \tag{12}$$

and F is a vector of components:

$$F(x_i) = \begin{cases} hf(x_i) & \text{if } x_i \notin \partial\Omega_k^+, x_i \in \Omega_k \\ A_I \left[\phi_0(x_i) \right] & \text{if } x_i \in \partial\Omega_k^+. \end{cases} \tag{13}$$

Because $\| P \| = 1$ the operator M is not a contractive operator and in the general case we cannot insure a unique solution of (11).

Markov chain associated to the process.

We consider a random process with finite states $y_i^{(\nu)}$, where ν is the time discrete variable, i the initial state and the probability evolution is given by the matrix P defined in (12). $P_{i,j}$ is the probability transition from node i to node j with the convention that $P_{i,i} = 1$ if $x_i \in \partial\Omega_k^+$.

In this Markovian system we can define a stochastic game problem with stopping times where the functional cost becomes:

$$J^k(i, \tau_1, \tau_2) =$$

$$\overline{\lim_{\nu \to \infty}} E \left(\sum_{\nu=1}^{\tau_1 \wedge \tau_2 \wedge \nu h \wedge \theta} F(y_i^\nu) + \phi_1(y(\tau_1)) \chi_{\tau_1 < \tau_2 \wedge \theta} + \phi_2(y(\tau_2)) \chi_{\tau_2 < \tau_1 \wedge \theta} \right.$$

$$\left. + \phi_0(y(\theta)) \chi_{\tau_2 \wedge \tau_1 > \theta} \right)$$

where, θ, τ_1, $\tau_2 \in hZ^+$ are the exit and stopping times associated with the process and F is defined in (13). For this discrete process the value function is:

$$V_h^k(i) = \inf_{\tau_2} \sup_{\tau_1} J^k(i, \tau_1, \tau_2) \qquad (14)$$

Theorem 4.1 *Equation* (11) *has at least one solution.* V_h^k, *given by* (14) *is one of its solutions.*

Again we omit the technical proof of this theorem, which does not contribute to the main purpose of this paper.

Equation (11) has in general more than one solution. We call S the set of its solutions. We are going to characterize S taking account of the fundamental structure of the graph associated with the Markov chain related to the discrete problem. We shall then identify in this set of solutions the one which represents the probabilistic formulation of the original problem.

We introduce the following equivalence relation in the set of nodes: $i \sim j$ if $i = j$ or if there exists $i_1, \ldots, i_\nu$, $\nu > 1$ such that $P_{i_m, i_{m+1}} > 0$ $\forall m = 1, \ldots, \nu$ $i_1 = i_\nu = i$ and there exists $0 < k < \nu$ with $i_k = j$.

This equivalence relation defines equivalence classes, C^r, $r = 1, \ldots, \hat{N}$. We are going to define an order relation between these classes. To that effect, define $S^\nu, \nu = 1, \ldots, \bar{\nu}$ such that:

$C^r \in S^1$ if there is no $\hat{r} \neq r$ such that $P_{i,j} > 0$, for any $i \in C^r$, $j \in C^{\hat{r}}$. $C^r \in S^\nu$ $\nu > 1$ if there exists $\hat{r} \neq r$ such that $P_{i,j} > 0$, $i \in C^r$, $j \in C^{\hat{r}}$, $C^{\hat{r}} \in S^{\nu-1}$.

If $C^r \in S^1$, $r = 1, \ldots, n_1$, we say that $C^r =: C_1^r$ is a **final class**. We call **transitory** all other classes that are not contained in S^1. If $C^r \in S^\nu$ we set $C^r = C_\nu^r$. Without loss of generality, we order the classes of S^1 putting in the first position classes $C_1^r = \{x_j\}$ corresponding to points $x_j \in \partial\Omega_k^+$. If $C_\nu^r \in S^\nu$ we denote the cardinality of C_ν^r by m_ν^r.

The analysis of (11) can be restricted to the analysis in each final class C_1^r (taking into account the fact that the values in transitory classes are a function of values in elements of S^1). In effect the following theorem holds:

Theorem 4.2 *If* $i \in C_\nu^r$ $\nu > 1$ *there exists a function* $Q : \mathbb{R}^t \to \mathbb{R}$, *with* $t = \cup_{r=1}^{n_1} m_1^r$, *where* $V_k^h(i) = Q\left(V_k^h(j)\ j \in C_1^r\right)$.

Proof: For each $C_1^r \in S^1$ $r = 1, \ldots n_1$ we obtain $v_1 \in M_{m_1^r x 1}$ with

$$v_1(i) = A_{[\phi_{11}(i), \phi_{21}(i)]}\left(P_1 V_1 + F_1\right)(i)$$

where $P_1 \in M_{m_1^r x m_1^r}$ is the restriction of matrix P considering only its action in $C_1^r \in S^1$. In the same way F_1, ϕ_{11}, $\phi_{21} \in M_{m_1^r x 1}$ are the restriction of F,

ϕ_1 and ϕ_2 respectively.

To compute the solution for $i \in C_\nu^r \in S^\nu$, $\nu > 1$ $r = 1, \ldots, n_\nu$ we compute in each class $C_\nu^r \in S^\nu$, $v^\nu \in M_{m_\nu^r x 1}$,

$$v_\nu(i) = A_{[\phi_{1\nu}(i),\phi_{2\nu}(i)]}\left(P_\nu V_\nu + \bar{F}_\nu\right)(i)$$

where $P_\nu \in M_{m_\nu^r x m_\nu^r}$ is the restriction of P considering only its action in $C_\nu^r \in S^\nu$, F_ν, ϕ_ν, $\phi_{2\nu} \in M_{m_\nu^r x 1}$, with $\bar{F}_\nu = \sum_{j=1}^{\nu-1} P_j v_j + F^\nu$. ∎

By Theorem 4.2 we obtain:

Theorem 4.3 *Let v, w be solutions of (11); if $v(x_i) = w(x_i) \forall x_i \in C_1^r \subset S^1$ then $v = w$.*

4.2 Study of equation (11) in final classes

1) Case $x_j \in \partial\Omega_k^+$. In this case the fully discrete solution is:

$$V_k^h(x_j) = A_I\left[\phi_0(x_j)\right] = \min\left[\phi_2(x_j), \max(\phi_1(x_j), \phi_0(x_j))\right].$$

2) Case $x_j \notin \partial\Omega_k^+$. We consider the restriction of matrix P to the final class C_1^r. Without loss of clarity we also denote this matrix by P. By definition of final classes, we know that P is an irreducible stochastic matrix. We call d the greatest common divisor of the length of all the circuits (closed path) determined by P. We define the following equivalence relation:

$$i \sim j \iff \exists \text{ a path between i and j with a length multiple of } d$$

We denote $c_i, i = 1, \ldots, d$, the d subclasses of equivalence. The matrix P has the following structure:

$$P = \begin{bmatrix} 0 & P^{1,2} & 0 & 0 \\ 0 & 0 & P^{2,3} & 0 \\ 0 & 0 & 0 & P^{d-1,d} \\ P^{d,1} & 0 & 0 & 0 \end{bmatrix} \tag{15}$$

where we denote $P^{i,j} = P^{c_i,c_j}$. The matrix P^d defines an ergodic Markov chain in each subclass c_i, then (see [18]) there exists a unique invariant measure $\mu^{c_i} = \mu^i$ of $(P^d)^{i,i}$, that is, $\mu^i(P^d)^{i,i} = \mu^i$ $\forall i = 1, ..., d$.

These invariant measures are related each other in the following way:

$$\mu^i P^{i,i+1} = \mu^{i+1} \quad \forall i = 1, ..., d \tag{16}$$

Then $\mu = \frac{1}{d}(\mu^1, \ldots, \mu^d)$ is the unique invariant measure of the Markov process P, i.e., $\mu = \mu P$.

The measure μ determines the characterization of the solutions of (11). In effect, we shall find the solution of the problem by analysing the cases $< \mu, f > \neq 0$ and $< \mu, f > = 0$.

a) Case $< \mu, f > \neq 0$

Theorem 4.4 *If* $< \mu, f > \neq 0$ *then the solution of* (11) *is unique and it is given by the limit point of the following convergent iteration,*

$$v^{\nu+1} = M^k v^{\nu}; \quad \forall \, initial \, v^0. \tag{17}$$

Outline of the proof: In this case we can prove that there exists $x_i \in X$ such that $V(x_i) = \phi_j(x_i)$ with $j = 1$ or 2. Then, as we can insure that the game finishes in a finite time, we can prove that there exists $\hat{n}$ such that $(M^k)^{\hat{n}}$ is a contractive operator. $\blacksquare$

Remark 4.1 : *To find numerically the solution of* (17) *we can use an adaptation of the accelerated algorithms presented in [9] and in [17].*

b) Case $< \mu, f > = 0$

In this case (11) can have more than one solution.

First, to prove the existence of solution of (11) (still restricted to a final class) we will obtain the solution, $\bar{\eta}$, of the linear system: $w(x_i) = (Pw + F)(x_i)$. In a second step, we will prove that the desired solution of (11) is found using the usual value iteration with initial condition $\bar{\eta}$.

Let $F = ((F^1)^t, \ldots, (F^d)^t)^t$; $F^i = F^{c_i}$ (where F^t is the transpose matrix of F), we are going to analyze the asymptotic behavior of the functions:

$$\psi_n(i) = E \left(\sum_{j=1}^{n} F(y_j) \right) = \left(\sum_{j=1}^{n} (P^{j-1} F)(i) \right). \tag{18}$$

The sequence $\psi_n(i)$ has the following properties

$$i) \psi_{md}(i) \to \eta_0(i) \quad \forall i \quad \text{if } m \to +\infty,$$

$$ii) \psi_{md+j}(i) \to \eta_j(i) \quad \forall i \quad \text{if } m \to +\infty.$$

Proof of i)

We have:

$$\psi_{md} = \sum_{q=1}^{d} P^{q-1} F + P^d \psi_{(m-1)d} = \hat{F} + P^d \psi_{(m-1)d} \tag{19}$$

where $\hat{F} = \sum_{j=1}^{d-1} P^j F$.

As P^d is block diagonal and each block defines a finite ergodic Markov chain, we have that (19) geometrically converges to a finite value. Then there exists $\eta_0(i)$ $\forall i$.

ii) is obvious by i) and by the relation $\psi_{md+j} = F + P\psi_{md+j-1}$.   ■

We want to compute $\overline{lim}_{n\to+\infty}\psi_n = \max_j \eta_j(i) = \bar{\eta}(i)$ because it represents the desired solution of the linear system. We have that $P^{md} \to G$ when $m \to +\infty$, where $G = \sum_j \hat{e}^j \hat{\mu}^j$, $\hat{e}^j = (0,\dots,1,\dots,1,0\dots,0)$ with 1 in the j-th cycle and $\hat{\mu}^j = (0,\dots,\mu^j,\dots,0)$ with μ^j the invariant measure vector restricted to subclass c_j.

By (18) and the properties of sequence ψ_n we have that η_j satisfies the following equation:

$$\eta_j = \eta_0 + \left(\sum_{r=1}^{d}\hat{e}^r\hat{\mu}^r\right)\left(\sum_{q=1}^{j-1}P^q F\right) = \eta_0 + \sum_{r=1}^{d}\hat{e}^r\sum_{q=1}^{j-1}\hat{\mu}^{r+q}F. \tag{20}$$

Calling $w_\rho = \hat{\mu}^\rho F$, we have, in each element of the cycle that:

$$\eta_j(s) = \eta_0(s) + \sum_{\rho=s}^{s+j-1} w_\rho.$$

Let $\bar{\rho}$ be such that $\sum_{r=1}^{\bar{\rho}} w_r = \max_\rho \sum_{r=1}^{\rho} w_r$ then we obtain $\overline{lim}_{n\to\infty}\psi_n(i) = \max_j \eta_j(i) = \eta_{\bar{\rho}}(i)$ $\forall i$. By (20) and the fact that $< \mu, f > = 0$ we have that $< \mu^j, \eta_0^j > = 0$, $i = 1,\dots,d$ where $\eta_0 = \left((\eta_0^1)^t,\dots,(\eta_0^d)^t\right)^t$. The last properties induce the following procedure to obtain $\bar{\eta}$.

Procedure to compute $\bar{\eta}$, the desired solution of the linear system

1. Consider c_p, compute n_p = cardinality of c_p and determine $\hat{p}$ such that $n_{\hat{p}} = \min_p n_p$.

2. Compute the invariant measure $\mu^{\hat{p}}$ and recursively μ^p for all p as a function of $\mu^{\hat{p}}$.

3. Compute $w_\rho = < \mu^\rho, F^\rho >$ and determine $\bar{\rho}$ such that $\sum_{r=1}^{\bar{\rho}} w_r = \max_\rho \sum_{r=1}^{\rho} w_r$.

4. Compute $\eta_0^{\bar{\rho}}$ through the equation: $\eta_0^{\bar{\rho}} = (P^d)^{\bar{\rho}}, \bar{\rho}\eta_0^{\bar{\rho}} + \tilde{F}^{\bar{\rho}}$ where $\tilde{F}^{\bar{\rho}} = \sum_{j=0}^{\bar{\rho}-1} P^j F$ with the additional condition $< \mu^j, \eta_0^j > = 0$.

5. Compute $\eta_{\bar{\rho}}(s) = \eta_0(s) + \sum_{\rho=s}^{s+\bar{\rho}-1} w_\rho$.

4.3 The main results

Accelerated algorithm to compute the solution of (11)
If $\bar{\eta}$ verifies that $\phi_1 \le \bar{\eta} \le \phi_2$ then it is precisely the solution of (11). If this is not the case, we present now an algorithm to obtain the solution of (11).

The procedure finds in a finite number of steps, $\bar{\bar{\eta}}$, the limit point of iterations $v^{\nu+1} = M^k v^\nu$, with $v^0 = \bar{\eta}$ and shows that $\bar{\bar{\eta}}$ is the solution of (11). The methodology is an extension of the one presented in [9] in the case where the operator M^k is not a contractive operator.

Recall that V is the solution of our continuous problem and define:

$$S_1 = \{j \in X \ / \ V(j) = \phi_1(j)\}$$

$$S_2 = \{j \in X \ / \ V(j) = \phi_2(j)\}$$

$$\tilde{S}_2^0 = \left\{j \ / \ \bar{\eta}(j) - \phi_2(j) = \max_i \left(\bar{\eta}(i) - \phi_2(i)\right)\right\} \bigcap \{j \ / \ \bar{\eta}(j) > \phi_2(j)\}$$

$$\tilde{S}_1^0 = \left\{j \ / \ \bar{\eta}(j) - \phi_1(j) = \max_i |\bar{\eta}(i) - \phi_1(i)|\right\} \bigcap \{j \ / \ \bar{\eta}(j) < \phi_1(j)\}.$$

Then we can prove:

Lemma 4.1 $\tilde{S}_1^0 \subset S_1$ *and* $\tilde{S}_2^0 \subset S_2$.

Proof: We are going to prove $\tilde{S}_2^0 \subset S_2$, we obtain $\tilde{S}_1^0 \subset S_1$ analogously. Let $\tilde{S}_2^0 \ne \emptyset$, we denote $\Delta = \max_i \left(\bar{\eta}(i) - \phi_2(i)\right)$, in the same way the vector with components equals to Δ and P the matrix defined in (11). Then, keeping in mind that $P\Delta = \Delta$, we have:
$\bar{\eta} - \Delta \le \phi_2 \quad \forall j = 1, \dots, N$
$\bar{\eta} - \Delta = \phi_2 \quad \forall j \in \tilde{S}_2^0$
$P(\bar{\eta} - \Delta) + F = P\bar{\eta} + F - P\Delta = \bar{\eta} - \Delta.$
As M^k is a monotone operator:

$$A_{[\phi_1,\phi_2]}\left(P(\bar{\eta} - \Delta) + F\right) = A_{[\phi_1,\phi_2]}\left(\bar{\eta} - \Delta\right) \ge \bar{\eta} - \Delta,$$

$$M^k(\bar{\eta} - \Delta) \ge \bar{\eta} - \Delta.$$

By induction

$$(M^k)^m(\bar{\eta} - \delta) \ge M^k(\bar{\eta} - \Delta) \ge \bar{\eta} - \Delta.$$

But

$$(M^k)^m(\bar{\eta} - \Delta)(i) \to \bar{w}(i), \text{when } m \to \infty.$$

Then

$$\bar{w}(i) \ge (M^k)^m(\bar{\eta} - \Delta)(i) \ge (M^k)(\bar{\eta} - \Delta)(i) \ge (\bar{\eta} - \Delta)(i) = \phi_2(i) \quad \forall i \in \tilde{S}_2^0.$$

This implies $i \in S_2$. ∎

By Lemma 4.1 we can deduce that we know the solution of (11) if $i \in \tilde{S}_1^0 \cup \tilde{S}_2^0$. We are going to use this information to find the discrete solution. We redefine the matrices P and F in the following way: $\quad P^0 = P$, $F^0 = F$, $w^0 = \bar{\eta}$ and $\forall n \geq 1$ w^n is the solution of

$$w^n = P^n w^n + F^n \tag{21}$$

where

$$P_{i,j}^n = \begin{cases} 0 & \forall i \in \tilde{S}_1^{n-1} \cup \tilde{S}_2^{n-1} \\ P_{i,j}^{n-1} & \text{if not,} \end{cases}$$

$$F^n(i) = \begin{cases} \phi_1(i) & \forall i \in \tilde{S}_1^{n-1} \\ \phi_2(i) & \forall i \in \tilde{S}_2^{n-1} \\ F^{n-1}(i) & \text{if not,} \end{cases}$$

with

$$\tilde{S}_2^n = \{j / w^n(j) - \phi_2(j) = \max_i (w^n(i) - \phi_2(i))\} \bigcap \{j / w^n(j) > \phi_2(j)\},$$
$$\tilde{S}_1^n = \{j / w^n(j) - \phi_1(j) = \max_i |w^n(i) - \phi_1(i)|\} \bigcap \{j / w^n(j) < \phi_1(j)\},$$
$$S_2^n = \bigcup_{m=0}^n \tilde{S}_2^m; \quad S_1^n = \bigcup_{m=0}^n \tilde{S}_1^m; \quad C^n = C(S_1^n \cup S_2^n).$$

Remark 4.2 : *We know that there exists a unique solution of (21) because P^n has a submatrix of norm smaller than 1.*

Lemma 4.2 *The following holds:*

$$w^n(i) = \begin{cases} \phi_1(i) & \text{if } i \in \tilde{S}_1^{n-1} \\ \phi_2(i) & \text{if } i \in \tilde{S}_2^{n-1}. \end{cases}$$

The proof is obvious by definitions of P^n and F^n given in (21).
We consider the following operator $M_n^k(w)$ defined by:

$$(M_n^k w)(j) = A_I (P^n w + F^n)(j)$$

with P^n and F^n defined in (21). It is easy to see that our solution is also the fixed point of operator M_n^k, that is:

$$\bar{w} = M_n^k \bar{w}. \tag{22}$$

The following Lemma holds:

Lemma 4.3 $\tilde{S}_1^n \cup \tilde{S}_2^n \neq \emptyset$ *or* $w^n = \bar{w}$

Proof: If $\tilde{S}_1^n \cup \tilde{S}_2^n \neq \emptyset$ and if $i \notin S_1^{n-1} \cup S_2^{n-1}$, we have:

$$w^n(i) \in [\phi_1(i), \phi_2(i)]$$

then

$$(M_n^k w^n)(i) = A_I (P^n w^n + F^n)(i) = A_I(w^n)(i) = w^n(i). \tag{23}$$

So, by (22) and (23) we have:

$$w^n = \bar{w}.$$

With Lemmas 4.2 and 4.3 we can now present the algorithm that computes the solution of (11).

ALGORITHM

Step 0: $n = 0$

Step 1: Compute w^n as in (21). Obtain $\tilde{S}_1^n$ and $\tilde{S}_2^n$

Step 2: If $\tilde{S}_1^n \cup \tilde{S}_2^n \neq \emptyset$ then stop.

If not, obtain $P^n(i,j)$ and $F^n(i)$ defined in (21).   Let $n = n + 1$ and go to step 1.

Theorem 4.5 *The algorithm finishes in a finite number of steps smaller than N.*

Proof: $\tilde{S}_1^n \cup \tilde{S}_2^n \neq \emptyset$ implies that cardinality of $C^n <$ cardinality C^{n-1}. But as cardinality of $C^1 \leq N$, there exists an index $\tilde{n}$ such that cardinality of $C^{\tilde{n}} =$ cardinality $C^{\tilde{n}-1}$, then $\tilde{S}_1^n \cup \tilde{S}_2^n = \emptyset$ and by Lemma 4.3 $w^{\tilde{n}} = \tilde{w}$. ∎

The preceding results are summarized in the following theorem.

Theorem 4.6 : Structure of the set of solutions

To characterize the structure of all the solutions of (11) in final classes we must take into account the following cases:

1. *There exists $\bar{\eta}$ the limit point of sequence ψ_n such that $\phi_1 \leq \bar{\eta} \leq \phi_2$. In this case $w = \bar{\eta} + ce$ is also a solution if $w(x) \in [\phi_1(x), \phi_2(x)]$ $\forall c \in [cmin, cmax]$ where $c, cmin, cmax$ are constants and e the identity vector.*

2. *$\bar{\eta}$ does not satisfy the last condition, then the iterative algorithm*

$$v^{n+1} = M^k v^n; \quad v^0 = \bar{\eta}$$

converges to the solution of problem (11) and the following situations can occur:

   - *$\bar{\bar{\eta}}$ is a solution of the linear system; then there exists a constant $a > 0$ such that $\bar{\bar{\eta}} + ce$ is a solution $\forall c \in [0, a]$ or $\forall c \in [-a, 0]$.*

   - *$\bar{\bar{\eta}}$ is not a solution of the linear system, then it is the "unique" solution of (11).*

5   Convergence properties

When we can insure that the game finishes in a finite time, we can easily prove the convergence property (of type $\sqrt{k}$) of the unique discrete solution of our problem. This is the case, for instance, for boundary problems or ergodic problems with a mean different from zero.
In these cases we can reduce the problem to a finite time problem (see [1], [2], [3], [4] and [5]) and we can prove the convergence, in the viscosity sense, with the techniques used in [3], [14], [15] and [16] for this type of problem.

6   Conclusion

We have studied an undiscounted zero sum differential game problem with stopping times. In this case we lose in general the regularity properties of the value function, as we have shown in an example. This fact implies the difficulty of its numerical resolution because the discrete problem may have infinitely many solutions and the algorithm might not converge.

We have totally characterized the set of solutions of the discrete problem. We have obtained the solution that represents the probabilistic formulation of the original problem and we have presented an algorithm to compute it. In regular cases we have proved the convergence of the discrete problem to the real solution, because in these cases the problem can be associated to a finite time problem and we can use the usual techniques to prove convergence properties.

REFERENCES

[1] Bardi M., Soravia P., "Approximation of differential games of pursuit-evasion by discrete-time games", Developments in modeling and computation, R.P. Hamalainen, H.K. Ethamo eds. Lectures Notes in Control and Information Sciences 156 (1991), Springer-Verlag, 131-143.

[2] Bardi M., Soravia P., "Hamilton-Jacobi equations with singular boundary conditions on a free boundary and applications to differential games", Trans. Amer. Math. Soc. 325 (1991), 205-229.

[3] Bardi M., Falcone M., Soravia P., "Fully discrete schemes for the value function of pursuit-evasions games", Proceedings of Fifth International Symposium on Dynamic Games and Applications, Grimentz, Switzerland, 15-18 July 1992.

[4] Bardi M., Falcone M., "Discrete approximations of the minimal time function for systems with regular optimal trajectories", Ricerca di Atenes. Ottimizzatione de Funcionali e Convergence Variazionali, 1990.

[5] Bardi M., Falcone M., "An approximation schemes for the minimum time function". Siam J. Control and Optimization. Vol 4 (1990), 950-965.

[6] Barles G., Perthame B., "Exit time problems in optimal control and vanishing viscosity method", Siam J. Control and Optimization, Vol 26:5 (1988), 1133, 1148.

[7] Capuzzo Dolcetta I., Ishii H., "Approximate solution of the Bellman equation of deterministic control theory", Appl. Math. Optim., Vol. 11 (1984), 161-181.

[8] Crandall M. G., Lions P. L., "Viscosity solutions of Hamilton-Jacobi equations", Tansactions of the American Mathematical Society, Vol 277:1 (1983), 1-42.

[9] Di Marco S., "Técnicas de descomposición-agregación en el tratamiento de la inecuación bilátera de Isaacs", Mecánica Computacional Vol 12 (1991), 509-518.

[10] Friedman A., "Differential Games", Wiley-Interscience, New York, 1971.

[11] Gonzalez R., "Sur la resolution de l'équation de Hamilton-Jacobi du côntrole déterministique", Cahiers de Mathématiques de la Decisión N 8029 and 8029 bis. Ceremade-Université de Paris-Dauphine, 1980.

[12] Gonzalez R., "Solución numérica de problemas de juegos diferenciales de suma nula con tiempo de detención", Proceedings of $1^e r$ Congreso Nacional de Informática y Teleinformática -USUARIA'83/13 JAIIO, Buenos Aires, 1983 4.1-4.17.

[13] Gonzalez R., Rofman E., "On deterministic control problems: an approximation procedure for the optimal cost. Part I and II", SIAM Journal on Control and Optimization, 23 (1985), 242-285.

[14] Gonzalez R., Tidball M., "On a Discrete Time Approximation of the Hamilton-Jacobi Equation of Dynamic Programming", Rapport de Recherche N1375, INRIA, 1990.

[15] Gonzalez R., Tidball M., "On the rate of convergence of fully discrete solutions of Hamilton-Jacobi equations", Rapport de Recherche , N 1376, INRIA, 1991.

[16] Gonzalez R., Tidball M., "Sur l'odre de convergence des solutions discrétisées en temps et en espace de l'équation de Hamilton-Jacobi" Comptes Rendus Acd. Sci., Paris, Tomo 314, Serie I, 1992, 479-482.

[17] Gonzalez R. L. V., Tidball M. M., "Fast solution of general nonlinear fixed point problems". System Modeling and Optimization, Proceedings of the 15th IFIP Conference on System Modeling and Optimization, Zurich, Switzerland, September 2-6, 1991, Lecture Notes in Control and Information Sciences, Vol.180 (1992), Springer Verlag, New York, 35-44.

[18] Ross S. M., "Applied Probability Models with Optimization Applications", Holden-Day, San Francisco, 1970.

[19] Soner M., "Optimal control with state-space constraints, II". SIAM J. of Control and Optim., 26 (1986), 1110-1122.

[20] Stettner L., "Zero-sum Markov Games with Stopping and Impulsive Strategies", Appl. Math. Optim., Vol. 9 (1982), 1-24.

[21] Strang G., Fix G., "An Analysis of the Finite Element Method", Prentice-Hall, Englewood Cliffs, NJ, 1973.

[22] Tidball M.M, Gonzalez R.L.V., "Zero sum differential games with stopping times. Some results about its numerical solution", Proceedings of Fifth International Symposium on Dynamic Games and Applications, Grimentz, Switzerland, 15-18 July 1992. Annals of Dynamics Games, Vol 1, 1993.

Guaranteed Result in Differential Games with Terminal Payoff

A. A. Chikrii and J. S. Rappoport *
Department of controlled processes optimization,
Cybernetics Institute NAS, 252187,
Glushkov Avenue, 40, Kiev, Ukraine

Abstract

We suggest a method of solving differential games with terminal payoff. The method consists of applying Fenchel-Moreau [1] duality ideas to the general scheme of the Resolving Functions method [2]. The gist of the suggested method is presentation of the Resolving Function by means of the function conjugate to the payoff, and employing of involutory property of the conjugation operator for closed convex functions to obtain guarantee estimation of terminal value of the payoff. This estimation can be represented by the initial payoff value and Resolving Function integral. This paper involves the ideas of [2], adjoins [3–9] and turns out to be a new aspect of Convex Analysis application to differential games.

We shall consider a differential game given by the equation

$$\dot{z} = Az + \varphi(u, v), \quad z \in R^n, u \in U, v \in V, \tag{1}$$

where A is a constant square matrix; U and V are given nonempty compact subsets of the Euclidean space R^n; $\varphi : U \times V \to R^n$ is a function, continuous in both arguments.

The payoff function $\sigma(z), \sigma : R^n \to R^1$ is a proper closed lower bounded function and its value determines the moment of the game termination.

If $z(t) = z(z_0, u_t(\cdot))$ is a trajectory of the system (1) resulting from initial position z_0 and chosen controls $u_t(\cdot) = \{u(s) : s \in [0, t]\}, v_t(\cdot)$, then we consider the game to be finished at the moment t_1 if

$$\sigma(z(t_1)) \leq 0 \tag{2}$$

The aim of the pursuer (u) is to finish the game as soon as possible and evader (v) has the opposite aim.

*all: Institute of Cybernetics, E-Mail: chik@d165.icyb.kiev.ua   Fax.: (044) 266-15-58
Tel.: (044) 266-21-58, 266-04-58

Suppose that throughout the game both pursuer and evader use as controls measurable functions $u(t) \in U, v(t) \in V$. We are going to take the side of the pursuer and find the guaranteed game finish time using the information about initial position z_0 and evader control history $v_t(\cdot)$.

Let us introduce some definitions and results connected with conjugate functions [1]. For the proper function $f(z), f : R^n \to R^1$, we define the conjugate function $f^*(p)$ and second conjugate function $f^{**}(z)$ in the following way:

$$f^*(p) = \sup_{z \in R^n} [(p, z) - f(z)], p \in R^n,$$
$$f^{**}(z) = \sup_{p \in R^n} [(p, z) - f^*(p)], z \in R^n.$$

Note that the following conditions are satisfied:

$$-f^*(0) = \inf_{z \in R^n} f(z),$$
$$f^{**}(z) \leq f(z), z \in R^n$$

Taking into account the Fenchel-Moreau Theorem [1] we can conclude that the proper function $f(z)$ is convex and closed if and only if $f^{**}(z) = f(z), z \in R^n$. From the above assumptions on the function $\sigma(z)$ we have

$$\sigma(z) = \sup_{p \in R^n} [(p, z) - \sigma^*(p)], \tag{3}$$

$$0 \in \mathrm{dom}\, \sigma^* = \{p \in R^n : \sigma^*(p) < +\infty\}.$$

Let π denote the orthogonal projection from R^n onto $L = \mathrm{lin}\, \mathrm{dom}\, \sigma^*$. Taking into account conditions (3), one can find

$$\sigma(z) = \sup_{p \in \mathrm{dom}\, \sigma^*} [(p, \pi z) - \sigma^*(p)] = \sigma(\pi z). \tag{4}$$

Now let us introduce the following Set-valued Mappings [2]

$$W(t, v) = \pi e^{At} \varphi(U, v),$$
$$W(t) = \bigcap_{v \in V} W(t, v)$$

and define

$$\mathrm{dom}\, W = \{t : W(t) \neq \emptyset\}.$$

Condition 1. $\mathrm{dom}\, W = [0, +\infty)$. *(See [9]).*

Inasmuch as $W(t)$ is upper semicontinuous [2], it contains at least one Borel selector [2]. Denote by Γ the set of all these selectors. Let us fix one selector $\gamma(\cdot) \in \Gamma$ and define

$$\xi(t, z, \gamma(\cdot)) = \pi e^{At} z + \int_0^t \gamma(t - \tau) d\tau.$$

Analogous to [2] we define the Resolving Function in the following way:

$$\alpha(t, \tau, z, v, \gamma(\cdot)) = \sup\Big\{ \alpha \geq 0 : \inf_{u \in U} \sup_{p \in \text{dom } \sigma^*} \big[(p, \pi e^{A(t-\tau)} \varphi(u, v) - \tag{5}$$

$$-\gamma(t - \tau)) + \alpha[(p, \xi(t, z, \gamma(\cdot))) - \sigma^*(p)]\big] \leq 0 \Big\},$$

where $t \geq \tau \geq 0, z \in R^n, v \in V$.

By Condition 1 we obtain

$$\min_{u \in U} \max_{p \in \text{dom } \sigma^*} \Big(p, \pi e^{A(t-\tau)} \varphi(u, v) - \gamma(t - \tau) \Big) \leq 0 \tag{6}$$

for all $t \geq \tau \geq 0, v \in V$.

It follows from Condition 1 that the inequality in definition (5) is satisfied at least when the Resolving Function equals zero. Also note that when $\sigma(\xi(t, z, \gamma(\cdot))) \leq 0$ the function $\alpha(t, \tau, z, v, \gamma(\cdot)) = +\infty$ for all $v \in V, \tau \in [0, t]$. But in the case when for some $t > 0, z \in R^n, \gamma(\cdot) \in \Gamma \quad \sigma(\xi(t, z, \gamma(\cdot))) > 0$, one can conclude that Resolving Function has finite values and is uniformly bounded with respect to $\tau \in [0, t]$ and $v \in V$. It follows directly from the following lemma.

Lemma 1 *Assume that Condition 1 is satisfied and for some $t > 0, z \in R^n$, $\gamma(\cdot) \in \Gamma$; $\sigma(\xi(t, z, \gamma(\cdot))) > 0$. Then the following inequality holds:*

$$\alpha(t, \tau, z, v, \gamma(\cdot)) \leq \inf_{p \in P(t,z)} \frac{\min_{u \in U} \Big(p, \pi e^{A(t-\tau)} \varphi(u, v) - \gamma(t - \tau) \Big)}{-\sigma(\xi(t, z, \gamma(\cdot)))} \tag{7}$$

where $t \geq \tau \geq 0, z \in R^n, v \in V$,

$$P(t, z) = \Big\{ p \in \text{dom } \sigma^* : \sigma(\xi(t, z, \gamma(\cdot))) = (p, \xi(t, z, \gamma(\cdot))) - \sigma^*(p) \Big\}.$$

Let us introduce another lemma which will be useful for the construction of pursuit control.

Lemma 2 *Suppose Condition 1 is satisfied and for some $t > 0, z \in R^n$, $\gamma(\cdot) \in \Gamma$; $\sigma(\xi(t, z, \gamma(\cdot))) > 0$. Then the Resolving Function (5) is Borel in (τ, v), $\tau \in [0, t], v \in V$.*

Let us consider function

$$T(z, \gamma(\cdot)) = \inf\Big\{ t \geq 0 : \int_0^t \inf_{v \in V} \alpha(t, \tau, z, v, \gamma(\cdot)) d\tau \geq 1 \Big\}. \tag{8}$$

If the inequality in curly brackets is not true for all $t \geq 0$, then we define $T(z, \gamma(\cdot)) = +\infty$.

Note that if $\sigma(\xi(t, z, \gamma(\cdot))) > 0$, then the function $\inf_{v \in V} \alpha(t, \tau, z, v, \gamma(\cdot))$ is measurable in τ [2] and inasmuch as it is uniformly bounded in τ, it is also summable on $[0, t]$. If, conversely, $\sigma(\xi(t, z, \gamma(\cdot))) \leq 0$, then $\inf_{v \in V} \alpha(t, \tau, z, v, \gamma(\cdot)) = +\infty$ for $\tau \in [0, t], t > 0$. In this case it is natural to set the integral to be equal to $+\infty$ and consequently the inequality in definition of $T(z, \gamma(\cdot))$ is satisfied automatically.

Theorem 1 *Let for the controlled process (1) Condition 1 be satisfied and for some $z_0 \in R^n$, $\gamma_0(\cdot) \in \Gamma$, $T(z_0, \gamma_0(\cdot)) < +\infty$. Then the game starting from initial position z_0 can be finished at the moment $T(z_0, \gamma_0(\cdot))$.*

Proof. Define $T = T(z_0, \gamma_0(\cdot))$. Let $v(\tau), v(\tau) \in V, \tau \in [0, T]$ be an arbitrary measurable function. We are going to indicate a method of pursuit control choice. Consider a case $\sigma(\xi(T, z_0, \gamma_0(\cdot))) > 0$. Let us introduce the control function

$$h(t) = 1 - \int_0^t \alpha(T, \tau, z_0, v(\tau), \gamma_0(\cdot))d\tau.$$

Function $h(t)$ is continuous, nonincreasing and $h(0) = 1$. It follows from the definition of T that there exists $t_* = t_*(v(\cdot)), 0 < t_* \leq T$, such that $h(t_*) = 0$.

Let us consider multivalued mappings

$$U_1(\tau, v) = \left\{ u \in U : \sup_{p \in \text{dom } \sigma^*} [(p, \pi e^{A(t-\tau)}\varphi(u, v) - \gamma_0(T - \tau) + \right.$$

$$\left. \alpha(T, \tau, z_0, v(\tau), \gamma_0(\cdot))[(p, \xi(T, z_0, \gamma_0(\cdot))) - \sigma^*(p)]] \leq 0 \right\},$$

$0 \leq \tau \leq t_*, v \in V$;

$$U_2(\tau, v) = \left\{ u \in U : \pi e^{A(t-\tau)}\varphi(u, v) - \gamma_0(T - \tau) = 0 \right\},$$

$t_* \leq \tau \leq T, v \in V$. Each of them is Borel in both of (τ, v) (see Lemma 2 and [2]). Then the selectors

$$u_1(\tau, v) = \text{lex min } U_1(\tau, \text{v}), 0 \leq \tau \leq t_*, \text{v} \in V,$$
$$u_2(\tau, v) = \text{lex min } U_2(\tau, \text{v}), t_* \leq \tau \leq T, \text{v} \in V,$$

are Borel in (τ, v) functions [2].

The control of the pursuer on the interval $[0, T]$ are defined in the following way:

$$u(\tau) = \left\{ \begin{array}{ll} u_1(\tau, v(\tau)), & \tau \in [0, t_*), \\ u_2(\tau, v(\tau)), & \tau \in [t_*, T]. \end{array} \right.$$

Function $u(\tau)$ is measurable [2].

Now consider a case when $\sigma(\xi(T, z_0, \gamma_0(\cdot))) \leq 0$. The control of the pursuer we define as $u(\tau) = u_2(\tau, v(\tau))$. This function is also measurable [2].

Now we are going to show that the chosen control guarantees the fairness of inequality (2) for the corresponding trajectories of the system (1). According to the Cauchy formula for process (1), we have

$$\pi z(T) = \xi(T, z_0, \gamma_0(\cdot)) + \int_0^T \left[\pi e^{A(T-\tau)}\varphi(u(\tau), v(\tau)) - \gamma_0(T - \tau)\right]d\tau. \quad (9)$$

Let $\sigma(\xi(T, z_0, \gamma_0(\cdot))) \leq 0$. Taking the chosen control of the pursuer into account, we obtain

$$\pi z(t) = \xi(T, z_0, \gamma_0(\cdot)).$$

It easily follows from relation (4) that (2) holds.

Now let $\sigma(\xi(T, z_0, \gamma_0(\cdot))) > 0$. By virtue of (4) and (9), we have

$$\sigma(z(T)) = \sup_{p \in \mathrm{dom}\ \sigma^*} \left[(p, \xi(T, z_0, \gamma_0(\cdot))) - \sigma^*(p) + \right.$$

$$\left. \int_0^T (p, \pi e^{A(T-\tau)}\varphi(u(\tau), v(\tau)) - \gamma_0(T - \tau))d\tau\right].$$

Adding and subtracting in brackets the value of

$$\left[(p, \xi(T, z_0, \gamma_0(\cdot))) - \sigma^*(p)\right] \int_0^{t_*} \alpha(T, \tau, z_0, v(\tau), \gamma_0(\cdot))d\tau,$$

we obtain

$$\sigma(z(T)) = \sup_{p \in \mathrm{dom}\ \sigma^*} \left[[(p, \xi(T, z_0, \gamma_0(\cdot))) - \sigma^*(p)]h(t_*) + \right.$$

$$+ \int_0^{t_*} [(p, \pi e^{A(T-\tau)}\varphi(u(\tau), v(\tau)) - \gamma_0(T - \tau)) +$$

$$+ \alpha(T, \tau, z_0, v(\tau), \gamma_0(\cdot))[(p, \xi(T, z_0, \gamma_0(\cdot))) - \sigma^*(p)]d\tau +$$

$$\left. + \int_{t_*}^T (p, \pi e^{A(T-\tau)}\varphi(u(\tau), v(\tau)) - \gamma_0(T - \tau))d\tau\right].$$

This presentation shows that the chosen control of the pursuer guarantees that the following relation holds at the moment T

$$\sigma(z(T)) \leq \sigma(\xi(T, z_0, \gamma_0(\cdot)))h(t_*) = 0.$$

Corollary 1 *Suppose Condition 1 holds. Then, if the pursuer uses the control described above, the following estimation holds for any $T, 0 < T < T(z_0, \gamma_0(\cdot))$*

$$\sup_{v(\cdot) \in \Omega_V} \sigma(z(T)) \leq \sigma(\xi(T, z_0, \gamma_0(\cdot)))\left[1 - \int_0^T \inf_{v \in V} \alpha(T, \tau, z_0, v, \gamma_0(\cdot))d\tau\right], \quad (10)$$

where Ω_V denotes the set of all measurable functions with their values in V.

Proof. The proof is similar to that of Theorem 1 with the following presentation of the Resolving Function taken into consideration

$$h(t) = \int_0^T \inf_{v \in V} \alpha(T, \tau, z_0, v, \gamma_0(\cdot))d\tau - \int_0^t \alpha(T, \tau, z_0, v(\tau), \gamma_0(\cdot))d\tau.$$

Suppose M^* is a convex set, S is a convex bounded set, $0 \in \text{int } S$. Then for all $z \in R^n$ the function of generalized distance can be defined [10] as

$$\sigma(z) = \inf\{p \geq 0 : z \in M^* + \rho S\} = d_S(z|M^*).$$

It is easy to show that this function meets the initial assumptions on $\sigma(z)$.

Let us find the conjugate function $\sigma^*(p), p \in R^n$. First of all note that the following formula is valid

$$d_S(z|M^*) = \inf\{\mu_S(z - m) : m \in M^*\},$$

where $\mu_S(x) = \inf\{\rho \geq 0 : x \in \rho S\}$ is a calibration function of the set S [1, 10].

Therefore according to the definition of the operation of infimal convolution [1] we have

$$d_S(z|M^*) = (f\square g)(z) = \inf\{f(z - y) + g(y) : y \in R^n\},$$

where $\square$ denotes the operation of infimal convolution [1], $f(x) = \mu_S(x)$, $x \in R^n$, $g(y) = \delta(y|M^*)$ is an indicator function of the set M^* [1].

According to the theorem of the duality of operations of addition and infimal convolution [1], we obtain the formula for conjugate function

$$d_{S^0}^*(p|M^*) = f^*(p) + g^*(p) = \begin{cases} C(M^*, p), p \in S^0, \\ +\infty, p \notin S^0, \end{cases}$$

where $C(M^*, p)$ is a supporting function of M^*, $f^*(p) = \delta(p|S^0)$, $g^*(p) = C(M^*, p)$, $S^0 = \{p \in R^n : (p, x) \leq 1, x \in S\}$ is a polar of S [1]. Here we use the fact that calibration function of S is a supporting function of the polar S^0 [1], as well as the property of duality of indicator and supporting function of convex closed set [1].

So we have

$$\text{dom } d_S^* = S^0,$$

$$d_S(z|M) = \sup_{p \in S^0} \lfloor (p, z) - C(M, p) \rfloor.$$

Taking this presentation into account, we can easily prove the following lemma.

Lemma 3 *Suppose X is a compact set, M^* is a convex closed set, S is a convex bounded set $0 \in \text{int } S$. Then $X \cap M^* \neq \emptyset$ if and only if*

$$\inf_{z \in X} \sup_{p \in S^0} \left[(p, z) - C(M^*, p)\right] \leq 0,$$

where S^0 is a polar of S.

Let us take as M^* a cylindrical set

$$M^* = M_0 + M,$$

where M_0 is a linear subspace in R^n, M is a convex compact from the orthogonal complement L of M_0 in R^n.

Using the relation (5) we can get the following expression for the Resolving Function $\alpha(t, \tau, z, v, \gamma(\cdot))$:

$$\sup\left\{\alpha \geq 0 : \inf_{u \in U} \sup_{p \in S^0} \left[\left(p, \pi e^{A(t-\tau)}\varphi(u,v) - \gamma(t-\tau)\right) + \right.\right.$$

$$\left.\left. +\alpha[(p, \xi(t, z, \gamma(\cdot))) - C(M, p)]\right] \leq 0\right\}.$$

Thus, by Lemma 3, we can show that this function coincides with the Resolving Function introduced in [2] according to the following formula

$$\sup\left\{\alpha \geq 0 : [W(t-\tau, v) - \gamma(t-\tau)] \cap \alpha[M - \xi(t, z, \gamma(\cdot))] \neq \emptyset\right\}.$$

REFERENCES

[1] Rockafellar R. T. *Convex Analysis.* Princeton University Press, 1970.

[2] Chikrii A. A. *Conflict-controlled processes.* Naukova Dumka, Kiev, 1992. (in Russian).

[3] Chikrii A. A. and Rappoport J. S. Group pursuit for controlled objects with different inertness. *Dokl. Acad. Nauk SSSR*, 321(3):486–490, 1991.

[4] Chikrii A. A. and Rappoport J. S. The first direct Pontryagin method in the problem of group pursuit for objects with different inertness. *Dokl. Acad. Nauk Ukrainy*, 2:12–15, 1992.

[5] Chikrii A. A. and Rappoport J. S. Calibration function in the differential games. *Avtomatika*, 5:9–17, 1992.

[6] Pshenichnii B. N., Chikrii A. A., and Rappoport J. S. An efficient method of solving differential games with many pursuers. *Dokl. Acad. Nauk SSSR*, 256(3):530–535, 1981.

[7] Pshenichnii B. N., Chikrii A. A., and Rappoport J. S. Pursuit by several controlled objects in the presence of phase constraints. *Dokl. Acad. Nauk SSSR*, 259(4):138–141, 1981.

[8] Pshenichnii B. N., Chikrii A. A., and Rappoport J. S. Group pursuit in differential games. *Technische Hochschule Leipzig*, 6:13–27, 1982.

[9] Pontryagin L. S. *Selected scientific works*, volume 2. Nauka, Moscow, 1988. (in Russian).

[10] Pshenichnii B. N. *Convex Analysis and Extremal Problems*. Nauka, Moscow, 1974. (in Russian).

PART IV
Nonzero sum games, theory

Lyapunov Iterations for Solving Coupled Algebraic Riccati Equations of Nash Differential Games and Algebraic Riccati Equations of Zero-Sum Games

T-Y. Li

Department of Mathematics, Michigan State University,
E. Lansing, MI 48824, USA

Z. Gajic

Department of Electrical and Computer Engineering, Rutgers University,
Piscataway, NJ 08855-0909, USA

Abstract

In this paper we study the symmetric coupled algebraic Riccati equations corresponding to the steady state Nash strategies. Under control-oriented assumptions, imposed on the problem matrices, the Lyapunov iterations are constructed such that the proposed algorithm converges to the nonnegative (positive) definite stabilizing solution of the coupled algebraic Riccati equations. In addition, the problem order reduction is achieved since the obtained Lyapunov equations are of the reduced-order and can be solved independently. As a matter of fact a parallel synchronous algorithm is obtained. A high-order numerical example is included in order to demonstrate the efficiency of the proposed algorithm. In the second part of this paper we have proposed an algorithm, in terms of the Lyapunov iterations, for finding the positive *semidefinite* stabilizing solution of the algebraic Riccati equation of the zero-sum differential games. The similar algebraic Riccati type equations appear in the H_∞ optimal control and related problems.

1 Coupled Algebraic Riccati Equations of Nash Differential Games

The solutions of the coupled algebraic Riccati equations produce the answers to some important problems of modern control theory, for example, the differential games with conflict of interest and simultaneous decision making (Nash strategies), (Starr and Ho, 1969; Basar, 1991), the H_∞ optimal control problems (Bernstein and Haddad, 1989; Basar and Bernhard, 1991), the optimal control of jump linear systems (Mariton, 1990).

In this paper we solve the symmetric coupled algebraic Riccati equations corresponding to the steady state Nash strategies of the linear-quadratic differential game problem. At the present time *there is no global efficient method for solving these coupled algebraic Riccati equations.* The obtained solution is stabilizing one, nonnegative (positive) definite and valid under the stabilizability-detectability assumptions imposed on the problem matrices. The proposed algorithm is of the reduced-order and can be implemented as a synchronous parallel algorithm (Bertsekas and Tsitsiklis, 1991). As a matter of fact, it will be clear from the convergence proof that this algorithm is based on the successive approximations technique of dynamic programming (Bellman, 1954, 1957, 1961; Larson, 1967; Bertsekas, 1987). The method of successive approximations is the main tool in solving the functional equation of dynamic programming. It has been used in several control theory papers, for example (Vaisbord, 1963; Mil'shtein, 1964; Leake and Liu, 1967; Kleinman, 1968; Levine and Vilas, 1973; Mageriou, 1977). This method can be used as a very powerful decomposition technique which simplifies computations. In the work of (Mil'shtein, 1964), an approximate convergent method for synthesis of the optimal control system is investigated. The approach is based on a combination of the ideas of Lyapunov's second method and Bellman's method of successive approximations. Convergent suboptimal control sequences were also obtained in (Bellman, 1954, 1961) and (Vaisbord, 1963; Kleinman, 1968; Mageriou, 1977).

A controlled linear dynamic system corresponding to the Nash differential game strategies is given by

$$\dot{x} = Ax + B_1 u_1 + B_2 u_2, \quad x(t_0) = x_0 \tag{1}$$

where $x \in \Re^n$ is a state vector, $u_1 \in \Re^{m_1}$ and $u_2 \in \Re^{m_2}$ are control inputs (for the reason of simplicity we limit our attention to two control agents), $A, B_1,$ and B_2 are constant matrices of appropriate dimensions. With each control agent a quadratic type function is associated

$$J_1(u_1, u_2, x_0) = \frac{1}{2} \int_{t_0}^{\infty} \left(x^T Q_1 x + u_1^T R_{11} u_1 + u_2^T R_{12} u_2 \right) dt \tag{2}$$

$$J_2(u_1, u_2, x_0) = \frac{1}{2} \int_{t_0}^{\infty} \left(x^T Q_2 x + u_1^T R_{21} u_1 + u_2^T R_{22} u_2 \right) dt \tag{3}$$

Weighting matrices are symmetric and

$$\begin{aligned}
Q_i &\geq 0, \quad i = 1, 2; \quad \text{(positive semidefinite)} \\
R_{ii} &> 0, \quad i = 1, 2; \quad \text{(positive definite)} \\
R_{ij} &\geq 0, \quad i = 1, 2; \quad j = 1, 2; \quad i \neq j
\end{aligned} \tag{4}$$

The optimal solution to the given problem leads to the so-called Nash optimal strategies u_1^* and u_2^* satisfying

$$J_1\left(u_1^*, u_2^*\right) \le J_1\left(u_1, u_2^*\right), \quad J_2\left(u_1^*, u_2^*\right) \le J_1\left(u_1^*, u_2\right) \tag{5}$$

It was shown in (Starr and Ho, 1969) that the *closed-loop* Nash optimal strategy is given by

$$u_i^* = -R_{ii}^{-1} B_i^T K_i x, \quad i = 1, 2 \tag{6}$$

where $K_i, i = 1, 2$, satisfy the coupled algebraic Riccati equations

$$\begin{aligned} K_1 A + A^T K_1 + Q_1 - K_1 S_1 K_1 - K_2 S_2 K_1 \\ -K_1 S_2 K_2 + K_2 Z_2 K_2 = \mathcal{N}_1\left(K_1, K_2\right) = 0 \end{aligned} \tag{7}$$

$$\begin{aligned} K_2 A + A^T K_2 + Q_2 - K_2 S_2 K_2 - K_2 S_1 K_1 \\ -K_1 S_1 K_2 + K_1 Z_1 K_1 = \mathcal{N}_2\left(K_1, K_2\right) = 0 \end{aligned} \tag{8}$$

with

$$S_i = B_i R_{ii}^{-1} B_i^T, \; i = 1, 2; \; Z_i = B_i R_{ii}^{-1} R_{ji} R_{ii}^{-1} B_i^T, \; i, j = 1, 2, \; i \ne j$$

The existence of the nonlinear optimal Nash strategies was established in (Basar, 1974), so that (6), in fact, are the best linear optimal feedback strategies. Since a linear control law is highly desirable from the practical point of view, the linear feedback strategies (6) attract the attention of many researchers.

The existence of Nash strategies (6) and a solution of the coupled algebraic Riccati equations (7)-(8) were studied in Papavassilopoulos et al., 1979, using the Brower fixed point theorem by imposing norm conditions on the given matrices. These conditions are, in general, difficult to test and they are not very useful from a practical point of view. We are interested in imposing the control oriented assumptions Wonham, 1968; Kucera, 1972, which accompanied with a convenient algorithm will lead to the required solution of (7)-(8). We will show that the obtained solution is a nonnegative (positive) definite and stabilizing one. In addition, the proposed algorithm operates only on two decoupled standard algebraic Riccati equations (initialization) and performs iterations on two algebraic Lyapunov equations; thus, it operates on the reduced-order problems, and from a computational point of view, the algorithm is extremely efficient.

So far the existing algorithms for solving (7)-(8) Krikelis and Rekasius, (1971); Tabak (1975) are of the local type, that is, they are faced with the problems of finding very good initial guesses. Furthermore, the algorithm proposed in (Tabak, 1975) does not necessarily converge even when the initial guesses are close to the optimal ones, as was pointed out in Olsder (1975).

On the other hand, it is not known how to generate the initial guesses for the Newton-type algorithm used in Krikelis and Rekasius (1971), such that the algorithm converges to the stabilizing solutions of (7)-(8). Note that the differential coupled Riccati equations of Nash differential games corresponding to the finite time optimization problems were studied in Jodar and Abou-Kandil, 1989; Abou-Kandil et al., 1993. In Abou-Kandil et al., 1993, the nonsymmetric coupled algebraic Riccati equations corresponding to the *open-loop* Nash strategies were also studied. Equations (7)-(8) have been studied for special classes of systems in (Khalil and Kokotovic, 1979; Khalil, 1980) - singularly perturbed systems, and (Ozguner and Perkins, 1977; Petrovic and Gajic, 1988) weakly coupled systems.

2 The Lyapunov Iterations for the Linear-Quadratic Nash Games

The considered algorithm is originally proposed by the authors in Gajic and Li, 1988; see also Gajic and Shen, 1993, pp. 359, where only the algorithm and simulation results were presented. In this paper we give the convergence proof. It is shown that the algorithm converges to the nonnegative (positive) definite stabilizing solution of (7)-(8) under the following control-oriented assumption.

Assumption 2.1 *Either the triple* $\left(A, B_1, \sqrt{Q_1}\right)$ *or* $\left(A, B_2, \sqrt{Q_2}\right)$ *is stabilizable-detectable.*

These conditions are quite natural since at least one control agent has to be able to control and observe unstable modes. Because the game is a noncooperative one, the assumption that their joint effect will take care of unstable modes seems to be very idealistic.

Let us suppose that $\left(A, B_1, \sqrt{Q_1}\right)$ is stabilizable-detectable. Then a unique positive definite solution of an auxiliary algebraic Riccati equation

$$K_1^{(0)} A + A^T K_1^{(0)} + Q_1 - K_1^{(0)} S_1 K_1^{(0)} = 0 \tag{9}$$

exists such that $\left(A - S_1 K_1^{(0)}\right)$ is a stable matrix. By plugging $K_1 = K_1^{(0)}$ in (8) we get the second auxiliary Riccati equation as

$$K_2^{(0)} \left(A - S_1 K_1^{(0)}\right) + \left(A - S_1 K_1^{(0)}\right)^T K_2^{(0)}$$
$$+ \left(Q_2 + K_1^{(0)} Z_1 K_1^{(0)}\right) - K_2^{(0)} S_2 K_2^{(0)} = 0 \tag{10}$$

Since $\left(A - S_1 K_1^{(0)}\right)$ is a stable matrix and $Q_2 + K_1^{(0)} Z_1 K_1^{(0)}$ is a positive semidefinite matrix, the corresponding closed-loop matrix

$\left(A - S_1 K_1^{(0)} - S_2 K_2^{(0)}\right)$ is stable.

In fact, the triple $\left(A - S_1 K_1^{(0)}, B_2, \sqrt{Q_2 + K_1^{(0)} S_1 K_1^{(0)}}\right)$ is stabilizable-detectable and the stabilizing $K_2^{(0)}$ is uniquely determined. In the following we will use the solutions of (9)-(10), that is, $K_1^{(0)}$ and $K_2^{(0)}$ to initialize our algorithm.

The following algorithm is proposed in Gajic and Li, 1988; see also Gajic and Shen, 1993, pp. 359 for solving the coupled algebraic Riccati equations (7)-(8)

Algorithm 1:

$$\left(A - S_1 K_1^{(i)} - S_2 K_2^{(i)}\right)^T K_1^{(i+1)} + K_1^{(i+1)} \left(A - S_1 K_1^{(i)} - S_2 K_2^{(i)}\right) = \tag{11}$$

$$= \overline{Q_1^{(i)}} = -\left(Q_1 + K_1^{(i)} S_1 K_1^{(i)} + K_2^{(i)} Z_2 K_2^{(i)}\right), \quad i = 0, 1, 2, \ldots$$

$$\left(A - S_1 K_1^{(i)} - S_2 K_2^{(i)}\right)^T K_2^{(i+1)} + K_2^{(i+1)} \left(A - S_1 K_1^{(i)} - S_2 K_2^{(i)}\right) = \tag{12}$$

$$= \overline{Q_2^{(i)}} = -\left(Q_2 + K_1^{(i)} Z_1 K_1^{(i)} + K_2^{(i)} S_2 K_2^{(i)}\right), \quad i = 0, 1, 2, \ldots$$

with initial conditions $K_1^{(0)}$ and $K_2^{(0)}$ obtained from (9)-(10).

This algorithm is based on the Lyapunov iterations. Even though it looks like this algorithm has the form of Kleinman (1968), it is quite easy to show that this is not the case. Note that Kleinman's algorithm is used to solve the regular algebraic Riccati equation. In our paper we are concerned with the problem of solving coupled algebraic Riccati equations, where the coupling comes through the nonlinear quadratic terms, so that our problem is much more complex. Also, it can be shown that the proposed algorithm (11)-(12) is not of the Newton type. As a matter of fact, the Kleinman algorithm is equivalent to the Newton method. Interestingly enough, it can be shown that Kleinman's algorithm can be obtained by using the successive approximations of dynamic programming, and for the regular algebraic Riccati equation this is equivalent to using the Newton method to solve it.

The algorithm (11)-(12) has the feature given in the following theorem.

Theorem 2.1 *Under Assumption 2.1 the unique nonnegative definite stabilizing solution of the coupled algebraic Riccati equations (7)-(8) exists. It is obtained by performing Lyapunov iterations (11)-(12).*

The proof of this theorem is based on the successive approximations technique of dynamic programming.

Proof: Consider the linear-quadratic optimal control problem of minimizing the following performance criteria

$$J_1\left(u_1, u_2, x\left(t\right)\right) \;=\; \frac{1}{2} \int_t^\infty \left(x^T Q_1 x + u_1^T R_{11} u_1 + u_2^T R_{12} u_2\right) d\tau \tag{13}$$

$$J_2\left(u_1, u_2, x\left(t\right)\right) \;=\; \frac{1}{2} \int_t^\infty \left(x^T Q_2 x + u_1^T R_{21} u_1 + u_2^T R_{22} u_2\right) d\tau \tag{14}$$

along the trajectories of dynamic system (1). Note that the optimization problem defined by (1) and (13)-(14) is more general than the one with fixed initial time (Kirk, 1970). Corresponding Hamiltonians for the Nash differential game for each control agent are given by Starr and Ho, 1969:

$$\begin{aligned}
H_1\left(x, u_1, u_2^*, \frac{\partial J_1^*}{\partial x}\right) &= \tfrac{1}{2}\left\{x^T Q_1 x + u_1^T R_{11} u_1 + u_2^{*T} R_{12} u_2^*\right\} \\
&\quad + \left(\frac{\partial J_1^*}{\partial x}\right)^T \left\{Ax + B_1 u_1 + B_2 u_2^*\right\}
\end{aligned} \tag{15}$$

$$\begin{aligned}
H_2\left(x, u_1^*, u_2, \frac{\partial J_2^*}{\partial x}\right) &= \tfrac{1}{2}\left\{x^T Q_2 x + u_1^{*T} R_{21} u_1^* + u_2^T R_{22} u_2\right\} \\
&\quad + \left(\frac{\partial J_2^*}{\partial x}\right)^T \left\{Ax + B_1 u_1^* + B_2 u_2\right\}
\end{aligned} \tag{16}$$

The necessary conditions for the Nash optimal strategies are (Starr and Ho, 1969)

$$\min_{u_1} H_1\left(x, u_1, u_2^*\right) = 0 \quad \Rightarrow \quad u_1^* = -R_{11}^{-1} B_1^T \left(\frac{\partial J_1^*}{\partial x}\right)^T \tag{17}$$

$$\min_{u_2} H_2\left(x, u_1^*, u_2\right) = 0 \quad \Rightarrow \quad u_2^* = -R_{22}^{-1} B_2^T \left(\frac{\partial J_2^*}{\partial x}\right)^T \tag{18}$$

The successive approximations technique of dynamic programming applied to (1) and (13)-(16) is composed of the following two steps.

Step 1. Take any stabilizable linear control law $u_1^{(0)}\left(x\left(t\right)\right)$ and $u_2^{(0)}\left(x\left(t\right)\right)$, for example $u_1^{(0)}\left(x\left(t\right)\right) = -R_{11}^{-1} B_1^T K_1^{(0)} x\left(t\right)$ and $u_2^{(0)}\left(x\left(t\right)\right) = -R_{22}^{-1} B_2^T K_2^{(0)} x\left(t\right)$ with $K_1^{(0)}$ and $K_2^{(0)}$ being symmetric, and evaluate the expression for the performance criterion

$$\begin{aligned}
J_1^{(0)} \;=&\; \frac{1}{2} \int_t^\infty [x^T\left(\tau\right) Q_1 x\left(\tau\right) + u_1^{(0)T}\left(x\left(\tau\right)\right) R_{11} u_1^{(0)}\left(x\left(\tau\right)\right) \\
&\quad + u_2^{(0)T}\left(x\left(\tau\right)\right) R_{12} u_2^{(0)}\left(x\left(\tau\right)\right)] d\tau \\[4pt]
=&\; \frac{1}{2} \int_t^\infty x^T\left(\tau\right) \left[Q_1 + K_1^{(0)} S_1 K_1^{(0)} + K_2^{(0)} Z_2 K_2^{(0)}\right] x\left(\tau\right) d\tau
\end{aligned} \tag{19}$$

along the trajectories of the system

$$\dot{x}(t) = Ax(t) + B_1 u_1^{(0)}(x(t)) + B_2 u_2^{(0)}(x(t)) = \left(A - S_1 K_1^{(0)} - S_2 K_2^{(0)}\right) x(t) \tag{20}$$

Similarly, evaluate for the second control agent

$$
\begin{aligned}
J_2^{(0)} &= \frac{1}{2} \int_t^\infty [x^T(\tau) Q_2 x(\tau) + u_1^{(0)^T}(x(\tau)) R_{21} u_1^{(0)}(x(\tau)) \\
&\quad + u_2^{(0)^T}(x(\tau)) R_{22} u_2^{(0)}(x(\tau))] d\tau \\
&= \frac{1}{2} \int_t^\infty x^T(\tau) \left[Q_2 + K_1^{(0)} Z_1 K_1^{(0)} + K_2^{(0)} S_2 K_2^{(0)}\right] x(\tau) d\tau
\end{aligned} \tag{21}
$$

From (19) and (21) we are also able to find the expressions for $\frac{\partial J_1^{(0)}}{\partial x}(t)$ and $\frac{\partial J_2^{(0)}}{\partial x}(t)$ along (20). These expressions will be determined later.

Step 2. For the known value of $\frac{\partial J_1^{(0)}}{\partial x}(t)$ find a new approximation for the control law by minimizing with respect to u_1 the "partially frozen" Hamiltonian

$$
\begin{aligned}
H_1\left(x, u_1, u_2^*, \left(\tfrac{\partial J_1}{\partial x}\right)^{(0)}\right) &= \tfrac{1}{2}\{x^T(t) Q_1 x(t) + u_1^T(t) R_{11} u_1(t) \\
&\quad + u_2^{*^T}(t) R_{21} u_2^*(t)\} + \left(\tfrac{\partial J_1}{\partial x}(t)\right)^{(0)^T}(Ax(t) + B_1 u_1(t) + B_2 u_2^*(t))
\end{aligned} \tag{22}
$$

The minimization produces a stabilizing control given by

$$u_1^{(1)}(t) = -R_{11}^{-1} B_1^T \left(\frac{\partial J_1}{\partial x}(t)\right)^{(0)} \tag{23}$$

Similarly, the minimization of

$$
\begin{aligned}
H_2\left(x, u_1^*, u_2, \left(\tfrac{\partial J_2}{\partial x}\right)^{(0)}\right) &= \tfrac{1}{2}\{x^T(t) Q_2 x(t) + u_1^{*^T}(t) R_{21} u_1^*(t) \\
&\quad + u_2^T(t) R_{22} u_2(t)\} + \left(\tfrac{\partial J_2}{\partial x}(t)\right)^{(0)^T}(Ax(t) + B_1 u_1^*(t) + B_2 u_2(t))
\end{aligned} \tag{24}
$$

with respect to u_2 produces the stabilizing control for the second agent as

$$u_2^{(1)}(t) = -R_{22}^{-1} B_2^T \left(\frac{\partial J_2}{\partial x}(t)\right)^{(0)} \tag{25}$$

Note that $\frac{\partial J_1}{\partial x}$ along the system trajectory can be calculated from (1), (14), namely, by using the indentity

$$\frac{dJ_1}{dt} = \frac{\partial J_1}{\partial x}\frac{dx}{dt} = -\frac{1}{2}\left(x^T(t) Q_1 x(t) + u_1^T(t) R_{11} u_1(t) + u_2^T(t) R_{21} u_2(t)\right) \tag{26}$$

Under the stabilizing control laws $u_1^{(0)}(x(t))$ and $u_2^{(0)}(x(t))$ the last equality produces

$$\frac{\partial J_1^{(0)}}{\partial x}\left(A - S_1 K_1^{(0)} - S_2 K_2^{(0)}\right) x(t)$$
$$= -\tfrac{1}{2} x^T(t)\left[Q_1 + K_1^{(0)} S_1 K_1^{(0)} + K_2^{(0)} S_2 K_2^{(0)}\right] x(t) \tag{27}$$

This simple partial differential equation has a solution of the form

$$J_1^{(0)} = \frac{1}{2} x^T(t) K_1^{(1)} x(t) \tag{28}$$

By using the fact that

$$\frac{\partial J_1^{(0)}}{\partial x} = K_1^{(1)} x(t) \tag{29}$$

we get

$$x^T(t) K_1^{(1)}\left(A - S_1 K_1^{(0)} - S_2 K_2^{(0)}\right) x(t)$$
$$= -\tfrac{1}{2} x^T(t)\left[Q_1 + K_1^{(0)} S_1 K_1^{(0)} + K_2^{(0)} S_2 K_2^{(0)}\right] x(t) \tag{30}$$

Using the standard symmetrization technique known from the derivations of the Riccati equation, that is

$$x^T M x = \frac{1}{2} x^T \left(M + M^T\right) x, \quad \text{for any square matrix} \quad M \tag{31}$$

we get

$$\left(A - S_1 K_1^{(0)} - S_2 K_2^{(0)}\right)^T K_1^{(1)} + K_1^{(1)}\left(A - S_1 K_1^{(0)} - S_2 K_2^{(0)}\right)$$
$$= -\left(Q_1 + K_1^{(0)} S_1 K_1^{(0)} + K_2^{(0)} S_2 K_2^{(0)}\right) \tag{32}$$

Due to the fact that $A - S_1 K_1^{(0)} - S_2 K_2^{(0)}$ is a stable matrix and the right-hand side of equation (32) is negative semidefinite, it follows that a unique nonnegative definite solution $K_1^{(1)}$ exists. *Note that if we assume that the penalty matrix Q_1 is positive definite then the corresponding solution of (32) will be also positive definite.* From (23) and (29) we have

$$\frac{\partial J_1^{(0)}}{\partial x}(t) = K_1^{(1)} x(t) \;\Rightarrow\; u_1^{(1)}(t) = -R_{11}^{-1} B_1^T K_1^{(1)} x(t) \tag{33}$$

Similarly, performing operations (26)-(33) for the second control agent we get

$$\frac{\partial J_2^{(0)}}{\partial x}(t) = K_2^{(1)} x(t) \Rightarrow u_2^{(1)}(t) = -R_{22}^{-1} B_2^T K_2^{(1)} x(t) \tag{34}$$

By repeating steps 1 and 2 now with $u_1^{(1)}(x(t))$ and $u_2^{(1)}(x(t))$ we get $u_1^{(2)}(x(t))$ and $u_2^{(2)}(x(t))$ as well as $K_1^{(2)}$ and $K_2^{(2)}$. Continuing the same procedure, we get the sequences of the solution matrices. It is easy to show (minimization technique in the negative gradient direction) that these sequences are convergent since $K_1^{(m)}, K_2^{(m)}, m = 1, 2, \ldots$ cause the corresponding Hamiltonians to tend to zero. In addition, let $K_1^{(\infty)}$ and $K_2^{(\infty)}$ be the limit points of the corresponding sequences; then from (11)-(12) we have

$$\left(A - S_1 K_1^{(\infty)} - S_2 K_2^{(\infty)}\right)^T K_1^{(\infty)} + K_1^{(\infty)}\left(A - S_1 K_1^{(\infty)} - S_2 K_2^{(\infty)}\right)$$
$$+ \left(Q_1 + K_1^{(\infty)} S_1 K_1^{(\infty)} + K_2^{(i)} Z_2 K_2^{(\infty)}\right) = 0 \tag{35}$$

$$\left(A - S_1 K_1^{(\infty)} - S_2 K_2^{(\infty)}\right)^T K_2^{(\infty)} + K_2^{(\infty)}\left(A - S_1 K_1^{(\infty)} - S_2 K_2^{(\infty)}\right)$$
$$+ \left(Q_2 + K_1^{(\infty)} Z_1 K_1^{(\infty)} + K_2^{(\infty)} S_2 K_2^{(\infty)}\right) = 0 \tag{36}$$

that is, $K_1^{(\infty)}$ and $K_2^{(\infty)}$ satisfy (7)-(8) so that they represent the sought solutions of these equations.

Note that the stronger result than the one stated in Theorem 2.1 can be similarly obtained by assuming that the penalty matrices Q_1 and Q_2 are positive definite (see (32) and the corresponding comment below it).

Assumption 2.2 *The state penalty matrices satisfy $Q_1 > 0$, $Q_2 > 0$.*

In that case we have the following theorem.

Theorem 2.2 *Under Assumptions 2.1 and 2.2 the unique positive definite stabilizing solution of the coupled algebraic Riccati equations (7)-(8) exists. It is obtained by performing Lyapunov iterations (11)-(12).*

Numerical Example 1: In order to demonstrate the efficiency of the proposed algorithm we have run a tenth-order example, which is in fact a system of 110 nonlinear algebraic equations. Matrices A, B_1, and B_2 have been chosen randomly, whereas the choice of matrices Q_1, Q_2, R_{11}, R_{12}, and R_{22} assures that Assumption 2.1 is satisfied. These matrices are given by

$$B_1^T = \begin{bmatrix} -2.036 & 1.560 & -0.907 & -1.214 & 0.813 & 0.044 & -0.750 & 0.901 & 0.913 & 0.084 \\ 0.637 & 0.447 & 1.154 & -1.091 & -0.575 & 0.729 & 0.690 & -1.826 & 0.635 & -0.209 \end{bmatrix}$$

$$B_2^T = \begin{bmatrix} -1.648 & 0.171 & -0.380 & -1.465 & 1.854 & 0.015 & 0.458 & 0.255 & 0.274 & -0.502 \\ -0.759 & 1.256 & -1.076 & -0.101 & 0.745 & 1.717 & -0.091 & -1.304 & -0.763 & 1.345 \end{bmatrix}$$

$$A = \begin{bmatrix} -1.944 & 0.572 & 1.446 & -0.576 & 0.736 & -0.601 & -0.722 & -0.088 & 0.977 & 0.380 \\ 1.440 & 0.393 & 1.023 & -0.711 & 1.282 & -0.679 & 0.010 & 0.588 & 1.281 & -1.414 \\ -0.881 & 1.058 & -1.492 & 1.113 & -1.728 & 0.498 & 0.313 & 1.509 & -1.536 & -0.264 \\ -1.170 & -1.055 & -0.058 & -0.723 & -0.939 & 1.453 & -1.087 & -0.486 & 1.066 & 0.235 \\ 0.736 & -0.569 & 1.449 & -1.383 & 0.116 & -0.052 & 1.387 & 0.659 & -1.658 & -1.437 \\ 0.014 & 0.658 & 0.586 & -0.850 & -0.074 & -1.335 & -0.261 & -1.021 & -0.449 & 1.444 \\ -0.734 & 0.621 & 0.422 & -0.369 & -0.395 & -0.453 & 1.228 & 0.213 & -1.380 & 1.307 \\ 0.820 & -1.746 & 0.178 & -0.860 & -1.235 & -0.902 & 0.390 & -0.656 & -1.658 & 1.329 \\ 0.831 & 0.569 & 1.408 & 1.500 & 1.396 & -0.605 & 0.387 & -0.729 & 1.717 & 1.309 \\ 0.051 & -0.224 & 1.394 & 0.104 & -1.742 & -0.386 & -0.047 & -0.505 & -1.135 & 1.392 \end{bmatrix}$$

$$R_{11} = \begin{bmatrix} 1 & 0 \\ 0 & 2 \end{bmatrix}, \quad R_{12} = \begin{bmatrix} 3 & 0 \\ 0 & 4 \end{bmatrix}, \quad R_{21} = \begin{bmatrix} 5 & 0 \\ 0 & 6 \end{bmatrix}, \quad R_{22} = \begin{bmatrix} 7 & 0 \\ 0 & 8 \end{bmatrix}, \quad Q_1 = I_{10}, \ Q_2 = I_{10}$$

The obtained results are really remarkable since only after 8 iterations we got very good convergence. These results are presented in Table 1. The errors are defined as the absolute values of the largest elements in matrices $\mathcal{N}_1\left(K_1^{(i)}, K_2^{(i)}\right)$ and $\mathcal{N}_2\left(K_1^{(i)}, K_2^{(i)}\right)$ where i stands for the number of iterations.

Simulation results presented in Table 1 are obtained by using the software package L-A-S (Bingulac and Vanlandingham, 1993).

Iteration	*error 1*	*error 2*
1	$1.5283 \times 10^{+2}$	$1.4193 \times 10^{+2}$
2	$1.6726 \times 10^{+1}$	$4.0585 \times 10^{+1}$
3	$3.1057 \times 10^{+0}$	$1.2188 \times 10^{+1}$
4	2.3207×10^{-1}	$2.4337 \times 10^{+0}$
5	1.2386×10^{-1}	7.6489×10^{-2}
6	4.1600×10^{-3}	6.9948×10^{-5}
7	7.0661×10^{-4}	3.0096×10^{-7}
8	2.4374×10^{-5}	9.2183×10^{-8}

Table 1: Simulation results for a system of 110 nonlinear scalar equations

3 Algorithm for Solving the Generalized Algebraic Riccati Equation

Consider *the generalized algebraic Riccati equation* appearing in the H_∞ optimal control problems and zero-sum differential games (Basar, 1991; Hewer, 1993; Zhou and Khargonekar, 1987; Bernstein and Haddad, 1989; Mageirou, 1976; Peterson, 1988; Basar and Bernhard, 1991)

$$A^T P + PA + Q - P\left(\frac{1}{\gamma^2}S - Z\right)P = 0 \tag{37}$$

where A, Q, S, Z, and P are real constant matrices of dimensions $n \times n$ and γ is a real positive parameter $0 < \gamma \le \infty$. In addition, matrices S, Z, *and* Q *are positive semidefinite.* An important feature of equation (37) which distinguishes this equation from the standard linear-quadratic optimal control algebraic Riccati equation is that the matrix $\frac{1}{\gamma^2}S - Z$ is in general indefinite. Equation (37) or its forms also appear in the stabilization of uncertain systems (Peterson and Hollot, 1986; Peterson, 1988), disturbance attenuation problems (Peterson, 1987), and decentralized stabilization (Mageirou and Ho, 1977).

In this paper we develop an elegant and simple algorithm which converges globally to the *positive semidefinite stabilizing* solution of (37) under stabilizability-detectability assumptions. The algorithm is given in terms of the standard algebraic Riccati equations, which have to be solved iteratively. We have also presented the Lyapunov iterations version of the proposed algorithm. Note that the Lyapunov iterations for finding the *positive definite* solution of (37), based on the "Bellman approximation in policy space," were obtained in Mageirou and Ho, 1977; Mageirou, 1977, see also Basar, 1991. However, that algorithm is different from the corresponding one presented in this paper and *it cannot be used* for finding the positive semidefinite stabilizing solution of (37), which is of interest for H_∞ optimal control problems.

The generalized algebraic Riccati equation (37) will be solved under the following assumption.

Assumption 3.1 *The triple* $\left(A, \sqrt{S}, \sqrt{Q}\right)$ *is stabilizable-detectable.*

We propose the following algorithm for solving (37).
Algorithm 2:

$$A^T P^{(i+1)} + P^{(i+1)}A - \frac{1}{\gamma^2}P^{(i+1)}SP^{(i+1)} + Q + P^{(i)}ZP^{(i)} = 0$$

$$\textit{with the initial condition obtained from} \tag{38}$$

$$A^T P^{(0)} + P^{(0)}A + Q - \frac{1}{\gamma^2}P^{(0)}SP^{(0)} = 0$$

The properties of this algorithm are stated in the next theorem.

Theorem 3.1 *The proposed algorithm converges globally under Assumption 3.1 to the positive semidefinite stabilizing solution of the generalized algebraic Riccati equation (37), assuming that such a solution exists.*

Note that Assumption 3.1 may be tightened so that the algorithm converges to the positive definite stabilizing solution of (37), see Remark 1.

Before we proceed with the proof of the algorithm's global convergence we need the following lemma which follows from Hewer, 1993; Peterson, 1988.

Lemma 3.1 *If it exists the positive semidefinite stabilizing solution of (37) is unique.*

In addition, it is easy to observe that if P is stabilizing, that is, $A - \frac{1}{\gamma^2}SP + ZP$ is stable, then the feedback matrix $A - \frac{1}{\gamma^2}SP$ is stable too.

Proof: Under Assumption 3.1 the unique positive semidefinite stabilizing solution of the standard algebraic Riccati equation given by

$$A^T P^{(0)} + P^{(0)} A - \frac{1}{\gamma^2} P^{(0)} S P^{(0)} + Q = 0 \tag{39}$$

exists. Subtracting (39) from (37) and assuming that the positive semidefinite stabilizing solution of (37) exists, we get

$$\begin{aligned}
\left(A - \tfrac{1}{\gamma^2}SP\right)^T \left(P - P^{(0)}\right) + \left(P - P^{(0)}\right)\left(A - \tfrac{1}{\gamma^2}SP\right) \\
= -\tfrac{1}{\gamma^2}\left(P - P^{(0)}\right)S\left(P - P^{(0)}\right) - PZP
\end{aligned} \tag{40}$$

Since $\left(A - \frac{1}{\gamma^2}SP\right)$ is stable and the right side of (40) is negative semidefinite, it follows that

$$P - P^{(0)} \geq 0 \;\Rightarrow\; P \geq P^{(0)} \tag{41}$$

Consider now the next iteration of (38), that is

$$A^T P^{(1)} + P^{(1)} A - \frac{1}{\gamma^2} P^{(1)} S P^{(1)} + \left(Q + P^{(0)} Z P^{(0)}\right) = 0 \tag{42}$$

Since

$$Q + \frac{1}{\gamma^2} P^{(0)} S P^{(0)} \geq Q \tag{43}$$

we have

$$P^{(1)} \geq P^{(0)} \tag{44}$$

Continuing the same procedure for $i = 2, 3, ...$, it can be shown by using the properties of the algebraic Riccati equation that

$$Q + P^{(i)} Z P^{(i)} \geq Q + P^{(i-1)} Z P^{(i-1)} \;\Rightarrow\; P^{(i+1)} \geq P^{(i)} \tag{45}$$

hence, we have from (38)

$$P^{(i+1)} \geq P^{(i)} \geq ...P^{(0)} \geq 0, \quad i = 0, 1, 2, ... \tag{46}$$

This monotonically nondecreasing sequence $\{P^{(i)}\}$ has the upper bound. To show this we subtract (38) from (37) which produces

$$\begin{aligned} \left(A - \tfrac{1}{\gamma^2}SP\right)^T \left(P - P^{(i+1)}\right) + \left(P - P^{(i+1)}\right)\left(A - \tfrac{1}{\gamma^2}SP\right) \\ = -\tfrac{1}{\gamma^2}\left(P - P^{(i+1)}\right)S\left(P - P^{(i+1)}\right) - \left(PZP - P^{(i)}ZP^{(i)}\right) \end{aligned} \tag{47}$$

Established relation (41) and the induction arguments imply

$$P \geq P^{(i+1)}, \quad i = 0, 1, 2, ... \tag{48}$$

Thus, the required positive semidefinite stabilizing solution of (37) represents the upper bound for the sequence (46). The bounded sequence defined by (46) and (48) is convergent by the monotonic convergence of positive operators (Wonham, 1968; Kantorovich and Akilov, 1964). Assuming that $P^{(\infty)}$ is the limit point of the sequence $\{P^{(i)}\}$ we have

$$A^T P^{(\infty)} + P^{(\infty)} A - \frac{1}{\gamma^2} P^{(\infty)} S P^{(\infty)} + Q + P^{(\infty)} Z P^{(\infty)} = 0 \tag{49}$$

Since equations (37) and (49) are identical, it follows that the proposed algorithm converges to the required solution of (37). Since by Lemma 3.1 the stabilizing solution of (37) is unique, the proposed algorithm is globally convergent. This completes the convergence proof of the proposed algorithm and proves stated Theorem 3.1.

Remark 1. By tightening Assumption 3.1 to the triple $\left(A, \sqrt{S}, \sqrt{Q}\right)$ is stabilizable-observable, the proposed algorithm produces *the positive definite* stabilizing solution of (37). This is important for zero-sum differential games where the required solution is positive definite.

Note that an algorithm for solving (37) of zero-sum differential games, in terms of the Lyapunov iterations, has been proposed in Mageirou and Ho, 1977; Mageirou, 1977 in the form

$$\left(A + QM^{(i)}\right)^T M^{(i+1)} + M^{(i+1)}\left(A + QM^{(i)}\right) = (S - Z) + M^{(i)}QM^{(i)} \tag{50}$$

with $M^{(0)}$ being any antistabilizing initial guess and

$$P^{(i+1)} = M^{(i+1)^{-1}}, \quad P^{(\infty)} = M^{(\infty)^{-1}} \to P \tag{51}$$

Remark 2. The first obvious drawback of algorithm (50)-(51) is that it produces only the positive definite solution (P must be invertible) so that it cannot be used for the H^∞ optimal control problems. Secondly, finding the stabilizing initial guess for high order problems is computationally involved.

Note that Algorithm 2 can be reformulated in terms of the Lyapunov iterations by having in mind that the solution of any standard algebraic Riccati equation can be obtained by performing iterations on the Lyapunov algebraic equations (Kleinman, 1968). The "linearized" version of Algorithm 2 is given by

Algorithm 3:

$$
\left(A - \tfrac{1}{\gamma^2}SL^{(i)}\right)^T L^{(i+1)} + L^{(i+1)}\left(A - \tfrac{1}{\gamma^2}SL^{(i)}\right)
$$
$$
= -\left(Q + \tfrac{1}{\gamma^2}L^{(i)}SL^{(i)} + L^{(i)}ZL^{(i)}\right) \tag{52}
$$
$$
\text{with} \quad A^T L^{(0)} + L^{(0)}A + Q - \tfrac{1}{\gamma^2}L^{(0)}SL^{(0)} = 0
$$

It should be pointed out that the sequence generated by Lyapunov iterations (52) is closer to the required positive semidefinite stabilizing solution than the corresponding sequence generated by the Riccati iterations (38). To see this, observe that both sequences start at the same initial point and that from (38) and (52) we have

$$
\left(A - \frac{1}{\gamma^2}SL^{(i)}\right)^T \left(L^{(i+1)} - P^{(i+1)}\right) + \left(L^{(i+1)} - P^{(i+1)}\right)\left(A - \frac{1}{\gamma^2}SL^{(i)}\right) =
$$
$$
-\frac{1}{\gamma^2}\left(P^{(i+1)} - L^{(i)}\right)S\left(P^{(i+1)} - L^{(i)}\right) - \left(L^{(i)}ZL^{(i)} - P^{(i)}ZP^{(i)}\right) \tag{53}
$$

Since $L^{(0)} = P^{(0)}$ it follows that $L^{(1)} \geq P^{(1)}$; then by induction, it is easy to establish that

$$
L^{(i+1)} \geq P^{(i+1)}, \quad i = 0, 1, 2, \ldots \tag{54}
$$

Other technical details establishing the complete convergence proof for Algorithm 3 can be obtained similarly to those of Algorithm 2.

Numerical Example 2: In order to demonstrate the efficiency of the proposed algorithm we solve the following example. The problem matrices for (37) are given by

$$
A = \begin{bmatrix} 1 & 0 & 3 & 0 \\ 0 & -2 & 3 & 0 \\ 0 & 1 & -3 & 0 \\ 1 & 0 & 0 & -4 \end{bmatrix}, \quad
Q = \begin{bmatrix} 1 & 0 & 0 & 0 \\ 0 & 1 & 0 & 0 \\ 0 & 0 & 0 & 0 \\ 0 & 0 & 0 & 0 \end{bmatrix}
$$

$$
S = B_1 R_1^{-1} B_1^T, \quad Z = B_2 R_2^{-1} B_2^T, \quad R_1 = I_2, \quad R_2 = I_1, \quad \gamma = 1
$$

$$
B_1^T = \begin{bmatrix} 0.0116 & 0.6020 & 0.9215 & 0.5565 \\ 0.5450 & 0.0730 & 0.5565 & 0.3834 \end{bmatrix}
$$

$$
B_2^T = \begin{bmatrix} 0.4814 & 0.3909 & 0.4087 & 0.5591 \end{bmatrix}
$$

The open-loop eigenvalues of the matrix A are $\{-4.3028, -4, -0.6972, 1\}$, which indicates that this system is open-loop unstable. The rank of the observability matrix is 3 so that this system is not observable. However, the system is both stabilizable and detectable. Note that matrices B_1 and B_2 are obtained by using a random number generator. Simulation results are obtained using MATLAB. The proposed Algorithm 2 has produced the positive *semidefinite* stabilizing solution with accuracy of $O\left(10^{-5}\right)$ after 16 iterations. It is interesting to point out that the same accuracy is obtained after the same number of iterations by using Algorithm 3 based on the Lyapunov iterations. Results for the trace of the matrix P per iteration for both the Riccati iterations algorithm (38) and the Lyapunov iterations algorithm (52) are given in Table 2. Note that the Lyapunov iterations results (column

Iteration	Algorithm 2 $traceP^{(i)}$	Algorithm 3 $traceP^{(i)}$
1	4.2911	4.4778
2	4.7192	4.8320
3	4.9459	5.0093
4	5.0678	5.1026
5	5.1337	5.1526
6	5.1694	5.1796
7	5.1886	5.1941
8	5.1990	5.2020
9	5.2046	5.2062
10	5.2078	5.2085
11	5.2093	5.2097
12	5.2101	5.2104
13	5.2106	5.2107
14	5.2109	5.2109
15	5.2110	5.2110
16	5.2111	5.2111
Optimal =	5.2111	5.2111

Table 2: Solution of the generalized algebraic Riccati equation

3) are closer to the desired solution in each iteration than those from the Riccati iterations (column 2), as established in (54). The obtained stabilizing

positive semidefinite solution is given by

$$P^{(16)} = \begin{bmatrix} 3.0103 & 0.3834 & 2.1315 & 0 \\ 0.3834 & 0.3875 & 0.5205 & 0 \\ 2.1315 & 0.5205 & 1.8134 & 0 \\ 0 & 0 & 0 & 0 \end{bmatrix} \geq 0$$

The eigenvalues of the closed-loop matrix are obtained as

$$\lambda\left\{ A - (S - Z)\, P^{(16)} \right\} = \left\{ \begin{array}{c} -4.1317 \\ -4.0000 \\ -1.3959 \pm j0.5865 \end{array} \right\}$$

4   Conclusions

An iterative algorithm leading to the nonnegative (positive) definite stabilizing solution of coupled algebraic Riccati equations is constructed. Computational requirements are reduced considerably since the problem decomposition is achieved and only reduced-order Lyapunov equations have to be solved. In addition, a simple and elegant algorithm is presented for solving the algebraic Riccati equation of H^∞ optimal control and zero-sum differential games. This algorithm, in fact, finds the desired positive semidefinite stabilizing solution, assuming that it exists. The algorithm's initial guess is easily obtained.

REFERENCES

[1]   H. ABOU-KANDIL, G. FREILING, and G. JANK, *Necessary conditions for constant solutions of coupled Riccati equations in Nash games*, Systems & Control Letters, 21 (1993), 295-306.

[2]   T. BAŞAR, *Generalized Riccati equations in dynamic games*, in *The Riccati Equation*, S. Bittanti, A. Laub, and J. Willems, eds., Springer-Verlag, 1991.

[3]   T. BAŞAR and P. BERNHARD, *H^∞ Optimal Control and Related Minimax Design problems: A Dynamic Game Approach*, Birkhäuser, Boston, 1991.

[4]   T. BAŞAR, *A counterexample in linear-quadratic games: existence of non-linear Nash strategies*, J. of Optimization Theory and Applications, 14 (1974), 425-430.

[5] R. BELLMAN, *Monotone approximation in dynamic programming and calculus of variations*, Proc. The National Academy of Science USA, 44 (1954), 1073-1075.

[6] R. BELLMAN, *Dynamic Programming*, Princeton University Press, 1957.

[7] R. BELLMAN, *Adaptive Control Processes: A Guided Tour*, Princeton University Press, 1961.

[8] D. BERNSTEIN and W. HADDAD, *LQG control with an H_∞ performance bound: A Riccati equation approach*, IEEE Trans. Automatic Control, 34 (1989), 293-305.

[9] D. BERTSEKAS, *Dynamic Programming: Deterministic and Stochastic Models*, Prentice Hall, Englewood Cliffs, 1987.

[10] D. BERTSEKAS and J. TSITSIKLIS, *Some aspects of parallel and distributed iterative algorithms - A survey*, Automatica, 27 (1991), 3-21.

[11] S. BINGULAC and H. VANLANDINGHAM, *Algorithms for Computer-Aided Design of Multivariable Control Systems*, Marcel Dekker, New York, 1993.

[12] Z. GAJIC and T-Y. LI, *Simulation results for two new algorithms for solving coupled algebraic Riccati equations*, Third Int. Symp. on Differential Games, Sophia Antipolis, France, June 1988.

[13] Z. GAJIC and X. SHEN, *Parallel Algorithms for Optimal Control of Large Scale Linear Systems*, Springer Verlag, London, 1993.

[14] G. HEWER, *Existence theorems for positive semidefinite and sign indefinite stabilizing solutions of H_∞ Riccati equations*, SIAM J. Control and Optimization, 31 (1993), 16-29.

[15] L. JODAR and H. ABOU-KANDIL, *Kronecker products and coupled matrix Riccati differential equations*, Linear Algebra and Its Applications, 121 (1989), 39-51.

[16] L. KANTOROVICH and G. AKILOV, *Functional Analysis in Normed Spaces*, Macmillan, New York, 1964.

[17] D. KLEINMAN, *On an iterative techniques for Riccati equation computations*, IEEE Trans. Automatic Control, 13 (1968), 114-115.

[18] H. KHALIL and P. KOKOTOVIC, *Feedback and well-posedness of singularly perturbed Nash games*, IEEE Trans. Automatic Control, 24 (1979), 699-708.

[19] H. KHALIL, *Multimodel design of a Nash strategy*, J. Optimization Theory and Application, (1980), 553-564.

[20] D. KIRK, *Optimal Control Theory*, Prentice Hall, Englewood Cliffs, 1970.

[21] N. KRIKELIS and A. REKASIUS, *On the solution of the optimal linear control problems under conflict of interest*, IEEE Trans. Automatic Control, 16 (1971), 140-147.

[22] V. KUCERA, *A contribution to matrix quadratic equations*, IEEE Trans. Automatic Control, 17 (1972), 344-347.

[23] R. LARSON, *A survey of dynamic programming computational procedures*, IEEE Trans. Aut. Control, 12 (1967), 767-774.

[24] R. LEAKE and R. LIU, *Construction of suboptimal control sequences*, SIAM J. Control, 5 (1967), 54-63.

[25] M. LEVINE and T. VILIS, *On-line learning optimal control using successive approximation techniques*, IEEE Trans. Aut. Control, 19, (1973) 279-284.

[26] E. MAGERIOU, *Values and strategies for infinite time linear quadratic games*, IEEE Trans. Automatic Control, 21 (1976), 547-550.

[27] E. MAGERIOU, *Iterative techniques for Riccati game equations*, J. Optimization Theory and Applications, 22 (1977), 51-61.

[28] E. MAGERIOU and H. HO, *Decentralized stabilization via game theoretic methods*, Automatica, 13 (1977), 393-399.

[29] M. MARITON, *Jump Linear Systems in Automatic Control*, Marcell Dekker, New York, Basel, 1990.

[30] G. MIL'SHTEIN, *Successive approximation for solution of one optimum problem*, Auto. and Rem. Control, 25 (1964), 298-306.

[31] G. OLSDER, *Comment on a numerical procedure for the solution of differential games*, IEEE Trans. Automatic Control, 20 (1975), 704-705.

[32] U. OZGUNER and W. PERKINS, *A series solution to the Nash strategy for large scale interconnected systems*, Automatica, 13 (1977), 313-315.

[33] G. PAPAVASSILOPOULOS, J. MEDANIC, and J. CRUZ, *On the existence of Nash strategies and solutions to coupled Riccati equations in linear-quadratic games*, J. of Optimization Theory and Applications, 28 (1979), 49-75.

[34] I. PETERSON, *Disturbance attenuation and H^∞ optimization: A design method based on the algebraic Riccati equation*, IEEE Trans. Automatic Control, 32 (1987), 427-429.

[35] I. PETERSON, *Some new results on algebraic Riccati equations arising in linear quadratic differential games and stabilization of uncertain linear systems*, Systems & Control Letters, 10 (1988), 341-348.

[36] I. PETERSON and C. HOLLOT, *A Riccati approach to the stabilization of uncertain linear systems*, Automatica 22 (1986), 397-411.

[37] B. PETROVIC and Z. GAJIC, *The recursive solution of linear quadratic Nash games for weakly interconnected systems*, J. Optimization Theory and Application, 56 (1988), 463-477.

[38] A. STARR and Y. HO, *Nonzero-sum differential games*, J. of Optimization Theory and Applications, 3 (1969), 184-206.

[39] D. TABAK, *Numerical solution of differential game problems*, Int. J. Systems Sci., 6 (1975), 591-599, 1975.

[40] E. VAISBORD, *An approximate method for the synthesis of optimal control*, Auto. and Rem. Control, 24 (1963), 1626-1632.

[41] W. WONHAM, *On a matrix Riccati equation of stochastic control*, SIAM J. on Control, 6 (1968), 681-697.

[42] K. ZHOU and P. KHARGONEKAR, *An algebraic Riccati equation approach to H^∞ optimization*, Systems & Control Letters, 11 (1987), 85-91.

A Turnpike Theory for Infinite Horizon
Open-Loop Differential Games
with Decoupled Controls *

D. Carlson
Dept. of Mathematics, University of Toledo,
Toledo, Ohio, USA.

A. Haurie
Dept. of Management Studies, University of Geneva,
102 Carl-Vogt, CH-1211, Geneva, Switzerland
and GERAD-Ecole des HEC, Montréal, Canada.

Abstract

This paper deals with a class of open-loop differential games played over an infinite time horizon. The equilibrium concept is defined in the sense of overtaking optimal responses by the players to the program choices of the opponents. We extend to this dynamic game framework the results obtained by Rosen for concave static games. We prove existence, uniqueness and asymptotic stability (also called the turnpike property) of overtaking equilibrium programs for a class of games satisfying a strong concavity assumption (strict diagonal concavity).

1   Introduction

This paper deals with a class of dynamic competitive process models defined over an infinite time horizon. We use overtaking optimality to deal with unbounded payoffs. In the present paper we consider a dynamic competitive process modeled in continuous time and we provide a theory of existence, uniqueness and asymptotic stability. We obtain these results under conditions very similar to Rosen's *strict diagonal concavity.*

This paper complements Ref. [9] in which we have developed a turnpike theory for discrete time competitive processes. Both papers use basically the same approach; however, the continuous time framework requires specific

---

*Research supported by FNRS-Switzerland, FCAR-Québec, NSERC-Canada, a Summer Research Fellowship from the University of Toledo, and by a travel grant from the Dept. of Management Studies of University of Geneva,.

developments which were not covered in [9], especially when we are proving existence of an equilibrium in the absence of discounting.

The theory of optimal control over an infinite time horizon with the *overtaking criterion* has been motivated by the study of optimal economic growth models. We refer the reader to the books [1] and [14] for a presentation of the economic models and to the book [10] for a comprehensive discussion of the optimal control problem. An extension of this theory to the case of open-loop differential games is natural both as an economic paradigm and as an optimization problem. Instead of considering a single decision maker accumulating production capacities it is possible to model several firms competing in a market through the accumulation of production capacity. One of the first models of competition among a few firms, over an infinite time horizon, is due to Brock [6]. This model was characterized by an assumption of decoupled dynamics. Indeed each firm is controlling its own accumulation of production capacity. The coupling between the firms was then essentially due to their interactions through the price determining demand law in the definition of the payoff functionals.

An attempt to extend the *global asymptotic stability (GAS)* conditions of state and co-state trajectories, known as the *turnpike property*, to open-loop differential games is also reported in [6]. A set of sufficient conditions for obtaining *GAS* results in open-loop infinite horizon differential games has been proposed by Haurie and Leitmann in [24]. In that work the equilibrium is defined in terms of overtaking optimality of the response of each player to the controls chosen by the opponents. Conditions for *GAS* are given in terms of a so-called *vector Lyapunov function* applied to the *pseudo Hamiltonian system* resulting from the necessary optimality conditions.

Knowing that such a *GAS* property holds permits in particular the development of numerical methods for solving infinite horizon differential games. The use of this property in the analysis of competition has been well illustrated in Ref. [21] where a transboundary fisheries model, with several nations exploiting the same biomass, has been studied and numerically solved, using the asymptotic steady state as terminal conditions.

More recently Haurie and Roche [25] have proposed a numerical analysis of a class of stochastic oligopolistic games which exploits this *GAS* property. More specifically they developed a theory for a *piecewise deterministic oligopoly*, where the solution is obtained through the analysis of a class of associated infinite horizon open-loop differential games. The numerical technique developed and illustrated in [25] exploits the *turnpike property* for these associated differential games.

In [25], the authors also give the elements of a theory of sufficiency and asymptotic stability of the coupled Hamiltonian systems characterizing open-loop equilibria which is close to those developed for the single controller case in [22] and [17]. This theory is based on the assumption that a *strong support property* of the optimal trajectories holds. In [25] it is observed that the support property used to provide n sufficient conditions for asymptotic stability is similar to the *strict diagonal concavity* condition introduced by Rosen in his important paper [33], where he proved existence and uniqueness of equilibria in static concave games. This similarity has been exploited in [23] and [27] where the coordination of an oligopoly under a *long term* global environmental constraint is considered.

These scattered results were still lacking two important components in order to provide a theory as complete as in the optimal control case, namely an existence result and a uniqueness result. This is provided in this paper which proposes a "complete" theory in the case of differential games with decoupled controls, i.e. where the players only interact through the state variables and not, directly through the control they use. This decoupling of controls is less restrictive than the decoupling of dynamics. For the sake of comprehensiveness we include also the theory of asymptotic stability, which is close to what was already done in [25] but which is now developed in a more general framework.

The paper is organized as follows: in section 2 we explore the *turnpike property* for a general class of differential games, permitting, as in [28] for the single player case, a nonautonomous dynamics; in section 3 we show that the conditions used to insure asymptotic stability also imply existence of *overtaking equilibria*; in section 4 we show that in addition we get uniqueness.

2 Turnpikes for Overtaking Equilibria

In this section we define the concept of an *overtaking equilibrium* for a class of games which represent infinite horizon dynamic competition among m firms. We give conditions under which all the overtaking equilibrium programs, emanating from different initial states, bunch together at infinity.

2.1 Competitive programs

We consider a competitive process defined by the following data:

- An infinite time horizon $t \in [0, \infty)$.

- A set $M \doteq \{1, ..., m\}$ of m players (or firms) represented at time t by a state $x_j(t) \in \mathbb{R}^{n_j}$, where n_j is a positive integer (we denote $n \doteq n_1 + ... + n_m$). This state is e.g., the production capacity of firm j.

- For each $j \in M$, a program for player j is defined as an absolutely continuous function $\mathbf{x}_j = (x_j(t) \in \mathbb{R}^{n_j} : t \geq 0)$. An M-program is defined as $\mathbf{x} = (x(t) : t \geq 0) \doteq ((x_j(t))_{j \in M} : t \geq 0)$.

- Along an M-program, the reward accumulation process for player j is defined as

$$\phi_j^T(\mathbf{x}) = \int_0^T e^{-\rho_j t} L_j(t, x(t), \dot{x}_j(t))\, dt, \quad T > 0, \tag{1}$$

where $\rho_j \geq 0$ is the discount rate for player j , $L_j : [0, \infty) \times \mathbb{R}^n \times \mathbb{R}^{n_j} \mapsto \mathbb{R} \cup \{-\infty\}$, $j \in M$ are given functions and $\dot{x}_j(t) = \frac{d}{dt} x_j(t)$. The expression $L_j(t, x(t), \dot{x}_j(t))$ represents e.g. the net income to firm j when the market price is a function of total supply $\sum_{j \in M} x_j(t)$ minus the cost for the capacity adjustment $\dot{x}_j(t)$. The control decoupling comes from the fact that only the velocity $\dot{x}_j(t)$ enters in the definition of the reward rate of player j.

Remark 2.1 *There is indeed no loss of generality in adopting this generalized calculus of variations formalism instead of a state equation formulation of each player's dynamics. For details on the transformation of a full fledged control formulation, including state and control constraints, into a generalized calculus of variation formulation we refer to [7], [17] or [10].*

2.2 Optimality

We now introduce a version of the Nash-equilibrium concept which is adapted to the consideration of an infinite time horizon. Given an M-program $\mathbf{x}^*$ we denote $[\mathbf{x}^{*(j)}; \mathbf{x}_j]$ the M-program obtained when player j unilaterally changes his program to $\mathbf{x}_j$.

Definition 2.1 *An M-program $\mathbf{x}^*$ is an **equilibrium** at x^o if*

1. $x^*(0) = x^o$

2. $\lim_{T \to \infty} \phi_j^T(\mathbf{x}^*) < \infty$ *for all* $j \in M$

3. $\liminf_{T \to \infty}(\phi_j^T(\mathbf{x}^*) - \phi_j^T([\mathbf{x}^{*(j)}; \mathbf{x}_j])) \geq 0$ *for all programs* $\mathbf{x}_j$ *such that* $x_j(0) = x_j^o$ *for all* $j \in M$.

If only the first and third conditions hold the M-program $\mathbf{x}^*$ is called an **over-taking-equilibrium** *at* x^o.

Remark 2.2 *The consideration of the* overtaking equilibrium *concept in dynamic games seems to have been introduced by Rubinstein in [32]. The concept has also been used in [24] and [26].*

2.3 Optimality Conditions

We recall the necessary optimality conditions for open-loop (overtaking) equilibrium. These conditions are a direct extension of the celebrated *Maximum Principle* established by Halkin [20] for infinite horizon control problems but written for the case of convex systems. Let's introduce for $j \in M$ and $p_j \in \mathrm{IR}^{n_j}$ the *Hamiltonians* $H_j : [0, \infty)\, \mathrm{IR}^n \times \mathrm{IR}^{n_j} \mapsto \mathrm{IR} \cup \{-\infty\}$, defined as

$$H_j(t, x, p_j) = \sup_{z_j}\{L_j(t, x, z_j) + p_j' z_j\}. \tag{2}$$

Here p_j is called a *j-supporting price* vector. A function

$$\mathbf{p} = ((p_j(t))_{j \in M}\; :\; t \geq 0)$$

will be called an *M-price schedule*.

Assumption 2.1 *We assume that the Hamiltonians H_j are concave in x_j, convex in p_j.*

For an (overtaking) equilibrium the following necessary conditions hold:

Theorem 2.1 *If $\mathbf{x}^*$ is an* **overtaking equilibrium** *at initial state x^o then there exists an absolutely continuous M-price schedule $\mathbf{p}^*$ such that*

$$\dot{x}_j^*(t) \;\in\; \partial_{p_j} H_j(t, x^*(t), p_j^*(t)) \tag{3}$$
$$\dot{p}_j^*(t) + \rho_j\, p_j^*(t) \;\in\; -\partial_{x_j} H_j(t, x^*(t), p_j^*(t)) \tag{4}$$

for all $j \in M$.

Proof: For a fixed set of programs $\mathbf{x}^{*(j)}$ chosen by the competitors, Player-j's optimal reply $\mathbf{x}_j^*$ is obtained as the solution of an infinite horizon optimal control problem where only the x_j dynamics matter. We now apply the necessary conditions given in [31] for control systems satisfying Assumption 2.1, and adapt the proof given by Halkin [20] in an obvious manner. $\bullet$

The equations (3) and (4) have been called *pseudo-Hamiltonian systems* in [24].

These conditions are incomplete since only initial conditions are specified for the M-programs and no transversality conditions are given for their associated M-price schedules.

In the single player case, this system is made complete by invoking the turnpike property which provides an asymptotic transversality condition. Due to the coupling among the players, the system (4) considered here does not fully enjoy the rich geometric structure found in the classical optimization setting (e.g. the saddle point behavior of Hamiltonian systems in the autonomous case). In the next two sections we provide conditions under which the turnpike property holds for these pseudo-Hamiltonian systems.

2.4   A Turnpike Result for the Undiscounted Case

Let us first consider the case where $\rho_j \equiv 0$ for all $j \in M$. Then, clearly, we have to deal with overtaking equilibria since the sequence of accumulated rewards can be unbounded.

We introduce below the fundamental assumption which underlines our developments. It is directly linked with the *Strict Diagonal Concavity Assumption* made by Rosen [33] in his study of concave static games. Recall first the definition given by Rosen. Let $x = (x_j)_{j=1,\ldots,m} \in \mathbb{R}^{n_1} \times \ldots \times \mathbb{R}^{n_m}$ and consider m continuously differentiable functions $\phi_j(x)$. Let $\nabla_j \phi_j(x)$ denote the gradient of $\phi_j(x)$ with respect to x_j. The sum

$$\sigma(x) = \sum_{j=1}^{m} \phi_j(x)$$

is said to be *diagonally strictly concave* if for every x^0 and x^1 we have

$$\sum_{j=1,\ldots,m} (x_j^1 - x_j^0)'(\nabla_j \phi_j(x^0) - \nabla_j \phi_j(x^1)) > 0.$$

Our assumption concerns the sum of the Hamiltonians. It is also formulated in a slightly more general way to take care of the possible nondifferentiability of these functions.

Assumption 2.2 *(Strict Diagonal Concavity-Convexity Assumption (SDC-CA)). We assume that the combined Hamiltonian $\sum_{j \in M} H_j(t, x, p_j)$ is strictly diagonally concave in x, convex in p. That is*

$$\sum_{j \in M} \left[(\hat{p}_j - \tilde{p}_j)'(\hat{\pi}_j - \tilde{\pi}_j) + (\hat{x}_j - \tilde{x}_j)'(\hat{\xi}_j - \tilde{\xi}_j) \right] > 0, \tag{5}$$

for all $(\hat{x}_j, \tilde{x}_j, \hat{p}_j, \tilde{p}_j)$ and $(\hat{\pi}_j, \tilde{\pi}_j, \hat{\xi}_j, \tilde{\xi}_j)$, such that

$$\hat{\pi}_j \;\in\; \partial_{p_j} H_j(t, \hat{x}, \hat{p}_j), \quad \tilde{\pi}_j \in \partial_{p_j} H_j(t, \tilde{x}, \tilde{p}_j) \tag{6}$$

$$\hat{\xi}_j \;\in\; -\partial_{x_j} H_j(t, \hat{x}, \hat{p}_j), \quad \tilde{\xi}_j \in -\partial_{x_j} H_j(t, \tilde{x}, \tilde{p}_j). \tag{7}$$

A direct consequence of Assumption 2.2 is the following lemma which gives insight into the definition of a Lyapunov function providing a sufficient condition for *GAS*.

Lemma 2.1 *Assume $\rho_j \equiv 0$ for all $j \in M$. Let $\hat{\mathbf{x}}$ and $\tilde{\mathbf{x}}$ be two (overtaking) equilibria at $\hat{x}^o$ and $\tilde{x}^o$ respectively, with their respective associated M-price schedules $\hat{\mathbf{p}}$ and $\tilde{\mathbf{p}}$. Then under Assumptions 2.1 and 2.2 the inequality*

$$\sum_{j \in M} \frac{d}{dt} \left[(\hat{p}_j(t) - \tilde{p}_j(t))'(\hat{x}_j(t) - \tilde{x}_j(t)) \right] > 0 \tag{8}$$

holds.

Proof: According to Theorem 2.1 we have

$$\frac{d}{dt}\hat{x}_j(t) \;\in\; \partial_{p_j} H_j(t, \hat{x}(t), \hat{p}_j(t)) \tag{9}$$

$$\frac{d}{dt}\hat{p}_j(t) \;\in\; -\partial_{x_j} H_j(t, \hat{x}(t), \hat{p}_j(t)) \tag{10}$$

$$\frac{d}{dt}\tilde{x}_j(t) \;\in\; \partial_{p_j} H_j(t, \tilde{x}(t), \tilde{p}_j(t)) \tag{11}$$

$$\frac{d}{dt}\tilde{p}_j(t) \;\in\; -\partial_{x_j} H_j(t, \tilde{x}(t), \tilde{p}_j(t)). \tag{12}$$

Then (8) follows directly from Assumption 2.2. $\quad\bullet$

We now prove the Turnpike theorem for (overtaking) equilibria under a strengthening of the inequality (8).

Definition 2.2 *We say that the (overtaking) equilibrium M-program $\hat{\mathbf{x}}$ is f bf strongly diagonally supported by the M-price schedule $\hat{\mathbf{p}}$ if, for every $\varepsilon > 0$ there exists a $\delta > 0$, such that for all $t \geq 0$, $\|x - \hat{x}(t)\| + \|p - \hat{p}(t)\| > \varepsilon$ implies*

$$\sum_{j \in M} \left[(\frac{d}{dt}\hat{p}_j(t) - \pi_j)'(\hat{x}_j(t) - x_j) + (\frac{d}{dt}\hat{x}_j(t) - \xi_j)'(\hat{p}_j(t) - p_j) \right] > \delta, \tag{13}$$

for all (x_j, p_j) and (π_j, ξ_j), such that

$$\pi_j \;\in\; \partial_{p_j} H(t, x, p_j) \tag{14}$$

$$\xi_j \;\in\; -\partial_{x_j} H_j(t, x, p_j). \tag{15}$$

Remark 2.3 *In the autonomous case, treated in more detail in section 2.7, the stricter inequality (13) is obtained as a consequence of Assumption (2.2) or inequality (8) when the state variable x remains in a compact set (this is known as the Atsumi lemma in the single player case). In the general nonautonomous case, the condition (13) is certainly more restrictive and not always easy to verify. The case of discounted payoffs is a special case of the general nonautonomous case and it exhibits a special structure that permits these strong support properties to be modified in such a way as to be useful. This is particularly true if this nonautonomy arises only as a result of the discounting. We pursue these developments in subsections 2.6 and 2.7 below.*

Theorem 2.2 *Assume $\rho_j \equiv 0$ for all $j \in M$. Let $\hat{\mathbf{x}}$ with its associated M-price schedule $\hat{\mathbf{p}}$ be a strongly diagonally supported (overtaking) equilibrium at $\hat{x}^o$, such that*

$$\limsup_{t \to \infty} \|(\hat{x}(t), \hat{p}(t))\| < \infty.$$

Let $\tilde{\mathbf{x}}$ *be another (overtaking) equilibrium at* $\tilde{x}^o$ *with* $\tilde{\mathbf{p}}$ *its associated M-price schedule such that*

$$\limsup_{t\to\infty} \|(\tilde{x}(t), \tilde{p}(t))\| < \infty.$$

Then

$$\lim_{t\to\infty} \|(\hat{x}(t) - \tilde{x}(t), \hat{p}(t) - \tilde{p}(t))\| = 0. \tag{16}$$

Proof: Assume (16) does not hold. Then, according to Equation (13) of Definition 2.2 we have

$$\lim_{T\to\infty} \int_0^T \sum_{j\in M} \left[(\frac{d}{dt}\hat{p}_j(t) - \frac{d}{dt}\tilde{p}_j(t))'(\hat{x}_j(t) - \tilde{x}_j(t)) \right.$$
$$\left. + (\hat{p}_j(t) - \tilde{p}_j(t))'(\frac{d}{dt}\hat{x}_j(t) - \frac{d}{dt}\tilde{x}_j(t)) \right] dt = \infty. \tag{17}$$

However, the left-hand-side of the above expression is also equal to $\lim_{T\to\infty} V(T)$ where

$$V(T) \doteq \sum_{j\in M} [(\hat{p}_j(T) - \tilde{p}_j(T))'(\hat{x}_j(T) - \tilde{x}_j(T))$$
$$- (\hat{p}_j(0) - \tilde{p}_j(0))'(\hat{x}_j(0) - \tilde{x}_j(0))]. \tag{18}$$

Since the M-programs and their associated M-price schedules are bounded $V(T)$ is bounded for all T. This contradicts (17). $\bullet$

Remark 2.4 *This turnpike result is very much in the spirit of McKenzie [28] since it is established for nonautonomous systems as well as a nonconstant turnpike. The special case of autonomous systems will be considered in more detail in subsection 2.7*

2.5 Conditions for SDCCA

We show below how the central assumption of SDCCA can be checked on the data of the differential game.

Lemma 2.2 *Assume* $L_j(t, x, z_j)$ *is concave in* (x_j, z_j) *and assume that the total reward function* $\sum_{j\in M} L_j(t, x, z_j)$ *is diagonally strictly concave in* (x, z), *i.e. verifies*

$$\sum_{j\in M} [(z_j^1 - z_j^0)'(\zeta_j^1 - \zeta_j^0) + (x_j^1 - x_j^0)'(\eta_j^1 - \eta_j^0)] < 0, \tag{19}$$

for all

$$\eta_j^1 \in \partial_{x_j} L_j(t, x^1, z_j^1) \quad \zeta_j^1 \in \partial_{z_j} L_j(t, x^1, z_j^1)$$
$$\eta_j^0 \in \partial_{x_j} L_j(t, x^0, z_j^0) \quad \zeta_j^0 \in \partial_{z_j} L_j(t, x^0, z_j^0).$$

Then Assumption 2.2 holds true.

Proof: The concavity of $L_j(t, x, z_j)$ in (x_j, z_j) implies that each Hamiltonian $H_j(t, x, p_j) = \sup_{z_j}\{L_j(t, x, z_j) + p_j' z_j\}$ defined in (2) is concave in x_j and convex in p_j, for $j \in M$. This property with (19) implies that (5) holds for all $(\hat{x}_j, \tilde{x}_j, \hat{p}_j, \tilde{p}_j)$ and $(\hat{\pi}_j, \tilde{\pi}_j, \hat{\xi}_j, \tilde{\xi}_j)$, satisfying (14) and (15). $\bullet$

Remark 2.5 *In the case when the functions $L_j(t, x, z_j)$ are smooth, explicit conditions for the total reward function, $\sum_{j \in M} L_j(t, x, z_j)$, to be diagonally strictly concave in (x, z) are given in Rosen [33] in Theorem 6, page 528 (for an explanation of the terminology please see pp. 524–528).*

2.6 The Turnpike Property with Discounting

The discounting of payoffs can be viewed as a particular form of nonautonomous behavior of the controlled system. It can however be treated more explicitly by introducing in the support property of Definition 2.2 a so-called "curvature condition." This terminology has been introduced by Rockafellar in [31]. It indicates the "amount" of strict concavity-convexity needed to obtain *GAS* when discounting is introduced. A way to check this property, when the functions L_j are autonomous, is proposed in subsection 2.7.

Theorem 2.3 *Suppose Assumptions 2.1 and 2.2 hold. Let $\hat{x}(\cdot)$ and $\tilde{x}(\cdot)$ be two overtaking equilibria at $\hat{x}^o$ and $\tilde{x}^o$ respectively, with associated M-price schedules $\hat{\mathbf{p}}(\cdot)$ and $\tilde{\mathbf{p}}(\cdot)$ such that*

$$\limsup_{t \to \infty} \| (\hat{x}(t), \hat{p}(t)) \| < \infty \ \ and \ \ \limsup_{t \to \infty} \| (\tilde{x}(t), \tilde{p}(t)) \| < \infty$$

and additionally satisfy the following property:

For each $\epsilon > 0$ there exists $\delta > 0$ so that whenever

$$\| (\hat{x}(t) - \tilde{x}(t), \hat{p}(t) - \tilde{p}(t)) \| > \epsilon,$$

one has

$$\sum_{j \in M} \left[(\hat{p}_j(t) - \tilde{p}_j(t))' (\hat{\pi}_j(t) - \tilde{\pi}_j(t)) + (\hat{x}_j(t) - \tilde{x}_j(t))' \left(\hat{\xi}_j(t) - \tilde{\xi}_j(t) \right) \right]$$

$$> \delta + \sum_{j \in M} \rho_j (\hat{x}_j(t) - \tilde{x}_j(t))' (\hat{p}_j(t) - \tilde{p}_j(t)) \tag{20}$$

for all

$$\hat{\pi}_j(t) \ \in \ \partial_{p_j} H_j(t, \hat{x}(t), \hat{p}_j(t)) \tag{21}$$

$$\hat{\xi}_j^k \ \in \ -\partial_{x_j} H_j(t, \hat{x}(t), \hat{p}_j(t)) \tag{22}$$

$$\tilde{\pi}_j(t) \ \in \ \partial_{p_j} H_j(t, \tilde{x}(t), \tilde{p}_j(t)) \tag{23}$$

$$\tilde{\xi}_j(t) \ \in \ -\partial_{x_j} H_j(t, \tilde{x}(t), \tilde{p}_j(t)). \tag{24}$$

Then

$$\lim_{t \to \infty} \| (\hat{x}(t) - \tilde{x}(t), \hat{p}(t) - \tilde{p}(t)) \| = 0.$$

Proof: The proof of this result is a straightforward adaptation of the proof of Theorem 2.2.

Remark 2.6 *The condition (20) is clearly a strengthening of (13). This condition indeed is analogous to the conditions found in the papers of Cass and Shell, Brock and Scheinkman, Rockafellar, et. al. which are nicely collected in [14].*

2.7 The Autonomous Case

We now specialize our study to the case when the functions $L_j(\cdot, \cdot, \cdot)$ are independent of the time t (i.e., $L_j(t, x, z_j) \equiv L_j(x, z_j)$). In this case, the optimality conditions become

$$
\begin{aligned}
\dot{x}_j(t) &\in \partial_{p_j} H_j(x(t), p_j(t)) \\
\dot{p}_j(t) + \rho_j\, p_j(t) &\in -\partial_{x_j} H_j(x(t), p_j(t)),
\end{aligned}
$$

where

$$H_j(x, p_j) = \sup_{z_j \in \mathbb{R}^{n_j}} \left\{ L_j(x, z_j) + p_j' z_j \right\}. \tag{25}$$

The above conditions define an autonomous pseudo-Hamiltonian system and the possibility arises that there exists a *steady-state equilibrium.* That is, a pair $(\bar{x}, \bar{p}) \in \mathbb{R}^n \times \mathbb{R}^n$ that satisfies

$$
\begin{aligned}
0 &\in \partial_{p_j} H_j(\bar{x}, \bar{p}_j) \\
\rho_j \bar{p}_j &\in -\partial_{x_j} H_j(\bar{x}, \bar{p}_j).
\end{aligned}
$$

When a unique steady-state equilibrium exists, the turnpike properties discussed above provide conditions when the pair $(\bar{x}, \bar{p})$ becomes an attractor for all bounded (overtaking) equilibria. Also, in this case, the support properties and curvature assumptions along a trajectory described on the nonautonomous case become simpler. Moreover, as indicated previously, when $\rho = 0$ the strong support property (26) is a direct consequence of SDCCA and boundedness of state variables (Atsumi's Lemma, see e.g. [7]).

Definition 2.3 *Let $(\bar{x}, \bar{p})$ be a steady state equilibrium. We say that the strong diagonal support property for $(\bar{x}, \bar{p})$ holds if for each $\epsilon > 0$ there exists $\delta > 0$ so that whenever $\|x - \bar{x}\| + \|p - \bar{p}\| > \epsilon$ one has*

$$\sum_{j \in M} \left[(p_j - \bar{p}_j)' \pi_j \ + \ (x_j - \bar{x}_j)' (\xi_j - \rho_j \bar{p}_j) \right]$$

$$> \ \delta + \sum_{j \in M} \rho_j \, (x_j - \bar{x}_j)' (p_j - \bar{p}_j), \tag{26}$$

for all $j \in M$ and pairs (π_j, ξ_j) satisfying

$$\pi_j \in \partial_{p_j} H_j(x, p_j) \text{ and } \xi_j \in -\partial_{x_j} H_j(x, p_j).$$

Remark 2.7 *If $\mathbf{x}$ is an M-program with an associated M-price schedule bfp such that*

$$\|(x(t) - \bar{x}, p(t) - \bar{p})\| > \epsilon,$$

then making the substitutions $x_j = x_j(t)$, $p_j = p_j(t)$, $\pi_j = \dot{x}_j(t)$, and $\xi_j = \dot{p}_j(t)$ in (26) immediately shows that the steady-state $(\bar{x}, \bar{p})$ satisfies (20). This leads immediately to the following result.

Theorem 2.4 *Assume that $(\bar{x}, \bar{p})$ is a unique steady state equilibrium that has the strong diagonal support property given by (26). Then for any M-program $\mathbf{x}$ with an associated M-price schedule $\mathbf{p}$ that satisfies*

$$\limsup_{t \to \infty} \| (x(t), p(t)) \| < \infty,$$

we have

$$\lim_{t \to \infty} \| (x(t) - \bar{x}, p(t) - \bar{p}) \| = 0.$$

Proof: Follows immediately from the above remark and Theorem 2.3. $\bullet$

In the case of an infinite horizon optimal control problem with discount rate $\rho > 0$, Rockafellar [31] and Brock and Scheinkman [8] have given easy-to-verify curvature conditions. Namely a steady-state equilibrium $\bar{x}$ is assured to be an attractor of overtaking trajectories by requiring the Hamiltonian of the optimally controlled system to be *a-concave* in x and *b-convex* in p for values of $a > 0$ and $b > 0$ for which the inequality

$$(\rho_j)^2 < 4ab$$

holds.

We conclude this section by extending this result to the case considered here.

Definition 2.4 *Let $a = (a_1, a_2, \ldots a_m)$ and $b = (b_1, b_2, \ldots b_m)$ be two vectors in $\mathbb{R}^m$ with $a_j > 0$ and $b_j > 0$ for all $j \in M$. We say that the combined Hamiltonian $\sum_{j \in M} H_j(x, p_j)$ is strictly diagonally a-concave in x, b-convex in p if $\sum_{j \in M} \left[H_j(x, p_j) + \frac{1}{2} \left(a_j \|x_j\|^2 - b_j \|p_j\|^2 \right) \right]$ is strictly diagonally concave in x, convex in p.*

Theorem 2.5 *Assume that there exists a unique steady state equilibrium, $(\bar{x}, \bar{p})$. Let $a = (a_1, a_2, \ldots a_m)$ and $b = (b_1, b_2, \ldots b_m)$ be two vectors in $\mathbb{R}^m$ with $a_j > 0$ and $b_j > 0$ for all $j \in M$, and assume that the combined Hamiltonian is strictly diagonally a-concave in x, b-convex in p. Further, let*

x *be a bounded equilibrium M-program with an associated M-price schedule* **p** *that also remains bounded. Then if the discount rates ρ_j, $j \in M$ satisfy the inequalities*

$$(\rho_j)^2 < 4a_j b_j, \tag{27}$$

the M-program **x** *converges to $\bar{x}$.*

Proof: Let **x** be a bounded M-program and let **p** be its associated $M-$price schedule (which is also bounded). Then we have the following inclusions for all $t \geq 0$ and $j \in M$

$$\dot{x}_j(t) - b_j p(t) \in \partial_{p_j}\Big[H_j(x(t), p_j(t)) \\ + frac12 \left(a_j \|x_j(t)\|^2 - b_j \|p_j(t)\|^2 \right) \Big]$$

$$\dot{p}_j(t) + \rho_j p_j(t) - a_j x_j(t) \in -\partial_{x_j}\Big[H_j(x(t), p_j(t)) \\ + \frac{1}{2} \left(a_j \|x_j(t)\|^2 - b_j \|p_j(t)\|^2 \right) \Big]$$

$$-b_j \bar{p} \in \partial_{p_j}\left[H_j(\bar{x}, \bar{p}_j) + \frac{1}{2}\left(a_j \|\bar{x}_j\|^2 - b_j \|\bar{p}_j\|^2 \right)\right]$$

$$\rho_j \bar{p}_j - a_j \bar{x}_j \in -\partial_{x_j}\left[H_j(\bar{x}, \bar{p}_j) + \frac{1}{2}\left(a_j \|\bar{x}_j\|^2 - b_j \|\bar{p}_j\|^2 \right)\right].$$

Thus, as a consequence of our strict diagonal concavity and convexity assumptions, we have for each $t \geq 0$ that

$$0 < \sum_{j \in M} \left[(p_j(t) - \bar{p}_j)' \left(\dot{x}_j(t) - b_j p(t) + b_j \bar{p} \right) \right.$$

$$\left. + (x_j(t) - \bar{x}_j)' \left(\dot{p}_j(t) + \rho_j p_j(t) - a_j x_j(t) - (\rho_j \bar{p}_j - a_j \bar{x}_j) \right) \right]$$

$$= \sum_{j \in M} \left[(p_j(t) - \bar{p}_j)' \dot{x}_j(t) + (x_j(t) - \bar{x}_j)' \dot{p}_j(t) \right] - \sum_{j \in M} \left[b_j \|p_j(t) - \bar{p}_j\|^2 \right.$$

$$\left. - \rho_j \left(p_j(t) - \bar{p}_j \right)' \left(x_j(t) - \bar{x}_j \right) + a_j \|x_j(t) - \bar{x}_j\|^2 \right].$$

Integrating from $t = 0$ to $t = T$ we may write

$$V(T) > \int_0^T \left\{ \sum_{j \in M} \left[b_j \|p_j(t) - \bar{p}_j\|^2 \right. \right.$$

$$\left. \left. - \rho_j \left(p_j(t) - \bar{p}_j \right)' \left(x_j(t) - \bar{x}_j \right) + a_j \|x_j(t) - \bar{x}_j\|^2 \right] \right\} dt$$

where, as before,

$$V(T) = \int_0^T \left\{ \sum_{j \in M} \left[(p_j(t) - \bar{p}_j)' \, \dot{x}_j(t) + (x_j(t) - \bar{x}_j)' \, \dot{p}_j(t) \right] \right\} dt.$$

This inequality may be equivalently written as

$$
\begin{aligned}
V(T) \quad &> \quad \int_0^T \left\{ \sum_{j \in M} b_j \| (p_j(t) - p_j) - \frac{\rho_j}{2b_j} (x_j(t) - \bar{x}) \|^2 \right. \\
&\qquad \left. + \left(a_j - \frac{\rho_j^2}{4b_j} \right) \| x_j(t) - \bar{x} \|^2 \right\} dt \\
&> \quad \int_0^T \left\{ \sum_{j \in M} \left(a_j - \frac{\rho_j^2}{4b_j} \right) \| x_j(t) - \bar{x} \|^2 \right\} dt.
\end{aligned}
$$

To conclude this proof we see, from above, that if the M-program $\mathbf{x}$ does not converge to $\bar{x}$, then $\lim_{T \to \infty} V(T) = \infty$ because of (27). This however is a contradiction since we may equivalently write

$$V(T) = \sum_{j \in M} \left((p_j(t) - \bar{p}_j)' (x_j(t) - \bar{x}_j) - (p_j(0) - \bar{p}_j)' (x_j(0) - \bar{x}_j) \right),$$

showing that $V(T)$ must remain bounded for all T.   •

Remark 2.8 *In the above results, we have extended the classical asymptotic turnpike theory to a dynamic game framework with separated dynamics. The fact that the players interact only through the state variables and not the control ones is essential. An indication of the increased complexities of coupled state and control interactions may be seen in Haurie and Leitmann [24].*

3 Existence of Equilibria

In this section we extend Rosen's approach to show existence of equilibria in dynamic autonomous competitive processes, under sufficient smoothness and compactness conditions. Basically we reduce the existence proof to a fixed-point argument for a point-to-set mapping constructed from an associated class of infinite horizon concave optimization problems.

3.1 Existence of Overtaking Equilibria in the Undiscounted Case

Our proof of existence of an overtaking equilibrium for undiscounted dynamic competitive processes uses extensively sufficient overtaking optimality conditions for single player optimization problems (see [10], chap. 2). For this appeal to sufficiency conditions, the existence of a bounded attractor to all good programs is important. This is the reason why our existence theory is restricted to autonomous systems, for which a steady-state equilibrium provides such an attractor.

Remark 3.1 *Existence of overtaking optimal control for autonomous systems (discrete or continuous time) can be established through a reduction to finite cost argument (see e.g. [10]). There is a difficulty in extending this approach to the case of dynamic open-loop games. It comes from the inherent time-dependency introduced by the other players' decisions. Our approach circumvents this difficulty by implementing a reduction to finite costs for an associated class of infinite horizon concave optimization problems.*

We first make the following assumptions:

Assumption 3.1 *The functions $L_j : I\!R^n \times I\!R^{n_j} \to I\!R$ are strictly concave in (x_j, z_j) and additionally we have that $\frac{\partial L_j}{\partial x_j}$ and $\frac{\partial L_j}{\partial z_j}$ are continuous on $I\!R^n \times I\!R^{n_j}$ for each $j \in M$.*

Remark 3.2 *While the above conditions are somewhat restrictive, we maintain that they are satisfied by a large useful class of models. Indeed, the concavity conditions are those typically satisfied in economic growth models. The desired differentiability conditions are also satisfied by a robust class of models. For example if the dynamic game is described in an ordinary control formulation the integrand, $L_j(\cdot, \cdot)$ is the optimal value function of a nonlinear mathematical programming problem such as*

$$L_j(x, z_j) = \max_{u_j}\{f^0(x, u_j)| \quad z_j = f(x, u_j) \text{ and } u_j \in U_j\}.$$

Under appropriate smoothness conditions on (f^0, f), uniqueness of the optimal solution, and constraint qualification assumptions it is well known by the envelope theorem, that $L_j(\cdot, \cdot)$ is continuously differentiable. Further, uniqueness is assured by assuming strict concavity conditions. In addition to these remarks we provide a simple class of examples in which these conditions are easily verified.

Assumption 3.2 *There exists a unique steady state equilibrium $\bar{x} \in I\!R^n$ and a corresponding constant M-price schedule $\bar{p} \in I\!R^n$ satisfying*

$$0 \in \partial_{p_j} H_j(\bar{x}, \bar{p}_j)$$

$$\tag{28}$$

$$0 \in \partial_{x_j} H_j(\bar{x}, \bar{p}_j).$$

We need a controllability assumption which says that the steady state equilibrium is locally reachable in a uniform time.

Assumption 3.3 *There exists $\epsilon_0 > 0$ and $S > 0$ such that for any $\tilde{x} \in I\!\!R^n$ satisfying $\|\tilde{x} - \bar{x}\| < \epsilon_0$ there exists an M-program $w(\tilde{x}, \cdot)$ defined on $[0, S]$ such that $w(\tilde{x}, 0) = \tilde{x}$ and $w(\tilde{x}, S) = \bar{x}$.*

In order to achieve our result we must assure that all admissible M-programs lie in a compact set. Additionally we must further assume that their rates of growth are not too large. Thus we make the following additional assumption.

Assumption 3.4 *For each $j \in M$ there exists a closed bounded set*

$$X_j \subset I\!\!R^{n_j} \times I\!\!R^{n_j}$$

such that each M-program, $\mathbf{x}$ satisfies $(x_j(t), \dot{x}_j(t)) \in X_j$ a.e. $t \geq 0$.

Additionally we introduce the following notation.

- We let Ω denote the set of all M-programs that start at x^o and converge to $\bar{x}$, the unique steady state equilibrium.

- We define the family of functionals $\theta^T : \Omega \times \Omega \to I\!\!R$, $T \geq 0$, by the formula

$$\theta^T(\mathbf{x}, \mathbf{y}) \doteq \int_0^T \sum_{j \in M} \left[L_j \left(\left[x^{(j)}(t), y_j(t) \right], \dot{y}_j(t) \right) \right] dt.$$

We view Ω as a subset of all bounded continuous functions in $I\!\!R^n$ endowed with the topology of uniform convergence on bounded intervals.

Definition 3.1 *Let $\mathbf{x}, \mathbf{y} \in \Omega$. We say that $\mathbf{y} \in \Gamma(\mathbf{x})$ if*

$$\liminf_{T \to \infty} \left(\theta^T(\mathbf{x}, \mathbf{y}) - \theta^T(\mathbf{x}, \mathbf{z}) \right) \geq 0,$$

for all M-programs $\mathbf{z}$ such that $z_j(0) = x_j^0$, $j \in M$. That is, $\mathbf{y}$ is an overtaking optimal solution of the infinite horizon optimization problem whose objective functional is defined by $\theta^T(\mathbf{x}, \cdot)$. Hence $\Gamma(\mathbf{x})$ can be viewed as the set of optimal responses by all players to an M-program $(\mathbf{x})$.

Theorem 3.1 *Under the above assumptions, there exists an overtaking equilibrium for the infinite horizon dynamic game.*

Proof: To prove our result we prove that the set-valued map $\Gamma : \Omega \to 2^\Omega$ has a fixed point using the Kakutani fixed point theorem. To do this we need to show the following

  1. For each $\mathbf{x} \in \Omega$, the set $\Gamma(\mathbf{x})$ is nonempty, convex and compact.

2. The map $\Gamma(\cdot)$ has a closed graph. That is, if $(\mathbf{y}_l, \mathbf{x}_l)$ is a sequence of functions in $\Omega \times \Omega$ that converges to $(\mathbf{y}, \mathbf{x})$ and additionally satisfies

$$\mathbf{y}_l \in \Gamma(\mathbf{x}_l)$$

for all $l = 1, 2, \ldots$, then $\mathbf{y} \in \Gamma(\mathbf{x})$.

We begin by first showing, for each $\mathbf{x} \in \Omega$, that $\Gamma(\mathbf{x})$ is nonempty. To see this we fix $\mathbf{x} \in \Omega$, and let $\hat{T} > 0$ be such that for all $t \geq \hat{T}$ we have $\|x(t) - \bar{x}\| \leq \epsilon_0$. Define for each $l = 1, 2, \ldots,$ the times $\tau_l = \hat{T} + lS$ and the M-programs $\mathbf{x}_l$ as

$$x_l(t) = \begin{cases} x(t) & \text{if } t < \tau_l \\ w(x_l(\tau_l), t - \tau_l) & \text{if } \tau_l \leq t < \tau_{l+1} \\ \bar{x} & \text{if } \tau_{l+1} \leq t \end{cases}$$

for $t \geq 0$. Now consider the problem of maximizing the functional $\gamma(\mathbf{x}_l, \cdot)$, defined by the formula

$$
\begin{aligned}
\gamma(\mathbf{x}_l, \mathbf{z}) \;&\doteq\; \int_0^\infty \sum_{j \in M} \left[L_j \left(\left[x_l^{(j)}(t), z_j(t) \right], \dot{z}_j(t) \right) - L_j(\bar{x}, 0) + \bar{p}_j \dot{z}_j(t) \right] dt \\
&=\; \int_0^{\tau_l} \sum_{j \in M} \left[L_j \left(\left[x_l^{(j)}(t), y_j(t) \right], \dot{y}_j(t) \right) - L_j(\bar{x}, 0) + \bar{p}_j \dot{y}_j(t) \right] dt \\
&\quad + \int_{\tau_l}^{\tau_{l+1}} \sum_{j \in M} \left[L_j \left(\left[\mathbf{w}_l^{(j)}(x(\tau_l), t - \tau_l), y_j(t) \right], \dot{y}_j(t) \right) \right. \\
&\qquad\qquad\qquad\qquad\qquad\qquad \left. - L_j(\bar{x}, 0) + \bar{p}_j \dot{y}_j(t) \right] dt \\
&\quad + \int_{\tau_{l+1}}^\infty \sum_{j \in M} \left[L_j \left(\left[\bar{x}^{(j)}, y_j(t) \right], \dot{y}_j(t) \right) - L_j(\bar{x}, 0) + \bar{p}_j \dot{y}_j(t) \right] dt,
\end{aligned}
$$

over all M-programs starting at x^o. Observe that for each $t > \tau_{l+1}$, the terms of $\gamma(\mathbf{x}_l, \mathbf{z})$ are nonpositive and in fact equal zero whenever we take $z_j(t) = \bar{x}_j$. From this it follows easily that there exists an M-program, say $\mathbf{z}$, for which $\gamma(\mathbf{x}_l, \mathbf{z}) > -\infty$ and moreover, since for each M-program the map $t \to (x(t), \dot{x}(t))$ lies in a compact subset of $\mathbb{R}^n$ it also follows that $\gamma(\mathbf{x}_l, \mathbf{z})$ is bounded above. Therefore, there exists an M-program $\mathbf{y}_l$ that maximizes $\gamma(\mathbf{x}_l, \mathbf{z})$ as desired. Furthermore, as a result of the strict concavity assumptions given above, it follows that $\mathbf{y}_l \in \Gamma(\mathbf{x}_l)$ and in addition we also have

$$\lim_{t \to \infty} y_l(t) = \bar{x}.$$

This asymptotic stability property has been proved by Brock and Haurie in [7] (see also [10]). In this way we generate the sequence $(\mathbf{y}_l)$ in Ω, a compact set. Thus we can assume that the sequence $(\mathbf{y}_l)$ converges to an M-program $\mathbf{y} \in \Omega$. Moreover, since the derivatives $(\dot{\mathbf{y}}_l)$ are bounded a.e. on $[0, +\infty)$ we can further assume that the sequence of derivatives converges weakly in $L^1_{loc}([0, +infty); \mathbb{R}^n)$. Additionally, we also know that the sequence $(\mathbf{x}_l)$ converges to our original M-program $\mathbf{x}$. We now show that $\mathbf{y} \in \Gamma(\mathbf{x})$. To this end we note that for each $l = 1, 2, \ldots$, there exists an M-price schedule $\mathbf{p}_l$ such that

$$\dot{y}_{lj}(t) \in \partial_{p_j} H_j\left([x_l^{(j)}(t), y_{lj}(t)], p_{lj}(t)\right)$$

$$\dot{p}_{lj}(t) \in -\partial_{x_j} H_j\left([x_l^{(j)}(t), y_{lj}(t)], p_{lj}(t)\right),$$

or equivalently that

$$p(t)_{lj} = \frac{\partial}{\partial z_j} L_j\left([x_l^{(j)}(t), y_{lj}(t)], \dot{y}_{lj}(t)\right) \tag{29}$$

$$\dot{p}_{lj}(t) = \frac{\partial}{\partial x_j} L_j\left([x_l^{(j)}(t), y_{lj}(t)], \dot{y}_{lj}(t)\right).$$

for almost all $t \geq 0$. As a result of our compactness assumptions on admissible M-programs and the continuity of the partial derivatives of $L_j(\cdot, \cdot)$, $j \in M$, it follows that the sequences $(p_{lj}(\cdot), \dot{p}_{lj}(\cdot))_{l=1}^{+\infty}$, for $j \in M$, are also bounded almost everywhere. Thus without loss of generality we may assume that there exists a locally absolutely continuous functions $p_j(\cdot) : [0, +\infty) \to \mathbb{R}^{n_j}$, $j \in M$, so that the sequences of M-price schedules $(p_{lj}(\cdot))_{l=1}^{+\infty}$ converge uniformly on compact subsets of $\mathbb{R}$ to $p_j(\cdot)$ and that the sequences $(\dot{p}_{lj}(\cdot))_{l=1}^{+\infty}$ converge weakly in $L^1_{loc}[0, +\infty)$ to $\dot{p}_j(\cdot)$. Clearly we have $p_j(t) \to \bar{p}_j$ as $t \to +\infty$. We now show that the limit function $\mathbf{p} = (p_j(\cdot))_{j \in M}$ is an M-price schedule for the trajectory $\mathbf{y} = (y_j(\cdot))_{j \in M}$. To see this we observe that for each $j \in M$ and each pair $(u_j, z_j) \in \mathbb{R}^{n_j} \times \mathbb{R}^{n_j}$ we have

$$L_j\left([x_l^{(j)}(t), u_j], z_j\right) - L_j\left([x_l^{(j)}(t), y_{lj}(t)], \dot{y}_{lj}(t)\right)$$
$$\leq \dot{p}_{lj}(t)'(u_j - y_{lj}(t)) + p_{lj}(t)'(z_j - \dot{y}_{lj}(t))$$

which holds for almost all $t \geq 0$. Thus for $s > 0$ and $h \geq 0$ we have

$$\int_s^{s+h} L_j\left([x_l^{(j)}(t), u_j], z_j\right) - L_j\left([x_l^{(j)}(t), y_{lj}(t)], \dot{y}_{lj}(t)\right) dt$$
$$\leq \int_s^{s+h} \left[\dot{p}_{lj}(t)'(u_j - y_{lj}(t))\right.$$

$$+ p_{lj}(t)'(z_j - \dot{y}_{lj}(t))\Big] \, dt$$

$$= \int_s^{s+h} \dot{p}_{lj}(t)' u_j \, dt + \int_s^{s+h} p_{lj}(t)' z_j \, dt$$

$$- [p_{lj}(s+h)' y_{lj}(s+h) - p_{lj}(s)' y_{lj}(s)]$$

which, upon letting $l \to +\infty$, gives us

$$\int_s^{s+h} L_j\left([x^{(j)}(t), u_j], z_j\right) \; - \; L_j\left([x^{(j)}(t), y_j(t)], \dot{y}_j(t)\right) \, dt$$

$$\leq \int_s^{s+h} [\dot{p}_j(t)'(u_j - y_j(t)) + p_j(t)'(z_j - \dot{y}_j(t))] \, dt.$$

The last inequality in the above follows from our concavity condition and the fact that for any $s \geq 0$ and $h \geq 0$ the integral functional

$$J_j(\psi, \eta) = \int_s^{s+h} L_j(\psi(t), \eta(t)) \, dt,$$

defined for all measurable functions $\psi : [s, s+h] \to \mathbb{R}^n$ and integrable functions $\eta : [s, s+h] \to \mathbb{R}^{n_j}$, is upper semicontinuous with respect to pointwise convergence in $\psi(\cdot)$ and weak L^1 convergence in $\eta(\cdot)$ (see e.g., Berkovitz [3]). To finish we now divide by h in this last inequality and let $h \to 0$ to get

$$L_j\left([x^{(j)}(s), u_j], z_j\right) \; - \; L_j\left([x^{(j)}(s), y_j(s)], \dot{y}_j(s)\right)$$

$$\leq \; \dot{p}_j(s)'(u_j - y_j(s)) + p_j(s)'(z_j - \dot{y}_j(s))$$

holding for almost all $s \geq 0$ since the set of Lebesgue points of a summable function is a set of full measure. Thus, since u_j and z_j are arbitrary in $\mathbb{R}^{n_j}$, our concavity and continuity assumptions allow us to conclude

$$p_j(t) \;=\; \frac{\partial}{\partial z_j} L_j\left([x^{(j)}(t), y_j(t)], \dot{y}_j(t)\right)$$

$$\dot{p}_j(t) \;=\; \frac{\partial}{\partial x_j} L_j\left([x^{(j)}(t), y_j(t)], \dot{y}_j(t)\right),$$

for almost all $t \geq 0$ and $j \in M$. That is, we can associate with $\mathbf{y} \in \Omega$ an M-price schedule $\mathbf{p}$ and moreover, since $\mathbf{x}, \mathbf{y} \in \Omega$ we have $\lim_{t \to \infty} p_j(t) = \bar{p}_j$ as well. Therefore, by appealing to standard sufficient conditions for overtaking optimality we obtain $\mathbf{y} \in \Gamma(\mathbf{x})$ as desired. Further, it is an easy matter to see that, as a result of our concavity assumptions on L_j, $\Gamma(\mathbf{x})$ is a convex set.

Since $\Gamma(\mathbf{x})$ is a subset of Ω, a compact set, the compactness of $\Gamma(\mathbf{x})$ follows immediately once we show that it is closed. To see this let $(\mathbf{y}_l)$ be a sequence in $\Gamma(\mathbf{x})$ that converges to $\mathbf{y}$ and let $(\mathbf{p}_l)$ be the sequence of corresponding M-price schedules. By a direct adaptation of the above argument it can be shown there is an M-price schedule $\mathbf{p}$ associated with $\mathbf{y}$ that converges to $\bar{p}$ as $t \to \infty$. Hence $\mathbf{y} \in \Gamma(\mathbf{x})$ giving us the desired compactness condition.

It remains to prove that the graph of Γ is closed. To see this let $(\mathbf{y}_l, \mathbf{x}_l) \to (\mathbf{y}, \mathbf{x})$ as $l \to \infty$ be such that $\mathbf{y}_l \in \Gamma(\mathbf{x}_l)$ for each $l \in \mathbb{N}$. Further we let $(\mathbf{p}_l)$ be a sequence of associated M-price schedules, satisfying (30). Then, once again, letting $l \to \infty$ one sees that there exists an M-price schedule $\mathbf{p}$, associated with $\mathbf{x}$ such that $\mathbf{p}_l \to bfp$. Thus proceeding as above we see that $\mathbf{y} \in \Gamma(\mathbf{x})$ giving us the closed graph property.

We have just shown that the conditions needed to apply Kakutani's fixed point theorem are satisfied. Therefore there exists $\mathbf{x}^* \in \Gamma(\mathbf{x}^*)$. The proof that the M-program $\mathbf{x}^*$ is an overtaking equilibrium now follows as in Rosen [33], Theorem 1.

$\bullet$

3.2 Existence of Equilibria in Discounted Competitive Processes

Under the assumptions made above to establish the existence of an overtaking equilibrium, it is an easy matter to treat the case of autonomous games with discounting. Indeed, by requiring all admissible M-programs to satisfy the compact constraints of Assumption 3.4 and letting $\rho_j > 0$ for all $j \in M$, it follows that the functionals $\theta : \Omega \times \Omega \to \mathbb{R}$, given by

$$\theta(\mathbf{x}, \mathbf{y}) \doteq \int_0^{+\infty} \sum_{j \in M} \left[e^{\rho_j t} L_j \left(\left[x^{(j)}(t), y_j(t) \right], \dot{y}_j(t) \right) \right] dt.$$

are bounded both above and below. Thus, the existence of an equilibrium can be assured by an easy modification of the above argument. To do this we enlarge Ω to be the set of all M-programs satisfying the initial condition x^o and now define $\Gamma : \Omega \to 2^\Omega$ as follows:

> For an M-program $\mathbf{x} \in \Omega$ we let $\Gamma(\mathbf{x})$ denote the set of all M-programs $\mathbf{y} \in \Omega$ such that

$$\theta(\mathbf{x}, \mathbf{y}) = \max_{\mathbf{z} \in \Omega} \theta(\mathbf{x}, \mathbf{z}).$$

Further, the positive discounting eliminates the need to define the associated problems using the functional $\gamma(\cdot, \cdot)$. Therefore we state the following result without proof.

Theorem 3.2 *Let $\rho_j > 0$ for $j \in M$ and let the Assumptions 3.1, 3.2, and 3.4 hold. Then there exists an overtaking equilibrium for the discounted autonomous infinite horizon dynamic game.*

4   Uniqueness of Equilibria

In Rosen's paper [33], the strict diagonal concavity condition was introduced essentially in order to assure uniqueness of equilibria. In our developments of section 2 we have introduced a similar assumption to get asymptotic stability of equilibrium programs. Indeed this condition also leads to uniqueness.

Theorem 4.1 *Suppose that the assumptions of Theorem 2.2 hold. Then, if there exists an overtaking equilibrium at x^o, it is unique.*

Proof: Assume that $\hat{\mathbf{x}}$ and $\tilde{\mathbf{x}}$ are two distinct equilibrium trajectories at x^o, with their respective M-price schedules $\hat{\mathbf{p}}$ and $\tilde{\mathbf{p}}$. According to the necessary conditions, for all $t \geq 0$, there exist

$$\hat{\pi}_j(t) \quad \in \quad \partial_{p_j} H_j(t, \hat{x}(t), \hat{p}_j(t)), \quad \tilde{\pi}_j(t) \in \partial_{p_j} H_j(t, \tilde{x}(t), \tilde{p}_j(t)) \tag{30}$$

$$\hat{\xi}_j(t) \quad \in \quad -\partial_{x_j} H_j(t, \hat{x}(t), \hat{p}_j(t)), \quad \tilde{\xi}_j(t) \in -\partial_{x_j} H_j(t, \tilde{x}, \tilde{p}_j). \tag{31}$$

such that

$$\sum_{j \in M} \left[(\hat{p}_j(t) - \tilde{p}_j(t))'(\hat{\pi}_j(t) - \frac{d}{dt}\hat{x}_j(t) \quad - \quad \tilde{\pi}_j(t) + \frac{d}{dt}\tilde{x}_j(t)) \right.$$

$$\left. + (\hat{x}_j(t) - \tilde{x}_j(t))'(\hat{\xi}_j(t) - \frac{d}{dt}\hat{p}(t)_j \quad - \quad \tilde{\xi}_j(t) + \frac{d}{dt}\tilde{p}(t)_j) \right]$$

$$= \quad 0. \tag{32}$$

Integrating over $[0, T]$ we get

$$\sum_{j \in M} \int_0^T \left[(\frac{d}{dt}\hat{p}_j(t) - \frac{d}{dt}\tilde{p}_j(t))'(\hat{x}_j(t) - \tilde{x}_j(t)) + \right.$$

$$\left. (\hat{p}_j(t) - \tilde{p}_j(t))'(\frac{d}{dt}\hat{x}_j(t) - \frac{d}{dt}\tilde{x}_j(t)) \right] dt =$$

$$\int_0^T \sum_{j \in M} \left[(\hat{p}_j(t) - \tilde{p}_j(t))'(\hat{\pi}_j(t) - \tilde{\pi}_j(t)) + (\hat{x}_j(t) - \tilde{x}_j(t))'(\hat{\xi}_j(t) - \tilde{\xi}_j(t)) \right] dt. \tag{33}$$

Now, when $T \to \infty$, the right-hand side of (33) tends to a strictly positive number, due to SDCCA, whereas the left-hand side, which is also equal to

$$\sum_{j \in M} [(\hat{p}_j(T) - \tilde{p}_j(T))'(\hat{x}_j(T) - \tilde{x}_j(T)) + (\hat{p}_j(0) - \tilde{p}_j(0))'(\hat{x}_j(0) - \tilde{x}_j(0))], \tag{34}$$

goes to 0, due to the turnpike property. This is a contradiction, hence the overtaking equilibrium is unique.   ●

5   Example

We consider the following dynamic game in which the accumulated reward for the j-th player is described by

$$\phi_j^T(x, u_j) = \int_0^T f^0(x(t), u_j(t)) \, dt$$

and the corresponding dynamics are described by the control system

$$\dot{x}_j(t) \;=\; f_j(x(t)) + u_j(t) \quad \text{for a.e. } t \geq 0$$

$$x_j(0) \;=\; x^0$$

$$u_j(t) \;\in\; U_j \subset \mathrm{I\!R}^{m_j} \text{ a.e. } t \geq 0$$

$$x_j(t) \;\in\; S_j \subset \mathrm{I\!R}^{n_j} \text{ for } t \geq 0.$$

Here we assume that $f_j^0 : \mathrm{I\!R}^n \times \mathrm{I\!R}^{m_j} \to \mathrm{I\!R}$ and $f_j : \mathrm{I\!R}^n \to \mathrm{I\!R}^{n_j}$ are continuously differentiable, U_j is a closed bounded convex set, and S_j is a closed bounded convex set. With this notation it is easy to see that the above problem may be expressed in the variational form considered here by defining $L_j : \mathrm{I\!R}^n \times \mathrm{I\!R}^{n_j} \to \mathrm{I\!R}$ through the formula

$$L_j(x, z_j) = f_j^0(x, z_j - f_j(x)).$$

It is now easy to impose additional restrictions on f_j^0 and f to insure the desired conditions are satisfied. Moreover, we observe that in the above control formulation, the dynamics of each player are indeed coupled with the dynamics of the other players, even though the controls are decoupled.

6   Conclusion

In this paper we have shown that, for infinite horizon open-loop differential games where the players only interact through the state variables and have decoupled controls, an assumption very similar to the one made by Rosen in his proof of uniqueness of equilibria in static concave games, brings together global asymptotic stability of equilibrium trajectories, existence and uniqueness of overtaking equilibria at each initial state. This extends the so-called *turnpike* theory of optimal economic growth to a competitive case. One extension of Rosen's work still remains to be extended to the infinite horizon differential game framework; this is the characterization of *normalized equilibria* for differential games with *coupled constraints*. The usefulness of the concept has been shown in [23] and [27] for the analysis of the policy coordination of an oligopoly facing a global environmental constraint.

REFERENCES

[1] ARROW K.J AND M. KURZ, *Public Investment, The Rate of Return, and Optimal Fiscal Policy*, The Johns Hopkins Press, 1970.

[2] BAŞAR T. AND G.J. OLSDER, *Dynamic Noncooperative Game Theory*, Academic Press: London, New York, 1982.

[3] BERKOVITZ, L. D., Lower Semicontinuity of Integral Functionals, *Transactions of the American Mathematical Society*, **192** (1974), 51–57.

[4] BRETON M., J.A. FILAR, A. HAURIE AND T.A. SCHULTZ, On the Computation of Equilibria in Discounted Stochastic Dynamic Games, in *Dynamic Games and Applications in Economics*, Lect. Notes Econ. Math. Syst., **265** (1986), Springer Verlag, 64–87.

[5] BROCK W.A., On Existence of Weakly Maximal Programmes in a Multisector Economy, *Review of Economic Studies*, **37**(1970), 275–280.

[6] BROCK W.A., Differential Games with Active and Passive Variables, in *Mathematical Economics and Game Theory: Essays in Honor of Oskar Morgenstern* Henn and Moeschlin (Eds.), Springer Verlag, 1977, 34–52.

[7] BROCK W.A. AND A. HAURIE, On Existence of Overtaking Optimal trajectories over an Infinite Time Horizon, *Mathematics of Operations Research*, **1**, (1976), 337–346.

[8] BROCK W.A. AND J.A. SCHEINKMAN, Global Asymptotic Stability of Optimal Control Systems with Application to the Theory of Economic Growth, *Journal of Economic Theory* , **12**(1976), 164–190.

[9] CARLSON D.A., HAURIE A., *A Turnpike Theory for Infinite Horizon Competitive Processes*, 1994, to appear.

[10] CARLSON D.A., HAURIE A. AND LEIZAROWITZ A., *Infinite Horizon Optimal Control: Deterministic and Stochastic Systems*, Springer Verlag, 1991.

[11] CARLSON D.A. AND HAURIE A. AND LEIZAROWITZ A., *Overtaking Equilibria for Switching Regulator and Tracking Games*, in *Advances in Dynamic Games and Applications*, T. Başar and A. Haurie (eds.), Annals of the International Society of Dynamic Games, **1** (1994), Birkhäuser, 247–268.

[12] CASS D., Optimum Growth in Aggregative Model of Capital Accumulation: A Turnpike Theorem, *Econometrica*, **34**(1965), 833–850.

[13] CASS D., Optimum Growth in Aggregative Model of Capital Accumulation, *Review of Economic Studies*, **32**(1965), 233–240.

[14] CASS D. AND SHELL K., *The Hamiltonian Approach to Dynamic Economics*, Academic Press, 1976.

[15] CASS D. AND SHELL K., The Structure and Stability of Competitive Dynamical Systems,
i Journal of Economic Theory,**12**(1976), 30–70.

[16] COURNOT A., *Recherches sur les principes mathématiques de la théorie des richesses*, Hachette, Paris, 1838.

[17] FEINSTEIN C.D. AND LUENBERGER D.G., Analysis of the Asymptotic Behaviour of Optimal Control Trajectories: the Implicit Programming Problem, *SIAM J. Control Optim.* **19** (1981), 561–585.

[18] FRIEDMAN J.W., *Oligopoly and the Theory of Games*, North-Holland, Amsterdam, 1977.

[19] FUDENBERG D. AND TIROLE J., *Game Theory*, The MIT Press, 1991.

[20] HALKIN H., Necessary Conditions for Optimal Control Problems with Infinite Horizon, *Econometrica*, **42**(1974), 267–273.

[21] HÄMÄÄLÄINEN, R.P., HAURIE A. AND KAITALA V., *Equilibria and Threats in a Fishery Management Game, Optimal Control Applications and Methods*, **6**(1985), 315–333.

[22] HAURIE A., Existence and Global Asymptotic Stability of Optimal Trajectories for a Class of Infinite-Horizon, Nonconvex Systems, *J. of Optimization Theory and Applications*, **31**: 4(1980), 515–533.

[23] HAURIE A., *Environmental Coordination in Dynamic Oligopolistic Markets Group Decision and Negotiation*, 4(1995), 39–57. to appear.

[24] HAURIE A. AND LEITMANN G., On the Global Stability of Equilibrium Solutions for Open-Loop Differential Games, *Large Scale Systems* 6(1984), 107–122.

[25] HAURIE A. AND ROCHE M.,   Turnpikes and Computation of Piecewise Open-Loop Equilibria in Stochastic Differential Games, 1993, *Journal Economic Dynamics and Control,* to appear.

[26] HAURIE A. AND TOLWINSKI B.,   Definition and Properties of Cooperative Equilibria in a Two-Player Game of Infinite Duration, *Journal of Optimization Theory and Applications,* **46**, No.4, 1985, 525–534.

[27] HAURIE A. AND ZACCOUR G.,   Differential Games Models of Global Environmental Management, Fondazione ENI Enrico Mattei, Nota di Lavoro 46.93, to appear in *Annals of Dynamic Games,* 1993.

[28] MCKENZIE L.W.,   Turnpike Theory, *Econometrica,* 1976, 841–866.

[29] NASH, J.F.,   Non-Cooperative Games, *Annals of Mathematics,* **54** (1951), 286–295.

[30] ROCKAFELLAR R.T.,   Saddle points of Hamiltonian Systems in Convex Problems of Lagrange, *Journal of Optimization Theory and Applications* , **12**(1973), 367–399.

[31] ROCKAFELLAR R.T.,   Saddle points of Hamiltonian systems in convex Lagrange problems having nonzero discount rate, *Journal of Economic Theory,* **12**(1976), 71–113.

[32] RUBINSTEIN, A.,   Equilibrium in Supergames with the Overtaking Criterion, *Journal of Economic Theory,* **21**(1979), 1–9.

[33] ROSEN J.B.,   Existence and uniqueness of equilibrium points for concave N-person games, *Econometrica,* 1965, 520–534.

[34] R. SELTEN, Rexamination of the Perfectness Concept for Equilibrium Points in Extensive Games, *International Journal of Game Theory,* Vol. 4, 1975, 25–55.

[35] VON WEIZÄCKER C.C.,   Existence of Optimal Programs of Accumulation for an Infinite Time Horizon, *Review of Economic Studies,* **32**(1965), 85–104.

Team-Optimal Closed-Loop Stackelberg Strategies for Discrete-Time Descriptor Systems

Hua Xu and Koichi Mizukami

Faculty of Integrated Arts and Sciences, Hiroshima University

1-7-1, Kagamiyama, Higashi-Hiroshima, Japan 724

Abstract

In this paper we investigate the team-optimal closed-loop Stackelberg strategies for discrete-time descriptor systems. We show that the closed-loop no-memory information on the descriptor variables is sufficient for the leader to design the team-optimal feedback closed-loop Stackelberg strategies for a general class of linear-quadratic Stackelberg games. Sufficient conditions for the existence of such strategies are derived. A recursive scheme is presented to determine the team-optimal feedback closed-loop Stackelberg strategies. A numerical example is solved to illustrate the validity of the sufficient conditions.

1   Introduction

In this paper, we consider the linear-quadratic closed-loop Stackelberg game problem for discrete-time descriptor systems $Ex_{k+1} = Ax_k + Bu_k + Cv_k$, where E is in general a singular matrix and the system structure is in general noncausal.

There is no doubt nowadays that the descriptor systems form a very important class of mathematical models that covers a large area of the modelling of natural phenomenon. They may arise in singular perturbation systems, but are not limited to only these systems. Applications of descriptor systems can also be found in economic systems, interconnected systems and many other areas (Verghese et al. 1981, Luenberger 1978 and Aplevich 1991).

For the descriptor system described above, if the system is causal, then it can be transformed to a regular full-order or reduced-order state space system depending on whether the matrix E is nonsingular or singular. It is well known that, under certain conditions, the linear-quadratic Stackelberg game for state space systems admits the closed-loop Stackelberg strategy which involves the memory of the state variable, for example, one-step memory closed-loop Stackelberg strategies (Başar and Selbuz 1979). However, if the system is noncausal, it cannot be transformed to a regular state space system as above for a causal descriptor system. Therefore, the methods developed for state space systems are not valid for solving the linear-quadratic Stackelberg game of descriptor systems.

In this paper, we develop a method to solve the linear-quadratic Stackelberg game for discrete-time descriptor systems. Sufficient conditions for the existence of the team-optimal feedback closed-loop Stackelberg strategies are obtained. A recursive scheme is presented to determine the closed-loop Stackelberg strategies. Different from the results for state space systems (Başar and Selbuz 1979, Tolwinsky 1981), we show that the closed-loop no-memory information on x_k is sufficient for the leader to find the team-optimal closed-loop Stackelberg strategies for a general class of linear-quadratic Stackelberg games. The closed-loop Stackelberg strategies for both the leader and the follower are realized in linear feedback form. Moreover, since the basic information structure for the construction of the closed-loop Stackelberg strategies in this paper is the closed-loop no-memory information on x_k, it is not necessary to assume that the follower does not take action at the last stage of the game, which is made in Başar and Selbuz (1979) and Tolwinski (1981) as a temporary assumption. Although such an assumption is not crucial and can be removed later, the realized game value for the leader will not be the original team value for the leader's team-optimal control problem.

The results in this paper are parallel in some ways to those for continuous-time descriptor systems in Xu and Mizukami (1994). However, the discrete-time problem has its own features. First, the existing methods for the optimal control of discrete-time descriptor systems (Bender and Laub 1987b, Mantas and Krikelis 1989) are not suitable for the special purpose of this paper. We extend the method developed by Wang *et al.* (1988) from the continuous-time descriptor system to the discrete-time descriptor system to solve the corresponding optimal control problems. Second, in contrast with the continuous-time problem, the team-optimal feedback closed-loop Stackelberg strategies can easily be calculated through a recursive process. Finally, it seems that discrete-time descriptor systems can find more applications in economic systems than continuous-time descriptor systems. One such application is the Leontief dynamic model of a multisector economy (Luenberger and Arbel 1977). In fact, the Leontief model was one of the original reasons for interest in descriptor systems (Bender and Laub 1987b).

This paper is organized as follows. In Section 2, we formulate the linear-quadratic team-optimal closed-loop Stackelberg games for the discrete-time descriptor systems. In Section 3, we extend the method used by Xu and Mizukami (1994) in continuous-time descriptor systems to discrete-time descriptor systems to solve the team-optimal control problem. Section 4 is concerned with the derivation of the Stackelberg strategies under the closed-loop no-memory information. Sufficient conditions for the existence of the team-optimal feedback closed-loop Stackelberg strategies are found. In Section 5, we present a recursive scheme to determine the sufficient conditions.

A numerical example is given to illustrate the viability of the conditions. Section 6 contains some conclusions.

2   Problem Formulation

Consider the linear discrete-time descriptor system

$$Ex_{k+1} = Ax_k + Bu_k + Cv_k, \quad x_0 \text{ is given,} \tag{1}$$

for $k = 0, 1, 2, ..., N - 1$, where

x_k is the n-dimensional vector of the descriptor variables,
u_k is the m-dimensional vector of the leader's control variables,
v_k is the l-dimensional vector of the follower's control variables.

The matrix E is a square matrix of rank $r \leq n$. The pencil $(sE - A)$ is assumed to be regular (i.e., $|(sE - A)| \not\equiv 0$). The cost functions for the leader and the follower are given by J_1 and J_2, respectively, where

$$J_1 = \frac{1}{2} x_N^T E^T Q_N^1 E x_N + \frac{1}{2} \sum_{k=0}^{N-1} \{ x_k^T Q^1 x_k + u_k^T R_{11} u_k + v_k^T R_{12} v_k \}, \tag{2a}$$

$$J_2 = \frac{1}{2} x_N^T E^T Q_N^2 E x_N + \frac{1}{2} \sum_{k=0}^{N-1} \{ x_k^T Q^2 x_k + u_k^T R_{21} u_k + v_k^T R_{22} v_k \}, \tag{2b}$$

with $R_{11} > 0$, $R_{12} > 0$, $R_{22} > 0$, and all other weighting matrices being nonnegative definite. The superscript ^{T} denotes the transpose of the matrix. The information structure of the problem is closed-loop no-memory information on x_k, under which the closed-loop strategy spaces for the leader and the follower at stage k, $k = 0, 1, ..., N - 1$, are denoted by Γ_k^1 and Γ_k^2, respectively. Γ_k^1 and Γ_k^2 are all composed of pure linear feedback strategies of x_k, denoted by γ_k^1 and γ_k^2, respectively. Their open-loop realizations are u_k and v_k respectively. Similar to Başar and Olsder (1982), let us introduce the notation

$$\gamma^i \in \Gamma^i := \{ \gamma_k^i \in \Gamma_k^i, \ k = 0, 1, ..., N - 1 \}, \ i = 1, 2, \tag{3}$$

and define the rational reaction set of the follower by

$$R(\gamma^1) := \{ \gamma_o^2 \in \Gamma^2, \ J_2(\gamma^1, \gamma_o^2) \leq J_2(\gamma^1, \gamma^2), \ \forall \ \gamma^2 \in \Gamma^2 \}. \tag{4}$$

Definition 2.1 $R(\gamma^1)$ *is called a realization singleton if the strategies in* $R(\gamma^1)$ *admit a unique open-loop realization.*

Definition 2.2 *A strategy* $\gamma^1 \in \Gamma_a^1 \subset \Gamma^1$ *is called an admissible strategy of the leader if* $R(\gamma^1)$ *is the realization singleton. Correspondingly,* Γ_a^1 *is called the admissible strategy space of the leader.*

Definition 2.3 *For the dynamic game posed above, an admissible strategy $\gamma^{1*} \in \Gamma_a^1$ constitutes the closed-loop Stackelberg strategy for the leader if*

$$J_1(\gamma^{1*}, R(\gamma^{1*})) \leq J_1(\gamma^1, R(\gamma^1)), \ \forall \ \gamma^1 \in \Gamma_a^1. \tag{5}$$

Remark 2.1 *Definition 2.3 is the simplified definition of the closed-loop Stackelberg strategy when the rational reaction set of the follower is a realization singleton. For the state space system, it is known that the rational reaction set of the follower will be a realization singleton if the leader's strategy is an affine one. However, that is not true for the discrete-time descriptor system. For the discrete-time descriptor system, the rational reaction set of the follower may not be a realization singleton even if the leader's strategy is a pure linear feedback one.*

Remark 2.2 *For the problem of state space systems, the follower is usually assumed not to take action at the last stage of the game. Otherwise, the attainable lower bound by the leader for J_1 will not be the original team value. In contrast with the state space system, such an assumption is not necessary for the descriptor system since the linear feedback strategy of the leader does not involve the memory information on x_k.*

3   Team-Optimal Solutions

As in the state space system, an indirect method is employed to solve the Stackelberg game problem formulated above. By this method, the leader can enforce his team-optimal solution on the follower. The related team-optimal control problem is formulated as follows.

$$\min_{\gamma^1 \in \Gamma^1} \min_{\gamma^2 \in \Gamma^2} J_1(\gamma^1, \gamma^2) \tag{6}$$

subject to (1).

The team-optimal control problem (6) is a multi-person extension of the standard linear-quadratic optimal control problem for discrete-time descriptor systems. There are two standard methods for solving such optimal control problems (Bender and Laub, 1987b; Mantas and Krikelis 1989). In this paper, we extend the method developed by Wang et al., (1988), which deals with the optimal regulator problem for continuous-time descriptor systems, to the discrete-time case. An advantage of the extension is that the feedback gains of the optimal control strategies can be expressed as the function of unknown matrices explicitly.

Suppose that the positive semidefinite weighting matrix Q^1 can be factored as $Q^1 = D^T D$ and $y_k = D x_k$ is the output of the system (1). Then, a basic assumption is made.

Assumption 3.1 *The descriptor system (1) is causally controllable and observable.*

Assumption 3.1 guarantees the uniqueness of the solution for the team-optimal control problem formulated above.

Lemma 3.1 *The descriptor system (1) is causally controllable and observable if and only if the rows of the matrices $[A_{22}\ B_2\ C_2]$ and $[A_{22}^T\ D_2^T]$ are independent respectively, where the matrices are defined below in equations (11a) and (11b).*

The necessary conditions for u_k and v_k, $0 \le k \le N-1$, to be the team-optimal solution are (Bender and Laub, 1987b; Mantas and Krikelis, 1989)

$$Ex_{k+1} = Ax_k + Bu_k + Cv_k, \tag{7a}$$

$$E^T \lambda_k = A^T \lambda_{k+1} + Q^1 x_k, \tag{7b}$$

$$0 = R_{11}u_k + B^T \lambda_{k+1}, \tag{7c}$$

$$0 = R_{12}v_k + C^T \lambda_{k+1}, \tag{7d}$$

with boundary conditions

$$x_{k=0} = x_0, \ E^T \lambda_N = E^T Q_N^1 E x_N. \tag{8}$$

Following Dai (1989), there exist nonsingular matrices M and H such that

$$MEH = \begin{bmatrix} I_r & 0 \\ 0 & 0 \end{bmatrix}, \ r = rank\ E. \tag{9}$$

Then, the necessary conditions can be transformed as

$$z_{k+1}^1 = A_{11}z_k^1 + A_{12}z_k^2 + B_1 u_k + C_1 v_k, \tag{10a}$$

$$0 = A_{21}z_k^1 + A_{22}z_k^2 + B_2 u_k + C_2 v_k, \tag{10b}$$

$$\lambda_k^1 = A_{11}^T \lambda_{k+1}^1 + A_{21}^T \lambda_{k+1}^2 + Q_{11}^1 z_k^1 + Q_{12}^1 z_k^2, \tag{10c}$$

$$0 = A_{12}^T \lambda_{k+1}^1 + A_{22}^T \lambda_{k+1}^2 + Q_{21}^1 z_k^1 + Q_{22}^1 z_k^2, \tag{10d}$$

$$0 = R_{11}u_k + B_1^T \lambda_{k+1}^1 + B_2^T \lambda_{k+1}^2, \tag{10e}$$

$$0 = R_{12}v_k + C_1^T \lambda_{k+1}^1 + C_2^T \lambda_{k+1}^2, \tag{10f}$$

where

$$MAH = \begin{bmatrix} A_{11} & A_{12} \\ A_{21} & A_{22} \end{bmatrix}, \ MB = \begin{bmatrix} B_1 \\ B_2 \end{bmatrix}, \ MC = \begin{bmatrix} C_1 \\ C_2 \end{bmatrix}, \tag{11a}$$

$$H^T Q^1 H = \begin{bmatrix} Q_{11}^1 & Q_{12}^1 \\ Q_{12}^{1T} & Q_{22}^1 \end{bmatrix} = \begin{bmatrix} D_1^T \\ D_2^T \end{bmatrix} \begin{bmatrix} D_1 & D_2 \end{bmatrix}, \tag{11b}$$

and

$$\begin{bmatrix} z_k^1 \\ z_k^2 \end{bmatrix} = H^{-1} x_k, \qquad \begin{bmatrix} \lambda_k^1 \\ \lambda_k^2 \end{bmatrix} = M^{-T} \lambda_k. \tag{12}$$

The boundary conditions become

$$H^{-1} x_0 = \begin{bmatrix} z_0^1 \\ z_0^2 \end{bmatrix}, \tag{13}$$

and

$$H^T E^T \lambda_N = (H^T E^T M^T)(M^{-T} \lambda_N) = \begin{bmatrix} I_r & 0 \\ 0 & 0 \end{bmatrix} \begin{bmatrix} \lambda_N^1 \\ \lambda_N^2 \end{bmatrix}$$

$$= \begin{bmatrix} I_r & 0 \\ 0 & 0 \end{bmatrix} \begin{bmatrix} Q_{11N}^1 & Q_{12N}^1 \\ Q_{12N}^{1T} & Q_{22N}^1 \end{bmatrix} \begin{bmatrix} I_r & 0 \\ 0 & 0 \end{bmatrix} \begin{bmatrix} z_N^1 \\ z_N^2 \end{bmatrix}, \tag{14}$$

where

$$M^{-T} Q_N^1 M^{-1} = \begin{bmatrix} Q_{11N}^1 & Q_{12N}^1 \\ Q_{12N}^{1T} & Q_{22N}^1 \end{bmatrix}. \tag{15}$$

From (10e) and (10f), we get

$$u_k^t = -R_{11}^{-1} B_1^T \lambda_{k+1}^1 - R_{11}^{-1} B_2^T \lambda_{k+1}^2, \tag{16a}$$

$$v_k^t = -R_{12}^{-1} C_1^T \lambda_{k+1}^1 - R_{12}^{-1} C_2^T \lambda_{k+1}^2. \tag{16b}$$

Substituting (16) into (10a) and (10b) yields

$$\begin{bmatrix} z_{k+1}^1 \\ \lambda_k^1 \end{bmatrix} = \begin{bmatrix} A_{11} & -S_{11} \\ Q_{11}^1 & A_{11}^T \end{bmatrix} \begin{bmatrix} z_k^1 \\ \lambda_{k+1}^1 \end{bmatrix} + \begin{bmatrix} A_{12} & -S_{12} \\ Q_{12}^1 & A_{21}^T \end{bmatrix} \begin{bmatrix} z_k^2 \\ \lambda_{k+1}^2 \end{bmatrix}, \tag{17a}$$

$$0 = \begin{bmatrix} A_{21} & -S_{12}^T \\ -Q_{12}^{1T} & -A_{12}^T \end{bmatrix} \begin{bmatrix} z_k^1 \\ \lambda_{k+1}^1 \end{bmatrix} + \begin{bmatrix} A_{22} & -S_{22} \\ -Q_{22}^1 & -A_{22}^T \end{bmatrix} \begin{bmatrix} z_k^2 \\ \lambda_{k+1}^2 \end{bmatrix}, \tag{17b}$$

where

$$S_{11} = B_1 R_{11}^{-1} B_1^T + C_1 R_{12}^{-1} C_1^T, \tag{18a}$$

$$S_{12} = B_1 R_{11}^{-1} B_2^T + C_1 R_{12}^{-1} C_2^T, \tag{18b}$$

$$S_{22} = B_2 R_{11}^{-1} B_2^T + C_2 R_{12}^{-1} C_2^T. \tag{18c}$$

Define

$$T_1 = \begin{bmatrix} A_{11} & -S_{11} \\ Q_{11}^1 & A_{11}^T \end{bmatrix}, \quad T_2 = \begin{bmatrix} A_{12} & -S_{12} \\ Q_{12}^1 & A_{21}^T \end{bmatrix},$$

$$\tag{19}$$

$$T_3 = \begin{bmatrix} A_{21} & -S_{12}^T \\ -Q_{12}^{1T} & -A_{12}^T \end{bmatrix}, \quad T_4 = \begin{bmatrix} A_{22} & -S_{22} \\ -Q_{22}^1 & -A_{22}^T \end{bmatrix}.$$

Lemma 3.2 *(Wang et al., 1988)* $|T_4| \neq 0$ *if and only if the rows of the matrices $[A_{22} \; B_2 \; C_2]$ and $[A_{22}^T \; D_2^T]$ are independent respectively.*

Since T_4 is nonsingular, we have

$$
\begin{bmatrix} z_k^2 \\ \lambda_{k+1}^2 \end{bmatrix} = -T_4^{-1} T_3 \begin{bmatrix} z_k^1 \\ \lambda_{k+1}^1 \end{bmatrix} = \begin{bmatrix} M_{11} & M_{12} \\ M_{21} & M_{22} \end{bmatrix} \begin{bmatrix} z_k^1 \\ \lambda_{k+1}^1 \end{bmatrix}, \tag{20}
$$

and

$$
\begin{bmatrix} z_{k+1}^1 \\ \lambda_k^1 \end{bmatrix} = (T_1 - T_2 T_4^{-1} T_3) \begin{bmatrix} z_k^1 \\ \lambda_{k+1}^1 \end{bmatrix} = \begin{bmatrix} A_0 & -S_0 \\ Q_0 & A_0^T \end{bmatrix} \begin{bmatrix} z_k^1 \\ \lambda_{k+1}^1 \end{bmatrix}, \tag{21}
$$

from (17), where the corresponding matrices are defined in Appendix A.

(21) gives the two-point boundary value problem with the boundary conditions (13) and (14). Let $\lambda_k^1 = P_k^1 z_k^1$, a matrix Riccati equation can be obtained from (21)

$$
P_k^1 = Q_o + A_o^T P_{k+1}^1 [I + S_o P_{k+1}^1]^{-1} A_o, \;\; P_N^1 = Q_{11N}^1. \tag{22}
$$

Using the solution of (22), we arrive at the following equations

$$
z_{k+1}^1 = Z_k^1 z_k^1, \tag{23}
$$

$$
\lambda_{k+1}^1 = L_k^1 z_k^1, \tag{24}
$$

where

$$
Z_k^1 = [I + S_o P_{k+1}^1]^{-1} A_o, \tag{25a}
$$

$$
L_k^1 = P_{k+1}^1 [I + S_o P_{k+1}^1]^{-1} A_o. \tag{25b}
$$

Furthermore, substituting λ_{k+1}^1 into (20) yields

$$
z_k^2 = Z_k^2 z_k^1, \tag{26}
$$

$$
\lambda_{k+1}^2 = L_k^2 z_k^1, \tag{27}
$$

where (also see Appendix A for the corresponding terms)

$$
Z_k^2 = M_{11} + M_{12} L_k^1, \tag{28a}
$$

$$
L_k^2 = M_{21} + M_{22} L_k^1. \tag{28b}
$$

Based on the derivations given above, we arrive at the following conclusion.

Lemma 3.3 *Suppose that Assumption 3.1 is satisfied. Then,*

(i) the team-optimal control problem defined above admits uncountably many linear feedback solutions given by

$$\gamma_k^{1t} = -R_{11}^{-1}B^T K_k^1 x_k, \ 0 \le k \le N - 1, \tag{29a}$$

$$\gamma_k^{2t} = -R_{12}^{-1}C^T K_k^2 x_k, \ 0 \le k \le N - 1, \tag{29b}$$

where

$$K_k^1 = M^T \begin{bmatrix} L_k^1 & 0 \\ L_k^2 - F_k^1 Z_k^2 & F_k^1 \end{bmatrix} H^{-1}, \tag{30a}$$

$$K_k^2 = M^T \begin{bmatrix} L_k^1 & 0 \\ L_k^2 - F_k^2 Z_k^2 & F_k^2 \end{bmatrix} H^{-1}, \tag{30b}$$

and F_k^1 and F_k^2 are arbitrary two $(n - r) \times (n - r)$ matrices making $A_{22} - B_2 R_{11}^{-1} B_2^T F_k^1 - C_2 R_{12}^{-1} C_2^T F_k^2$ invertible;

(ii) the open-loop realizations of γ_k^{1t} and γ_k^{2t} are unique, given by

$$u_k^t = -R_{11}^{-1}[B_1^T \ B_2^T] \begin{bmatrix} L_k^1 \\ L_k^2 \end{bmatrix} z_k^{1t}, \ 0 \le k \le N - 1, \tag{31a}$$

$$v_k^t = -R_{12}^{-1}[C_1^T \ C_2^T] \begin{bmatrix} L_k^1 \\ L_k^2 \end{bmatrix} z_k^{1t}, \ 0 \le k \le N - 1, \tag{31b}$$

respectively, where z_k^{1t} is the unique solution of (23) with the initial condition $z_{k=0}^1 = z_0^1$.

Proof. The proof follows the derivations prior to the statements of Lemma 3.3 and the reasoning similar to the one employed in the implementation of the optimal feedback control (Wang et al., 1988). $\qquad\square$

4   Feedback Closed-Loop Stackelberg Strategy

It is obvious that (29a,b) provide the sets of the linear feedback team-optimal strategies for the leader and the follower respectively, where, the superscript 't' represents the terms related to the team-optimal solution. In terms of dynamic games, the strategies in (29) can also be called the feedback representations of the open-loop team-optimal solution. In the following, we shall find a special feedback representation among (29a) to constitute the closed-loop Stackelberg strategy for the leader.

By substituting (29a) into (2b) and (1), we formulate an optimal control problem for the follower's cost function J_2, that is, minimizing

$$J_2 = \frac{1}{2}x_N^T E^T Q_N^2 E x_N + \frac{1}{2} \sum_{k=0}^{N-1} \{x_k^T[Q^2 + K_k^{1T} B R_{11}^{-1} R_{21} R_{11}^{-1} B^T K_k^1]x_k + v_k^T R_{22} v_k\},$$

$$\tag{32}$$

under the constraint

$$Ex_{k+1} = [A - BR_{11}^{-1}B^T K_k^1]x_k + Cv_k, \quad x_0 \ is \ given, \tag{33}$$

for $k = 0, 1, 2, ..., N - 1$.

The solution of the above problem constitutes the rational reaction of the follower when a strategy in (29a) is announced to him. In the following, we assume for the moment that the follower adopts the same transformation as that of the leader in solving the optimal control problem. Then (32) and (33) can be rewritten as

$$J_2 = \frac{1}{2}z_N^{1T}Q_{11N}^2 z_N^1 + \frac{1}{2}\sum_{k=0}^{N-1}\{[z_k^{1T}\ z_k^{2T}]\begin{bmatrix} \hat{Q}_{11k}^2 & \hat{Q}_{12k}^2 \\ \hat{Q}_{12k}^{2T} & \hat{Q}_{22k}^2 \end{bmatrix}\begin{bmatrix} z_k^1 \\ z_k^2 \end{bmatrix} + v_k^T R_{22}v_k\},$$

$$\tag{34}$$

and

$$z_{k+1}^1 = \hat{A}_{11k}z_k^1 + \hat{A}_{12k}z_k^2 + C_1 v_k, \tag{35a}$$

$$0 = \hat{A}_{21k}z_k^1 + \hat{A}_{22k}z_k^2 + C_2 v_k, \tag{35b}$$

respectively, where the reader is referred to Appendix B for the definitions of the related terms.

Similar to the derivations in Section 3, the necessary condition for (34) to be minimized may take the form

$$\begin{bmatrix} z_{k+1}^1 \\ \hat{\lambda}_k^1 \end{bmatrix} = \begin{bmatrix} \hat{A}_{11k} & -\hat{S}_{11} \\ \hat{Q}_{11k}^2 & \hat{A}_{11k}^T \end{bmatrix}\begin{bmatrix} z_k^1 \\ \hat{\lambda}_{k+1}^1 \end{bmatrix} + \begin{bmatrix} \hat{A}_{12k} & -\hat{S}_{12} \\ \hat{Q}_{12k}^2 & \hat{A}_{21k}^T \end{bmatrix}\begin{bmatrix} z_k^2 \\ \hat{\lambda}_{k+1}^2 \end{bmatrix},$$

$$\tag{36a}$$

$$0 = \begin{bmatrix} \hat{A}_{21k} & -\hat{S}_{12}^T \\ -\hat{Q}_{12k}^{2T} & -\hat{A}_{12k}^T \end{bmatrix}\begin{bmatrix} z_k^1 \\ \hat{\lambda}_{k+1}^1 \end{bmatrix} + \begin{bmatrix} \hat{A}_{22k} & -\hat{S}_{22} \\ -\hat{Q}_{22k}^2 & -\hat{A}_{22k}^T \end{bmatrix}\begin{bmatrix} z_k^2 \\ \hat{\lambda}_{k+1}^2 \end{bmatrix}, \tag{36b}$$

where

$$\hat{S}_{11} = C_1 R_{22}^{-1} C_1^T, \quad \hat{S}_{12} = C_1 R_{22}^{-1} C_2^T, \quad \hat{S}_{22} = C_2 R_{22}^{-1} C_2^T.$$

The optimal control of the follower is

$$v_k^o = -R_{22}^{-1}C_1^T\hat{\lambda}_{k+1}^1 - R_{22}^{-1}C_2^T\hat{\lambda}_{k+1}^2. \tag{37}$$

Define

$$\hat{T}_{1k} = \begin{bmatrix} \hat{A}_{11k} & -\hat{S}_{11} \\ \hat{Q}_{11k}^2 & \hat{A}_{11k}^T \end{bmatrix}, \quad \hat{T}_{2k} = \begin{bmatrix} \hat{A}_{12k} & -\hat{S}_{12} \\ \hat{Q}_{12k}^2 & \hat{A}_{21k}^T \end{bmatrix},$$

$$\tag{38}$$

$$\hat{T}_{3k} = \begin{bmatrix} \hat{A}_{21k} & -\hat{S}_{12}^T \\ -\hat{Q}_{12k}^{2T} & -\hat{A}_{12k}^T \end{bmatrix}, \quad \hat{T}_{4k} = \begin{bmatrix} \hat{A}_{22k} & -\hat{S}_{22} \\ -\hat{Q}_{22k}^2 & -\hat{A}_{22k}^T \end{bmatrix}.$$

Lemma 4.1 *A strategy given by (29a) constitutes an admissible strategy of the leader if $\hat{T}_{4k}$ is invertible at each stage $0 \leq k \leq N - 1$.*

Proof. According to Definition 2.2, for a strategy (29a) to be an admissible strategy, there must exist a unique open-loop solution to the follower's optimal control problem. The given condition will ensure the uniqueness of the open-loop solution for the follower's optimal control problem. Therefore, the leader's corresponding strategy is admissible. $\square$

The admissible strategy may not be unique since there exist uncountably many F_k^1, $0 \leq k \leq N - 1$, in (29a), such that the condition of Lemma 4.1 holds.

Now, suppose that $\hat{T}_{4k}$ is invertible at each stage $0 \leq k \leq N - 1$. Then, we have

$$\left[\begin{array}{c} z_k^2 \\ \hat{\lambda}_{k+1}^2 \end{array} \right] = -\hat{T}_{4k}^{-1}\hat{T}_{3k} \left[\begin{array}{c} z_k^1 \\ \hat{\lambda}_{k+1}^1 \end{array} \right] = \left[\begin{array}{cc} \hat{M}_{11k} & \hat{M}_{12k} \\ \hat{M}_{21k} & \hat{M}_{22k} \end{array} \right] \left[\begin{array}{c} z_k^1 \\ \hat{\lambda}_{k+1}^1 \end{array} \right], \qquad (39)$$

and

$$\left[\begin{array}{c} z_{k+1}^1 \\ \hat{\lambda}_k^1 \end{array} \right] = (\hat{T}_{1k} - \hat{T}_{2k}\hat{T}_{4k}^{-1}\hat{T}_{3k}) \left[\begin{array}{c} z_k^1 \\ \hat{\lambda}_{k+1}^1 \end{array} \right] = \left[\begin{array}{cc} \hat{A}_{0k} & -\hat{S}_{0k} \\ \hat{Q}_{0k} & \hat{A}_{0k}^T \end{array} \right] \left[\begin{array}{c} z_k^1 \\ \hat{\lambda}_{k+1}^1 \end{array} \right], \qquad (40)$$

from (36) where the corresponding matrices are defined in Appendix C.

Remark 4.1 *In the practical computation of the problem, it is not necessary to calculate the corresponding matrices in (39) and (40) by following the definitions given in Appendix C. With the help of some computer algebra system (for example, REDUCE, Copyright (c) The RAND Corporation 1985, 1993), one can calculate the corresponding matrices directly.*

(40) constitutes the two-point boundary value problem with the boundary conditions $z_{k=0}^1 = z_0^1$ and $\hat{\lambda}_N^1 = Q_{11N}^2$. Let $\hat{\lambda}_k^1 = P_k^2 z_k^1$, $0 \leq k \leq N - 1$, a matrix Riccati equation can be obtained from (40)

$$P_k^2 = \hat{Q}_{ok} + \hat{A}_{ok}^T P_{k+1}^2 [I + \hat{S}_{ok} P_{k+1}^2]^{-1} \hat{A}_{ok}, \quad P_N^2 = Q_{11N}^2. \qquad (41)$$

Then we have

$$z_{k+1}^1 = \hat{Z}_k^1 z_k^1, \qquad (42)$$

$$\hat{\lambda}_{k+1}^1 = \hat{L}_k^1 z_k^1, \qquad (43)$$

where

$$\hat{Z}_k^1 = [I + \hat{S}_{ok} P_{k+1}^2]^{-1} \hat{A}_{ok}, \qquad (44a)$$

$$\hat{L}_k^1 = P_{k+1}^2 [I + \hat{S}_{ok} P_{k+1}^2]^{-1} \hat{A}_{ok}. \qquad (44b)$$

Furthermore, substituting $\hat{\lambda}^1_{k+1}$ into (39) gives

$$z^2_k = \hat{Z}^2_k z^1_k, \tag{45}$$

$$\hat{\lambda}^2_{k+1} = \hat{L}^2_k z^1_k, \tag{46}$$

where(also see Appendix C for the corresponding terms)

$$\hat{Z}^2_k = \hat{M}_{11k} + \hat{M}_{12k}\hat{L}^1_k, \tag{47a}$$

$$\hat{L}^2_k = \hat{M}_{21k} + \hat{M}_{22k}\hat{L}^1_k. \tag{47b}$$

From (37), the optimal control for the follower's problem is

$$v^o_k = -R^{-1}_{22}(C^T_1 \hat{L}^1_k + C^T_2 \hat{L}^2_k)z^{1o}_k, \ 0 \le k \le N - 1, \tag{48}$$

where z^{1o}_k is the unique solution of (42) with the initial condition $z^1_{k=0} = z^1_0$. Obviously, the value of v^o_k depends on the choice of the matrix sequence F^1_k, $0 \le k \le N - 1$, which is contained in the parameter matrices of the follower's optimal control problem. In the following, through the selection of the matrix sequence F^1_k, $0 \le k \le N - 1$, we will find a specific admissible strategy γ^{1*}_k, $0 \le k \le N - 1$ to constitute the linear feedback closed-loop Stackelberg strategy for the leader.

Condition 4.1 *There exists at least one matrix sequence F^{1*}_k, $0 \le k \le N - 1$, such that*

$$Z^1_k = \hat{Z}^{1*}_k, \tag{49a}$$

$$R^{-1}_{12}(C^T_1 L^1_k + C^T_2 L^2_k) = R^{-1}_{22}(C^T_1 \hat{L}^{1*}_k + C^T_2 \hat{L}^{2*}_k), \tag{49b}$$

where the matrices with the superscript $*$ represent the corresponding matrices obtained when F^1_k is substituted by F^{1*}_k.

Theorem 4.1 *For the closed-loop Stackelberg game formulated in Section 2, assume that system (1) is causally controllable and observable. Then the following holds.*

(i) If there exists one matrix sequence F^{1*}_k, $0 \le k \le N - 1$, such that Condition 4.1 is satisfied, then the admissible strategy

$$\gamma^{1*}_k = -R^{-1}_{11}B^T K^{1*}_k x_k, \ 0 \le k \le N - 1, \tag{50}$$

constitutes the team-optimal feedback closed-loop Stackelberg strategy of the leader, where

$$K^{1*}_k = M^T \begin{bmatrix} L^1_k & 0 \\ L^2_k - F^{1*}_k Z^2_k & F^{1*}_k \end{bmatrix} H^{-1}.$$

(ii) (29b) constitutes the feedback closed-loop Stackelberg strategies for the follower.

Proof. (i) First, let us consider the case when the follower adopts the same transformation as the leader. Since the system (1) is regular, its solution will be determined uniquely by u_k, v_k, $0 \leq k \leq N-1$. Based on the derivations before this theorem, if Condition 4.1 is satisfied, we have $z_k^{1t} = z_k^{1*}$ and $v_k^t = v_k^*$ by observing (49a,b), where the terms with the superscript $*$ denote the corresponding ones obtained from (42) and (48) when the strategy (50) is announced by the leader. Therefore, the strategy obtained by the leader constitutes the team-optimal closed-loop Stackelberg strategy.

Now, suppose that the follower adopts a different transformation. Since the strategy (50) is admissible, we can prove that the open-loop optimal solution for the follower's optimal control problem is unique no matter what transformation is used to find it by contradiction; that is, if the statement is not true, we will arrive at the conclusion that the solution of (42) is not unique, that is false. The detail is omitted. Different transformations will only affect the feedback representation of the open-loop solution. Furthermore, since (49a,b) are satisfied, the unique open-loop optimal solution for the follower's problem is equal to v_k^t, the team-optimal solution of the follower. At the same time, the open-loop realization of (50) is the team-optimal solution of the leader. Therefore, (50) still constitutes the closed-loop Stackelberg strategy.

(ii) This follows from the result that the open-loop representation of (29b) coincides with the optimal control of the follower's optimal control problem when (50) is announced to the follower. $\qquad\square$

Remark 4.2 *Condition 4.1 is composed of two sets of nonlinear algebraic equations of F_k^1. In order to find their solutions, we need some computer algebra system as stated in Remark 4.1.*

Corollary 4.1 *In terms of z_0^1, the optimal cost for the leader and the follower are equal to*

$$J_1^* = \frac{1}{2} z_0^{1T} P_0^1 z_0^1, \tag{51a}$$

$$J_2^* = \frac{1}{2} z_0^{1T} P_0^{2*} z_0^1, \tag{51b}$$

respectively. Where P_k^{2*}, $0 \leq k \leq N-1$, is obtained from (41), in which F_k^1, $0 \leq k \leq N-1$ is replaced by F_k^{1*}, $0 \leq k \leq N-1$.

5   Recursive Scheme for Determination of F_k^{1*}

In Section 4, it would be more convenient to provide the sufficient conditions in terms of the matrices involved in the system description (1),(2). However, it seems difficult or impossible to get such general conditions because of the complicated matrix manipulations. An important feature for the discrete-time problem is that the strategy (50) can be obtained through a recursive process, that is, the coefficient matrices F_k^{1*} can be computed recursively. In the following, we first present a recursive scheme for the computation of F_k^{1*}, and then give a numerical example to illustrate the validity(nonvoidness) of the condition.

The recursive scheme starts from $k = N - 1$ and goes backward, where $P_N^{2*} = Q_{11N}^2$ is given.

(a) Use (49) to determine F_k^{1*}, where P_{k+1}^{2*} has been determined by the previous iteration.

(b) Use (39),(40) to calculate the corresponding matrices $(\hat{M}_{11k}^*,...,\hat{A}_{ok}^*,...)$.

(c) Use (41) to compute P_k^{2*}.

(d) Set $k = k - 1$, go back to (a) and iterate.

When the coefficient matrices $F_k^{1*}, 0 \le k \le N - 1$, are determined, we can obtain the team-optimal feedback closed-loop Stackelberg strategy γ_k^{1*} by substituting F_k^1 in (29a) with F_k^{1*}.

A Numerical Example. Consider a two-dimension discrete-time descriptor system.

$$\begin{bmatrix} 1 & 0 \\ 0 & 0 \end{bmatrix} \begin{bmatrix} z_{k+1}^1 \\ z_{k+1}^2 \end{bmatrix} = \begin{bmatrix} 2 & 1 \\ 1 & 0 \end{bmatrix} \begin{bmatrix} z_k^1 \\ z_k^2 \end{bmatrix} + \begin{bmatrix} 2 \\ 1 \end{bmatrix} u_k + \begin{bmatrix} 1 \\ 1 \end{bmatrix} v_k, \; n \le k \le N-1. \tag{52}$$

The cost functions J_1, J_2 are, respectively

$$J_1 = \frac{5}{2}(z_N^1)^2 + \frac{1}{2} \sum_{k=n}^{N-1} \{x_k^T x_k + 3u_k^2 + v_k^2\}, \tag{53a}$$

$$J_2 = \frac{3}{2}(z_N^1)^2 + \frac{1}{2} \sum_{k=n}^{N-1} \{x_k^T x_k + u_k^2 + 2v_k^2\}, \tag{53b}$$

where $x_k^T = [z_k^1 \; z_k^2]$. For this example, the relevant terms for the team-optimal solution of J_1 are determined from

$$Z_k^1 = 3/(5P_{k+1}^1 + 4),$$

$$L_k^1 = 3P_{k+1}^1/(5P_{k+1}^1 + 4),$$

$$Z_k^2 = -3P_{k+1}^1/(5P_{k+1}^1 + 4),$$

$$L_k^2 = 3/(5P_{k+1}^1 + 4),$$

$$U_k = -(2P_{k+1}^1 + 1)/(5P_{k+1}^1 + 4),$$

$$V_k = -(3P_{k+1}^1 + 1)/(5P_{k+1}^1 + 4),$$

$$P_k^1 = (11P_{k+1}^1 + 7)/(5P_{k+1}^1 + 4), P_N^1 = 5.$$

Using (49), F_k^{1*} is determined from

$$F_k^{1*} = 9(P_{k+1}^{2*} - P_{k+1}^1)/(3P_{k+1}^{2*} + 4P_{k+1}^1 + 5),$$

where

$$P_k^{2*} = \hat{Q}_{ok}^* + (\hat{A}_{ok}^*)^2 P_{k+1}^{2*}/(1 + \hat{S}_{ok}^* P_{k+1}^{2*}), \quad P_N^{2*} = 3.$$

The other relevant terms are determined from

$$|\hat{T}_{4k}^*| = -(3(F_k^{1*})^2 + 9)/18,$$

$$\hat{Z}_k^{1*} = \hat{A}_{ok}^*/(1 + \hat{S}_{ok}^* P_{k+1}^{2*}),$$

$$\hat{L}_k^{1*} = P_{k+1}^{2*} \hat{A}_{ok}^*/(1 + \hat{S}_{ok} P_{k+1}^{2*}),$$

$$\hat{Z}_k^{2*} = \hat{M}_{11k}^* + \hat{M}_{12k}^* \hat{L}_k^{1*},$$

$$\hat{L}_k^{2*} = \hat{M}_{21k}^* + \hat{M}_{22k}^* \hat{L}_k^{1*},$$

$$V_k^* = -(\hat{L}_k^{1*} + \hat{L}_k^{2*})/2,$$

where $\hat{A}_{ok}^*, ..., \hat{M}_{11k}^*, ...$ are expressed in Appendix D. Some important numerical values for forming the strategy are given in Table 1 for $k = N - 1, ..., N - 10$.

k	F_k^{1*}	$\hat{T}_{4k}^*$	P_k^1	L_k^1	L_k^2	Z_k^2	P_k^{2*}
N	-	-	5	-	-	-	3
N-1	-0.5294	-0.5467	2.1379	0.5172	0.1034	-0.5172	2.2140
N-2	0.0339	-0.5001	2.0774	0.4366	0.2042	-0.4366	2.2333
N-3	0.0701	-0.5008	2.0748	0.4331	0.2085	-0.4331	2.2367
N-4	0.0727	-0.5008	2.0747	0.4330	0.2087	-0.4330	2.2369
N-5	0.0729	-0.5008	2.0747	0.4330	0.2087	-0.4330	2.2369
N-6	0.0729	-0.5008	2.0747	0.4330	0.2087	-0.4330	2.2369
N-7	0.0729	-0.5008	2.0747	0.4330	0.2087	-0.4330	2.2369
N-8	0.0729	-0.5008	2.0747	0.4330	0.2087	-0.4330	2.2369
N-9	0.0729	-0.5008	2.0747	0.4330	0.2087	-0.4330	2.2369
N-10	0.0729	-0.5008	2.0747	0.4330	0.2087	-0.4330	2.2369

Table 1: Numerical calculations of relevant terms

Therefore, we obtain the team-optimal feedback closed-loop Stackelberg strategy in the following form.

$$\gamma_k^{1*} = -\frac{1}{3}[2\ 1]\begin{bmatrix} L_k^1 & 0 \\ L_k^2 - F_k^{1*} Z_k^2 & F_k^{1*} \end{bmatrix}\begin{bmatrix} z_k^1 \\ z_k^2 \end{bmatrix}.$$

Furthermore, $|\hat{T}_{4k}^*| \neq 0$ in Table 1 verifies that the strategy obtained is admissible.

6 Conclusions

In this paper, we have investigated the linear-quadratic Stackelberg game for discrete-time descriptor systems. The sufficient conditions for the existence of the linear feedback closed-loop Stackelberg strategy is found. The numerical example shows the exact existence of such a strategy. In contrast with some closed-loop Stackelberg strategies for state space systems (Başar and Selbuz 1979 and Tolwinsky 1981), the most important feature is that there exists the team-optimal linear feedback closed-loop Stackelberg strategy for descriptor systems. Since a linear feedback strategy does not involve the memory information on x_k, we do not need to assume that the follower does not take action at the last stage of the game. As the result, a complete team-optimal value can be achieved by the leader. It is expected that such a linear feedback closed-loop Stackelberg strategy could find some applications in coordination problems of multilevel systems.

Appendix A. Suppose that Assumption 3.1 is satisfied, then the matrix T_4 is invertible. However, since A_{22} is singular, it is necessary to introduce an auxiliary matrix Σ, which makes $A_{22} - S_{22}\Sigma$ invertible, to find the inverse of T_4.

$$\begin{bmatrix} A_{22} & -S_{22} \\ -Q_{22}^1 & -A_{22}^T \end{bmatrix} =$$

$$\begin{bmatrix} I & 0 \\ \Sigma & I \end{bmatrix} \begin{bmatrix} A_{22} - S_{22}\Sigma & -S_{22} \\ -\Sigma A_{22} - A_{22}^T\Sigma + \Sigma S_{22}\Sigma - Q_{22}^1 & \Sigma S_{22} - A_{22}^T \end{bmatrix} \begin{bmatrix} I & 0 \\ -\Sigma & I \end{bmatrix}. \tag{54}$$

Obviously, the inverse of T_4 will not depend on the choice of Σ. For the purpose of simplicity, we choose Σ to satisfy the equation

$$- \Sigma A_{22} - A_{22}^T\Sigma + \Sigma S_{22}\Sigma - Q_{22}^1 = 0. \tag{55}$$

Since Assumption 3.1, the solution of (55) can exist which makes $A_{22} - S_{22}\Sigma$ invertible. Hence,

$$\begin{bmatrix} A_{22} & -S_{22} \\ -Q_{22}^1 & -A_{22}^T \end{bmatrix}^{-1} = \begin{bmatrix} I & 0 \\ \Sigma & I \end{bmatrix} \begin{bmatrix} \tilde{A}_{22}^{-1} & -\tilde{A}_{22}^{-1}S_{22}\tilde{A}_{22}^{-T} \\ 0 & -\tilde{A}_{22}^{-T} \end{bmatrix} \begin{bmatrix} I & 0 \\ -\Sigma & I \end{bmatrix}. \tag{56}$$

where $\tilde{A}_{22} = A_{22} - S_{22}\Sigma$. Using (20), (21) and (56), we can obtain

$$M_{11} = -\tilde{A}_{22}^{-1}A_{21} - \tilde{A}_{22}^{-1}S_{22}N_2^T,$$

$$M_{12} = \tilde{A}_{22}^{-1}S_{12}^T + \tilde{A}_{22}^{-1}S_{22}N_1^T,$$

$$M_{21} = \Sigma M_{11} - N_2^T,$$

$$M_{22} = \Sigma M_{12} + N_1^T,$$

$$A_0 = A_{11} + N_1 A_{21} + N_1 S_{22} N_2^T + S_{12} N_2^T,$$

$$S_0 = S_{11} + N_1 S_{12}^T + N_1 S_{22} N_1^T + S_{12} N_1^T,$$

$$Q_0 = Q_{11}^1 - N_2 A_{21} - N_2 S_{22} N_2^T - A_{21}^T N_2^T, \tag{57}$$

where

$$N_1 = -\tilde{A}_{12}\tilde{A}_{22}^{-1}, \ N_2^T = \tilde{A}_{22}^{-T}\tilde{Q}_{12}^{1T},$$

$$\tilde{A}_{12} = A_{12} - S_{12}\Sigma, \ \tilde{Q}_{12}^1 = Q_{12}^1 + A_{21}^T\Sigma. \tag{58}$$

Appendix B.

$$\begin{bmatrix} \hat{Q}_{11k}^2 & \hat{Q}_{12k}^2 \\ \hat{Q}_{12k}^{2T} & \hat{Q}_{22k}^2 \end{bmatrix} :=$$

$$\begin{bmatrix} Q_{11}^2 & Q_{12}^2 \\ Q_{12}^{2T} & Q_{22}^2 \end{bmatrix} + \begin{bmatrix} L_k^{1T} & L_k^{2T} - Z_k^{2T} F_k^{1T} \\ 0 & F_k^{1T} \end{bmatrix} \begin{bmatrix} S_1 & S_2 \\ S_2^T & S_3 \end{bmatrix} \begin{bmatrix} L_k^1 & 0 \\ L_k^2 - F_k^1 Z_k^2 & F_k^1 \end{bmatrix},$$

$$M^{-T} Q_N^2 M^{-1} = \begin{bmatrix} Q_{11N}^2 & Q_{12N}^2 \\ Q_{12N}^{2T} & Q_{22N}^2 \end{bmatrix}, \ H^T Q^2 H = \begin{bmatrix} Q_{11}^2 & Q_{12}^2 \\ Q_{12}^{2T} & Q_{22}^2 \end{bmatrix},$$

$$S_1 = B_1 R_{11}^{-1} R_{21} R_{11}^{-1} B_1^T,$$

$$S_2 = B_1 R_{11}^{-1} R_{21} R_{11}^{-1} B_2^T,$$

$$S_3 = B_2 R_{11}^{-1} R_{21} R_{11}^{-1} B_2^T,$$

$$\hat{A}_{11k} = A_{11} - B_1 R_{11}^{-1}[B_1^T L_k^1 + B_2^T(L_k^2 - F_k^1 Z_k^2)],$$

$$\hat{A}_{12k} = A_{12} - B_1 R_{11}^{-1} B_2^T F_k^1,$$

$$\hat{A}_{21k} = A_{21} - B_2 R_{11}^{-1}[B_1^T L_k^1 + B_2^T(L_k^2 - F_k^1 Z_k^2)],$$

$$\hat{A}_{22k} = A_{22} - B_2 R_{11}^{-1} B_2^T F_k^1. \tag{59}$$

Appendix C. Similar to the derivation of Appendix A, for each $0 \leq k \leq N - 1$, we get

$$\hat{M}_{11k} = -\tilde{A}_{22k}^{-1} A_{21k} - \tilde{A}_{22k}^{-1} \hat{S}_{22} \hat{N}_{2k}^T,$$

$$\hat{M}_{12k} = \tilde{A}_{22k}^{-1} \hat{S}_{12}^T + \tilde{A}_{22k}^{-1} \hat{S}_{22} \hat{N}_{1k}^T,$$

$$\hat{M}_{21k} = \hat{\Sigma}_k \hat{M}_{11k} - \hat{N}_{2k}^T,$$

$$\hat{M}_{22k} = \hat{\Sigma}_k \hat{M}_{12k} + \hat{N}_{1k}^T,$$

$$\hat{A}_{0k} = \hat{A}_{11k} + \hat{N}_{1k} \hat{A}_{21k} + \hat{N}_{1k} \hat{S}_{22} \hat{N}_{2k}^T + \hat{S}_{12} \hat{N}_{2k}^T,$$

$$\hat{S}_{0k} = \hat{S}_{11} + \hat{N}_{1k} \hat{S}_{12}^T + \hat{N}_{1k} \hat{S}_{22} \hat{N}_{1k}^T + \hat{S}_{12} \hat{N}_{1k}^T,$$

$$\hat{Q}_{0k} = \hat{Q}^2_{11k} - \hat{N}_{2k}\hat{A}_{21k} - \hat{N}_{2k}\hat{S}_{22}\hat{N}^T_{2k} - \hat{A}^T_{21k}\hat{N}^T_{2k}, \tag{60}$$

where

$$\hat{N}_{1k} = -\tilde{A}_{12k}\tilde{A}^{-1}_{22k}, \quad \hat{N}^T_{2k} = \tilde{A}^{-T}_{22k}\tilde{Q}^{2T}_{12k},$$

$$\tilde{A}_{12k} = \hat{A}_{12k} - \hat{S}_{12}\hat{\Sigma}_k, \quad \tilde{Q}^2_{12k} = \hat{Q}^2_{12k} + \hat{A}^T_{21k}\hat{\Sigma}_k. \tag{61}$$

$\hat{\Sigma}_k$ satisfies

$$-\hat{\Sigma}_k\hat{A}_{22k} - \hat{A}^T_{22k}\hat{\Sigma}_k + \hat{\Sigma}_k\hat{S}_{22}\hat{\Sigma}_k - \hat{Q}^2_{22k} = 0, \tag{62}$$

such that $\tilde{A}_{22k} = \hat{A}_{22k} - \hat{S}_{22}\hat{\Sigma}_k$ invertible.

Appendix D.

$$\hat{A}^*_{ok} = -(4(F^{1*}_k)^2 P^1_{k+1} - 4(F^{1*}_k)^2 - 3F^{1*}_k P^1_{k+1} - 15F^{1*}_k - 27P^1_{k+1} - 27)/$$
$$3(5(F^{1*}_k)^2 P^1_{k+1} + 4(F^{1*}_k)^2 + 15P^1_{k+1} + 12),$$

$$\hat{S}^*_{ok} = ((F^{1*}_k)^2 - 6F^{1*}_k + 9)/3((F^{1*}_k)^2 + 3),$$

$$\hat{Q}^*_{ok} = (152(F^{1*}_k)^2(P^1_{k+1})^2 + 200(F^{1*}_k)^2 P^1_{k+1} + 80(F^{1*}_k)^2 - 72F^{1*}_k(P^1_{k+1})^2 -$$
$$90F^{1*}_k P^1_{k+1} + 432(P^1_{k+1})^2 + 720P^1_{k+1} + 315)/3(25(F^{1*}_k)^2(P^1_{k+1})^2 +$$
$$40(F^{1*}_k)^2 P^1_{k+1} + 16(F^{1*}_k)^2 + 75(P^1_{k+1})^2 + 120P^1_{k+1} + 48),$$

$$\hat{M}^*_{11k} = -F^{1*}_k(3F^{1*}_k P^1_{k+1} - 4P^1_{k+1} - 5)/(5(F^{1*}_k)^2 P^1_{k+1} + 4(F^{1*}_k)^2 + 15P^1_{k+1} + 12),$$

$$\hat{M}^*_{12k} = (F^{1*}_k - 3)/((F^{1*}_k)^2 + 3),$$

$$\hat{M}^*_{21k} = -2(5(F^{1*}_k)^2 P^1_{k+1} + 4(F^{1*}_k)^2 - 9F^{1*}_k P^1_{k+1} + 27P^1_{k+1} + 27)/$$
$$3(5(F^{1*}_k)^2 P^1_{k+1} + 4(F^{1*}_k)^2 + 15P^1_{k+1} + 12),$$

$$\hat{M}^*_{22k} = -(5(F^{1*}_k)^2 - 6F^{1*}_k + 9)/3((F^{1*}_k)^2 + 3).$$

REFERENCES

[1] APLEVICH, J.D., 1991, Implicit linear systems. Lecture Notes in Control and Information Sciences, Edited by M.Thoma and A.Wyner, Springer-Verlag, Berlin.

[2] BAŞAR, T. and SELBUZ, H., 1979, Closed-loop Stackelberg strategies with applications in the optimal control of multilevel systems. IEEE Transactions on Automatic Control, **24**, 166-179.

[3] BAŞAR, T. and OLSDER, G.J., 1982, Dynamic noncooperative game theory, Academic Press, New York.

[4]   BENDER, D.J. and LAUB, A.J., 1987a, The linear-quadratic optimal regulator for descriptor systems. IEEE Transactions on Automatic Control, **32**, 672-688.

[5]   BENDER, D.J. and LAUB, A.J., 1987b, The linear-quadratic optimal regulator for descriptor systems: discrete-time case. Automatica, **23**, 71-85.

[6]   DAI, L., 1989, Singular control systems. Lecture Notes in Control and Information Sciences, Edited by M.Thoma and A.Wyner, Springer-Verlag, Berlin.

[7]   HEARN, A. C., 1993, REDUCE version 3.5, User's Manual, RAND Publication.

[8]   LUENBERGER, D.G., 1977, Dynamic equations in descriptor form. IEEE Transactions on Automatic Control, **22**, 312-321.

[9]   LUENBERGER, D.G. and ARBEL, A., 1977, Singular dynamic Leontief systems. Econometrica, **45**, 991-995.

[10] LUENBERGER, D.G., 1978, Time-invariant descriptor systems. Automatica, **14**, 473-480.

[11] MANTAS, G.P. and KRIKELIS, N.J., 1989, Linear quadratic optimal control for discrete descriptor systems. Journal of Optimization Theory and Applications, **61**, 221-245.

[12] MIZUKAMI, K. and XU, H., 1992, Closed-loop Stackelberg strategies for linear -quadratic descriptor systems, Journal of Optimization Theory and Applications. **74**, 151-170.

[13] TOLWINSKI, B., 1981, Closed-loop Stackelberg solution to multistage linear quadratic game. Journal of Optimization Theory and Applications, **34**, 485-501.

[14] VERGHESE, G.C., LEVY, B.C., and KAILATH, T., 1981, A generalized state-space for singular systems. IEEE Transactions on Automatic Control, **26**, 811-831.

[15] WANG, Y.Y., SHI, S.J., and ZHANG, Z.J., 1988, A descriptor-system approach to singular perturbation of linear regulators. IEEE Transactions on Automatic Control, **33**, 370-373.

[16] XU, H., and MIZUKAMI, K., 1994, New sufficient conditions for linear feedback closed-loop Stackelberg strategy of descriptor systems. IEEE Transactions on Automatic Control, **39**, 1097-1102.

On Independence of Irrelevant Alternatives and Dynamic Programming in Dynamic Bargaining Games

Harri Ehtamo and Jukka Ruusunen
Systems Analysis Laboratory
Helsinki University of Technology
Espoo, Finland

Abstract

We consider bargaining games where the players control a dynamic system and the feasible set consists of utility gains over multiple time periods. It is assumed that the players can make binding contracts over the time horizon in question and that they are allowed to monitor the contract as the game evolves. A property of the bargaining solution arising in this situation is its dynamic consistency. A weaker form of it was recently proposed by Ehtamo and Ruusunen, who also showed that it is equivalent to the independence of irrelevant alternatives property. Here it is shown that independence of irrelevant alternatives property serves as Bellman's principle of optimality for multi-period bargaining games. Namely, if the bargaining solution is independent of irrelevant alternatives, then the contract and the corresponding cooperative policies can be obtained backwards by applying dynamic programming.

1 Introduction

The multi–period (or continuous time) bargaining game consists of a feasible set S defined by feasible utility *gains* evaluated over a fixed time horizon, and a bargaining solution F. In a dynamic world the different periods are further linked to each other by a state equation. In dynamic games the way that the contract $F(S)$ is implemented depends on whether the parties can make binding contracts or not. If binding contracts are possible, $F(S)$ defines the players' actions for the whole duration of the game (see, e.g., Ehtamo *et al* [6]). So, from the game theoretical point of view, this kind of contracting does not bring anything new conceptually to the traditional bargaining models. The properties, or axioms, that can be used to represent rationality, or fairness, in such bargaining games are the same as those for bargaining games in normal form.

If the players cannot make binding agreements other approaches are required. Haurie [8] has raised the problem of reopening of negotiations at an intermediate state that has been generated by the cooperative actions. In these negotiations the players solve a new dynamic bargaining problem at the state in question as if no cooperation has taken place thus far except that the initial state of the new problem has been generated by the previous cooperative actions. We say that F is *dynamically consistent* if in this case the solution for the remaining periods is the same as the original solution for these periods. The dynamic consistency requirement is very strong: Of the well–known bargaining solutions only the utilitarian solution that maximizes the sum of the players' utilities is dynamically consistent.

Another drawback of a cooperative solution is that the solution is not necessarily an *equilibrium*, so that a player can achieve a better result by unilaterally deviating from the agreement. Consequently, when the players know that they cannot make a binding agreement, they must take into account possible deviations from the cooperative policy and the possibility of renegotiations in the future when they design the present contract.

Tolwinski [19] has proposed a *cooperative equilibrium* for games without binding agreements. In his model the cooperative strategies are constructed stage by stage backwards in time as a function of the system state, assuming that cooperative strategies will also be applied in the future. The strategies are chosen at each stage by using a traditional axiomatic bargaining model. The solution is made an equilibrium by using so-called memory strategies that act as a threat against the violation of the contract. The price that the players have to pay for not being able to make a binding contract is that the solution does not, in general, satisfy the axioms of the bargaining model in the original game. Yet, in the original game there might be solutions that would give considerable positive gains for each player.

For binding contracts, Ehtamo and Ruusunen [3], [4] and [5] have proposed a weaker form of dynamic consistency. This property can be stated as follows. The contract $F(S)$ made initially does not change as a consequence of rebargaining at an intermediate stage, provided the contract has been followed thus far and the realized past gains are explicitly taken into account in rebargaining, i.e., the reference point in rebargaining is kept fixed at the original one. Thus, if a_1, $a_2, \ldots,$ a_T are the cooperative actions defining $F(S)$, and if at period t the feasible set is evaluated by restricting the first $t-1$ actions to be $a_1, \ldots, a_{t-1}$, then F applied to this new game, say S', must equal $F(S)$.

An interesting question now is how *weak dynamic consistency* (WDC) is related to the other properties that have been proposed to bargaining solutions in the literature. An almost straightforward observation is the fact that WDC is implied by the *independence of irrelevant alternatives* (IIA) property of the bargaining solution (Ehtamo and Ruusunen [4], [5]). Thus, for example, the Nash bargaining solution [15] satisfies WDC while the Kalai-Smorodinsky bargaining solution [11] does not. That IIA implies WDC is not surprising if we consider the definition of the latter as given above. The game S' has fewer options than the original game S. Yet $F(S)$ is in S' because one option that we still have is to continue with the original plan to achieve $F(S)$. Further, slightly generalizing the class of bargaining games considered in this paper, e.g., by including isoperimetric control constraints, makes these two properties equivalent (Ehtamo and Ruusunen [4]).

In this paper we show that WDC serves as *Bellman's principle of optimality* for multiperiod bargaining games. We show that if F satisfies IIA then $F(S)$ and the corresponding cooperative policies can be obtained backwards by correctly decomposing S into subgames and applying dynamic programming. The original bargaining game thus decomposes into a sequential bargaining game of T stages where the feasible set at stage t depends, in addition to the system state x_t and the set of admissible actions for that stage, on the state y_t of the cumulative gains up to that stage. To simplify presentation we shall consider only two-period problems. Nevertheless, the analysis will contain the basic incredients of the multi-stage dynamic programming procedure.

In Sections 2 and 3 we define a two-period bargaining game with additively time separable total utilities and discuss the relation between WDC and IIA. We also review the recent literature on related axiomatic bargaining models. In Sections 4 and 5 we show that $F(S)$ can be obtained by using dynamic programming provided F satisfies IIA. Section 6 concludes.

2 Two-Period Additive Model

We shall consider a group of N players who are in a position to bargain for a mutual benefit over two time periods. The players are indexed by i and the time periods by t.

We first suppose that the players' utility functions over the two periods are additively time separable, i.e., of the form

$$u_{i1} + u_{i2}$$

where u_{it} denotes the utility level of player i at period t, $t = 1, 2$.

The gains from cooperation are measured relative to a reference (status quo) outcome $u_{i1}^{\delta} + u_{i2}^{\delta}$, $1 \leq i \leq N$, that the players would get provided no agreement were reached. We assume that the utilities u_{it} directly measure the *gains* from cooperation, either positive or negative, with respect to the reference utility levels u_{it}^{δ}. Hence the reference outcome for player i is the sum of the zero gains from the two periods.

The vectors $(u_{1t}, u_{2t}, \ldots, u_{Nt})$ are denoted by u_t, and the set of all such vectors, i.e., the set of *feasible gains* that can be achieved from period t, is denoted by U_t, $t = 1, 2$. Thus $U_t \subset R^N$ for each t.

We suppose that the vector $u_t \in U_t$ represents the gains that the players realize by taking an *admissible* action $a_t \in A_t$, where $A_t \subset R^m$.

We may further assume that a_t contains the admissible actions of all the players, i.e., $a_t = (a_{1t}, a_{2t}, \ldots, a_{Nt})$ where $a_{it} \in A_{it} \subset R^{m_i}$ is the admissible action of player i at period t, $1 \leq i \leq N$; consequently $A_t = A_{1t} \times \cdots \times A_{Nt}$ and $\sum_{1=1}^{N} m_i = m$.

Thus, U_t is given by

$$U_t = \{u_t(a_t)|\ a_t \in A_t\}, \quad t = 1, 2.$$

Finally, the bargaining game, denoted by S_1, consists of all possible *total* gains. It is given by

$$\begin{aligned} S_1 &= U_1 + U_2 \\ &= \{u_1 + u_2|\ u_t \in U_t,\ t = 1, 2\} \\ &= \{u_1(a_1) + u_2(a_2)|\ a_t \in A_t,\ t = 1, 2\}. \end{aligned} \tag{1}$$

The status quo actions, denoted by $\delta_t \in A_t, t = 1, 2$, correspond to the status quo outcome $\bar{0} \in S_1$, i.e., $u_1(\delta_1) + u_2(\delta_2) = \bar{0}$.

In order for S_1 to be well defined the following assumptions are made:

Assumption 2.1 *A_t is compact and convex and u_{it} is continuous and concave on A_t for $1 \leq i \leq N$, $t = 1, 2$.*

The continuity of the utility functions implies that U_1, U_2 and S_1 are compact sets. Denote $\Lambda = \{\lambda \in R^N | \lambda_i \leq 0,\ \forall i\}$. We say that a nonempty set $S \subset R^N$ is Λ-*convex* if the set $S + \Lambda$ is convex. Observe that any convex set is Λ-convex. Also observe that the set of Pareto optimal points of a set $S \subset R^N$ is the same as that of $S + \Lambda$. The concavity of the utility functions is needed to ensure that U_1 and U_2, and hence S_1, are Λ-convex.

In this paper, a *bargaining game* is formally defined to be a compact Λ-convex subset of R^N. The set of all bargaining games is denoted by B. Hence, under Assumption 2.1, U_1, U_2 and S_1 belong to B. Although for all our bargaining games the status quo outcome is fixed at the origin, we do not necessarily require that it belong to the bargaining game in question. Nevertheless, by definition it belongs to S_1.

The *bargaining solution* is a mapping $F : B \to R^N$ which assigns an outcome $F(S) \in S$, called the *contract*, to every $S \in B$.

We list the additional rationality assumptions that are usually made (see Moulin [13]):

(a) For $S \in B$ there is $z \in S$ such that $z_i > 0$ for all i.

(b) For $S \in B$, $F_i(S) \geq 0$ for all i.

(c) For $S \in B$, $F(S)$ is a Pareto optimal outcome. That is, there is no $z \in S$ such that $z_i \geq F_i(S)$ for all i, where $z_i > F_i(S)$ for at least one i.

Assumption (a) guarantees that there is proper total gain for each player in a game $S \in B$. Assumption (b) says that the bargaining solution satisfies individual rationality.

The contract corresponding to S_1 can be written as

$$
\begin{aligned}
F(S_1) &= F(U_1 + U_2) \\
&= u_1(\tilde{a}_1) + u_2(\tilde{a}_2),
\end{aligned}
\tag{2}
$$

where $\tilde{a}_1$ and $\tilde{a}_2$ are the actions defining $F(S_1)$.

It is not, in general, possible to implement an efficient two-period contract via two spot contracts where the players would bargain over the sets U_1 and U_2 separately. This is due to the fact that a sequence of spot contracts does not allow for intertemporal utility exchanges. In this paper we assume a cooperative setting where *binding* contracts over two time periods are possible. In the case of cooperation the players *will* successively take the actions $\tilde{a}_1$ and $\tilde{a}_2$ to obtain the total gains defined by (2). Otherwise, in the case no unanimous agreement is reached, the status quo outcome $\bar{0}$ is the result.

Our aim is to find conditions under which $F(S_1)$ can be defined by using a suitable dynamic programming procedure. This is done in Section 4. Before that we present a rationality property of a bargaining solution which seems to be very essential for the dynamic programming procedure to work.

3   WDC and IIA in Additive Games

3.1   Definition

Suppose that the contract is binding but that the the players have the possibility to *monitor* its evolution. This means that the players have the possibility to *check* the fairness of the contract in the course of the game. Suppose the reference point is kept fixed at the original one, i.e., at the origin. Thus, at period 1 the players realize the gain $\tilde{u}_1$, $\tilde{u}_1 = u_1(\tilde{a}_1)$, and at the beginning of period 2 they face the game

$$
S_2 = \tilde{u}_1 + U_2.
$$

Suppose the rationality assumptions do not change in the course of the game (i.e., the same F which was used initially applies also later on). We then define:

Definition 3.1 *The bargaining solution F satisfies WDC if $F(\tilde{u}_1 + U_2) = F(U_1 + U_2)$ for any $\tilde{u}_1$, $\tilde{u}_2$ such that $F(U_1 + U_2) = \tilde{u}_1 + \tilde{u}_2$, and for any U_1, $U_2 \in B$.*

WDC thus means that the contract made initially does not change as a consequence of rebargaining at period 2 provided the initial contract has been realized thus far, and the gains from the first period *are taken into account in rebargaining*. WDC is a natural property since in a deterministic environment there is essentially *nothing* that could change the solution. Hence, the players will have no rationale to change the contract in the course of the game.

Since $\tilde{u}_1 + U_2 \subset U_1 + U_2$, and $F(U_1 + U_2) \in \tilde{u}_1 + U_2$, dynamic consistency of F is implied by the following assumption:

Assumption 3.1 *F satisfies IIA, i.e., if $S' \subset S$ and $F(S) \in S'$, then $F(S') = F(S)$.*

The IIA property was first introduced by John Nash [15] in his famous paper. Although it has played a central role in bargaining theory, it has been controversial. Here IIA arises naturally from WDC which can be easily motivated. The equivalence of these two properties can be proven by considering a larger class of two–period games with individual total utilities not necessarily time separable (see Ehtamo and Ruusunen [4]). One such larger class consists of games with isoperimetric control constraints (see Section 5).

Let us next review recent literature on related bargaining models. Kalai [10] proposes an axiomatic *step–by–step negotiation* model which allows the solution of a static (one-period) bargaining game stage by stage. Kalai shows that a solution satisfies the step–by–step negotiation condition if and only if it is *proportional*. At each period the players negotiate over a subset of the original feasible set using the agreement reached at the previous stages as a reference point at the current stage. In intertemporal problems that we are studying the feasible set itself changes in the course of the game, whereas the reference point remains fixed.

Peters [17] has studied the problem of simultaneous bargaining over two issues. He considers games of the form $S = T + U$ where T and U are the utility gains from the two issues, respectively. Peters studies a spot contract over S while the very essential feature in our formulation is the dynamic nature of the problem that includes monitoring and rebargaining. Nevertheless, it is interesting to note that in Peters' formulation an alternative characterization of the nonsymmetric Nash bargaining solutions can be obtained by replacing IIA with two axioms: an additivity property of the bargaining solution, called *restricted additivity*, and *Pareto continuity* property of the bargaining solution. Restricted additivity means that if T and U have unique lines of support at the points $F(T)$ and $F(U)$, respectively, and $F(T + U)$ is a Pareto optimal point of $T + U$, then $F(T + U) = F(T) + F(U)$.

An additive structure also arises in situations where the bargaining game S is of the form $\lambda T + (1 - \lambda)U$, $\lambda \in (0, 1)$, i.e., S consists of a lottery on two games T and U with probabilities λ and $1 - \lambda$, respectively. A problem arising in this class of games is the timing of the agreement; see Myerson [14], and Perles and Maschler [16]. For example, Myerson defines concavity of F on the class of games in question which then implies that the players prefer to reach the agreement before the outcome of the lottery is available. It should be noted that although timing of choices plays an important role in these games they are static in nature since the players act *only once*.

Finally, the intertemporal bargaining game considered in this paper also differs essentially from sequential, or noncooperative bargaining models: possibly defined over several issues, see Fershtman [7] and the references in that paper, or over several time periods, see Houba and de Zeeuw [9]. These models describe the bargaining process itself, i.e., the way the players can reach a particular outcome by a sequence of moves (offers and replies) over time. In our model the players take a sequence of (arbitrary) actions over multiple time periods and they have the possibility to monitor these actions and to check the fairness of the contract as the game evolves.

3.2 Example

As an example we consider a two player game over two time periods. The game at period 1 is $S_1 = U_1 + U_2$, see Figure 1, where U_1 is the convex hull of the set $\{(0,0),(0,4),(4,0)\}$, and U_2 is the convex hull of the set $\{(0,0),(0,3),(5,2),(5,0)\}$.

The *Nash bargaining solution* [15], denoted by F^N, maximizes the product of the individual gains on the feasible set in question. Figure 1 shows the contract $F^N(S_1)$ which is the sum of two gains $u_1^N \in U_1$ and $u_2^N \in U_2$. Figure 1 also shows the game $S_2 = u_1^N + U_2$. Since F^N satisfies WDC (i.e., since $F^N(S_1)$ maximizes the product of the individual total gains on S_1, it does so in S_2) we have $F^N(S_2) = F^N(S_1)$.

The *Kalai-Smorodinsky bargaining solution* [11], $F^{KS}(S_1) = u_1^{KS} + u_2^{KS}$, is the Pareto optimal point of S_1 satisfying

$$\frac{u_{11}^{KS} + u_{12}^{KS}}{u_{21}^{KS} + u_{22}^{KS}} = \frac{z_1(S_1)}{z_2(S_2)},$$

where $z(S_1) = (z_1(S_1), z_2(S_1))$ is the ideal point of S_1; see Figure 2. Note that the contract *explicitly* depends on the feasible set S_1 (via the ideal point of S_1). Figure 2 also illustrates the Kalai-Smorodinsky solution at period 2. Since the ideal points of S_2 and S_1 are different we have $F^{KS}(S_2) \neq F^{KS}(S_1)$. Hence the Kalai-Smorodinsky solution does not satisfy WDC.

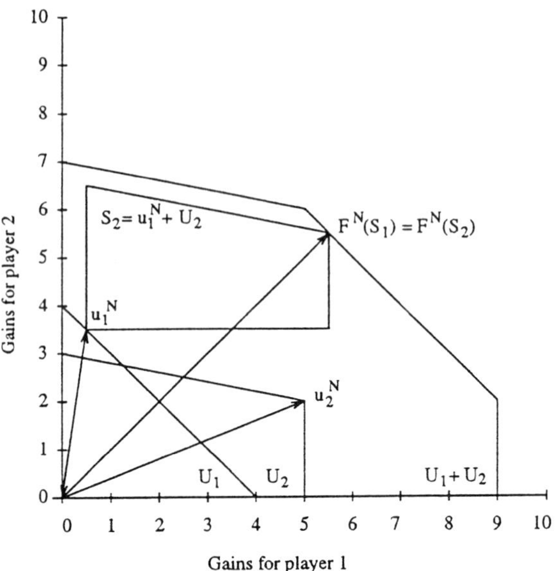

Figure 1. The Nash bargaining solution

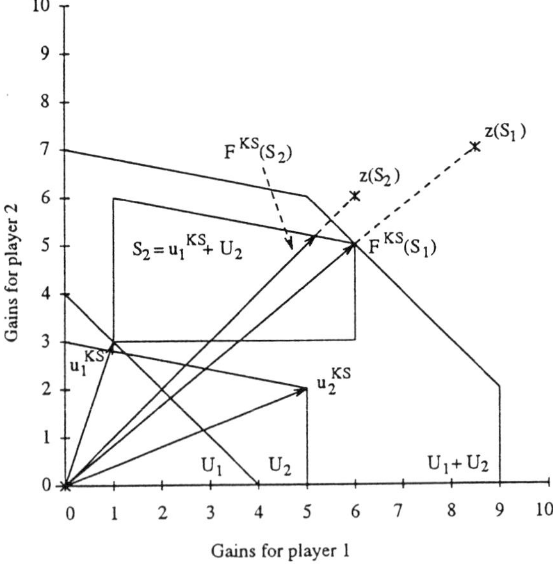

Figure 2. The Kalai-Smorodinsky bargaining solution

4 Dynamic Programming in Additive Games

In this section we show how $F(S_1)$, and the cooperative actions defining it, can be defined backwards by correctly applying dynamic programming. The sufficient condition for this procedure to work is the IIA assumption. We first consider the game defined in the two previous sections and in the next section we generalize our model to include the state equation.

First, the dynamic programming procedure is defined as follows. Let $Y_2 \subset R^N$ describe an appropriate set of gain histories from period 1, and for $y_2 \in Y_2$ define the set $S_2^*(y_2)$ by

$$
\begin{aligned}
S_2^*(y_2) &= y_2 + \{u_2(a_2)|a_2 \in A_2\} \\
&= y_2 + U_2.
\end{aligned} \tag{3}
$$

Let the strategy $\sigma_2^* : Y_2 \to A_2$ define $F(S_2^*(y_2))$, i.e., let

$$
F(S_2^*(y_2)) = y_2 + u_2(\sigma_2^*(y_2)). \tag{4}
$$

Then denote $v_2 = u_2 \circ \sigma_2^*$ and define the set S_1^* by

$$
S_1^* = \{y_2 + v_2(y_2)|y_2 = u_1(a_1), \ a_1 \in A_1'\}, \tag{5}
$$

where $A_1' \subset A_1$ is such that $u_1(a_1) \in Y_2$ for all $a_1 \in A_1'$. Let $\sigma_1^* \in A_1'$ define $F(S_1^*)$, i.e., let

$$
F(S_1^*) = u_1(\sigma_1^*) + v_2(u_1(\sigma_1^*)).
$$

As earlier, denote $F(U_1 + U_2) = u_1(\tilde{a}_1) + u_2(\tilde{a}_2) = \tilde{u}_1 + \tilde{u}_2$. Of course, the set A_1', and hence Y_2, must be sufficiently large in order that $\tilde{a}_1 \in A_1'$ (one possible choice, although not the most economical one from the computational point of view, is $Y_2 = U_1$, $A_1' = A_1$). Therefore we make the following assumption.

Assumption 4.1 $\tilde{a}_1 \in A_1'$.

We then have:

Theorem 4.1 *Let Assumptions 3.1 and 4.1 hold, and let the strategy pair (σ_1^*, σ_2^*) be defined by the dynamic programming procedure above. Then (σ_1^*, σ_2^*) defines $F(S_1)$, i.e., $F(S_1) = F(S_1^*)$.*

Proof. We have

$$
\begin{aligned}
F(U_1 + U_2) &= F(\tilde{u}_1 + U_2) = F(S_2^*(\tilde{u}_1)) \\
&= \tilde{u}_1 + u_2(\sigma_2^*(\tilde{u}_1)) \\
&= \tilde{u}_1 + v_2(\tilde{u}_1)
\end{aligned}
$$

where the first equality follows from WDC and the other equalities from definitions (3), (4) and the definition of v_2, respectively. Since $\tilde{u}_1 = u_1(\tilde{a}_1)$ and $\tilde{a}_1 \in A'_1$ it follows that $F(S_1) \in S_1^*$.

Since $u_1(a_1) \in U_1$ and $v_2(u_1(a_1)) \in U_2$ for all $a_1 \in A'_1$, we have $S_1^* \subset S_1$. Hence, by IIA,

$$
\begin{aligned}
F(S_1) &= F(S_1^*) \\
&= y_2^* + v_2(y_2^*) \\
&= y_2^* + u_2(\sigma_2^*(y_2^*)),
\end{aligned}
$$

where $y_2^* = u_1(\sigma_1^*)$. $\quad\square$

Remarks. (a) Although $F(S_1)$ is unique by definition the action pair $(\tilde{a}_1, \tilde{a}_2)$ defining it need not be unique. Thus $(\sigma_1^*, \sigma_2^*(y_2^*))$ need not equal $(\tilde{a}_1, \tilde{a}_2)$.

(b) IIA was used in the proof twice. First, IIA was used in the form of Definition 3.1 to show that $F(S_1) \in S_1^*$, and second, it was used in the form of Assumption 3.1 to deduce that $F(S_1) = F(S_1^*)$. In the latter case IIA reveals itself in a form of Bellman's [1] principle of optimality: *An optimal path must be composed of optimal subpaths.* Here this principle can be stated by saying that a fair path consists of fair subpaths.

(c) To simplify notation and presentation we have only considered two-period models. Nevertheless, the generalization to multiple time periods, $T > 2$, can be done by induction. The general case will have the same basic steps as the two-period case.

(d) In (5) the function v_2 describes the value function, "the cost to go", which is a function of the current state value y_2. In (3) the state y_2 describes the possible cumulative gains up to period 2, and in (5) the state equation $y_2 = u_1(a_1)$ is used in the usual way when moving from period 2 to period 1. In the general case the state equation is updated according to

$$
y_{t+1} = y_t + u_t(a_t), \quad y_1 = \bar{0}, \tag{6}
$$

and the value function according to

$$
v_t(y_t) = u_t(\sigma_t^*(y_t)) + v_{t+1}(y_{t+1}^*)
$$

where y_{t+1}^* is defined by (6) by applying $a_t = \sigma_t^*(y_t)$.

(e) Especially note that it is *not* possible to nullify the additive gain history part y_2 in (3) and (5). Tolwinski [19] used this kind of procedure when he defined cooperative threat strategies. Such strategies do not, however, define $F(S_1)$ (except in the case of the utilitarian solution that maximizes the sum of the players' utilities).

(f) It is clear that $S_2^*(y_2)$ is well defined, i.e., it belongs to B. This is not automatically so in the case of S_1^*. Compactness of S_1^* is clear if, e.g.,

$F(y_2 + U_2)$ is continuous with respect to y_2. Instead, it is difficult to give conditions under which S_1^* is Λ-convex. Assumption 2.1 does not guarantee it; however, see Ehtamo and Ruusunen [4]. One way to avoid this difficulty is to consider the convex hull $H(S_1^*)$ of S_1^*, and to define $F(S_1^*) = F(H(S_1^*))$. This gives a correct result since $F(S_1) \in S_1^*$ in any case.

5 Dynamic Programming with State Equation

In this section we generalize the previous model by including the state equation

$$x_{t+1} = f_t(x_t, a_t), \quad t = 1, 2, \quad x_1 = \tilde{x}_1 \quad \text{given}, \tag{7}$$

where $x_t \in R^n$ for $t = 1, 2, 3$. Let the total payoff for player i corresponding to action pair $(a_1, a_2) \in A_1 \times A_2$ be given by

$$g_{i3}(x_3) + \sum_{t=1}^{2} g_{it}(x_t, a_t), \quad 1 \le i \le N, \tag{8}$$

where the functions g_{it} are the players' one–period payoffs and the state trajectory is generated by (7). Suppose the appropriate continuity and concavity assumptions hold. We further assume that the players' utility functions are linear functions of their one-period payoffs so that (8) can be considered to represent directly the total utilities of the players.

Let $(\delta_1, \delta_2) = A_1 \times A_2$ be the status quo action pair, and let x_t^δ, $t = 1, 2, 3$, be the corresponding status quo trajectory. Define the utility gains from the different periods by

$$\begin{aligned}
u_{it}(x_t, a_t) &= g_{it}(x_t, a_t) - g_{it}(x_t^\delta, \delta_t), \quad 1 \le i \le N, \quad t = 1, 2, \\
u_{i3}(x_3) &= g_{i3}(x_3) - g_{i3}(x_3^\delta), \quad 1 \le i \le N,
\end{aligned}$$

denote $u_t = (u_{1t}, u_{2t}, \ldots, u_{Nt})$, $t = 1, 2, 3$, and define the set of feasible total gains by

$$S_1 = \Big\{ u_3(x_3) + \sum_{t=1}^{2} u_t(x_t, a_t) | a_t \in A_t, \ t = 1, 2 \Big\}.$$

Observe that since u_t depends on previous actions through the state equation, S_1 can no more be written in the form $U_1 + U_2$ as earlier.

Let $\tilde{a}_1, \tilde{a}_2$ define $F(S_1)$, i.e., let

$$F(S_1) = u_3(\tilde{x}_3) + \sum_{t=1}^{2} u_t(\tilde{x}_t, \tilde{a}_t), \tag{9}$$

where $\tilde{x}_t$ is the trajectory corresponding to $\tilde{a}_1, \tilde{a}_2$. WDC can then be stated as follows (Ehtamo and Ruusunen [5]).

Definition 5.1 *Define*

$$S_2 = u_1(\tilde{x}_1, \tilde{a}_1) + \{u_2(x_2, a_2) + u_3(x_3) | a_2 \in A_2\}.$$

Then F satisfies WDC if $F(S_2) = F(S_1)$. If F satisfies WDC, $F(S_2)$ can be represented by (9).

Since $S_2 \subset S_1$, and since $\tilde{a}_2 \in A_2$ so that $F(S_1) \in S_2$, WDC is again implied by IIA. The equivalence of WDC and IIA can be proven, e.g., by adding isoperimetric control constraints of the form $a_1 + a_2 \in A$ for some appropriate $A \subset R^m$ to the above model and by showing that for every pair $(S, S') \in B \times B$ for which $S' \subset S$, $F(S) \in S'$, there exists a pair (S_1, S_2) such that $S = S_1$ and $S' = S_2$ (for details, see Ehtamo and Ruusunen [4]).

We next define the dynamic programming procedure to the bargaining problem above.

Let $X_2 \subset R^n$, $Y_2 \subset R^N$ be appropriate sets of the state and gain histories from period 1, respectively. For $(x_2, y_2) \in X_2 \times Y_2$ define the set $S_2^*(x_2, y_2)$ by

$$S_2^*(x_2, y_2) = y_2 \quad + \quad \{u_2(x_2, a_2) | a_2 \in A_2\}$$
$$+ \quad \{u_3(x_3) | x_3 = f_2(x_2, a_2),\ a_2 \in A_2\}.$$

Let the strategy $\sigma_2^*(x_2, y_2)$ define $F(S_2^*(x_2, y_2))$. Then define

$$\begin{aligned} v_2(x_2, y_2) &= u_2(x_2, \sigma_2^*(x_2, y_2)) + u_3(x_3^*), \\ x_3^* &= f_2(x_2, \sigma_2^*(x_2, y_2)), \end{aligned}$$

and define the set S_1^* by

$$S_1^* = \{y_2 + v_2(x_2, y_2) | x_2 = f_1(\tilde{x}_1, a_1),\ y_2 = u_1(\tilde{x}_1, a_1),\ a_1 \in A_1'\},$$

where $A_1' \subset A_1$ is such that $f_1(\tilde{x}_1, a_1) \in X_2$, $u_1(\tilde{x}_1, a_1) \in Y_2$ for all $a_1 \in A_1'$. Let $\sigma_1^* \in A_1'$ define $F(S_1^*)$, i.e., let

$$F(S_1^*) = y_2^* + v_2(x_2^*, y_2^*),$$

where

$$\begin{aligned} x_2^* &= f_1(\tilde{x}_1, \sigma_1^*), \\ y_2^* &= u_1(\tilde{x}_1, \sigma_1^*). \end{aligned}$$

Following the proof of Theorem 4.1 it is straightforward to show that if $\tilde{a}_1 \in A_1'$, and if F satisfies IIA, then $F(S_1) = F(S_1^*)$, so that (σ_1^*, σ_2^*) also defines $F(S_1)$.

6 Conclusion

We have studied bargaining games with the feasible set consisting of utility gains evaluated over multiple time periods. We have shown how the contract for such games can be obtained by using the dynamic programming technique provided the bargaining solution satisfies the IIA property. This property can further be motivated by the equivalent WDC property. Thus, in the case of multi-period bargaining games, IIA reveals itself in a form of Bellman's principle of optimality.

In the literature various generalizations of Bellman's [1] original formulation of dynamic programming have been presented. Consequently, various refinements of the optimality principle have also been stated. In most cases the optimality principle has been replaced by various *monotonicity* assumptions (see for example Carraway *et al*, [2], Mitten [12]). Tauxe *et al.*, [18] were the first who studied multiobjective dynamic programming. The dynamic programming equations of our paper are similar to those of Tauxe *et al.* [18], in that the subproblems depend on the state variable of the cumulative gains from the past. In multiobjective problems the subproblems are optimization problems, whereas in bargaining games the solutions to subgames are obtained by applying the bargaining solution to these one-period games.

Our formulation in this paper was done for two-period games with additively time separable total utilities. It was shown how the basic steps of the dynamic programming procedure can be generalized to the multi-period case. Explicit derivation can be found in Ehtamo and Ruusunen [3] where preliminary results for multi-period problems where contracting takes place under exogenous uncertainty are also presented.

REFERENCES

[1] R.E. Bellman, *Dynamic Programming*, Princeton University Press, Princeton, NJ, 1957.

[2] R.L. Carraway, T.L. Morin, and H. Moskowitz, Generalized dynamic programming for multicriteria optimization, *European J. of Operational Research*, **44**, 1990, pp. 95-104.

[3] H. Ehtamo and J. Ruusunen, A theory of intertemporal bargaining, Helsinki University of Technology, Systems Analysis Laboratory, Research Reports A35, 1990.

[4] H. Ehtamo and J. Ruusunen, Intertemporal bargaining and dynamic

consistency, Helsinki University of Technology, Systems Analysis Laboratory, Research Reports A48, 1993.

[5] H. Ehtamo and J. Ruusunen, Contracting in dynamic games, *Group Decision and Negotiation*, **4**, 1995, pp. 59–69.

[6] H. Ehtamo, J. Ruusunen, V. Kaitala, and R.P. Hämäläinen, Solution for a dynamic bargaining problem with an application to resource management, *J. Optimiz. Theory Appl.*, **59**, 1988, pp. 391-405.

[7] C. Fershtman, The importance of the agenda in bargaining, *Games and Economic Behavior*, **2**, 1990, pp. 224-238.

[8] A. Haurie, A note on nonzero-sum differential games with bargaining solution, *J. Optimiz. Theory Appl.*, **18**, 1976, pp. 31-39.

[9] H. Houba and A. de Zeeuw, Strategic bargaining for the control of a dynamic system in state-space form, *Group Decision and Negotiation*, **4**, 1995, pp. 69–95.

[10] E. Kalai, Proportional solutions to bargaining situations: intertemporal utility comparisons, *Econometrica*, **45**, 1977, pp. 1623-1630.

[11] E. Kalai and M. Smorodinsky, Other solutions to Nash's bargaining problem, *Econometrica*, **43**, 1975, pp. 513-518.

[12] L.G. Mitten, Preference order dynamic programming, *Management Sci.*, **21**, 1974, pp. 43-46.

[13] H. Moulin, *Axioms of Cooperative Decision Making*, Cambridge University Press, New York, 1988.

[14] R.B. Myerson, Utilitarianism, Egalitarianism, and the timing effect in social choice problems, *Econometrica*, **49**, 1981, pp. 883-897.

[15] J. Nash, The bargaining problem, *Econometrica*, **18**, 1950, pp. 883-897.

[16] M.A. Perles and M. Maschler, The super-additive solution for the Nash bargaining game, *Int. J. Game Theory*, **10**, 1981, pp. 163-193.

[17] H. Peters, Simultaneity of issues and additivity in bargaining, *Econometrica*, **54**, 1986, pp. 153-169.

[18] G.W. Tauxe, R.R. Inman, and D.M. Mades, Multiobjective dynamic programming: a classic problem redressed, *Water Resources Research*, **15**, 1979, pp. 1398-1402.

[19] B. Tolwinski, A concept of cooperative equilibrium for dynamic games, *Automatica*, **18**, 1982, pp. 431-441.

The Shapley Value for Differential Games

Leon A. Petrosjan
Faculty of Applied Mathematics
St. Petersburg University, 198904
Bibliotechnaya pl. 2, Petrodvorets,
St. Petersburg, Russia

Abstract

Let $N = \{1, \ldots, n\}$ be the set of players and $\Gamma(x_0, T - t_0)$ the differential n-person cooperative game with prescribed duration, $V(S; x_0, T - t_0)$ $S \subset N$ a characteristic function, $x^*(\tau)$, $\tau \in [t_0, T]$ the "optimal" trajectory in $\Gamma(x_0, T - t_0)$ maximizing the sum of the players' payoffs. The "refinement" of c.f. $V(S; x_0, T - t_0)$ is defined by the formula

$$\overline{V}(S; x_0, T - t_0) = - \int_{t_0}^{T} V(S; x^*(\tau), T - \tau) \frac{V'(N; x^*(\tau), T - \tau)}{V(N; x^*(\tau), T - \tau)} d\tau.$$

$\overline{V}(S; x_0, T - t_0)$ is also a c.f. in $\Gamma(x_0, T - t_0)$. It is proved that the Shapley value defined for c.f. $\overline{V}(S; x_0, T - t_0)$ is time consistent and the set of all imputations $L(x_0, T - t_0)$ is strongly time consistent.

1　Introduction

The Shapley value [1] is an optimality principle from the static cooperative game theory. For using this optimality principle in the cooperative differential games, a special imputation distribution procedure (IDP) on the time interval $[t_0, T]$ is to be defined to provide the time consistency (dynamic stability) of Shapley value [2]. Unfortunately this is not always possible. For any given characteristic function, we propose a "refinement" of this function which is also a characteristic function, and for which the corresponding Shapley value is time consistent.

Consider n-person cooperative differential game $\Gamma(x_0, T - t_0)$

$$\dot{x} = f(x, u_1, \ldots, u_n), \quad u_i \in U_i \subset Comp\ R^l, \quad x \in R^n, \tag{1}$$

with integral payoffs

$$K_i(x_0, T - t_0, u_1, \ldots, u_n) = \int_{t_0}^{T} h_i(x(t)) dt,$$

$$h_i > 0, \quad i = 1, \ldots, n.$$

where $x(t)$ is a solution of (1) when the open loop controls $u_1, \ldots, u_n$ are used by the players. We assume that all conditions which guarantee the existence, uniqueness and prolongability of the solution $x(t)$ on the time interval $[t_0, T]$ for any n-tuples of measurable open-loop controls are satisfied. Suppose that there exists such an n-tuple of controls

$$u^*(t) = \{u_1^*(t), \ldots, u_n^*(t)\}, \quad t \in [t_0, T],$$

that the following condition holds

$$K(x_0, T - t_0; u_1^*(t), \ldots, u_n^*(t)) =$$

$$= \max_{u_1, \ldots, u_n} \sum_{i=1}^{n} K_i(x_0, T - t_0; u_1(t), \ldots, u_n(t)) =$$

$$= \sum_{i=1}^{n} \int_{t_0}^{T} h_i(x^*(t)) dt = V(N; x_0, T - t_0). \tag{2}$$

The solution $x^*(t)$ of (1) corresponding to $u^*(t)$ is called an optimal trajectory. In the cooperative n-person game theory [2] it is assumed that before starting the game the players agree to play $u^*(t) = \{u_1^*(t), \ldots, u_n^*(t)\}$ and thus the cooperative differential game $\Gamma(x_0, T - t_0)$ always develops along an optimal trajectory $x^*(t)$.

2   The characteristic function

Let $N = \{1, \ldots, i, \ldots, n\}$ be the set of players, $S \subset N$. We introduce the characteristic function of $\Gamma(x_0, T - t_0)$ axiomatically as a real valued function defined on the set of all coalitions $S \subset N$ (subsets of the set N), with the following properties:

1. $V(\emptyset; x_0, T - t_0) = 0$
2. $V(S_1 \cup S_2; x_0, T - t_0) \geq V(S_1; x_0, T - t_0) + V(S_2; x_0, T - t_0)$, for

$$S_1 \subset N, \ S_2 \subset N, \ S_1 \cap S_2 = \emptyset.$$

3. $V(N; x_0, T - t_0) = K(x_0, T - t_0; u_1^*(t), \ldots, u_n^*(t))$, where $u^*(t) = (u_1^*(t), \ldots, u_n^*(t))$ is defined by formula (2).

Consider the family of subgames of the game $\Gamma(x_0, T - t_0)$ along the optimal trajectory $\Gamma(x^*(t), T - t)$; i.e. the family of cooperative differential games from the initial position $x^*(t)$, and defined on the time interval $[t, T], \quad t \in [t_0, T]$ with the payoff functions

$$K_i(x^*(t), T - t; u_1, \ldots, u_n) = \int_{t}^{T} h_i(x(\tau)) d\tau, \quad i = 1, \ldots, n, \tag{3}$$

where $x(t)$ is the solution of (1) from the initial position $x^*(t)$ when the controls $u_1, \ldots, u_n$ are used by the players. Let $V(S; x^*(t), T - t)$, $S \subset N$, $t \in [t_0, T]$ be the characteristic function of the subgame $\Gamma(x^*(t), T - t)$. We suppose that the function $V(S; x^*(t), T - t)$ for every fixed $S \subset N$ is continuous on the time interval $[t_0, T]$ (it is always true if the c.f. is defined as the value of the associated zero-sum game played between the coalitions S and $N \setminus S$). For the function $V(N; x^*(t), T - t)$ $(S = N)$ the Bellman's equation along $x^*(t)$ is satisfied, i.e.,

$$V(N; x_0, T - t_0) = \int_{t_0}^t \sum_{i=1}^n h_i(x^*(\tau))d\tau + V(N; x^*(t), T - t). \qquad (4)$$

We get from (4),

$$V'(N; x^*(t), T - t) = -\left[\sum_{i=1}^n h_i(x^*(t))\right].$$

Define the new function $\overline{V}(S; x_0, T - t_0)$, $S \subset N$ by the formula

$$\overline{V}(S; x_0, T - t_0) = -\int_{t_0}^T V(S; x^*(\tau), T - \tau) \frac{V'(N; x^*(\tau), T - \tau)}{V(N; x^*(\tau), T - \tau)} d\tau. \qquad (5)$$

In the same manner for $t \in [t_0, T]$

$$\overline{V}(S; x^*(t), T - t) = -\int_t^T V(S; x^*(\tau), T - \tau) \frac{V'(N; x^*(\tau), T - \tau)}{V(N; x^*(\tau), T - \tau)} d\tau. \qquad (6)$$

Theorem 2.1 $\overline{V}(S; x_0, T - t_0)$, $S \subset N$ *is a characteristic function in the game* $\Gamma(x_0, T - t_0)$.

Proof. Condition 1 follows from the definition of $\overline{V}(\emptyset; x_0, T - t_0)$. Condition 2 follows from the following inequality

$$\overline{V}(S_1 \cup S_2; x_0, T - t_0) =$$

$$= -\int_{t_0}^T V(S_1 \cup S_2; x^*(\tau), T - \tau) \frac{V'(N; x^*(\tau), T - \tau)}{V(N; x^*(\tau), T - \tau)} d\tau \geq$$

$$\geq -\int_{t_0}^T V(S_1; x^*(\tau), T - \tau) \frac{V'(N; x^*(\tau), T - \tau)}{V(N; x^*(\tau), T - \tau)} d\tau -$$

$$-\int_{t_0}^T V(S_2; x^*(\tau), T - \tau) \frac{V'(N; x^*(\tau), T - \tau)}{V(N; x^*(\tau), T - \tau)} d\tau =$$

$$= \overline{V}(S_1; x_0, T - t_0) + \overline{V}(S_2; x_0, T - t_0)$$

for $S_1 \subset N$, $S_2 \subset N$, $S_1 \cap S_2 = \emptyset$. Condition 3 follows from

$$\overline{V}(N; x_0, T - t_0) = V(N; x_0, T - t_0) = \sum_{i=1}^{n} \int_{t_0}^{T} h_i(x^*(\tau)) d\tau.$$

The theorem is proved.

Let $L(x_0, T - t_0)$ be the set of imputations defined in $\Gamma(x_0, T - t_0)$ with the help of characteristic function $V(S; x_0, T - t_0)$, $S \subset N$, i.e.

$$L(x_0, T - t_0) = \{\xi = \{\xi_i\} : \sum_{i=1}^{n} \xi_i =$$

$$= V(N; x_0, T - t_0), \ \xi_i \geq V(\{i\}; x_0, T - t_0)\}. \tag{7}$$

In the same way define the sets of imputations $L(x^*(t), T - t)$, $t \in [t_0, T]$ in subgames $\Gamma(x^*(t), T - t)$

$$L(x^*(t), T - t) = \{\xi(t) = [\xi_i(t)] : \sum_{i=1}^{n} \xi_i(t) =$$

$$= V(N; x^*(t), T - t), \ \xi_i(t) \geq V(\{i\}; x^*(t), T - t), \ i \in N\}. \tag{8}$$

Denote the set of imputations defined by characteristic functions $\overline{V}(S; x_0, T - t_0)$, $\overline{V}(S; x^*(t), T - t)$, $t \in [t_0, T]$ by $\overline{L}(x_0, T - t_0)$ and $\overline{L}(x^*(t), T - t)$ correspondingly. Let $\xi(t) \in L(x^*(t), T - t)$ be an integrable selector, $t \in [t_0, T]$ ($\xi(t)$ is bounded, and if it is measurable on $[t_0, T]$, then it is integrable), define

$$\overline{\xi} = -\int_{t_0}^{T} \xi(t) \frac{V'(N; x^*(t), T - t)}{V(N; x^*(t), T - t)} dt, \tag{9}$$

$$\overline{\xi}(t) = -\int_{t}^{T} \xi(\tau) \frac{V'(N; x^*(\tau), T - \tau)}{V(N; x^*(\tau), T - \tau)} d\tau, \tag{10}$$

$t \in [t_0, T]$.

Theorem 2.2 *The set* $\overline{L}(x_0, T - t_0)$ *consists of vectors defined by (9) for all possible integrable selectors* $\xi(t), t \in [t_0, T]$ *with values in* $L(x^*(t), T - t)$.

Proof. Remember that $\sum_{i=1}^{n} \overline{\xi}_i = V(N; x_0, T - t_0)$,

$$\sum_{i=1}^{n} \overline{\xi}_i(t) = V(N; x^*(t), T - t),$$

$$\overline{\xi}_i \geq -\int_{t_0}^{T} V(\{i\}; x^*(t), T - t) \frac{V'(N; x^*(t), T - t)}{V(N; x^*(t), T - t)} dt = \overline{V}(\{i\}; x_0, T - t_0)$$

in the same way

$$\overline{\xi}_i(t) \geq \overline{V}(\{i\}; x^*(t), T - t), \ i = 1, \ldots, n, \ t \in [t_0, T].$$

This means that the vectors $\overline{\xi} = \{\overline{\xi}_i\}$ and $\xi(t) = \{\xi_i(t)\}$ are imputations in the games $\Gamma(x_0, T - t_0)$, $\Gamma(x^*(t), T - t)$ correspondingly under the characteristic function $\overline{V}$. We have that $\overline{\xi} \in \overline{L}(x_0, T - t_0)$, $\overline{\xi}(t) \in \overline{L}(x^*(t), T - t)$. But the converse statement is also true. Any imputation $\xi \in \overline{L}(x_0, T - t_0)$ may be represented in the form (9) (correspondingly (10)) for some integrable selector.

Suppose $\overline{\xi} \in L(x_0, T - t_0)$; then it is sufficient to show that we may find a function $\widehat{\xi}(t)$ with values in $L(x^*(t), T - t)$ such that

$$\overline{\xi}_i = -\int_{t_0}^{T} \widehat{\xi}_i(t) \frac{V'(N; x^*(t), T - t)}{V(N; x^*(t), T - t)} dt$$

Define $\widehat{\xi}_i(t)$ by formula

$$\widehat{\xi}_i(t) = V(\{i\}; x^*(t), T - t) + \delta_i(t),$$

where $\delta_i(t) \geq 0$ satisfies the conditions

$$\overline{\xi}_i + \int_{t_0}^{T} V(\{i\}; x^*(t), T - t) \frac{V'(N; x^*(t), T - t)}{V(N; x^*(t), T - t)} dt =$$

$$= -\int_{t_0}^{T} \delta_i(t) \frac{V'(N; x^*(t), T - t)}{V(N; x^*(t), T - t)} dt.$$

The theorem is proved.

3   The time-consistency problem

Let $\xi \in L(x_0, T - t_0)$ and the functions $\beta_i(t), \ i = 1, \ldots, n, \ t \in [t_0, T]$ satisfy the condition

$$\int_{t_0}^{T} \beta_i(t) dt = \overline{\xi}, \quad \beta_i(t) \geq 0.$$

The function $\beta(t) = \{\beta_i(t)\}$ we call the imputation distribution procedure (IDP). Define

$$\int_{t_0}^{\Theta} \beta_i(t) dt = \overline{\xi}_i(\Theta), \ i = 1, \ldots, n.$$

Let $C(x_0, T - t_0) \subset L(x_0, T - t_0)$ be any of the known classical optimality principles from the cooperative game theory (core, NM-solution, Shapley value or any other OP). Consider $C(x_0, T - t_0)$ as an optimality principle in $\Gamma(x_0, T - t_0)$. In the same manner let $C(x^*(t), T - t)$ be an optimality principle in $\Gamma(x^*(t), T - t)$, $t \in [t_0, T]$.

Definition 3.1 *The optimality principle (OP) $C(x_0, T - t_0)$ is time consistent (TC) if there exists such an IDP $\beta(t) = \{\beta_i(t)\}$ that*

$$\overline{\xi} - \overline{\xi}(\Theta) \in C(x^*(\Theta), T - \Theta) \tag{11}$$

for all $\Theta \in [t_0, T]$. The (OP) $C(x_0, T - t_0)$ is called strongly time consistent (STC) if there exists such an IDP $\beta(t) = \{\beta_i(t)\}$ that

$$\overline{\xi}(\Theta) + C(x^*(\Theta), T - \Theta) \subset C(x_0, T - t_0), \tag{12}$$

for all $\Theta \in [t_0, T]$ (here $a + B$, where $a \in R^n$, $B \subset R^n$, means the set of vectors $a + b$, $b \in B$).

The STC of the OP means that if an imputation $\xi \in C(x, T - t_0)$ and an IDP $\beta(t) = \{\beta_i(t)\}$ of ξ are selected, then after getting by the players, on the time interval $[t_0, \Theta]$, the amount

$$\xi_i(\Theta) = \int_{t_0}^{\Theta} \beta_i(t)dt, \quad i = 1, \ldots, n,$$

the optimal income (in the sense of the OP $C(x * (\Theta), T - \Theta)$) on the time interval $[\Theta, T]$ in the subgame $\Gamma(x^*(\Theta), T - \Theta)$ together with $\xi(\Theta)$ constitutes the imputation belonging to the OP in the original game $\Gamma(x_0, T - t_0)$. The condition is stronger than time consistency, which means only that the part of the previously considered "optimal" imputation belongs to the OP in the corresponding current subgame $\Gamma(x^*(\Theta), T - \Theta)$.

Theorem 2.2 implies that the set of all imputations $\overline{L}$ if considered as OP in $\Gamma(x_0, T - t_0)$ is a strongly time consistent optimality principle.

Suppose $C(x_0, T - t_0) = \overline{L}(x_0, T - t_0)$ and $C(x^*(t), T - t) = \overline{L}(x^*(t), T - t)$. From theorem 2.2 we have

$$\overline{L}(x_0, T - t_0) \supset \overline{\xi}(\Theta) + \overline{L}(x^*(\Theta), T - \Theta)$$

for all $\Theta \in [t_0, T]$. Suppose that the set $C(x_0, T - t_0)$ consists of the unique imputation — the Shapley value. In this case from time consistency the strong time consistency follows immediately. Condition (11) can be rewritten in the form

$$Sh(x_0, T - t_0) = \int_{t_0}^{\Theta} \beta(t)dt + Sh(x^*(\Theta), T - \Theta)$$

(here by $Sh(x^*(\Theta), T - \Theta)$ we denote the Shapley value for the game $\Gamma(x^*(\Theta), T - \Theta))$, which gives us the expression for $\beta(t)$

$$\beta(\Theta) = -Sh'(x^*(\Theta), T - \Theta),$$

or, if we suppose the differentiability of $V(S; x^*(\tau), T - \tau)$, $\quad \tau \in [t_0, T]$ along $x^*(\tau)$

$$\beta_i(\Theta) = -\sum_{\substack{S:i\in S \\ S\subset N}} \frac{(s-1)!(n-s)!}{n!} \times$$

$$\times [V'(S; x^*(\Theta), T - \Theta) - V'(S \setminus \{i\}; x^*(\Theta), T - \Theta)].$$

The above expression shows that condition $\beta_i(\Theta) \geq 0$, may not take place, since the differences in brackets may take negative values. Thus $\beta_i(\Theta)$ may not be an IDP, which means that the Shapley value may be time inconsistent.

Theorem 3.1 *The Shapley value defined for the "refined" c.f.* $\overline{V}(S; x_0, T - t_0)$ *is time consistent.*

Proof. For every integrable selected $\xi(\tau) \in C(x^*(\tau), T - \tau)$ define the IDP $\beta(\tau)$, $\tau \in [t_0, T]$ by the formula

$$\beta_i(\tau) = \frac{\xi_i(\tau) \sum_{i=1}^{n} h_i(x^*(\tau))}{V(N; x^*(\tau), T - \tau)} = -\frac{\xi_i(\tau) V'(N; x^*(\tau), T - \tau)}{V(N; x^*(\tau), T - \tau)} \geq 0,$$

if the optimality principle $C(x^*(\tau), T - \tau)$ consists of the unique imputation — the Shapley value, i.e.

$$C(x^*(\tau), T - \tau) = Sh(x^*(\tau), T - \tau) = \xi^\tau,$$

where

$$Sh_i(x^*(\tau), T - \tau) =$$

$$= \sum_{\substack{S:i\in S, \\ S\subset N}} \frac{(s-1)!(n-s)!}{n!} [V(S; x^*(\tau), T - \tau) - V(S \setminus \{i\}; x^*(\tau), T - \tau)],$$

the formula for $\beta_i(\tau)$ gives us

$$\beta_i(\tau) = \sum_{\substack{S:i\in S, \\ S\subset N}} \frac{(s-1)!(n-s)!}{n!} \times$$

$$\times [V(S; x^*(\tau), T - \tau) - V(S \setminus \{i\}; x^*(\tau), T - \tau)] \frac{\sum_{i=1}^{n} h_i(x^*(\tau))}{V(N; x^*(\tau), T - \tau)} =$$

$$= -\sum_{\substack{S:i\in S, \\ S\subset N}} \frac{(s-1)!(n-s)!}{n!} \times$$

$$\times [V(S; x^*(\tau), T - \tau) - V(S \setminus \{i\}; x^*(\tau), T - \tau)] \frac{V'(N; x^*(\tau), T - \tau)}{V(N; x^*(\tau), T - \tau)}.$$

At the same time we have

$$\overline{V}(S; x^*(t), T - t) = - \int_t^T \frac{V(S; x^*(\tau), T - \tau)V'(N; x^*(\tau, T - \tau))}{V(N; x^*(\tau), T - \tau)} d\tau.$$

The Shapley value computed for this c.f. $\overline{V}$ is equal in every subgame $\Gamma(x^*(t), T - t)$, $t \in [t_0, T]$ to

$$Sh_i(x^*(t), T - t) = \int_t^T \beta_i(\tau) d\tau$$

and trivially we have

$$Sh_i(x_0, T - t_0) = \int_{t_0}^t \beta_i(\tau) d\tau + Sh_i(x^*(t), T - t),$$

which is equivalent to (11), (12) ($Sh(x_0, T - t_0) = C(x_0, T - t_0) = \xi$, $\int_{t_0}^t \beta_i(\tau) d\tau = \xi(t)$, $Sh(x^*(t), T - t) = C(x^*(t), T - t)$), which means the time consistency of the Shapley value for the "refined" c.f. $\overline{V}$. The theorem is proved.

In the case under consideration the IDP $\{\beta_i(\tau)\} \geq 0$ has a natural interpretation as a Shapley value in the instantenous game ("small game") with the c.f. equal to $\overline{V}'(S; x^*(\tau), T - \tau)$, $S \subset N$ (see (4), (5)).

At the same time $\beta_i(\tau)$ divides the instantaneous common payoff

$$\sum_{i=1}^n h_i(x^*(\tau)) d\tau = -V'(N; x^*(\tau), T - \tau)$$

proportional to the Shapley value for the subgame $\Gamma(x^*(\tau), T - \tau)$ starting from $x^*(\tau)$ and with the duration $T - \tau$ and c.f. $V(S; x^*(\tau), T - \tau)$. Thus the "refined" c.f. and the corresponding Shapley value may be considered as differential optimality principles in cooperative differential games (see [2]).

It is easily seen that the Banzhaf index [3] will also be time consistent for the refined c.f. $\overline{V}$.

4 Conclusion

In cooperative differential games not only "optimal" imputation sets (such as core, or NM-solution), or imputations (such as the Shapley value and Banzhaf index) have to be found, but also the additional imputation distribution procedures (IDP) β, to define the earnings of the players on the time intervals $[t_0, \tau]$, $\tau \in [t_0, T]$. To follow the optimal trajectory $x^*(\tau)$, the players must be sure that the future earnings on the time interval $[\tau, T]$ remain optimal in the sense they were in the initial game $\Gamma(x_0, T - t_0)$. This is the time consistency condition. If we do not require $\beta \geq 0$, the time consistency problem can be easily solved, as it was in the case of the Shapley value, by putting $\beta(\theta) = -Sh'(x^*(\theta), T - \Theta)$. But negative β does not have much sense, since no one from the players would like to give back his earnings. In this paper we proposed a new approach for constructing time consistent optimality principles based on the idea of locally optimal behaviour.

REFERENCES

[1] Shapley, Lloyd S., "A value for n-persons Games" in *Contributions to the theory of games.* vol. II, H. W. Kuhn and A.W. Tucker, editors, Ann. Math. Studies 28, Princeton University Press, Princeton, New Jersey, 1953.

[2] Petrosjan Leon A., *Differential Games of Pursuit*, World Scientific, Singapore, London, 1993.

[3] Banzhaf, John F., *Weighted Voting Doesn't Work: A Mathematical Analysis*, Rutgers Law Review, Rutgers Univers, New Brunswick, 19, 1965.

PART V
Nonzero sum games, applications

Dynamic Game Theory
and Management Strategy

Steffen Jørgensen
Department of Management
Odense University
5230 Odense M, Denmark

Abstract

The paper deals with some issues of dynamic game modelling and management strategy, in particular pertaining to a firm's market strategy. The interface between the dynamic game formalism and the design of market strategies is emphasized. The paper provides a characterization of dynamic game models and proceeds to discuss the relative merits of repeated games, differential (difference) games, and games of incomplete information, with a view to assess their potential use as support in management strategy decisions.

1   Introduction

It was a pleasure and an honor to address the very first plenary session of the symposium. The title of the presentation could signal an intention to survey a series of models and results, but instead of engaging in such a potentially boring enterprise, I will deal more broadly with a number of issues in dynamic game modelling, management strategy and the interface between the two. There are no mathematics at all. I decided to proceed rather informally, hoping that such a violation of the rules of the game can be forgiven. My aim is to convey a number of observations I find encouraging, but also some I find troublesome.

"Management strategy is a subset of the field of management that combines ideas about competition and organizations with lessons learned from practical business experience", [27] p. 355. Management strategy focuses on the *manager's* formulation of a plan of action – a strategy – to satisfy the objectives of the firm. Management strategy draws from work in the firm's various functional areas, but the main issue is decision-making which cuts across functional lines. Thus, insights from a variety of disciplines must be integrated. Historically, management strategy has its roots in financial accounting, marketing, and corporate planning. Over the years, a wide variety of "management strategy paradigms" have been proposed; some have survived and are now among the standard inventory in the textbooks while others have fallen into oblivion. There are, however, no generally accepted theoretical methods for sorting out these various disjointed approaches, [27] p. 358.

Two areas of management strategy are of major importance: the firm's *market strategy* and its *organizational strategy*. I shall touch upon both, but with special emphasis on the former as the development of marketing strategy to achieve competitive advantage is the key component of management strategy.

Game theory can be described as a collection of analytical methods designed to help us understand and predict the outcome of conflicts or cooperation between decision-makers (players). Basic assumptions are the *rationality* of players and their *strategic reasoning*.

The theory employs two general types of "solution concepts": *cooperative* (e.g., the core, the bargaining set, the Shapley value), and *noncooperative* (Nash equilibrium and its extensions as well as the max-min solution for zero-sum games). A cooperative solution is a natural notion in situations where contracts can be made legally binding. Where distrust prevails, and no external enforcement mechanisms can sustain a cooperative solution, one would look for a noncooperative solution. However, even in basically noncooperative environments, individual self-interest can lead to cooperative behavior (implicit collusion). Sometimes such cooperation can be sustained by credible threats of punishment of defectors – despite the fact that no player legally can commit to retaliate. In the literature, noncooperative games have found much more application in the managerial sciences and industrial organization than cooperative games.

The rest of the presentation is organized as follows. Section 2 provides some useful characterizations of dynamic game models. Section 3 discusses the relative merits of repeated games, differential (difference) games, and games of incomplete information. Section 4 addresses some issues in the modelling of dynamic competition while Section 5 deals with the assumptions of dynamic game theory. Section 6 concludes.

2   Characterizations of Dynamic Game Models

The term "dynamic" has no precise meaning in game theory but – as we shall see – the players' access to, and use of, information is an important key to understanding the difference between various games labelled "dynamic".

Broadly speaking, a player's information is what the player knows when he makes a move. In a game of *perfect* information, players' never move simultaneously; when a player moves, he knows everything that every player did (and observed) at every past decision point. In games of perfect information this is not the case. Typically, the focus has been on imperfect public information. In a game of *complete* information, all players know the rules of the game (including the payoffs, feasible actions sets etc.). If this is not the case, the game is of incomplete information; typically, one player does not know all the details of the rivals' payoffs. A game of imperfect information may be transformed into one with imperfect information by introducing a player "Nature" who moves first and the move is unobserved by at least one player ("Harsanyi's trick"). For more details, see [22], [8].

In differential (or difference) games, information usually refers to how the players' condition their actions on the state variable $x(t)$, $t \geq 0$. A player may base his action at time t on the initial state $x(0)$ (open-loop strategy), on the current state $x(t)$ (Markov or feedback strategy), or on the state-history $x(s)$ for $s \in [0, t]$ (closed-loop strategy). Other specifications are available; see [3].

Some people would say that all games involving time explicitly are dynamic, referring to the fact that "variables" appear at different instants of time. However, even if a game involves time explicitly, it is sometimes assumed that no player uses any information gained in the course of the play. For some mysterious reason – and in some mysterious way – the players agree from the outset not to base their future actions on incoming information. Essentially such a game is static, but is often counted among dynamic games. The standard example is a differential game played with open-loop strategies.

A definition having gained recognition is the following: A game is *dynamic* if a player can use a strategy that depends on previous actions; at least one player increases his information as the game evolves. The intuition behind the usage of the term "dynamic" is that a player – when making a move – has the possibility of observing and reacting to the rivals' previous moves.

The industrial organization literature contains a lot of games played over two periods, incorporating some kind of incomplete information. According to the definition, such a game is dynamic. To illustrate, in the typical *game of entry*, a monopolist makes a strategic action in the first period. (The number of possible strategic actions studied in the literature is quite impressive). The other player is a potential entrant who must decide whether he should enter or stay out, in view of the incumbent's decision. The entrant does not know for sure the incumbent's type (e.g., his cost structure). If the entrant decides to enter, a Cournot or Bertrand game is played in period two. The competitive situation in period two is a static duopoly, although the game in that period certainly can be influenced by the action taken in period one. Time is not involved, apart from the simple fact that period one precedes period two. It does not matter whether the monopoly period is 10 minutes, and the duopoly game is played over 40 years, or vice versa.

Such models are often chosen for their tractability. Nevertheless, two periods may be sufficient to highlight (i) the strategic importance of the timing of actions, (ii) the effects of incomplete information, and (iii) the importance of the monopolist's strategic action in period one.

Other writers prefer to see dynamic games as the "combination" of game theory and *optimal control*, e.g., [3]. A basic reference here is [14]. This stream of literature employs analytical tools that are influenced by those of optimal control. Along the same lines, a theory of stochastic games has been developed. A dynamic game is here just another name for a differential (or difference) game.

All textbooks in the area insiders have canonized *"mainstream game theory"* provide an extensive coverage of repeated games and games of incomplete information. Differential games are often completely left out. In [20] Myerson devotes four pages to continuous-time games, but is mainly concerned with the problems of discrete vs. continuous-time modeling. He concludes that the general approach in game theory is to work with discrete-time models. This observation is probably valid for what Myerson perceives is game theory, but the existence of a theory of differential games certainly provides a counterexample. [8] deals with mainstream game theory but includes a chapter on games with state variables. This book is a notable attempt to put together mainstream dynamic game theory and differential game theory.

It is useful to distinguish three types of dynamic interaction: structural, behavioral (or strategic), and informational.

We have *structural dynamics (structural time-dependence)* if payoffs at time t depend not only on the players' actions at that time but also previous actions. The latter dependence may be indirect, through a state variable. Feasible action sets may change over time and can depend on the current state of the game. Most differential (difference) games are structurally dynamic. Indeed, some of them have a rather rich structure. Repeated games – where the same game (the constituent game) is played in each repetition – are structurally independent. With respect to structural dynamics, a repeated game is an extreme modelling approach.

Behavioral dynamics (behavioral time-dependence) refers to the choice of strategy spaces, the players' use of information. In general, it is fair to assume that each player will condition his current action on the game history – or part of that history (provided he can recall the history). A game is behaviorally dynamic if players condition their current move on the game history.

Repeated games rely heavily on behavioral dynamics. (This could perhaps be seen as a "compensation" for the lack of structural dynamics). Repeated games are increasingly often extended with incomplete information, for instance, with respect to the players' costs or the demand situation. Technical difficulties, however, often prohibit the study of more than a two-stage game.

For similar reasons, differential games have tended to ignore the behaviorally dynamic aspects of the players' interaction. Indeed, the choice of strategy spaces in differential games has drawn a fair amount of criticism on the use of these games. In games with open-loop strategies, players precommit to fixed time functions for their actions throughout the game. A player has only one decision point and no revisions of plans are possible. Strategically, the game is static. In [17] it is rightly argued that precommitment should not be allowed "to enter by the back door". Rather, the rules of the game should admit precommitment as possible actions the players can take. An interesting study of the effects of commitment is [23].

Remark. In practice, plans may be so complex and costly to change that some precommitment is unavoidable. There may also be a "point of no return" after which a fixed plan has to be carried out.

A game is *informationally dynamic* if players can use the history of the game to learn about parameters they do not know for sure. In games of incomplete information, history conveys information about some unknown characteristic of other players. Sometimes a modeler wishes to choose either a game of incomplete information or a repeated game (of complete information). It has been suggested that it may be easier to decide which types of incomplete information that can reasonably be assumed in a specific setting, rather than having to choose between the multiple equilibria that typically occur in a repeated game. Although this may sound reasonable it does not eliminate the problem of multiple equilibria; many games of incomplete information also suffer from nonuniqueness of equilibria.

In principle, the specification of all three types of dynamics should be derived from the *institutional characteristics* of the problem under study. These features should also govern the specification of incomplete information and the equilibrium concept. Different institutional settings lead to different games.

3 Three Classes of Dynamic Games

Dynamic competition has often been studied by repeated games, differential (difference) games, and multi-stage games of incomplete information. Repeated games have often been employed in the study of stationary, *mature* industries, differential games in non-stationary, *immature* industries. A substantial literature has been accumulated in all three areas, although the stock of differential game applications is growing at a slower rate than the two others. Surprisingly enough, there is only insignificant interaction between the three streams of research.

Before becoming mature, any industry was growing and generated the history of the mature industry. This clearly indicates a demand for the study of industry dynamics in itself, and the effects of industry history on the mature stages of industrial competition. The dynamics of immature industries cannot be studied properly in a repeated game framework. Although they have tended to ignore behavioral dynamics, differential games are a better approach to the study of those highly non-stationary environments that characterize immature industries.

In mature-industry competition, it is a basic notion that past behavior influences future performance through the expectations of the members of the industry; the current and future behavior of firms depend on the history of the industry. This has a long verbal history in industrial economics (the predecessor of industrial organization) and is intuitively appealing too.

The repeated game studies of competition typically assume rivalry to take place in prices or in quantities. In practice, one often observes "battles for market share" as the market saturates, but these battles are fought with more sophisticated artillery than just prices or quantities. The use of a repeated game approach may be precluded by the problem's institutional features, requiring structural dynamics in the model. An example is learning-by-doing in production which causes the unit production cost in each period to decrease as a function of accumulated output. Production decisions in one period affect the cost functions, and therefore the game in later periods. (Other examples are natural resource exploitation, capital accumulation, and R&D). The general problem with repeated games could be that the approach is too simplistic to describe and predict oligopolistic behavior in a more definitive way. The predictions of such games are by no means definitive: for the case of observable actions the Folk Theorem asserts that any strictly individually rational payoff vector can be supported by an equilibrium of the repeated game with sufficiently little discounting.

Introducing imperfect information [9] considers repeated games in which players – quite realistically – observe only a public outcome that is a random function of the actions played. The framework applies in Cournot oligopoly where firms sell output unobservably and the market price is a random function of total supply, cf. [10]. To establish the Folk Theorems (Nash-threat, minimax) various hypotheses are needed with respect to the way the probability distribution over public outcomes depends on the players' actions. Roughly speaking, the hypotheses amount to saying that the game should have enough observable outcomes.

In fact, that players have "enough information" is a crucial condition for the sustainability of *implicit collusion*, in repeated games or in differential games. (Implicit collusion is a stable non-binding agreement to secure monopolistic profits for the parties). It seems, however, that the informational requirements this literature imposes on the players are not chosen to reflect the institutional features of the problem. Sometimes the assumptions regarding available information are made rather mechanically, in other cases the assumptions are precisely those that will enable one to prove a desired result.

Implicit collusion is an important issue – if not *the* issue – in repeated games, but repeated game collusive equilibria only hold together because of the *bootstrap* phenomenon. A firm conditions its current action on history only because the rivals do so; history matters only because players threaten to make it matter.

Collusive arrangements are often sustained by threats and punishments based on *trigger strategies:* As long as everybody colludes, stick to the collusive strategy. If somebody cheats, punish him by reverting to a "grim" strategy. (Grim strategies often amount to "overkill" – it is efficient, but rather excessive to punish overtime parking by a life-time sentence). A good reference on the design of punishments is [1].

Also the differential games literature has studied the problem of collusion sustained by trigger strategies, [11], [4]. In [6] continuous strategies are proposed as an alternative to the rather dramatic consequences of the discontinuous trigger strategies. Corresponding to the repeated games literature on collusion, there is a body of literature of differential games dealing with dynamic cooperation. The noncooperative feedback Nash equilibrium is often used as a threat to be implemented should cooperation fail, [16]. In [13] a stochastic differential game of a commercial fishery is studied. Due to randomness in the stock evolution equation, each player is unable to detect with certainty a possible deviation by the opponent from an agreed cooperative harvesting policy. See also [10].

As in repeated games, nonuniqueness of equilibria is not uncommon in differential games. Multiple equilibria occur in games of collusion, but also in other contexts. In differential games, closed-loop Nash equilibria are nonunique, [3]. An explanation of the multiplicity of equilibria is that an increase in information to at least one player creates new equilibria, but does not eliminate the equilibria obtained under the original information structure. For this reason such solutions are called informationally nonunique equilibria. Sometimes also feedback Nash equilibria are nonunique; see, e.g., [5]. In [15] the multiplicity of equilibria is exploited to construct "new" equilibria that are Pareto-efficient.

The last decade has seen a tremendous increase in the literature on incomplete information games. Often the models are very simple and there is quite a step from the highly stylized, two-period models that dominate the literature to models with richer institutional structure and more satisfactory dynamics. On the other hand, these games have made it possible to capture the important idea that decision-makers use history to predict future behavior, intentions and capabilities. The *descriptive significance* of the notion of incomplete information is obvious and so is the strategic importance of *signalling* and *reputation*. But the developments did not come without costs. There may be considerable technicalities in the analysis, equilibria are sensitive to the specification of uncertainties, multiple equilibria often occur, and the assumptions about out-of-equilibrium behavior are quite crucial.

4   Modelling Dynamic Competition

A usable theory of strategic behavior should be oriented toward real-life economic institutions and competitive processes. Then it is the job of the modeler to translate these real-life processes and institutions into the formal rules of the game. Usually, this process is based on the modeler's perception of the situation. In [20] it is suggested that we try to indentify the *players'* perceptions of the game they are to play. Real-life players may very well have other perceptions of a competitive encounter than the theorist.

Modeling is notoriously difficult, and it is easy but rather trivial to point out that there are many areas where our institutional descriptions (models) need improvement. Aumann and Shapley do have a point in [2] when they state that the ability of dynamic game theory to handle real applications is still far from satisfactory. They see the trouble lying less with the descriptive modelling than with the choice of a solution concept. I do not subscribe to this point of view: a sound representation of the institutional environment should be just as important as the choice of solution concept.

Having a sound representation of the game environment does not exclude "No-Fat Modelling", [22] p. 14. The heart of this (often used) approach is to employ the simplest assumptions needed to generate an interesting conclusion to a relatively narrow question. An important component of no-fat modelling is "blackboxing" which treats unimportant elements in a cursory way. Thus a modeler might choose to include only *payoff-relevant* features in the model. The assumption is that real-life players would not care about features of the game environment that have no (or only little) influence on their payoffs.

Although no-fat modelling has generated interesting answers, the general applicability of such models may be limited by the – purposedly strong – model assumptions. To illustrate, many industrial organization studies of market strategies of firms employ the following simplifying assumptions. The model deals with a *consumer product* market. Business-to-business and services markets have attracted surprisingly little research activity although more than twice the dollar volume of yearly transactions takes place between businesses than in consumer markets; service industries account for about 60–80% of GNP in many western economies. The model contains a *single decision variable*, where real-life firms use a broad range of tactic and strategic instruments both in growing and mature markets. The model's description of *consumer behavior* is rather abstract (simplistic).

Remark. Models in marketing science tend to give a richer description of consumer demand and distribution channel characteristics, and deal with a broader range of managerial decisions, e.g., retailing systems, various types of advertising, product lines, and sales promotions. Indeed, marketing is of critical importance to the development of management strategy [27].

As they stand, game-theoretic oligopoly theory and industrial organization are – by and large – collections of formalized, relatively simple stories. However, a number of strategic "principles" have been established. Critics of this literature have said that it merely retells informal anecdotes in a formal manner. Nevertheless, the formal studies help to pinpoint and understand the crucial features of the anecdotes. It has also been emphasized that the outcome (the predictions of the model) frequently depends rather heavily on the context. But this is not necessarily a drawback. To a large extent, optimal competitive strategies must be situation-specific; the study of strategic competition may not require the development of a counterpart to general equilibrium.

Firm-specific normative conclusions are only implicit in many game-theoretic studies in industrial organization. Quite often we are not given prescriptions for firm-specific optimal strategies This may be due to the tradition of industrial organization where the industry is the primary unit of analysis and main emphasis is placed on market structure and welfare implications. There have not been many attempts to translate these results into a setting that could be useful for management strategy. On the other hand, the *management strategy* (corporate strategy) literature takes the individual firm as the unit of analysis. The work in [21] has been pathbreaking in this area. The management strategy literature diverges from industrial organization in an attempt to proceed from the basic questions (industry structure, conduct and performance) to *normative implications for firm-specific strategies*. This stronger emphasis on prescriptions has created a need for more detailed analysis on the level of the individual firm, i.e., more specific descriptions of the firm as an organization, and of the decision-making behavior in a firm. When dealing with competition, however, the literature has often employed a more informal "scenario approach" (the rivals of a firm react rather mechanically in one of a few predetermined ways).

Remark. This approach is similar to the "defensive strategies" that have enjoyed some popularity in marketing studies. A defensive strategy is a firm's best reponse to a conjectured rival strategy.

The potential usefulness of dynamic games in management strategy depends crucially on the "picture" we have of the firm. Traditionally, the basic elements of our view of the firm come from microeconomics and management science.

A cohesive theory of the firm is lacking in *microeconomics*. This may be explained by the fact that microeconomic theory does not take (and perhaps should not take) an explicit management perspective. Milgrom and Roberts write: "Economists have too long ignored the study of how firms and economic systems actually operate in a dynamic, tumultuous environment. Those who have studied these matters, and those who have managed organizations in these environments, have too long labored without the benefits of useful theories to guide their investigations and their decisions" [21]. Economists have not paid very much attention to the nature of managerial decision-making, let alone the management strategy literature. Managerial decision-making behavior, however, should be a cornerstone of the economist's microfoundations and is crucial for the study of industrial organization and the market behavior of firms. [26] and [27] are two stimulating surveys on the interface between economic analysis and management strategy.

Management science has mainly focused upon operational and tactical planning, for instance, applied to problems in production scheduling, inventory planning, waiting lines, and distribution. The tendency has been to handle strategic, external relationships as exogenous constraints in a one-person optimization problem, disregarding that seemingly "internal" decisions often interact with the external environment of the firm. It is fair to add that a stream of recent literature in management science does take a strategic point of view and pays due respect to the firm's competitive environment.

My recommendation would be that the study of competition takes a more explicit *management perspective*. To this end it is necessary to integrate different theories of the firm, to get a more comprehensive description of the firm as well as a richer description of the competitive environment and the competitive behavior. There will also be a need for introducing organizational issues that are currently missing in microeconomic and industrial organization theories of the firm. These are surely big tasks – with plenty of opportunities for research. Nobody would expect a resolution to be just around the corner.

A management strategy perspective would put greater emphasis on firm-specific prescriptions but to resolve the players' fundamental problem of strategic uncertainty one would like the "solution of the game" to supply a unique recommended outcome and a course of action. Unfortunately, many of the currently popular models yield multiple equilibria as the rule rather than the exception. As already said, the Folk Theorems provide the background paradigm for the emergence of collusion in supergames, but we do not get a basis for predicting a specific cooperative outcome. Games of incomplete information and differential games frequently admit multiple equilibria. What can be done?

Common-sense prescriptions such as *symmetric equilibria* in symmetric situations or Schelling's ideas of *focal point* equilibria have been employed. A formal – and rather ambitious – approach is to try to develop a procedure that allows the players to select a unique equilibrium. *Equilibrium selection theory*, [12], attempts to cope with the problem, but applications to specific problems are rare. The description of how *preplay communication* works has been somewhat mysterious, but in recent years formal analysis has been applied to understand the role of cheap talk (e.g., the number of messages that can be sent, their timing, and possible restrictions on the content of messages). The basic idea of the *refinements* of the Nash equilibrium (cf. [29]) is that some equilibria are flawed as "rationally acceptable" equilibria. (The standard example is an equilibrium which rests on an incredible threat). A refinement defines a subclass of equilibria satisfying stronger rationality requirements than Nash equilibrium. On the refinement approach, James Friedman writes: "An impressive cottage industry in refinements has arisen with many of the contenders seeming to the dreamed up to yield a unique outcome in a special context" [7].

Some parts of the literature show a growing skepticism as to whether sophisticated strategic behavior – satisfying, for example, sequential equilibrium – is the natural end-product of introspection of economic agents. A fundamental question is why, and under what conditions, should we then believe in the Nash equilibrium and, in particular, its refinements? Addressing such questions should be a promising occupation for researchers in economic philosophy.

5   Assumptions of Dynamic Game Theory

This section deals – although rather briefly – with some issues in managerial decision-making behavior and rationality as well as the informational requirements of dynamic game theory. We all know that game theory imposes strong assumptions on the players. Many of us also have a feeling that real-life competition is less tidy.

The rationality assumptions of game theory have often been attacked. Strict *maximization of a single objective* has been questioned both by "bounded rationality" supporters, stressing the limits to managerial information collection and processing, and by the proponents of a multicriteria approach to optimization. The majority of large business firms do not have a single owner, they are not being managed by their owners, and the managers often do not share the objectives of the owners. Even if the owners were the managers, they would not necessarily agree on a common objective. Information is not universally distributed over the organization and is not necessarily truthfully transmitted. Such *asymmetries in information* lie at the heart of principal-agent modelling. Still, many research papers in the economic sciences assume that business firms make decisions as a single, monolithic entity where all relevant information is truthfully and immediately transmitted to the top. At this Olympic level we meet the single, all-knowing and universal decision-maker, *The Boss.*

Nobody who observes managerial behavior, even on a casual basis, can avoid noticing a considerable number of *managerial blunders*, even quite substantial ones. It could be an interesting task for game theorists to try to "rationalize" ("irrationalize"?) this kind of observed behavior.

Remark. Robert Townsend, former president of The Avis Corporation confesses: "Two out of every three decisions I made were wrong. But my mistakes were discussed openly and most of them corrected with a little help from my friends" [28].

Our formal models do not treat the firm's internal organization in a satisfactory way and if we wish to model the behavior of organizations such as business firms, the *microstructure of the firm* should be given more priority. Some progress is being made, in particular in organizational theory. It might be beneficial to look in this direction. The "Garbage Can Model" (developed by organizational researchers Cohen, March and Olsen) is a recent model of organizational decision processes. It deals with decision-making in organizations working under conditions of rapid change and high uncertainties. Such a scenario is met in many real-life industries. In these organizations goals and actions often are ill-defined, cause-and-effect relationships difficult to identify, and the participation in a specific decision fluid and limited. The overall pattern of organizational decision making can be thought of as a large garbage can, in which problems, solutions, decision makers and decisions are "mixed". Then one will see problems that arise but are not solved, solutions that are tried but do not work, and even solutions suggested where no problem exists. This point of view certainly runs counter to our standard perception of a firm as *a well-oiled machine.* See also [27], in particular Section 3.

It has not been costless to abandon the static games of *The Bad Old Days* in favor of more plausible dynamic versions. We need to impose rather heavy *information requirements* on the players. In incomplete information games the computation of equilibria proceeds with full rigor, but considerable strain is put upon the player's abilities to make probabilistic assessments. The currently acceptable assumption is "domesticated uncertainty". Such uncertainty is particularly well-behaved and can be satisfied with a diet of probabilities. But it is advisable to keep in mind that strategic decisions almost generically involve unknown, and unknowable, dimensions of the present and the future. Many of these dimensions present difficulties that are currently intractable.

6 Conclusions

Dynamic games deal with situations in which the fortunes of the agents, in some way or another, are interdependent over time. In management strategy, interdependence is a central theme, too. Managers need to deal with interdependence between the firm and its competitors (actual as well as potential), between the firm and the consumers in a market, between the firm and its suppliers, between the individuals and groups within the organization, between the firm and its suppliers of financing (lenders and potential investors), and between the firm and government regulatory agencies.

The *terminology* of game theory is in fact very suggestive of applications to management strategy. Concepts such as strategies, bargaining power, threats, commitments, asymmetric information, and credibility are not at all unknown in business life, although they are certainly not formalized to the degree seen in game theory. In management strategy, [21] emphasizes the need for understanding the strategies of existing and potential competitors, as well as their goals, assumptions and capabilities. Translating this into the terms payoffs, feasible actions, information and technology of game theory should indicate that game theory can make a contribution to the analysis and design of management strategy.

What would be the conditions for dynamic game modelling to become a more prominent element in strategic management? It seems to me that an important initial condition would be that the rationality postulates should not strain the *limits of plausibility*. Furthermore, our models should represent problems that occur, or at least are likely to occur in *real-life competition*. ("The desk is a dangerous place from which to watch the world". John Le Carré).

In [25] it is advocated that "... in terms of application and value to management at the highest levels, *Conversational Game Theory* which consists of advice, suggestions and counsel as to how to think strategically is of considerable worth. It shows how to understand the presence of paradoxes in so-called rational behavior when there are two or more players".

At the present state-of-the-art in our profession, it is not generally advisable to make "literal" applications of game-theoretic models to provide fine-tuned prescriptions for managerial decision-making. The importance of game-theoretic reasoning to management lies more in the possibility of providing managers with a broader *qualitative understanding* of the repercussions of their actions and to focus their attention on key *strategic components*. What our models can do here is to demonstrate the existence of certain "strategic factors" that a manager is well advised to take into consideration when analyzing the repercussions of real-life competition and making his strategic plans. Many of these factors are not unknown to managers – although a manager sometimes uses a different terminology than the game theorist – but their effects and interactions may not be well understood. ("A competitor is the guy who goes in a revolving door behind you and comes out ahead of you" [G. Romney, former CEO, American Motors].

Many of the games we study are simple and striking enough to serve as *illustrations* of basic problems in conflict, coordination and cooperation. For instance, in problems of entry and exit, game theory can point out which strategies are feasible and rational in a range of different situations. We have theoretical explanations for important real-life concepts such as credible threats, commitments, expectations, and reputation. Game theory offers strategic, information-based explanations for real-life phenomena such as price wars, apparently uninformative advertising, product variety, R&D races, strategic investments in excess capacity, and limit pricing.

The game-theoretic study of competition is a fast-growing business, occupied with the dynamics of competition, the tradeoffs faced by firms in pursuing competitive strategies, and the role of information in shaping those strategies. To help the potential users of these results we need to *communicate* better the intended *interpretations* of our analytic framework: for instance, which model assumptions are more or less speculation, what conclusions hold under fairly general conditions, and what pertains to specific models only. To improve this communication we could turn to specific cases, exploring the benefits of computer representations and simulations. The role of dynamic game models would be like a *decision support* tool.

Decision support models (e.g., in product mix decisions or media scheduling) have had some success in marketing management. Managers' subjective estimates are used to support the marketing decisions. The method is not rigorous in a formal sense but can contribute to extract and structure expert knowledge. In general, however, the impact of marketing modeling on practice has been well below its potential. Lilien writes in [18]: "Few topics concern marketing model practitioners and academics alike as much as the low level of impact new developments have on practice".

It is also true that managers have made little explicit use of game theory, either being completely unaware of it or rejecting it as esoteric and largely irrelevant. Although game theory and management strategy may have many similarities as far as problems and terminology are concerned, most managers seem to defend themselves against the insights offered by game theory. One reason could be that the claims we make for our theory may not be easily verifiable. The products of game theory are abstract and often invisible.

It is interesting to notice that a part – but only a part – of this story also applies to management or corporate strategy. In these areas managers also meet mysterious gurus and their gimmicks. The products (e.g., core competence, lean management, business reengineering) are also abstract and the claims for the "theories" largely unverifiable. The proposals seem to change almost like the fashion of the *haute couture*. Why then have management consultants and gurus succeeded in attracting so much attention at the top levels in business firms, and game theory so little?

Acknowledgements

The paper has benefited from a stimulating plenary discussion at the symposium as well as the constructive remarks of an associate editor and two anonymous reviewers. Earlier versions of the paper were presented at CentER, Tilburg University and Dept. of Business, University of Vienna. I am grateful to Kristian R. Miltersen for transforming the manuscript into LaTeX. This work has been supported in part by the Danish Research Council for the Social Sciences under Grant 5.20.31.02.

REFERENCES

[1]   D. Abreu, "Extremal Equilibria of Oligopolistic Supergames", *Journal of Economic Theory*, Vol. 39, pp. 191–225, 1986.

[2]   R.J. Aumann and L.S. Shapley, "Long-Term Competition – A Game Theoretic Analysis", in *Essays in Game Theory in Honor of Michael Maschler* (N. Megiddo, ed.). Springer, 1994.

[3]   T. Basar and G.J. Olsder, *Dynamic Noncooperative Game Theory*. Academic Press, 1982.

[4]   J. Benhabib and R. Radner, "The Joint Exploitation of a Productive Asset: A Game-Theoretic Approach", *Economic Theory*, Vol. 2, pp. 155–190, 1992.

[5]   E.J. Dockner, J.M. Hartwick and G. Sorger, "The Great Fish War Revisited". Mimeo, 1992.

[6]   H. Ehtamo and R.P. Hämäläinen, "A Two-Country Dynamic Game Model of Whaling", *Journal of Economic Dynamics and Control*, Vol. 17, pp. 659–678, 1993.

[7] J.W. Friedman, "A Review of Refinements, Equilibrium Selection, and Repeated Games", in *Problems of Coordination in Economic Activity* (J. W. Friedman, ed.). Kluwer, 1994.

[8] D. Fudenberg and J. Tirole, *Game Theory*. MIT Press, 1992.

[9] D. Fudenberg, D. Levine and E. Maskin, "The Folk Theorem with Imperfect Public Information", *Econometrica*, Vol. 62, pp. 997–1039, 1994.

[10] E.J. Green and R.H. Porter, "Noncooperative Collusion under Imperfect Price Information", *Econometrica*, Vol. 52, pp. 87–100, 1984.

[11] R.P. Hämäläinen, A. Haurie and V. Kaitala, "Equilibria and Threats in a Fishery Management Game", *Optimal Control Applications & Methods*, Vol. 6, pp. 315–333, 1985.

[12] J.C. Harsanyi and R. Selten, *A General Theory of Equilibrium Selection in Games*. MIT Press, 1988.

[13] A. Haurie, J.B. Krawczyk and M. Rocher, "Monitoring Cooperative Equilibria in a Stochastic Differential Game", *Journal of Optimization Theory and Applications*, Vol. 81, pp. 73–95, 1994.

[14] R. Isaacs, *Differential Games*. Wiley, 1965.

[15] V. Kaitala, "Nonuniqueness of No-memory Feedback Equilibria in a Fishery Resource Game", *Automatica*, Vol. 25, pp. 587–592, 1989.

[16] V. Kaitala and M. Pohjola, "Optimal Recovery of a Shared Resource Stock: A Differential Game Model with Efficient Memory Equilibria", *Natural Resource Modeling*, Vol. 3, pp. 91–119, 1988.

[17] D.M. Kreps and M. Spence, "Modelling the Role of History in Industrial Organization and Competition" in *Issues in Contemporary Microeconomics and Welfare* (G. Feiwel, ed.). MacMillan, 1985.

[18] G.L. Lilien, "Marketing Models: Past, Present and Future", in *Research Traditions in Marketing* (G. Laurent et al., eds.). Kluwer, 1994.

[19] P.J. Milgrom and J. Roberts, *Economics, Organization and Management*. Prentice-Hall, 1992.

[20] R. Myerson, *Game Theory: Analysis of Conflict*. Harvard University Press, 1991.

[21] M.E. Porter, *Competitive Strategy*. Free Press, 1980.

[22] E. Rasmusen, *Games and Information. An Introduction to Game Theory*. Blackwell, 1989.

[23] J.F. Reinganum and N.L. Stokey, "Oligopoly Extraction of a Common Property Natural Resource: The Importance of the Period of Commitment in Dynamic Games", *International Economic Review*, Vol. 26, pp. 161–173, 1985.

[24] A. Rubinstein, "Comments on the Interpretations of Game Theory", *Econometrica*, Vol. 59, pp. 909–924, 1991.

[25] M. Shubik, "What is an Application and When is Theory a Waste of Time?", *Management Science*, Vol. 33, pp. 1511–1522, 1987.

[26] D.F. Spulber, "Economic Analysis and Management Strategy: A Survey", *Journal of Economics & Management Strategy*, Vol. 1, pp. 535–574, 1992.

[27] D.F. Spulber, "Economic Analysis and Management Strategy: A Survey Continued", *Journal of Economics & Management Strategy*, Vol. 3, pp. 355–406, 1994.

[28] R. Townsend, *Up the Organization*. Knopf, 1974.

[29] E. van Damme, *Stability and Perfection of Nash Equilibria*. Springer, 1987.

Endogenous Growth as a Dynamic Game*

Simone Clemhout and Henry Wan, Jr.
Dept. of Economics, College of Arts and Sciences
Cornell University, Uris Hall, Ithaca
New York 14853-7701, USA

Abstract

Dissimilar growth patterns often prevail for economies similar in taste, technology and initial endowment. This may arise from the nature of 'knowledge capital', a durable public input, privately accumulated. Its external effect spawns a dynamic game. By a game-theoretic analysis, one obtains a continuum of Markovian-Nash equilibria, some being Pareto-ranked: better coordination means higher growth. The underlying strategical complementarity is shown to be equivalent to the externality of 'fish war'. The notion of splicing equilibrium is introduced and some open issues are isolated for future studies.

1   Motivation

'Knowledge capital' is the favorite growth engine in the literature of endogenous growth. Created at private risks, and serving as a durable public input for production [Shell (1966)], this concept is offered as the basis to theorize the observed facts – the widely varying growth rates across economies and over time.

In the past eight years, many competing hypotheses have appeared, through regression studies and analytical models to explain the mechanism for growth, ranging from R. and D. [e. g., Romer (1986)], education, learning-by-doing [e. g., Stokey (1988), Young (1991)], the investment-growth nexus [e. g., DeLong and Summers (1991)], the trade promotes growth thesis, to appropriate fiscal and financial regimes. So far, no consensus has emerged to explain why similar economies end up with dissimilar performances [e. g., The Philippines – Korea comparison of Lucas (1993)]. Skeptics begin to question whether the 'new' theory has any more explanatory power than its neoclassical predecessor [See e. g., Pack (1994) and Solow (1994)].

---

*   Earlier versions of this paper have been presented at Cornell University, Academia Sinica, The Hong Kong University of Science and Technology, Kobe University and the Sixth Conference for the International Association of Dynamic Games. We acknowledge the helpful comments received from the audience in these occasions, but we alone are responsible for the remaining shortcomings.

We believe that in the search for the true cause for growth, the nature of the knowledge capital is both a hindrance and a remedy. This becomes clear when we focus on two basic issues: (a) what explanations economists can possibly offer and (b) how can theoretical economics be applied to yield decisive insight. We base our reasoning on dynamic game theory.

To begin with, knowledge capital is a public good privately supplied. Its accumulation is a game among its 'investors'. This game is 'dynamic', and not 'one-shot', by the durability of knowledge. Under a subgame-perfect solution, the player of a dynamic game chooses a credible contingent plan which is the best reply against such plans of all others. The (memory-less) feed-back equilibrium is the simplest framework for presenting our argument.

We maintain that even if all individuals are identical, externality makes a multi-person economy fundamentally different from the economy of an isolated Crusoe. It will be made intuitively clear that the mutual expectations between individuals give rise to a continuum of Pareto-ranked equilibria. Thus, social cohesion matters no less than technology, initial wealth, intention, and information. In general, it plays a critical role in economic development. In particular, it is decisive for pairwise comparisons between the Philippines and Korea. Yet, for empirical studies, there is no satisfactory and observable proxy for individuals' beliefs about each other. Moreover, what shapes mutual expectations may include factors like history, politics, and culture – over which we enjoy no comparative advantage as professional economists. It is no coincidence that founders of growth theory (like Solow) would concentrate on economies with mature institutions. Other things being equal, individuals in such economies would cherish mutual expectations which favor growth.

Since the greatest differences in growth rates are observed among economies with low and middle income per capita [Lucas (1988)], it is only natural that economists today are eager to aim their artillery at where 'the big actions are'. But among these economies, part of the observed differences are the consequences of multiple equilibria. By omitting the effect of mutual expectations, one can easily be misled by spurious causality. Our task of identifying the fundamental mechanism for development is thus greatly complicated.

We believe a much more promising approach is to use the concept of 'social cohesion' as an 'organizing principle', even though we lack suitable proxies for it in our regressions. Borrowing a leaf from the study of frontier production functions, we may concentrate on those economies which are outliners of success, for example, the East Asian economies. By factoring in some degree of 'social cohesion' which can be reasonably expected in reality, there is better prospect of delineating some of the economic forces and policy environments which are conducive to high performance in growth. A step in that direction is given by Lau and Wan (1993).

The next sections are devoted to the following topics: (a) the need to reformulate endogenous growth as a dynamic game, (b) the conceptual basis and the method of proof for multiple equilibria, (c) an analogy with the models of 'fish war' for which the multiplicity of solutions is well established, and (d) certain open issues waiting for resolution. Some final remarks are provided in the concluding section.

2 Why a Reformulation is Needed for Endogenous Growth

Much of the endogenous growth literature follows Romer (1986). This is an analog of Marshall's static analysis of the industrial supply. In a Marshallian equilibrium, firms anticipate the concurrent 'actions' (i. e., outputs) of other firms. For endogenous growth, individuals may anticipate each others' beliefs, which decide the investment responses to the state of the system. In contrast, Romer's approach specifies that:

(i) In an economy with identical individuals, the output of each depends upon the levels of knowledge capital of both one's own and the economy-wide average. Both are state variables which influence agents' investment.

(ii) Each agent operates under the belief that the economy-wide average of investment may vary with time at the most, but not in response to the state of the system[1] (there is, therefore, the *asymmetric myopia*), and

(iii) At the equilibrium, the level of knowledge capital of each agent is identical to the economy-wide average.

We maintain that (i) is an over-simplification, leaving no scope to study issues about the 'structure, behavior and performance' of an industry. Yet these are needed to assess the impact of industrial concentration on R. and D[2].

(ii) is even more problematic in its *asymmetric myopia*. It denies that mankind can be cognizant of the fact that *other persons like oneself would respond to the system exactly like oneself.* After all, 'putting oneself into others' shoes' underlies the Golden Rule which is ubiquitous among diverse cultures[3]. Moreover, with all individuals being identical, there is no room for the uncertainties in the 'joy of fish' dialog [Zhuang Zhou (369B.C. – 286 B. C.): 'You are not I, how do you know that I do not know fish is joyful?']. Finally, introspection severs that Gordian knot of infinite regress: 'I know you know that I know...'.

---

[1]This resembles the *open-loop* equilibrium of dynamic games, in which players are not allowed to take account of other players' decision rules [cf. Hansen et al. (1985), cited by Romer, ibid].

[2]The industry-wide average does not reflect the second and higher moments for the distribution of knowledge capital among the agents.

[3]From Vedic text, Confucian Analects, the teachings of the Prophet Hillel, to the classical treatise of Plato and Seneca.

In fact, this formulation affects the study of transitional dynamics [See Benhabib and Perli (1994) and Xie (1994)] in two ways:

(i) One finds a continuum of equilibrium paths only in models with two or more state variables and under certain parameter values, and

(ii) There is no intuitive explanation of why a multiple solution prevails.

In contrast, it will be shown that once the problem is reformulated in terms of feedback strategies, the cause of multiplicity is seen to be the strategical complementarity between the players[4].

Like physical capital, the formation of knowledge capital is also costly in terms of consumption foregone. For the classical Ramsey problem, optimality requires that Along the optimal time path, the rate of return of the

(discounted) marginal utilityof consumption must offset the marginal return on investment. (The Euler-Lagrange condition).

The novelty here is that knowledge capital is a public 'investment good'. As various types of knowledge may complement each other, the returns on investments are no longer fixed, but can be lifted by bootstraps.

By knowledge capital we refer to specialized expertise in interdependent industries, no less than patentable discoveries. Thus the bottleneck for the final goods industries is often the paucity of reliable local suppliers for parts and repair services. The viability of these supporting industries again depends on the success of their clients, i.e., local manufacturers. Such an agglomerative effect explains that for long periods, the more mature industrialized economies (e.g., Japan, Taiwan today) can hold their own against competitors with much lower wages (e.g., Mainland China).

The upshot is that the more ready are other individuals to accumulate (complementary) knowledge, the more justifiable it is for an individual to accumulate knowledge capital on one's own part, out of self interest. This is how mutual expectations matter[5]. Such incremental calculation can proceed by infinitesimal degrees, giving rise to uncountably many Pareto-ranked equilibria. Clearly, such reasoning needs no special restrictions on either the number of state variables or the range of parameter values. Nor has *asymmetric myopia* any role to play in this scenario.

---

[4]The concept, 'strategic complementarity', originates in supermodular games [Topkis (1979), also the survey in Fudenberg and Tirole (1991)] where players select finite-dimensional vectors in one-shot games. In contrast, here players select functions. Thus, theorems from supermodular games cannot be invoked.

[5]Ironically, the only possible exception seems to be the two-period, discrete time example of Romer, ibid, where a player observes what has happened in period 1 only in period 2, which is the last period in that example. By then, it is too late to make any positive investment. Such an investment can only matter in period 3, which lies beyond the assumed horizon.

3 The Nature and Demonstration of Multiple Equilibria

According to Lucas (1987), each of his major contributions in macro-economics is the consequence of reformulating the problem as a dynamic game. The intuitive demonstration of multiple equilibria in the last section illustrates once more the power of game-theoretic reasoning. Yet, it will be shown below that to clarify *what* sort of multiple solutions one may obtain and *how* this multiplicity is established, one must carry out the analysis formally and with care.

Consider for the time being a *symmetric game*: where all players share the same intention and capabilities, excepting possibly some player-specific state variables.

Let

$N = \{1, \ldots, N\}$ be the set of players,

x be the state vector in the state space X, which is contained in R^M,

c_i be the control vector in the (common) control space C, a compact subset of R^M, $i = 1, \ldots, N$

$u(c_i)$ be the (common) *concave* felicity index for all individuals,

r be the (common) time preference rate,

$$v_i = \int u(c_i)e^{-rt}dt \tag{1}$$

be i's payoff integral over the infinite horizon,

$$dx/dt = f[x, (c_i)_{i\in N}] \tag{2}$$

be the state equation, where f denotes production technology.

Let S be the class of piece-wise continuously differentiable functions: $X \longrightarrow C$ where there are at most jump discontinuities. Members of S are referred to as strategies which include the following three types, singled out to facilitate our exposition:

1. Continuously differentiable strategies.

2. The 'extreme strategy': $c_i(x) = \max C$, for $x \neq 0$; $c_i(0) = 0$.

3. Splicing strategies which are continuously differentiable over some interval in the state space, but coincide with 1 or 2 or other splicing strategies over its complement.

Definition 3.1 $c^* = (c_{j^*})_{j\in N} = (c_{i^*}, c_{-i^*})$, *where* (c_{-i}^*) *refers to* $(c_{j^*})_{j\in N}$, $j \neq i$, *is an equilibrium for the game, if for each i, given (i) the state equation, (ii) the initial state $x(t_0)$ and (iii) the strategies of the others:* (c_{-i^*}), c_{i^*} *maximizes v_i over the class S, for all initial state x.*

Remark 3.1 *In c^*, each player i adopts c_i^* as the best reply against (c_{-i}^*).*

Remark 3.2 *Since* c_{i*} *is the solution of a 'conditional optimal control problem', it satisfies both: (i) the Euler-Lagrange first order condition necessary for an optimum and (ii) some sufficient condition for a maximum like the Mangasarian criterion.*

Definition 3.2 *The evolution of a game under a particular equilibrium from an initial state is referred to as an equilibrium play.*

Definition 3.3 *For a symmetric game, a symmetric equilibrium is an equilibrium where every player uses the same strategy.*

Definition 3.4 *A symmetric state is a state symmetric with respect to all players.*

Remark 3.3 *With a single state variable, all states are symmetric.*

Definition 3.5 *A symmetric equilibrium play is the equilibrium evolution from a symmetric state under a symmetric equilibrium.*

For illustration, the graphs of four symmetric strategies in a 'simple' fish war (to be discussed in the next section) are displayed in Figure 1 below. The arrow signs indicate the evolution under a *symmetrice quilibrium play*: f/N and c_i stand for per capita supply and demand of the replenishable resource. x rises (falls) if $(f/N) - c_i$ is positive (negative). e, ε and e' depict alternative steady states.

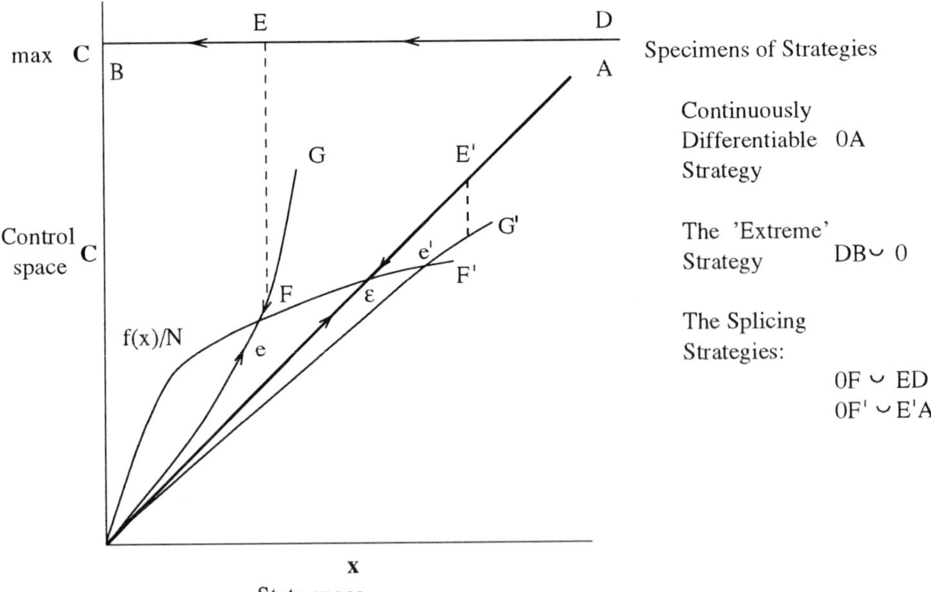

Figure 1: The example of the 'simple' fish war: $dx/dt = f(x) - \sum c_i$

Traditionally, attention is focused on the solution corresponding to the interior, 'saddle' path, i.e. OA in Figure 1, even though the 'extreme solution' is well known in the discrete time models (exhausting all supplies once and for all in one period). In continuous time, this corresponds to the situation where any one player can run down the stock of the 'public asset' single-handedly at some maximum speed. The latter is decided by physical or engineering considerations. By so doing, this player will induce all other players to follow suit. In that light, one can always expect the coexistence of at least two solutions.

However, this is not quite our current concern. It is the coexistence of a continuum of equilibria which can cause problems in analyzing the differences in growth performance. The typical strategy we consider coincides with a 'non-saddle' path like OFG, up to some point F. Not being the saddle path, this latter locus will bend backward at some point G, as Shimomura (1991) pointed out. Thus a strategy for the solution cannot coincide with it beyond G. At the same time, under an equilibrium play which starts from some initial state, any strategy must be defined over the entire state space, including the 'unreachable' states[6]. This calls for *splicing*, a procedure employed by Benhabib and Radner (1991). This yields OFED as the strategy for the symmetric equilibrium. There is also another strategy, OF'E'A, where the splicing is done with the help of the saddle path, OA.

In this illustration, the saddle path is decided by the Euler-Lagrange condition. It only qualifies as an equilibrium when some sufficient criterion is met. On the other hand, if such a criterion is met in the strict sense, then in its suitably defined neighborhood, there must be additional paths which also qualify as the basis of our splicing equilibria [Clemhout and Wan (1993)]. Thus, in general, we can demonstrate the multiplicity of equilibria by a 'constructive' proof. The steps involved are schematically depicted in Figure 2.

Conceptually, we focus attention on all solutions of the Euler-Lagrange equation. The latter is necessary for any solution path associated with an equilibrium. A subset of these also satisfies the sufficient conditions and therefore they correspond to equilibria for the game. Among these, we then verify that one constructed solution ('the verified solution') should correspond to an equilibrium, where the sufficient criterion is satisfied in the strict sense. By an argument in Clemhout and Wan, op. cit., all solution paths in some (appropriately defined) neighborhood must also meet the sufficient criterion, over some interval in the state space which contains the steady state. They can then be extended by splicing to form an equilibria. The

---

[6] Any equilibrium should specify what happens if an individual player deviates from one's equilibrium strategy. For that, one must identify the best replies of all other players in all states which are reachable after that individual's deviation. Since we are considering a symmetric equilibrium, it is desirable that the *common* strategy is defined at *every* state.

essence of our work shall be to construct a 'verified equilibrium'.

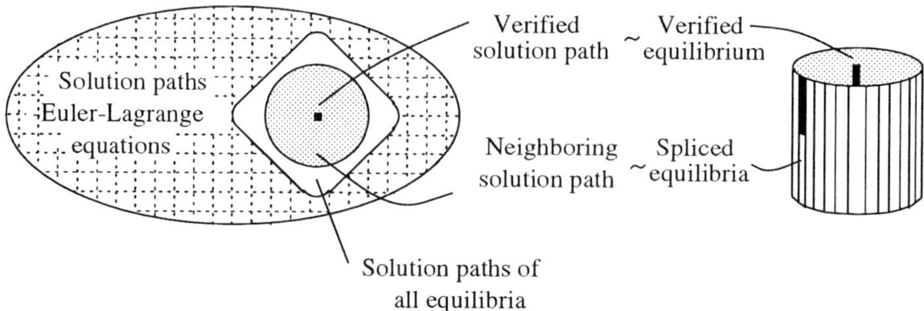

Figure 2: Constructive proof for multiplicity

Note that earlier discussions of multiplicity tend to be incomplete. Clemhout and Wan, ibid. point out the importance of the sufficient criterion, without which what is established is only the multiplicity of *candidates for equilibrium* and not the multiplicity of equilibria. On the other hand, that paper has discussed the need for splicing for the asymmetric equilibria, but not for the symmetric equilibria.

We must now face the meta-game-theoretical question, does 'continuous differentiability' constitute a suitable *selection principle* ? If so, the saddle path is re-established as the unique solution. We do not think so, not for players facing an initial state below the steady state point e, in any case, if splicing only happens beyond the point e. Splicing at states unreachable along an equilibrium play is likely to be the least of their concerns for ordinary individuals. It is their behavior that we try to capture in our game theoretical formulation. Thus, continuous differentiability will be too restrictive a criterion to apply.

4 Fish War and Endogenous Growth: The Simple Versions

We revisit the findings on fish war for two reasons: (i) Its properties are well known, and it is a general framework, under which several important models (including the endogenous growth game) may be subsumed as special cases, and (ii) We shall derive past findings by a new approach, which is applicable to the multi-state case that is crucial to future researches on the endogenous growth game.

We shall first demonstrate the structural affinity between various models before analyzing the simpler cases.

(A) Fish war–the 'general' case.

Consider now the game of fish war, formulated by Levhari and Mirman (1980) for the one species, discrete time model, and by Clemhout and Wan (1985) in the many species, continuous time version. For the latter case, (3.2) is specialized to the following additively separable form:

$$\begin{aligned} d\mathbf{x}/\mathbf{dt} &= \mathbf{f}[\mathbf{x}, (\mathbf{c_i})_{i\in\mathbf{N}}] \\ &= \mathbf{f}(\mathbf{x}) - \sum_{i\in\mathbf{N}} \mathbf{c_i} \end{aligned} \tag{3}$$

We shall now study the relationship of this model with various other models.

(B) The Ramsey model.

To begin with, the Ramsey model may be regarded as the special case with: $N = 1$.

On the other hand, at an equilibrium, $c^*(x)$, define:

$$\mathbf{f_i}(\mathbf{x}) = \mathbf{f}(\mathbf{x}) - \sum_{i\neq j\in\mathbf{N}} \mathbf{c_j^*}(\mathbf{x});$$

then (4.1) becomes:

$$d\mathbf{x}/\mathbf{dt} = \mathbf{f_i}(\mathbf{x}) - \mathbf{c_i} \tag{4}$$

Thus, from the viewpoint of an individual player in an equilibrium, the model of fish war also becomes an N-fold coupled 'conditional Ramsey model', conditioned upon the equilibrium strategies of all other players. The Euler-Lagrange condition applies to c_i in (4.2).

(C) Fish war – separated fishery.

Another relevant special case concerns the fish war with interrelated national fisheries. For example, by mutual agreement, both the Russian and the Finnish fishing fleets would only harvest a particular type of fish in their own territorial portion of the Gulf of Finland, yet the man-made demarcation line at sea means little for the fish population: a fixed proportion of fish in each national fishery would migrate to the other at each instant[7]. A simple formulation for this problem is to add to (4.1) an additional restriction:

$$\mathbf{c_i} = \mathbf{c_i}\mathbf{e_i}, \qquad \textbf{for every i} \tag{5}$$

where $\mathbf{e_i}$ is the ith unit vector of $\mathbf{R^M}$, and $M = N$. $\mathbf{x_i}$ now represents the particular fish population in the territorial waters of player i.

(D) Fish war – the 'simple' case.

---

[7]They may also interbreed in the open sea before returning to their native shore to spawn. Alternatively, they may prey on each others' young.

We are also interested in another special case, where in (4.1), $M = 1$ (one species, no non-separated fishery):

$$dx/dt = f(x) - \sum_{i \in \mathbf{N}} c_i \tag{6}$$

(E) Endogenous growth – 'general version'.

Next we return to the literature of endogenous growth. For the present purpose, consider a two-person economy, $M = 1$, $N = 2$. In the formulation of Romer, we have:

$$
\begin{aligned}
dx_i/dt &= F[x_i, (x_i + x_j)/2] - c_i \\
&= f(x_i, x_j) - c_i \, .
\end{aligned}
\tag{7}
$$

by defining a new function f. Specifying further that

$$\mathbf{f}(\mathbf{x}) = (\mathbf{f}(\mathbf{x_i}, .\mathbf{x_j}), \mathbf{f}(\mathbf{x_j}, .\mathbf{x_i})), \ \mathbf{c_i} = (\mathbf{c_i}, \mathbf{0}), \qquad \mathbf{c_j} = (\mathbf{0}, \mathbf{c_j}) \, , \tag{8}$$

in the formulation of (4.1), we have now seen the equivalence between the case of 'endogenous growth – general version' and the case of 'fish war – separated fishery'.

Differentiating F totally with respect to x_i in (4.5), we get:

$$dF/dx_i = (F_1) + (1/2)(F_2),$$

where the two terms at the right hand side are referred to in the literature as the 'internal effect' and the 'external effect' respectively. The 'spill over' aspect of the knowledge capital is due to the second term. If that term is absent, then knowledge capital is analytically no different from physical capital, and we are back to the Ramsey model.

(F) Endogenous growth – 'simple version'

Suppose now the internal effect is absent. Then

$$F[x_i, (x_i + x_j)/2] = G[(x_i + x_j)/2] = F[x_i, (x_i + x_j)/2], \tag{9}$$

for all x_i, x_j and for some function G. We can now introduce a new state variable x and redefine f as follows:

$$x = x_1 + x_2, \ \ f(x) = G(x/2) + G(x/2) = 2G(x/2), \tag{10}$$

so that we have:

$$dx/dt = f(x) - c_i - c_j, \ \ i, j = 1, 2, \tag{11}$$

which is equivalent to the fish war – the 'simple' case in (D), with $N = 2$. Figure 3 displays the interrelationship between these models.

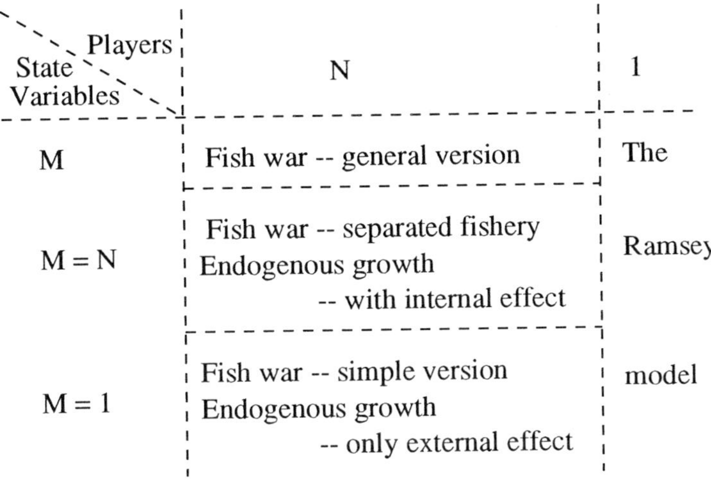

Figure 3: Relationship between the models

The multiplicity of solutions in the 'simple fish war' has been studied in Clemhout and Wan (1993). To make our discussions self contained, we sketch below a simple example.

Let

$$M = 1 \ , \ N = 2 \ , \quad \text{(Levhari-Mirman)}$$

$$u(c) = \log c \ , \quad \text{(Bernouilli)}$$

$$f(x) = (a - b \log x)x \quad \text{(Gompertz)}$$

Choosing an appropriate unit for time, we set $r = 1$, $\max C > e^{(a/b)-1}$. From the last condition, any determined player can single-handedly run down the stock of x toward 0, by setting the consumption rate to $\max C$.

4.1 Characterization of Equilibrium

At any equilibrium, $\mathbf{c}^* = (\mathbf{c_{1*}}, \mathbf{c_2^*})$, and an initial state–time pair (x_0, t_0), let the maximum attainable payoff for player $i = 1, 2$, be $V(x_0, t_0; c_j^*)$, $j = 2, 1$, then c_i^* must satisfy the Hamilton-Jacobi equation:

$$\partial V/\partial t = \max_{c_i}\{e^{-t}\log c_i + (\partial V/\partial x)[(a - b\log x)x - c_i - c_j^*(x)]\} \ ;$$

$$c_i^*(x) = argmax\{e^{-t}\log c_i + (\partial V/\partial x)[(a - b\log x)x - c_i - c_j^*(x)]\}.$$

In fact, there is a solution,

$$V = e^{-t_0}W(x_0; c_j^*) \qquad \text{for some } W(.), \text{ with}$$

$$\partial V/\partial x = e^{-t}/c_i, \quad \text{(The Maximum Principle)}$$

so that

$$\partial V/\partial t = -e^{-t}W, \ \partial V/\partial x = e^{-t}/c_i = e^{-t}W',$$

and thus,

$$W' = 1/c_i. \tag{12}$$

By substitution and rearrangement,

$$W(x) = \log c_i^*(x) + W'(x)[(a - b\log x)x - c_i^*(x) - c_j^*(x)].$$

Taking the total time derivative, we have

$$
\begin{aligned}
W'(x)[dx/dt] &= \{(d/dc_i)[\log c_i - (W')c_i]\}c_i^{*\prime}(x)[dx/dt] \\
&\quad + \{(d/dt)[W'(x)]\}[dx/dt] \\
&\quad + W'(x)\{(d/dx)[(a - b\log x)x - c_j^*(x)]\}[dx/dt] \\
&= \{0\}c_i^{*\prime}(x)[dx/dt] \\
&\quad + \{(d/dt)[W'(x)] \\
&\quad + W'(x)\{(d/dx)[(a - b\log x)x - c_j^*(x)]\}[dx/dt]
\end{aligned}
$$

which means either

$$dx/dt = 0,$$

or

$$
\begin{aligned}
d\,W'/dt &= \{1 - (d/dx)[(a - b\log x)x - c_j^{*\prime}(x)]\}W' \\
&= [(1 + b - a) + b\log x + c_j^{*\prime}(x)]\}W'
\end{aligned}
$$

By (4.10),

$$
\begin{aligned}
[a - b - 1 - b\log x - c_j^{*\prime}(x)]c_i^*(x) &= c_i^{*\prime}(x)(dx/dt) \\
&= c_i^{*\prime}(x)[(a - b\log x)x - c_i^*(x) - c_j^*(x)].
\end{aligned}
$$

Dropping the * sign to simplify notations and collecting terms, one has a simultaneous system of two first-order, inhomogeneous, ordinary differential equations in implicit form:

$$
\begin{aligned}
[(a - b\log x)x - c_1 - c_2]c_1' + c_1 c_2' &= (a - b\log x - b - 1)c_1 \\
c_2 c_1' + [(a - b\log x)x - c_1 - c_2]c_2' &= (a - b\log x - b - 1)c_2
\end{aligned}
\tag{13}
$$

The solution to (4.11) characterizes any candidate path corresponding to an equilibrium, be it symmetric or asymmetric. From (4.11), one can easily verify that there exists a symmetric equilibrium where strategies are linear in the state variable:

$$
\begin{aligned}
c_i^*(x) &= (1+b)x \\
&= x c_i^{*\prime}(x), \qquad i = 1, 2. \quad \text{(The 'verified equilibrium')} \quad (14)
\end{aligned}
$$

The strict concavity of $f(x)$ and the linearity of $c_j^*(x)$ implies the strict concavity of

$$
f_i(x) = f(x) - c_j^*(x),
$$

and thus, the satisfaction of the sufficient criterion for an equilibrium, both for $c^*(x)$ and for neighboring paths (defined with a *twice-continuously differentiable norm*) and sharing some singularity point.

For completeness, we substitute (4.12) back to (4.10) and get:

$$
W'(x) = 1/(1+b)x,
$$

for the verified equilibrium, thus

$$
W(x) = (\log x)/(1+b) + K, \quad \text{with} \quad K = \log(1+b) + a/(1+b) - 2.
$$

4.2 Stability Analysis

For stability analysis, it is convenient to change variables, with

$$
z = \log x \qquad \text{and} \qquad w_i(z) = c_i/x, \quad i = 1, 2,
$$

as the new state and control variables, where

$$
c_i'(x) = w_i(z) + w_i'(z).
$$

Thus, dividing (4.11) through by x and making substitutions, one obtains

$$
\begin{aligned}
{}[(a - bz) - w_1 - w_2](w_1 + w_1') + w_1(w_2 + w_2') &= (a - bz - b - 1)w_1, \\
w_2(w_1 + w_1') + [(a - bz) - w_1 - w_2](w_2 + w_2') &= (a - bz - b - 1)w_2,
\end{aligned}
$$

or,

$$
\begin{aligned}
{}[(a - bz) - w_1 - w_2]w_1' + w_1 w_2' &= [-(b + 1) + w_1]w_1, \\
w_2 w_1' + [(a - bz) - w_1 - w_2]w_2' &= [-(b + 1) + w_2]w_2.
\end{aligned}
$$

By Kramer's Rule, one obtains

$$
w_i'/w_i = \{[(a - bz) - w_i - w_j][-(b + 1) + w_i] - [-(b + 1) + w_j]\}/D,
$$

$$
i, j = 1, 2, \qquad i \neq j, \qquad D = [(a - bz) - w_i - w_j]^2 - w_i w_j,
$$

which is useful for evaluating the asymmetric equilibrium[8].

On the symmetric plane: $\{(w_1, w_2) : w_1 = w_2\}$, we have

$$w_i' = w_i[w_i - (b+1)]/(a - bz - w_i) \tag{15}$$

which may be transformed into a second order ordinary differential equation for the unknown 'current-value' value function $U(z)$. This is done by means of the transformed relation in (4.10):

$$U'(z) = 1/w_i.$$

Specifically,

$$U''(z) = [(b+1)U'(z) - 1]U'(z)/[(a - bz)U'(z) - 1] . \tag{16}$$

A steady state means

$$a - bz = 2w_i = 2/U'(z) ,$$

or,

$$U'(z) = 2/(a - bz) . \tag{17}$$

Near the steady state,

$$(z, w_1, w_2) = ((a - 2)b^{-1} - 2, b + 1, b + 1),$$

we can rewrite (4.13) to yield divergent solutions:

$$\begin{aligned}
w_i'/w_i &= [w_i - (b+1)]/(a - bz - w_i) \\
&\approx [w_i/(b+1)] - 1.
\end{aligned}$$

For reference, this *verified symmetric equilibrium* has the constant-value property:

$$w_i \equiv (b+1) \equiv w_j \tag{18}$$

as well as a linear 'current-value' value function:

$$U(z) = z/(b+1) + K . \tag{19}$$

For the verified equilibrium, Figure 4 displays the 'current value' value function, the 'asset value' for the state variable and the control variable.

---

[8] $D < 0$ near the steady state of the 'verified equilibrium' where the term in brackets, [.], is near zero but the term $w_i w_j$ is close to the value, $(b+1)^2 > 0$. Therefore, for asymmetric equilibriua, interior paths exist only near the steady state. There we have,

$(d/dz)log(w_i/w_j) \approx (w_i - w_j)/D$

which has a sign opposite to the expression, $w_i - w_j$, near the verified equilibrium $(b + 1, b + 1)$.

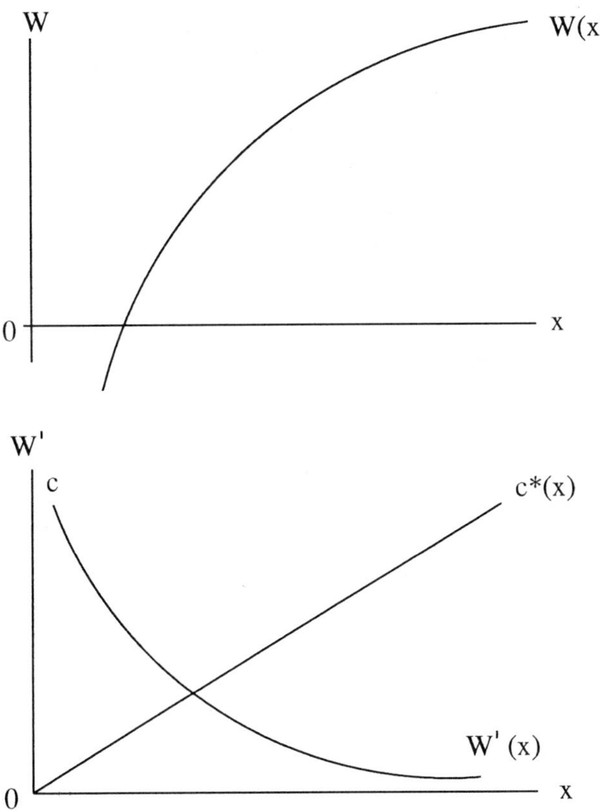

Figure 4: Value Function. Asset price and Consumption

Figure 5 provides the phase portraits of solution paths for various equilibria. The paths for symmetric equilibria are shown in the symmetric plane in both the transformed and the original versions.

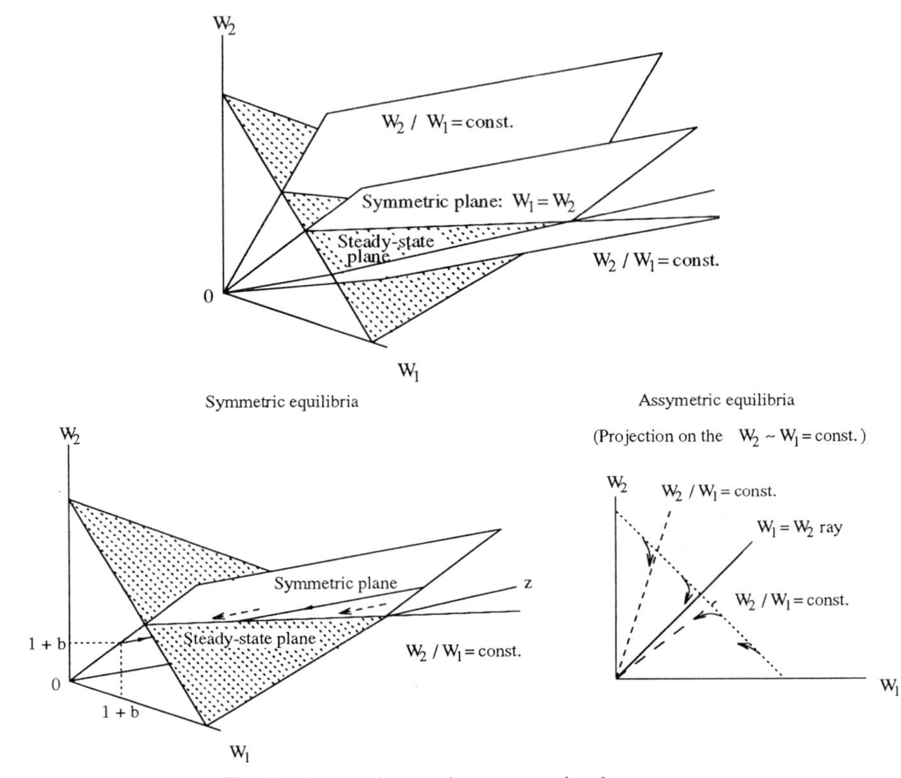

Figure 5: Global analysis of equilibrium paths

For the symmetric paths, note that at any symmetric steady state [for the *time-paths*: $\mathbf{x(t)}, \mathbf{c(t)}$ or $\mathbf{z(t)}, \mathbf{w(t)}$], the Lipschitz condition is satisfied for that differential equation in the *state-control space* (with the solution $c_i^*(x)$ or $w_i^*(z)$). Thus, the existence and uniqueness theorems for the solution of a differential equation may be used to extend the paths from the steady state positions, both (i) to lower values of the state variable, toward the singularity at the source: $x = 0$ (or $z = -\infty$), and (ii) to higher values of the state variables, until such a path does not correspond to an equilibrium any more.

The configuration in the phase diagram in original variables indicates clearly that all such extensions must go back to a singular point at $\mathbf{x = 0}$. In contrast, by (4.13), at some large value of z the nonlinear solution path may reach an impasse when $a - bz - w_i = o$, causing the absolute value of the growth rate for w_i to become infinity. Splicing now becomes necessary.

4.3   Splicing

Splicing introduces a 'decomposition' of the state space into (i) an open interval, containing the steady state and (ii) its complement, for higher values of the state variable (i. e., z, or x). For the ease of interpretation, we work with the original state space, $\mathbf{X}$, which is partitioned at some 'splicing point' x_p, between the steady state point along that solution path and some point where the latter path can no longer correspond to any equilibrium, because either (a) dc_i/dx is no longer finite, or (b) the implied function $f_i(.)$ may fail the sufficient criterion for equilibrium. The choice of $\mathbf{x_p}$ is arbitrary, introducing another level of multiplicity.

Splicing means that over the set $\mathbf{X_a} = \{x \in \mathbf{X} : \mathbf{x} \geq \mathbf{x_p}\}$, each player in an equilibrium faces a conditional variational problem of the variable endpoint type. There is the terminal payoff, $e^{-t_p} W(x_p)$, where $W(x_p)$ is decided by solving the variational problem over $\mathbf{X_b} = \mathbf{X} \backslash \mathbf{X_a}$, and the value of t_p is decided as a transversality condition.

Over X_a, the 'spliced on' equilibrium strategy may be the 'extreme strategy', $c_i \equiv \max C$. In fact, the value function $V(x_0)$ over $\mathbf{X_a}$ is

$$(1 - e^{-t_p}) \log(\max \mathbf{C}) + e^{-t_p} W(x_p),$$

and t_p is decided implicitly by the condition

$$\xi(t_p) = x_p,$$

where $\xi(.)$ is the solution of the differential equation

$$dx/dt = x(a - b \log x) - 2(\max \mathbf{C}), \mathbf{x(0)} = \mathbf{x_0} \,.$$

We note though that for certain paths with a steady state larger than that of the *continuously differentiable equilibrium*, then it may also be the latter. In such a case, the arbitrariness in the choice of the 'spliced on' strategy constitutes a further level of multiplicity.

At x_p, the discontinuity of the first kind for c_i^* satisfies the restriction:

$$
\begin{aligned}
\lim_{x \uparrow x_p} c_i^*(x) \quad &< \quad \lim_{x \downarrow x_p} c_i^*(x) \\
&= \quad \max \mathbf{C},
\end{aligned}
$$

which means, by symmetry, an induced discontinuity for $f_i(x)$ via the discontinuity in $c_j^*(x)$. This would cause an upward shift in the 'desired value' for $c_i^*(x)$ to some $c_d \geq \max \mathbf{C}$. Since,

$$
c_i^*(x) = \min\{c_d, \max \mathbf{C}\},
$$

this makes the condition

$$
c_d \leq \max \mathbf{C},
$$

binding. Thus, by the maximum principle over $\mathbf{X_b}$,

$$
\begin{aligned}
\lim_{x \uparrow x_p} \partial V/\partial x \quad &= \quad \lim_{x \uparrow x_p} c_i^*(x) \\
&> \quad 1/\max \mathbf{C} \\
&\geq \quad 1/c_d \\
&= \quad \lim_{x \downarrow x_p} \partial V/\partial x \, ,
\end{aligned}
$$

which implies a kink in the value function $W(x)$.

4.4 Additional Remarks

As discussed in Clemhout and Wan (1993), the presence of a linear verified equilibrium is convenient but not crucial for our result. Note that technically, we have derived the multiplicity of solution in a manner different from the past. The key step is in the derivation of (4.11). Previously, in Clemhout and Wan (1989, 1993), we have followed Uzawa and Koopmans. After substituting the adjoint variable with the control variable in the adjoint equation (by using the maximum principle), we divide the time derivative of the state variable into the time derivative of the optimal control . This approach cannot be generalized to the multi-state variable framework, since in general, the division operation between two vectors is undefined. The method we used here generalizes naturally.

5   Open Issues

Three open issues are not yet entirely resolved at this point.

First, one of Romer's concerns is the accelerating growth of income per capita. In contrast, the fish war example in the last section evolves under some 'limit for growth'[9] and its equilibrium time path converges toward some steady state. Regardless whether unbounded expansion is cosmologically possible in real life, it is conceptually desirable to extend our analysis to admit a positive and increasing rate of growth over the entire horizon. While the reasons causing multiplicity remain, there are technical matters requiring further study.

Second, when there is an 'internal effect', the equilibrium policy becomes a function of two or more variables. Its characterization involves partial derivatives of both the value function and the policy function. Any proof of multiplicity must involve partial differential equations and hence additional analytical challenge. However, the reasons causing multiplicity remain. More study is needed.

Third, it is often believed that when the number of players becomes large, the 'open-loop' solution (and perhaps the procedure used by Romer, Benhabib and Perli, Xie, etc.) approaches the feedback solution. Such a claim must be checked with care.

To do full justice, each of these three issues takes more space and time than is possible here. Nonetheless, we shall sketch the main questions below and report our preliminary findings such as they are. Our discussion is necessarily somewhat tentative in this section.

5.1 Perpetual Expansion

We must assume in this connection that:

(i) In contrast to our fish war example, function f is unbounded and perhaps convex as well. This calls for a somewhat more complicated sufficient condition for an equilibrium path. Now the attainable range of x is unbounded, one must show that the equilibrium value of the attainable x is exponentially bounded at the time preference rate. This is satisfied if the product of the growth rate of attainable x and the elasticity of marginal utility are bounded by the time preference rate. Following Xie (1991), an illustrative example is:

$$r = 1, \qquad N = 2, \quad u(c) = \sqrt{c}, \quad f(k) = 3x(1 - e^{-x}).$$

But this matter has to be pursued more thoroughly elsewhere.

(ii) The control space $\mathbf{C}$ must be either unbounded above at each state, or with an upper bound which is state-dependent. This may call for a different sort of splicing strategy, perhaps using the saddle point path like OF'E' A in Figure 1, Section 3.

---

[9]Specified both by the fact that $f(x) \leq 0$ for $x \geq e^{a/b}$, and $c \leq \max \mathbf{C}$.

5.2 Internal Effect of Knowledge Capital

We now illustrate the situation by specializing (4.5) as follows:

$$f(x_i, .x_j) = x_i[a - (b + \theta)\log x_i + bx_j], \ i, j = 1, 2, \ i \neq j.$$

Thus the state equations:

$$dx_i/dt = f(x_i, .x_j) - c_i, \quad dx_j/dt = f(x_i, .x_j) - c_j . \tag{20}$$

Further, we continue to specify that $u_i = \log c_i$ and $r = 1$. This is clearly a variation of the simple fish war model we studied earlier. At the same time, the strategical complementarity between the two players is less direct. No matter how high player j sets the level of c_j, it is not expected that player i would have to follow suit at the same instant. Instead, the path of the verified equilibrium may be used for splicing for some of the other strategies, analogous to OF'E'A in Figure 1 before.

5.2.1 General Considerations, A First Pass

Let the maximum attainable payoff for player $i = 1, 2$ be

$$V(x_{0i}, x_{0j}, t_0; c_j^*) = e^{-t_0}W(x_{0i}, x_{0j}; c_j^*), \ j = 2, 1 ,$$

in a symmetric equilibrium $c^* = (c^*(x_i, x_j), c^*(x_j, x_i))$. As before, again we change variables by setting:

$$z_i = \log x_i; \ w = c_i/x_i,$$

in order to obtain:

$$\begin{aligned} e^{-t_0}W(x_{0i}, x_{0j}; c_j^*) &= e^{-t_0}W(e^{z_{0i}}, e^{z_{oj}}; w_j^* x_j) \\ &= e^{-t_0}U(z_{0i}, z_{0j}; w_j^*), \end{aligned}$$

say, with the state equations taking the form:

$$dz_i/dt = a - (b + \theta)z_i + bz_j - w_i, \qquad i, j = 1, 2, \quad i \neq j,$$

and the payoff taking the form

$$v_i = \int (z_i + \log w_i)e^{-t}dt .$$

Here, w_i^* must satisfy the Hamilton-Jacobi equation:

$$-e^{-t}U \quad = \quad \max_{w_i}\{e^{-t}(z_i + \log w_i) + e^{-t}(U_i)[a - (b + \theta)z_i + bz_j] - w_i)$$

$$+e^{-t}(U_j)([a - (b + \theta)z_j + bz_i] - w_j^*(z_j, z_i))\} \, ,$$

where $U_i = \partial W/\partial x_i \quad$ and $\quad U_j = \partial W/\partial x_j$.

Again we have:

$$\begin{aligned} U_i &= 1/w_i \\ &= 1/w^*(z_i, z_j) \end{aligned} \tag{21}$$

Dividing through by e^{-t}, taking partial derivative with respect to z_i and using the facts:

$$U_{ji} = U_{ij},$$

and

$$dU_i/dt = (U_{ii})dz_i/dt + (U_{ij})dz_j/dt \, , \tag{22}$$

we have,

$$dU_i/dt = (1 + b + \theta)U_i - (b + \partial w_j^*/\partial z_i)U_j - 1 \, . \tag{23}$$

Likewise, we can also get

$$dU_j/dt = -bU_i + (1 + b + \theta + \partial w_j^*/\partial z_j)U_j \tag{24}$$

5.2.2 A Verified Equilibrium

This model is a variation of the fish war model of the last section. That model has a verified equilibrium with a linear strategy, with $w_i = constant$. Hence, a natural trial solution is again

$$\partial w_i^*/\partial z_i = 0 = \partial w_j^*/\partial x_j.$$

This is consistent with:

$$\begin{aligned} U_i &\equiv (1 + b + \theta)/(1 + \theta)(1 + 2b + \theta) = U_i^v; \\ U_j &\equiv b/(1 + \theta)(1 + 2b + \theta) = U_j^v, \end{aligned}$$

say. By (5.2), for the verified equilibrium,

$$w_i^* \equiv (1 + \theta)(1 + 2b + \theta)/(1 + b + \theta) \equiv w_j^* \, .$$

Substituting back into the Hamilton-Jacobi equation, we find:

$$U = \{[a - (U_i^v) - 1][U_i^v + U_j^v] - \log U_i^v\} + U_i^v z_i + U_j^v z_j,$$

which is a linear function of z_i and z_j. One can substitute back to obtain W which is loglinear in x_i and x_j.

Although we are solving for a symmetric equilibrium in a symmetric game, the equilibrium play need not start from a symmetric state. From the game theoretic point of view, W and c^* (or U and w^*) must be solved over the entire state space and not just for the symmetric states.

5.2.3   General Considerations, Once More

Generally speaking, in any equilibrium and at any state, symmetric or asymmetric,

$$dz_1/dt = a - (b + \theta)z_1 + bz_2 - 1/U_i(z_1, z_2), \tag{25}$$

$$dz_2/dt = a + bz_1 - (b + \theta)z_2 - 1/U_i(z_2, z_1), \tag{26}$$

where $U_i(z_1, z_2)$ denotes $\partial U(z_i, z_j)/\partial z_i$, i.e., the partial derivative with respect to the first argument.

As in the derivation of (4.11), we would like to characterize the policy function, or the value function, with no explicit reference to time. For this, we now collect all the above information together to transform the equations (5.4-5). Using (5.6-7) to eliminate dz_1/dt and dz_2/dt in (5.3) and using that result to eliminate any explicit reference of t in the left-hand side of (5.4-5), and using (5.2) to eliminate any reference of w in the right-hand side of (5.4-5), we end up with two equations involving only the first and second order partial derivatives of the unknown value function U.

This is then the analog of (4.13a) for the two independent variable case.

What is special in this problem is that for asymmetric states, the partial derivatives of the two functions f and c^* have to be evaluated simultaneously at the two points (x_i, x_j) and (x_j, x_i), or (z_i, z_j) and (z_j, z_i).

Figure 6 presents the evolution under the verified equilibrium from both a symmetric and an asymmetric initial state toward the steady state, (z_∞^v, z_∞^v), via symmetric and asymmetric equilibrium plays, respectively. We display the value function for reference. Next we also show how a spliced alternative equilibrium should be, if such a one exists.

As in the last section, we now focus attention to the steady state of an equilibrium, verified or otherwise, which is also a symmetric state.

At any steady state,

$$dz_1/dt = 0 = dz_2/dt,$$

so that, after denoting U_i as the partial derivative of U with respect to its first argument, we have

$$
\begin{aligned}
(b + \theta)z_1 \quad - \quad bz_2 &= a - w^*(z_1, z_2) &= a - 1/U_i(z_1, z_2) \\
-bz_1 \quad + \quad (b + \theta)z_2 &= a - w^*(z_2, z_1) &= a - 1/U_i(z_2, z_1)\,.
\end{aligned}
$$

At any *symmetric* steady state, where

$$z_1 = z = z_2,$$

we obtain by rearrangement:

$$U_1(z_1, z_2) = a - \theta z = U_1(z_2, z_1). \quad \text{[An analog to (4.14)]} \tag{27}$$

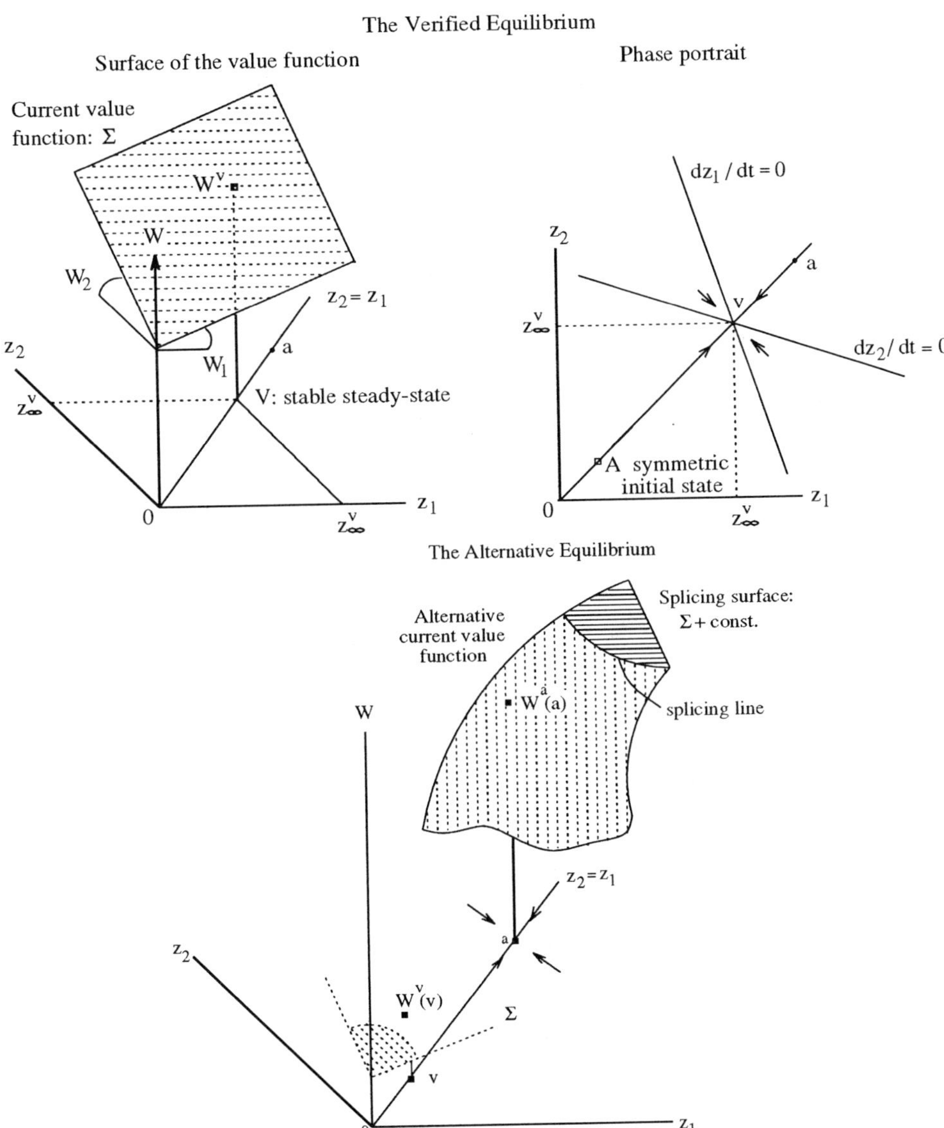

Figure 6: The verified and alternative equilibria

This is true at $v = (z_\infty^v, z_\infty^v)$ under the verified equilibrium as well as at steady state a, under an alternative equilibrium, as displayed in Figure 6. For the former, that corresponds to:

$$
\begin{aligned}
w^*(z_1, z_2) &\equiv w^*(z_\infty^v, z_\infty^v) \\
&= (1 + \theta)(1 + 2b + \theta)/(1 + b + \theta) ,
\end{aligned}
$$

where $z_\infty^v = (a - w*)/(2b + \theta)$ is the unique value for solving the above equations.

Now for the alternative equilibrium, the matter is more complex. In principle, once when we have solved that partial differential equation we mentioned after (5.7), we get a new value function U, as shown in Figure 6. Splicing with the verified equilibrium is also demonstrated graphically. Yet, as we shall see, some issues remain open.

In practice, one may attempt to derive the new surface 'by quadrature', up to an additive scaler, as follows:

(i) initiate with both (a) the initial states, $z_1 = z_2 = z_\infty^a$ as chosen, and (b) the initial slopes of U, $U_1(z_1^a, z_2^a)$ and $U_1(z_2^a, z_1^a)$ from (5.8),.....

(ii) select an increment, Δt, and all ratios of $\Delta z_2/\Delta z_1$ as well,

(iii) use the equations (5.2-7), starting from the new steady state, to track backward from the steady state as a 'sink', and

(iv) conduct splicing when needed in a manner analogous to what is done in the last section.

The remaining issue is, that for the alternative equilibrium, those 'interacting terms' ($\partial w_j^*/\partial z_i$ and $\partial w_j^*/\partial z_j$) which distinguish the feedback solution (from the open loop) do not vanish from (5.4 - 5) even at the steady state. The use of (5.8) reduces these two equations at the steady state into two algebraic equations for the three variables, U_j, $\partial w_j^*/\partial z_i$ and $\partial w_j^*/\partial z_j$, thus causing one degree of (new) indeterminacy.

In contrast, for the verified equilibrium, with the vanishing interacting terms, these two equations are used to decide both and EAU, and the value of the latter is used to determine the state variable at the steady state. Here the value of z_∞^a is imposed by our arbitrary choice.

Since this is a study of multiplicity of solutions, one can have 'two schools of thought' at this point. Those who prefer to use this 'new' indeterminacy as a *selection principle* may conclude that the verified equilibrium should be the unique equilibrium worthy of that appellation. Those like us who believe that mutual expectations matter, the perception by player i about player j's response to the state (for that is precise the meaning of $\partial w_j^*/\partial z_i$ and $\partial w_j^*/\partial z_j$) should not be surprised at all that such perception affects both the 'asset price(s)' of the state variable(s), U_j (and U_i), and hence the value of the state at the steady state as well. One more degree of indeterminacy would only bring such indeterminacy in line with the dimension of the state space. Thus, assigning arbitrary pairs of steady state values of $\partial w_j^* / \partial z_i$ and $\partial w_j^* / \partial z_j$ will yield commensurate pairs of steady state values of U_i and U_j by (5.4 - 5), and the former has to agree with the value of z_∞^v by (5.8).

Technically, the possible need for splicing, etc. remains to be explored in more detail. Hence, some future study is required.

5.3 The Behavior of the Model as the Number of Players Becomes Large

We do not concur with the often expressed view that, as the number of players becomes large, the open-loop, and the Romer approach will converge to the limit of the feedback equilibrium approach. Our results in the exhaustible resource game (formally a special case of the fish war) are against that: the open loop solution will always be Pareto efficient, and so will be its limit; the feedback solution is inefficient, becoming more and more so as the number of players becomes large [Cf. Clemhout and Wan (1989)].

At the same time, the above discussion points to the fact that one cannot meaningfully increase the number of players without specifying the particular manner under which the structure of the model (e.g., the production technology where externality occurs) changes as the number of players approaches infinity.

6 Concluding Observations

At this point, some economists work with discrete time models alone, others with only continuous time models, and still others, like us, with both. The issue is partly a matter of esthetics, in particular, whether one is ready to adopt certain sufficient conditions for the existence of solutions in continuous time models. Partly, it is also a matter of analytic convenience, since workers with continuous time models have access to the results of differential equations which have been well developed over the centuries. Our stand is eclectic. To reach the widest audience, one would like to develop results cherished by all, i.e., in terms of discrete time models. But to develop analytic intuition, there is an advantage to tackling the problem first in continuous time, and then trying to prove it in the discrete time version as well. This is especially true for problems with important implications in a substantive field, like economics. The game-theoretic reformulation of endogenous growth is precisely such a case.

REFERENCES

[1] J. Benhabib and R. Radner, The joint exploitation of a productive asset: a game-theoretic approach, *Economic Theory* **2** (1991), 155-190.

[2] J. Benhabib and R. Perli, Uniqueness and indeterminacy: transitional dynamics in a model of endogenous growth, *Journal of Economic Theory*, **63**(1994), 113-142.

[3] S. Clemhout and H. Y. Wan, Jr., Dynamic common property resources and environmental problems, *Journal of Optimization, Theory and Applications* **46** (1985), 471-481.

[4] S. Clemhout and H. Y. Wan, Jr., On games of cake-eating, in *Dynamic Policy Games in Economics*, (F. van der Ploeg and A. de Zeeuw eds.), New York: North Holland, 1989.

[5] S. Clemhout and H. Y. Wan, Jr., The non-uniqueness of Markovian strategy equilibrium: the case of continuous time models for non-renewable resources, T. Basar and A. Haurie eds., Advances in Dynamic Games and Applications, Boston: Birkhäuser, 1993.

[6] J. B. DeLong and L. Summers, Equipment investment and economic growth, Quarterly Journal of Economics, **106,** 445-502, 1991.

[7] D. Fudenberg and J. Tirole, *Game Theory,* Cambridge, MA: MIT Press, 1991.

[8] L. P. Hansen et al., Linear-quadratic duopoly models of resource depletion, in *Energy, Foresight and Strategy,* T. Sargent ed., Washington, D.C.: Resource for the Future, 1985.

[9] M. L. Lau and H. Wan, Jr., On the mechanism of catching-up, *European Economic Review,* **38,** 952-963, 1993.

[10] D. Levhari and L. J. Mirman, The great fish war: an example using the Cournot-Nash solution, *Bell Journal of Economics,* **11,** 322-334, 1980.

[11] R. Lucas, Jr., *Models of Business Cycles,* New York: Blackwell, 1987.

[12] R. Lucas, Jr., On the mechanism of economic development, *Journal of Monetary Economics,* **22,** 3-42, 1988.

[13] R. Lucas, Jr., Making a miracle, *Econometrica,* **61,**251-272, 1993.

[14] H. Pack, Endogenous growth theory: intellectual appeal and empirical shortcomings, *Journal of Economic Perspectives,* **8,** 55-72, 1994.

[15] P. M. Romer, Increasing returns and long-run growth, *Journal of Political Economy,* **94**, 1002-1037, 1986.

[16] K. Shell, Toward a theory of inventive activity and capital accumulation, *American Economic Review,* **56**, 62-68, 1966.

[17] K. Shimomura, The feedback equilibria of a differential game of capitalism, *Journal of Economic Dynamics and Control,* **15**, 317-338, 1991

[18] R. M. Solow, Perspectives on growth theory, *Journal of Economic Perspectives,* **8**, 45-54, 1994.

[19] N. L. Stokey, Learning by doing and the introduction of new goods, *Journal of Political Economy,* **96**, 701-717, 1988.

[20] D. Topkis, Equilibrium points in non-zero-sum, n-person submodular games, *SIAM Journal of Control and Optimization,* **17**, 773-787, 1979.

[21] D. Xie, Increasing returns and increasing rate of growth, *Journal of Political Economy,* **99**, 429-435, 1991.

[22] D. Xie, Divergence in economic performance: transitional dynamics with multiple equilibria, *Journal of Economic Theory,* **63**, 97-112, 1994.

[23] A. Young, Learning by doing and the dynamic effects of international trade, *Quarterly Journal of Economics,* **106**, 369-406, 1991.

Searching for Degenerate Dynamics in Animal Conflict Game Models involving Sexual Reproduction

W. G. S. Hines

Mathematics and Statistics, University of Guelph
Guelph, Ontario Canada, N1G 2W1

Abstract

The initial appeal of using conventional game theory to model and to analyze conflicts in biological populations is somewhat muted by the realization that most such populations reproduce sexually. By mixing parental genetic material, sexual reproduction introduces complications about the transmissions of strategies, if genetically determined, not present in standard game theoretic models of populations. If not addressed, these complications leave the relevance of game theoretic analyses to biological populations in considerable question.

In this paper, we discuss recent results concerning the effects of sexual reproduction, with its potential for mixing characteristics of parents in the offspring. To do so, we must consider the nature of the set of possible frequency distributions of genes for a given population, the nature of the mapping of those distributions into the distributions of competitive strategies of the corresponding populations, and the nature of evolutionarily stable populations. Central to this study are both equations describing the evolutionary effects of contests between individuals and a suitable measure of distance in gene-frequency space. (An added complication is that that measure proves to be discontinuous as genes disappear from the population, an often-present possibility.) Insights into the analysis and the nature of possible equilibrium populations are illustrated by a numeric example and some corresponding graphs.

1 Introduction

Can biologists safely rely on game theoretic analyses of conflicts in biological populations? In particular, given the central role of genetics in evolution, does the absence of similar mechanisms in standard game theory models invalidate such analyses? Unlike some characteristics such as adaptation, potential information utilization, population structure, or finiteness of population size, which might all be considered reasonably excluded from the analyses of at least some biological populations, sexual reproduction occurs at least occasionally in the life cycles of almost all biological species. As a result, like does not necessarily breed like, but only similar - and not even invariably that. The questions posed above are therefore both challenging and important. They are also far from being fully resolved.

This does not mean that the potential value of game-theoretic analyses to biology is generally unrecognized, or that no significant progress has been made. Some theoretical biologists have enthusiastically embraced and developed the idea of modelling the conflicts which occur in real biological populations by variants of the formalized competitions of game theory. Other biologists, concerned about the many complexities of the populations they study and about the complications introduced by sexual reproduction in particular, have been reluctant to trust analyses which ignore those complications, however attractive the results.

It can be argued that, understandably, some of those not ready to accept the game theoretic approach to biology have paid little continuing attention to its development and increased relevance. For example, one special class of game theoretic models that does incorporate an important and common set of assumptions about the genetics involved assumes the existence of multiple relevant loci, with additive effects both within and among loci. (The term 'locus' refers to a location in an individual's chromosome at which genetic coding for a particular gene exists. The coding at that locus which is received from a given parent is called an 'allele'.) Hines and Turelli [13] have recently demonstrated that for this class, the results of the (biologically-naive) game theoretic analysis are in fact quite relevant to fairly detailed models of sexually reproducing biological populations. While important, however, this one class of genetic models is far from the only one of interest to the biologist, who is also concerned about the possible effects of non-additivity, such as occurs in the case of dominant genes.

This paper discusses recent work on that case reported in such papers as Hines [8][9] and Cressman, Hofbauer and Hines [5], for the case of a single locus at which individuals might differ genetically. In conventional game theory models, it is natural to characterize individuals in the population by those strategies which they possess and which directly affect their fortunes in competition. In contrast, the more relevant characterization of sexually reproducing individuals is by the alleles that they carry, which are assumed to determine the strategies they use and the consequences of those strategies they experience in competitions. If individuals in the population mate at random, say with frequencies determined by their relative numbers and prior competitive successes, we can replace consideration of the detailed composition of the population of individuals by consideration on the comparatively simple frequency of the various alleles present in the gene pool from which the population was drawn.

The space of vectors of possible allelic frequencies contains some important structure. One obvious component is the set of surfaces on each of which all frequency vectors of alleles give rise to populations with a common mean strategy. Another is the set of frequency vectors for which the mapping from allelic frequency vector to mean strategy is singular in the following sense: for a population with such an allelic frequency vector, the directions in which its mean strategy can be moved by small perturbations of the frequency vector are restricted, which can result in equilibria not anticipated by the standard game-theoretic formulation. To explore the question of whether populations with initial frequencies sufficiently close to equilibria in the set of singular frequencies or close to surfaces corresponding to certain equilibrium mean strategies will converge to these equilibria, we will need an appropriate measure of distance, and will need to cope with the fact that it behaves discontinuously as any given allele moves from being rare in the gene pool to being totally absent.

2 Background

In pioneering work, Maynard Smith[15] developed the idea of the Evolutionarily Stable Strategy, commonly described as a competitive strategy with the property that its use by virtually all of a given biological population would place the users of a common deviant strategy at a strict selective disadvantage. For some competitive situations, such strategies in near-universal use do indeed confer considerable stability upon the population, not only against a single invasion by a group of identical individuals but also against a continuing series of invasions by a variety of individuals. In other situations, the protection provided is less complete: the presence of more than one type of invader can result in an evolution of the population to a point of equilibrium in which a diversity of equally rewarding strategies persists. Fortunately, the analysis of this more complex situation proved tractable, even after the introduction of some simple genetic models to explore the effects of sexual reproduction. Hines [7] provides an overview of much of that analysis.

Early studies of special cases of the sexual-reproduction model (Hines [6], Treisman [23], Maynard Smith [16]) focused on the simplest possible situations genetically - a single locus, a pair of possible alleles (or types of gene at that locus), a simple rule determining the strategy corresponding to each possible pair of alleles (or genotype) at the locus, and a simple model of a contest in which the strategy was to be employed. That work established for the cases considered that sexual populations would tend to evolve so that the average strategy of the population approached the ESS. It was possible however, that the population might never attain the ESS, for example because of an absence of any possible genotype which had the ESS as its strategy, or because no mix of the various genotypes existed which both had the ESS as its mean strategy and persisted from generation to generation as sexual reproduction continually reshuffled the various allele pairs present. (A similar but simpler situation occurred for the asexual case as well.) Hines and Bishop [10][11][12] and Thomas [20][21][22] later explored more general single-locus multi-allele cases. This work confirmed that population compositions could exist which were stable from generation to generation and which had mean strategies which were locally but not globally closest to the ESS: the continuous or continuing small changes in allele frequencies possible under gradually evolving population composition models could not bring the resulting mean strategy closer to the ESS.

Focusing on the use of statistical summaries (means, covariance matrices) of populations, the method used by Hines and Bishop was somewhat less conventional than that employed by Thomas. As such, the mean-covariance approach gave rise both to improved intuitive understandings of the processes involved and to concerns about limitations of the method. Lessard, for example, has commented in a personal communication that for a fully rigorous treatment of the situation using the mean-covariance approach, the evolution of the covariance matrix needed to be studied as well as that of the mean. While the intuitive nature of the mean-covariance approach was useful and welcome, a more rigorous approach remained clearly desirable.

Rigorous approaches do exist in various important special cases considered by Cressman (and described in Cressman, Hofbauer and Hines [5]) and by Hofbauer and Sigmund [14]. This present paper outlines and expands on another, more general, approach, which is as yet incomplete. Taken together, the various results obtained suggest, but do not yet establish that, under the conditions to be described, stability does occur in all cases.

3   The analysis

3.1   The contests

Simple animals exist in a featureless environment, engaging in contests with randomly selected opponents drawn from an infinite homogeneous population. At the end of each generation, individuals contribute to a common gene pool from which the offspring forming the next generation are drawn. In addition to a common level of contribution, individuals further contribute to the gene pool in amounts proportional to their winnings or losses, so that lineages using persistently superior strategies will become increasingly represented - assuming such strategies to exist.

Assume a finite number n_s of possible choices of behaviour, and let $A = (a_{ij})$, a payoff matrix with the entry a_{ij} being the expected return resulting if choice i is used against an opponent making choice j. Individual strategies can be summarized as n_s- dimensional probability (column) vectors denoted here as r, s,... . Under the assumption that choices are made independently by the two players, the return to an s-user of a contest against an r- using opponent is s · A r. The classic conditions obtained by Maynard Smith [1974] for a given strategy s to be an ESS are that for any other strategy r differing from s,

- s · A s $\geq$ r · A s, and

- for r with s · A s $=$ r · A s,

$$s \cdot As > r \cdot Ar.$$

(Maynard Smith [15]). We will use the notation s* to indicate a strategy which is an ESS, and if the mean strategy of a given population of dissimilar individuals is identical to such a strategy s*, we will denote that mean strategy by μ^*. We will assume without further comment that all components of μ^* are positive. (See Hines [7] for a discussion of various implications of that assumption.)

The ESS conditions, their derivations and their implications have been discussed in numerous reviews and other articles and in several texts; the interested reader is directed to various surveys and reviews such as Maynard Smith [17][18], Riechert and Hammerstein [19], Hines [7], or Hofbauer and Sigmund [14].

3.2   The anatomy of the allelic frequency simplex

In the original ESS formulation, an individual could be considered fully described by its strategy. With sexual reproduction, however, it is the genetic composition of the individual that matters, with this composition affecting both its strategy and its possible genetic contributions to the gene pool of the next generation. For definiteness, the strategy used by an individual carrying the alleles u and v at the locus being considered will be denoted by s(u, v) ($=$ s(v, u)). Let the frequency of allele u in the gene pool from which a population is drawn be denoted by F(u). If individuals are produced by the random pairing of alleles from that pool, then the probability of a given pair of alleles, say u and v, being drawn in a particular order is F(u) F(v) (so that the probability ignoring order is then double that product if u and v differ). The mean strategy present in the resulting population is then

$$\mu(F) = \sum_u \sum_v s(u, v) F(u) F(v)$$

while the mean strategy associated with an individual who carries allele u as, say, its 'first-drawn' allele is

$$\mu(F|u) = \sum_v s(u,v)F(v)$$

In the absence of variability in the mean strategies associated with the various alleles, all alleles result in equal average returns from contests and, in an infinite population model, this implies that no evolution in allelic frequencies will occur. A measure of the relevant variability (the 'additive genetic variability') in the population is given by the covariance matrix determined by F, C(F), where

$$C(F) = \sum_u \left(\mu(F|u) - \mu(F)\right)\left(\mu(F|u) - \mu(F)\right)^T F(u)$$

and calculations show that, if the subscript t denotes a quantity defined for generation t,

$$\frac{d\mu_t}{dt}\alpha C(F_t)A\left(\mu_t - \mu*\right)$$

where μ^* denotes the ESS for the payoff matrix A. This immediately implies that if $C(F_t)$ is not a zero matrix, if μ_t differs from μ^*, and if A $(\mu_t - \mu^*)$ is not a null eigenvector of $C(F_t)$, then the mean strategy of the population will evolve and, a fortiori, the composition of the population will also evolve. (The second ESS condition can be used to show that if μ_t differs from μ^*, then the vector A $(\mu_t - \mu^*)$ is not the zero vector.)

The function μ: F $\to \mu(F)$ maps the probability simplex Δ_a of all possible n_a-dimensional allelic frequency vectors to the probability simplex Δ_s of all possible n_s-dimensional strategy vectors. Even if $n_a = n_s$, the mapping is not necessarily 1-1 or onto. (In particular, if $n_a > n_s$, then necessarily for each allele frequency F with more than n_s positive components, there exist perturbations to it, say to F + ϵ x, where x is of unit norm and with components summing to unity, such that $\mu(F + \epsilon\ x) = \mu(F)$, or such that $\mu(F + \epsilon\ x) = \mu(F - \epsilon\ x) \neq \mu(F)$.) This mapping is investigated in some detail in Hines [8] as a preliminary to a consideration of the possible evolution of the allelic frequency in Hines [9]. In that first paper, two types of surface in the allelic simplex are considered: those on which the resulting mean strategy is constant and those on which the mapping μ is singular. Here, singularity is considered to occur if the dimension of the tangent space of μ for a given allelic frequency F is less than its maximum possible value (i.e., Min(n_a, n_s) - 1, the "-1" term resulting since all allelic frequencies and all strategy vectors each have components which necessarily sum to unity). Together, those two types of surfaces provide important information about the mechanisms linking the various possible genetic compositions of biological populations and the strategies present in the populations, and are central to an understanding of possible evolutions of those populations. In order to determine if evolution will work so as to take populations towards or away from those surfaces, we need an appropriate measure of distance.

3.3 Measuring distances in the allelic frequency simplex

A variant of the Euclidean norm - called the Shahshahani metric [24] by Akin [1][2][3] and others, although in prior use by Antonelli and various coworkers (e.g. [4]), proves appropriate after some modification. The square of the original Shahshahani metric defined for measuring the difference between two frequency vectors, say F' and F'' using weights determined by a third frequency vector, say F is given by

$$\|F','' \|_F^2 = \sum_u \frac{(F'(u) - F''(u))^2}{F(u)}$$

Our interest is in measuring the distance from a particular frequency F to nearby frequencies F', even though some such frequencies might give positive weight to some alleles u for which F(u) is zero. Consider the behaviour of that expression if F(u) in the denominator is replaced by $F^\epsilon(u)$ for some nearby frequency vector F^ϵ and if F''(u) is then replaced by F(u) in the numerator. Calculation shows that we can write the expression in three parts: one which diverges to infinity as F^ϵ tends to F, another which tends to zero, and a third component, $G(F', F)^2$, defined by

$$G(F', F)^2 = \sum_{u, F(u) > 0} \frac{(F'(u) - F(u))^2}{F(u)} + 2 \sum_{u, F(u) = 0} F'(u).$$

This last expression serves our needs as the square of a suitable measure of distance from a frequency vector to other frequency vectors (but not to more general vectors since it can then be negative). The distance from a set of frequency vectors to, say, an initial frequency vector can then be taken to be the minimum of the distances from individual frequency vectors in that set, or rather from that subset of those vectors which do not assign zero frequency to any alleles which are given positive frequencies by both other frequency vectors in the set and by the initial frequency vector. (This device avoids problems with the discontinuity associated with the measure as alleles move from being rare in a population to being totally absent.) With that measure of set-to-point distance, we can then explore questions of convergence to local equilibria under the conflict-induced population dynamics, even when the mapping μ: F $\to \mu$(F) is singular.

3.4   Equilibria and dynamics

The expression for the change in mean strategy indicates that the population
cannot be in equilibrium unless

- $\mu_t = \mu^*$ (i.e., the population has an ESS as its mean strategy),

- $C(F_t)$ is zero (i.e., the population lacks the variability necessary for
  evolution to occur), or

- $A\,(\mu_t - \mu^*)$ is different from zero and is a null eigenvector of $C(F_t)$,
  which occurs generically for isolated frequency vectors.

These possibilities include the one expected by an analysis of animal
conflicts which ignores the complications introduced by sexual reproduction,
the self-evident statement that a population unable to evolve will not evolve,
and the possibility anticipated by the sceptical biologist that genetic considerations might render a population which may be capable of at least some
evolution nonetheless incapable of evolution in directions which bring it closer
to the state predicted by game theory analyses which ignore sexual reproduction. (The last possibility includes the previous one as a limiting case.)
As noted above, all of these possibilities had been recognized previously by
the various authors who have addressed the problem.

Will these various possible equilibria prove to be stable? An explicit
expression for the rate of change of the above measure of squared distance
from various equilibria proves to be tractable and instructive. The following
conclusions can be reached from inspection of that measure and its properties, combined with information about the anatomy of the allele simplex (as
described in Hines [8][9]):

- If none of the set of all allelic frequencies which give rise to the ESS
  as a mean strategy is singular, then that set is locally attracting, in
  agreement with a result in Hofbauer and Sigmund [14].

- As well, if an allelic frequency gives rise to a mean strategy μ differing
  from μ^* which is strictly extreme in the direction $A\,(\mu - \mu^*)$ in the
  sense that any sufficiently small perturbation of frequencies leads only
  to strategies μ' such that $(\mu' - \mu) \cdot A\,(\mu - \mu^*) < 0$ (the other case
  considered by Hines and Bishop) that allelic frequency will be locally
  stable.

- If at least one of the set of all allelic frequencies which give rise to the
  ESS as a mean strategy is singular, the situation is more complicated,
  as described below.

Let F denote a singular frequency with $\mu(F) = \mu^*$, and consider the set X of directions x with the property that $\mu(F + \delta x) = \mu(F - \delta x)$ different from μ^* for δ different from 0. Let F' be some initial allelic frequency sufficiently close to the set of allelic frequencies which each give rise to the ESS as the population mean strategy; restrict attention to those frequencies among that set which give positive probability to all alleles given positive probability by both F and by F'; and let F be the frequency closest to F' among such frequencies. This can be shown to imply that if F' is written as F + ϵy for some suitable ϵ and y, then y is in or very close to the set X. For y in X, if A and the set of strategy vectors s(u, v) have been suitably chosen, then the distance of the allelic frequencies from F (and from the set of alleles giving rise to μ^*) increases rather than decreases, at least momentarily. (One of the requirements is that A have a relatively large skew-symmetric component and a relatively small symmetric one.) The region about the set F + X in which this increase occurs becomes increasingly tightly concentrated about that set as ϵ decreases.

We distinguish two cases depending on whether the evolution of the population's allelic frequencies is transverse to F + X or tangential to it. If the evolution is transverse, then the allelic frequency soon passes to a region in which the distance can be shown to be again decreasing. If the evolution is tangential, then it is conceivable that that evolution could remain so close to F + X that the distance continues to increase, implying an evolution of the population's mean strategy away from the ESS. That possibility currently appears to be unlikely, in light of a result in Hofbauer and Sigmund [14] which says that if all possible perturbations of allelic frequency lie in the set X then evolution will move the population's mean strategy to the ESS. Nonetheless, the possibility was the basis for constructing a numeric example which appeared to produce the sort of pattern of instability being sought.

4   A numeric example

The analysis by Hines [8][9] suggested various necessary conditions to be satisfied if an example in which the ESS was to be unstable as a population's mean strategy, and it was possible to choose the parameters of the example so that, for example, both the allelic frequency vector corresponding to the ESS and the ESS itself were vectors of constant components, say the 3-vector with all components (1/3). The strategy vectors for the various genotypes were chosen to be $(1, 0, 0)^T$, $(0, 1, 0)^T$ and $(1/2, 1/2, 0)^T$ for the first, second and third possible homozygotes, and $(1/4, 1/4, 1/2)^T$ for all possible heterozygotes. The payoff matrix A was chosen to be

$$A = \begin{pmatrix} 0 & -1 & 1 \\ 1 & 0 & -1 \\ -1 & 1 & 0 \end{pmatrix} + \epsilon \begin{pmatrix} 0 & 0 & 1 \\ 1 & 0 & 0 \\ 0 & 1 & 0 \end{pmatrix}$$

with ϵ positive, as required by the second ESS condition. As noted previously, that analysis by Hines had indicated that a necessary condition for instability of the ESS to occur was that there be relatively little symmetry present in A, a condition which corresponded to taking ϵ small. In early experimental numerical work, the value used for ϵ was 0.01, and graphs of the resulting evolution were produced. Graphically, that value of ϵ appeared to produce the anticipated instability.

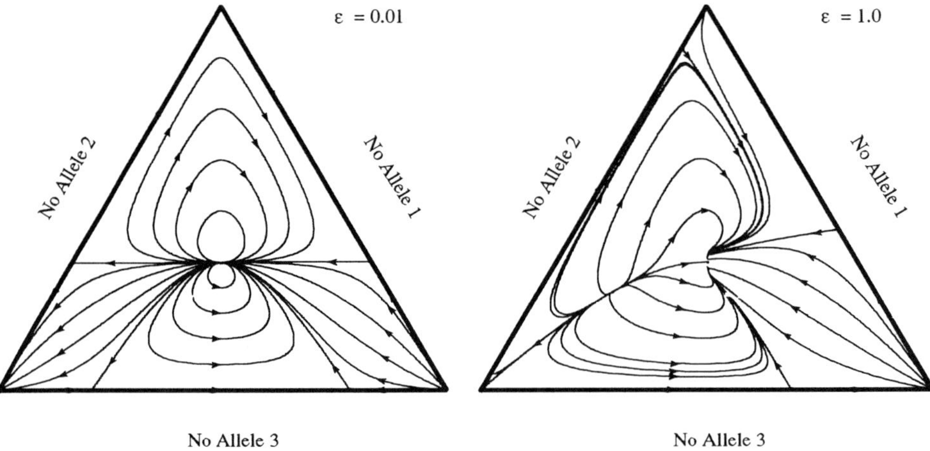

Figure 1: Changes in allele frequency induced by contest-derived fitness in diploid populations. Carriers of two copies of allele 1, 2 or 3 use tactic 1 only, tactic 2 only or tactics 1 and 2 equally often, respectively. Carriers of two different alleles use tactics 1, 2 and 3 in the ration 1:1:2. Parts a) and b) correspond to little and moderate symmetry in the payoff matrix ($\epsilon = 0.01$ and $\epsilon = 1.0$, respectively).

Figure 1a shows the evolution of allelic frequencies under various starting conditions - chosen so that the trajectories pass through frequencies indicated by small circles in that figure. (In fact, the indicated locations were used as initial frequencies for both evolution and evolution in reversed time.) The figure appears to imply that while populations starting at almost any allelic frequency close to that producing the ESS as a mean strategy do eventually converge in mean strategy to that strategy, exceptional frequencies do exist in a cusp shaped region just to the left of the ESS-producing frequency which evolve towards a different stable population composition, as shown. The indicated evolution of allelic frequencies implies a corresponding evolution of mean strategy. These evolutions must lie within 90^0 of the flow given by

$$\frac{d\mu_t}{dt} = \Pi A(\mu_t - \mu*)$$

where Π is a projection matrix with 2/3 along its diagonal and -1/3 elsewhere. This flow is shown in Figure 2a, along with the image of the allelic frequency simplex Δ_a under the mapping μ: $F \to \mu(F)$ (Hines [9] for example).

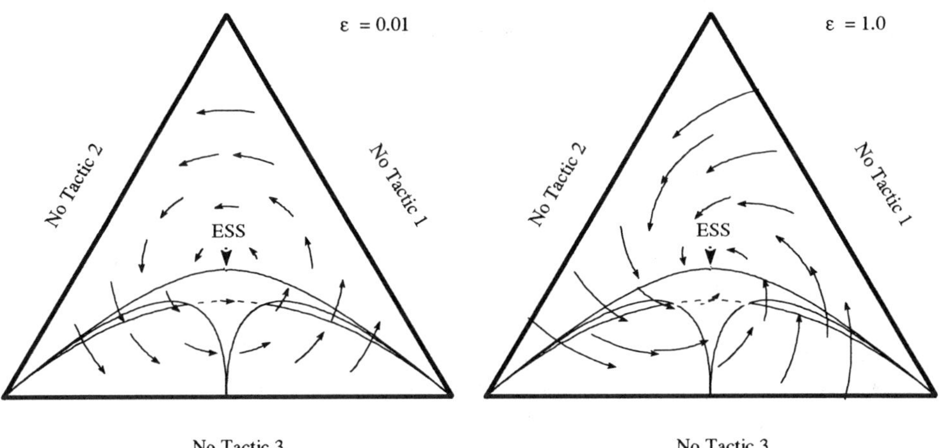

Figure 2: A vector field of optimal gradient changes in mean strategy in the simplex of possible mean strategy, superimposed on the image of the allelic simplex under the mapping from allelic frequency to resulting population mean strategy. Parts a) and b) correspond to $\epsilon = 0.01$ and $\epsilon = 1.0$, respectively. While actual changes in mean strategy deviate from those indicated, as a consequence of the potential of the population to evolve in various ways, those actual changes lie with $90°$ of the indicated flow.

Figure 3a shows the evolutions of mean strategies corresponding to the allelic frequency evolutions shown in Figure 1a.

Given these figures and the requirement on evolutions of mean strategy that they lie within 90^0 of the flow field in Figure 2a, the apparent result that the indicated ESS frequency is unstable is plausible.

Given the numeric simplicity of the example, however, Cressman decided to use central manifold calculations on it - and discovered that the ESS-producing frequency should in fact be stable. (This result was later ex-

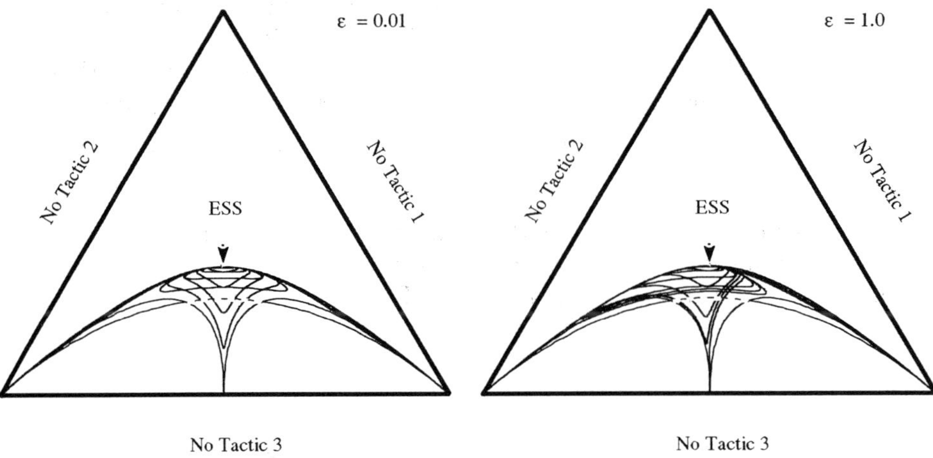

Figure 3: The actual evolutions in mean strategy for $\epsilon = 0.01$ and 1.9.

tended by him to the more general situation of arbitrary payoff matrices for which internal ESSs exist (i.e., those with all components positive) for arbitrary numbers of alleles and tactics - provided that the central manifold be one-dimensional, which can only happen if $n_a < n_s$. See Cressman, Hofbauer and Hines [5] for details.) This result by Cressman suggested the presence of a second, unstable, equilibrium frequency close to the ESS-producing frequency. To explore this, similar plots of frequency evolutions were produced for larger values of ϵ ($\epsilon = 1.0$ is illustrated in Figures 1b, 2b, 3b), and that hypothesis was confirmed. As ϵ was increased from its initial value of 0.01, the unstable equilibrium which is apparent in Figure 1b moved smoothly from the centre of the simplex down to the lower left vertex, and reached it by the time ϵ was 2.0.

The reason for the discrepancy between the results of the approaches of Hines and of Cressman is subtle and important. Hines focuses on the behaviour of the distance metric on parts of tangent planes to the surface of alleles all giving rise to the ESS, and that plane can lie close to the surface on which the relevant evolution occurs. When high-order effects, however, are important, that good approximation is insufficiently adequate. The central manifold approach involves consideration of a surface which better approximates that on which the evolution in question is occurring. As such, it is a more acute instrument, but subject to the practical limitation that central manifolds of more than a single dimension are not readily analyzed. It should be remarked that the analysis by Cressman shows that counterexamples to a claim of local stability of ESSs in this model will necessarily need to have the space generated by the elements of X to be more than one dimensional, while the results of Hines indicate that convergence will occur if population frequencies evolve away from those very near to the exceptional set noted above - and yet a result in [14] indicates that if all allele frequency vectors are singular, the ESS is again locally stable. Together, these results suggest that for the model under consideration, the ESS does indeed at least possess local stability. As the numeric example indicates, the radius of the domain of attraction for that stability need not be large.

5   Conclusions

As with most successful models, the description of competition in biological populations used in formulating the Evolutionarily Stable Strategy concept is, of necessity, incomplete. The work described here and elsewhere demonstrates, however, that the ESS determined strategy remains relevant when the effects of sexual reproduction are taken into account, for the class of competitive models and one-locus model considered. While populations with that strategy as a mean strategy are at equilibrium, other populations can be as well, and stably so, if genetic constraints prevent them from moving further towards that same mean strategy, but the direction of the flow towards that particular mean strategy remains extremely relevant - the population moves 'as far as it can' towards the ESS. So much was known, and expected from the analysis reported on here.

That analysis confirmed that the statistical mean-covariance approach, while intuitively instructive, could prove to be inadequately precise for some situations. The analysis also determined that the equilibria just mentioned were the only possibilities and started to explore the question of what happens when two of those possibilities collide. If a population is subject to genetic constraints that preclude evolution in some direction and if, as well, the population has the strategy identified as an ESS as its mean strategy, can effects conspire to produce instability? The possibility arises since there are different dominant terms that determine stability in the cases where the population mean is the ESS and where the population is constrained genetically: if both of those dominant terms are zero when both conditions are met, higher order terms that are usually unimportant demand consideration. Consideration of the most simple of the combined ESS and genetic constraints cases has demonstrated that in that case, as in other particular and relatively tractable cases, stability does indeed occur. If exceptions are yet to be found, the necessary conditions they must satisfy are now better understood, which should make the search for them somewhat more directed, if less promising. The original form of the numeric example considered, however, with $\epsilon = 0.01$, demonstrates that populations with strategies very close to ESSs can still be subject to persistent evolution away from the ESS under some circumstances.

In practical terms, it appears unlikely that any given model for conflict in a sexually reproducing biological population which uses simply the constant payoff matrix form considered above will suffer from the defect that its nominal ESS will prove to be unstable since the population composition there is singular. Although that danger seems slight, however, there is another: the existence of a truly stable ESS in such a population does not automatically imply that populations with initial mean strategies even modestly different from the ESS will converge in mean strategy to the ESS. While the use of game-theoretic modelling to determine possible locally-stable equilibria for biological populations is useful, and informative even when the population in question is not able to achieve those equilibria, the resulting picture of the possible evolutions of the populations can be incomplete. As with many powerful tools, care should be taken by the user.

Acknowledgements

The Figures of this paper incorporate improvements suggested by E. Akin, R. Cressman and J. Hofbauer.

REFERENCES

[1] Akin, E. (1979), The Geometry of Population Genetics, *Lecture notes in Biomathematics* 31, Springer Verlag, Berlin.

[2] Akin, E. (1982), Exponential Families and Game Dynamics. *Cdn. J. Math* 34, 374-405.

[3] Akin, E. (1990), The Differential Geometry of Population Genetics and Evolutionary Games, in *Mathematical and Statistical Developments of Evolutionary Theory* (S. Lessard, Ed.), Kluwer Academic, Dordrecht, 1-93.

[4] Antonelli, P. and Strobeck, C. (1977), The Geometry of Random Drift I Stochastic Distance and Diffusion. *Adv. in Appl. Prob.* 9, 238-249.

[5] Cressman, R., Hofbauer, J. and Hines, W.G.S (1994), Evolutionary Stability in Strategic Models in Single-Locus Frequency-Dependent-Viability Selection, *J. Math. Biol.*, in press.

[6] Hines, W.G.S. (1980), An Evolutionarily Stable Strategy Model for Randomly Mating Sexual Populations, *J. Theor. Biol.* 87, 379-384.

[7] Hines, W.G.S. (1987), Evolutionary Stable Strategies: A Review of Basic Theory *Theor. Pop. Biol.* 31, 195-272.

[8] Hines, W.G.S. (1994), ESS Modelling of Diploid Populations I: Anatomy of one-locus allelic frequency simpleces *Adv. Appl. Prob.* 26, 341-360

[9] Hines, W.G.S. (1994), ESS Modelling of Diploid Populations II: Stability analysis of possible equilibria, *Adv. Appl. Prob.* 26, 361-376.

[10] Hines, W.G.S. and Bishop, D.T. (1983), Evolutionarily Stable Strategies in Diploid Populations with General Inheritance Patterns, *J. Appl. Prob.* 20, 395-399.

[11] Hines, W.G.S. and Bishop, D.T. (1984), On the Local Stability of Evolutionarily Stable Strategies in Diploid Populations, *J. Appl. Prob.* 21, 215-224.

[12] Hines, W.G.S. and Bishop, D.T. (1984), Can and will a Sexual Diploid Population attain an Evolutionarily Stable Strategy? *J. Theor. Biol.* 111, 667-686.

[13] Hines, W.G.S. and Turelli, M. (1995), Multi-locus ESS Models: Additive effects. Submitted *J. Theor. Biol.*.

[14] Hofbauer, J. and Sigmund, K. (1988), *The Theory of Evolution and Dynamical Systems*, Cambridge University Press.

[15] Maynard Smith, J. (1974), The theory of games and the evolution of animal conflicts, *J. Theor. Biol.* 47, 209-221.

[16] Maynard Smith, J. (1981), Will a sexual population evolve to an ESS? *Amer. Naturalist* 177, 1015-1018.

[17] Maynard Smith, J. (1982), *Evolution and the Theory of Games*, Cambridge University Press.

[18] Maynard Smith, J. (1989), *Evolutionary Genetics*, Oxford University Press.

[19] Riechert, S. and Hammerstein, P. (1983), Game Theory in the ecological context, *Ann. Rev. Ecol. Syst.* 14, 377-409.

[20] Thomas, B. (1985), Genetical ESS-models I. Concepts and basic model, *Theor. Pop. Biol.* 28, 18-32.

[21] Thomas, B. (1985), Genetical ESS-models II. Multi-strategy models and multiple alleles, *Theor. Pop. Biol.* 28, 33-49.

[22] Thomas, B., (1985), Evolutionarily Stable Sets in Mixed Strategist Models, *Theor. Pop. Biol.,* 28, 332-34.

[23] Treisman, M. (1981), Evolutionary limits to the frequency of aggression between related or unrelated conspecifics in diploid populations with simple mendelian inheritance. *J. Theor. Biol.* 93, 97-124.

[24] Shahshahani, S. (1979), *A New Mathematical Framework for the Study of Linkage and Selection,* Amer. Math. Soc., 211, AMS, Providence, RI

Annals of the International Society of Dynamic Games

Series Editor
Tamer Başar
Coordinated Science Laboratory
University of Illinois
1308 West Main Street
Urbana, IL 61801
U.S.A.

This series publishes volumes in the general area of dynamic games and its applications.
It is an outgrowth of activities of "The International Society of Dynamic Games," *ISDG,*
which was founded in 1990. The primary goals of *ISDG* are to promote interactions
among researchers interested in the theory and applications of dynamic games; to facilitate
dissemination of information on current activities and results in this area; and to enhance
the visibility of dynamic games research and its vast potential applications.

The *Annals of Dynamic Games* Series will have volumes based on the papers presented
at its biannual symposia, including only those that have gone through a stringent review
process, as well as volumes of invited papers dedicated to specific, fast-developing topics,
put together by a guest editor or guest co-editors. More information on this series and on
volumes planned for the future can be obtained by contacting the Series Editor,
Tamer Başar, whose address appears above.

We encourage the preparation of manuscripts in LaTeX using Birkhäuser's macro.sty
for this volume.

Proposals should be sent directly to the editor or to: Birkhäuser Boston,
675 Massachusetts Avenue, Cambridge, MA 02139, U.S.A.

Volumes in this series are:

Advances in Dynamic Games and Applications
Tamer Basar and Alan Haurie

Control and Game-Theoretic Models of the Environment
Carlo Carraro and Jerzy A. Filar

New Trends in Dynamic Games and Applications
Geert Jan Olsder